Advanced Ceramics for Energy and Environmental Applications

T0256399

Editor
Akshay Kumar
Department of Nanotechnology
Sri Guru Granth Sahib World University
Fatehgarh Sahib, Punjab
India

CRC Press
Taylor & Francis Group
Boca Raton London New York

CRC Press is an imprint of the
Taylor & Francis Group, an **informa** business

A SCIENCE PUBLISHERS BOOK

First edition published 2022
by CRC Press
6000 Broken Sound Parkway NW, Suite 300, Boca Raton, FL 33487-2742

and by CRC Press
2 Park Square, Milton Park, Abingdon, Oxon, OX14 4RN

© 2022 Taylor & Francis Group, LLC

CRC Press is an imprint of Taylor & Francis Group, LLC

Library of Congress Cataloging-in-Publication Data

Names: Kumar, Akshay, 1982- editor.
Title: Advanced ceramics for energy and environmental applications /
    editor, Akshay Kumar, Department of Nanotechnology, Sri Guru Granth
    Sahib World University, Fatehgarh Sahib, Punjab, India.
Description: First. | Boca Raton : CRC Press, Taylor & Francis Group, 2021.
    | "A science publishers book." | Includes bibliographical references and
    index.
Identifiers: LCCN 2021016128 | ISBN 9780367436742 (hardcover)
Subjects: LCSH: Electronic ceramics. | Detectors--Materials. | Energy
    storage--Materials. | Ceramic powders.
Classification: LCC TK7871.15.C4 A35 2021 | DDC 621.042028/4--dc23
LC record available at https://lccn.loc.gov/2021016128

ISBN: 978-0-367-43674-2 (hbk)
ISBN: 978-1-032-02553-7 (pbk)
ISBN: 978-1-003-00515-5 (ebk)

DOI: 10.1201/9781003005155

Typeset in Times
by TVH Scan

Preface

Although traditional ceramics can be dated back to more than 25000 BCE, the 20th century has seen a boom in the research in modern ceramics known as Advanced Ceramics. Advanced Ceramics possess various unique properties and are gifted to withstand harsh environments. Modern materials science and technology highly relies on structural and functional ceramics. These have applications in the fields demanding high temperature capability, high corrosion and wear resistance, low electrical conductivity etc. Due to their application specific properties, research in advanced ceramics is increasing day by day. When these advanced ceramics are of nano-size, their properties drastically improved than bulk materials. These nanostructured ceramics can be used in various applications with improved efficiency. Nanostructured ceramics have great potential for photocatalysis, gas sensing, energy storage, biomedical applications, etc.

The aim of this book is to cover various aspects of advanced ceramics such as synthesis, properties, structure, mechanisms for energy and environment related applications. The increase in worldwide energy consumption, shortage of resources, and global warming immensely increased the demand for energy conversion, high efficiency with minimum resources, energy storage and with less or no emission of greenhouse gases and other pollutants. Also, the significance of environmental technologies is increasing day by day in correspondence to the need for energy due to the dramatic increase in air, water and soil pollution. Advanced ceramics with additional functionality propose significant potential for more impact in the field of energy and environmental technologies. The role of advanced ceramics in energy related applications such as energy conversion, photovoltaics, energy storage, supercapacitors, Lithium ion batteries, heat absorption, etc. is discussed in this book. Environment related applications of nanostructured ceramics are mainly categorized into sensing and removal of pollutants in this book. In sensing, gas sensors, humidity sensors, micro/bio sensors etc. are elaborated. Removal of heavy metal ions, degradation of dyes, removal of air pollutants, removal of pathogens, etc. are also discussed as environmental applications of advanced ceramics.

This book is highly focused on the nanostructured ceramics: synthesis, properties, structure-property relation and application in the area of energy and environment. It covers the high impact work from around 50 leading researchers throughout the world working in this field. This will help metallurgists, biologists, mechanical engineers, ceramicists, material scientists and researchers working in nanotechnology field with inclusion of every aspect of advanced ceramics for energy and environmental applications.

Akshay Kumar

Acknowledgements

First of all, I thank God for providing this valuable opportunity to complete this book successfully.

In this journey of book, I would like to express my gratitude to Ms. Manjot Kaur, Senior Research Fellow, Department of Nanotechnology, SGGSWU, for her continuous support in this work. This work would not have been possible without her involvement and her efforts on daily basis from the start of the project.

My sincere thanks to all the Authors and Reviewers for their valuable contributions, hard work and support to complete this book. While they live and work in different countries, the proposal of this book is a common goal to all of us. Also, we highly prefer to acknowledge various authors and publishers who allowed us the copyright to use their figures and tables. If unknowingly, someone's right has been being infringed, we would like to offer our deep request for forgiveness.

My heartfelt thanks to publishing house and editorial office for accepting this book proposal and their continuous support.

Akshay Kumar

Contents

1

Progress in Advanced Ceramics: Energy and Environmental Perspective

Kulwinder Singh, Manjot Kaur and Akshay Kumar[*]

1.1 INTRODUCTION

The term ceramics comes from a Greek word "Keramicos" which signifies 'burnt materials' or *'potter's clay'*. There are many compounds having no clay content that also come under ceramics compounds. In history, archaeologists have found that ceramics made by human belongs to approximately 26,000 BC in certain areas of Czechoslovakia (Boch and Niepce 2001). Main constituents of these discovered ceramics include animal fat as well as bone containing fined clay-like materials. The evidence of the usage of functional pottery vessels was related to back in 9,000 BC. Mainly, functional pottery vessels were used in houses for different purposes such as food storage as well as holding different grains. Apart from this, ancient glass making process is similar to the process of pottery making and the origin was found to be related to Egypt in about 8,000 BC (Boch and Niepce 2001). In earlier times, coloured ceramic pots are mainly made at high temperatures from calcium oxide comprising sand and soda. Researchers have found the origin of making glass-based products back to 1,500 BC which were used in different items.

Now coming to advanced ceramics: basically, advanced ceramics are the elaboration of ceramic materials with unique and improved properties than traditional ceramics. Advanced ceramics are high-performance ceramics. In other words, advanced ceramic materials are processed and engineered ceramic materials, i.e. composition controlling and functional manufacturing, to get the desirable properties (Somia 2013). There are different terminologies for defining this type of ceramics such as engineered, technical, fine, advanced etc. (Kulik 1999). In the American literatures, it is generally referred as 'technical' or 'advanced' ceramics, whereas in Japan these are cited as 'fine' ceramics. The term 'technical' ceramics is more commonly used in Europe (Kulik 1999). Generally, advanced ceramics are crystalline in nature. Being an integral part of modern technology, advanced ceramics nowadays play a crucial role in technologies such as energy, environment, transport, communication, life sciences, etc. (Greil 2002).

In traditional ceramics, main constituents are natural minerals like clay quartz and structure of a material can be determined from the content of clay in a material. These are mainly used for daily work and for building different structures. Advanced ceramics are made up of processed high-quality powders which alter their functional properties and are simple in terms of structure and composition. As these materials are processed in controlled manner, high quality of materials

---

Department of Nanotechnology, Sri Guru Granth Sahib World University, Fatehgarh Sahib, Punjab-140 406, India.
[*] Corresponding author: akshaykumar.tiet@gmail.com

can be maintained alongwith uniform structure. Different mouldings are suitable with the addition of organic ingredients to high purity powder raw materials. Further processing is necessary such as sintering at high temperature depending upon materials. Processes used for advanced ceramic materials overcome the previous limits of ceramics. Different processes such as sintering, hot pressing, and high-temperature isostatic pressing are used for preparation of advanced ceramic materials. Due to high quality than traditional ceramics, advanced ceramic materials possess various properties leading to their different applications such as abrasives, cutting tools, electrical insulation, catalysis, sensing, aerospace, thermal conductors and electronic devices etc.

1.2 CLASSIFICATION OF ADVANCED CERAMICS

Ceramics can be categorized into two classes broadly, which includes: (i) Traditional ceramics, and (ii) Advanced ceramics, as shown in Fig. 1.1. Traditional ceramics are mainly comprised of glasses, abrasives, insulating materials, refractories and different enamels.

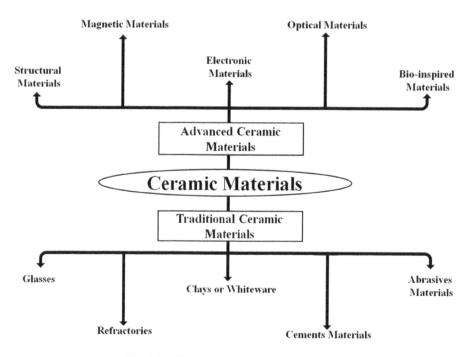

Fig. 1.1   Classification of ceramic materials

Advanced ceramic materials can be classified into two main categories: (1) Advanced structural ceramic materials, (2) Advanced functional ceramic materials. This categorization is shown in Fig. 1.2. Some other advanced ceramics are bio-ceramics, special glasses and ceramic coatings.

1.2.1   Advanced Structural Ceramics

Advanced structural ceramics are used for mechanical, structural, tribological, thermal, or chemical load applications due to their chemical inert nature, high strength, toughness, high corrosion and thermal shock resistance. Some examples of above said are zirconia (Aza et al. 2002), mullite, alumina (Schwentenwein and Homa 2015), titanium carbide (Kormann et al. 2008), boron carbide (Singh et al. 2014), silicon carbide (Shi et al. 2006), boron nitride (Singh et al. 2016), silicon nitride (Petzow and Herrmann 2002), aluminium nitride (Kaur et al. 2020b), and composites (Singh et al.

2020a, b). At high temperatures, these structural ceramics can carry significant levels of load. But they need to be properly fabricated to avoid defects during processing.

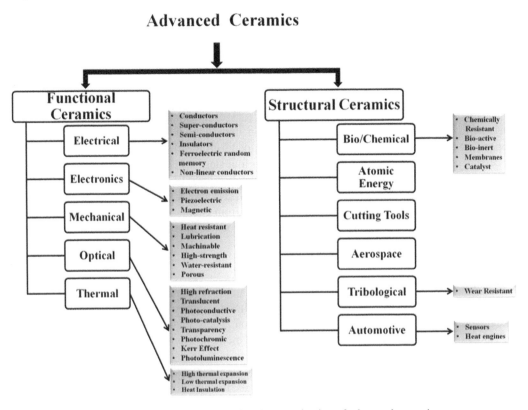

Fig. 1.2 Structural and functional categorization of advanced ceramics

1.2.2 Advanced Functional Ceramics

Advanced functional ceramics include materials that are used for several applications based on their functional capabilities. Microstructural effects are used in these ceramics which involve semiconducting, piezoelectric, ferroelectric, pyroelectric, and superconducting properties. On the basis of functions, these ceramics are further classified into electrical, magnetic, electronic, optical, thermal, mechanical, chemical, nuclear, and biological functions (Otitoju et al. 2020).

Also, depending upon the materials developed, the advanced ceramics can also be divided into two major categories: (1) Oxides, (2) Non-oxides (Weimar 1997). The typical oxides based advanced ceramics include alumina, zirconia, mullite, titania, ferrites, and perovskites. The non-oxide ceramics include nitrides like boron nitride (BN), aluminium nitride (AlN), titanium nitride (TiN), silicon nitride (Si_3N_4), carbides like boron carbide (B_4C), tungsten carbide (WC), silicon carbide (SiC), borides like titanium diboride (TiB_2), etc.

Advanced non-oxide ceramics can be distinguished by their bonding types and resulting properties as: (a) hard elements such as boron and carbon (diamond), (b) compounds of transition metals of groups IVa and VIa with B, C, N and Si, (c) binary and ternary compounds of elements of Groups IIIb to Vb, and (d) some salt – like chalcogenides and halogenides (Heimann 2010a).

1.3 PROPERTIES OF ADVANCED CERAMICS

In general, advanced ceramics outshine due to their high temperature stability, up to 2500 °C. and beyond, high corrosion resistance, high hardness, low coefficient of thermal expansion, and a huge variation of electrical properties (from insulators to semiconductors to highly conductive ceramics). Also, the development of reinforced advanced ceramics such as modern nanostructured ceramics and partially stabilized zirconia results in high performance materials even under hostile mechanical conditions. Technical ceramics are mainly inorganic – non-metallic and can be highly specialized with the slight manipulation of their unique electric, optical, magnetic, mechanical, thermal, biological, and environmental properties (Heimann 2010b). Due to their advanced processing technology, high value end products can be achieved.

1.4   SYNTHESIS OF ADVANCED CERAMICS

The traditional ceramics processing or fabrication undergoes following stages: (a) the powder preparation from raw materials, (b) moulding or shape forming, (c) sintering, and (d) finishing. The processing of advanced ceramics follows the similar four stages. The advanced ceramics industry is concerned with manufacturing of powders and application specific large-scale ceramic production which can work in particular environments. Traditional ceramics are generally derived from natural raw materials, while advanced ceramics are obtained from chemical synthesis routes or from high refinement of naturally found materials. Various methods used for the synthesis of ceramics are discussed briefly in this section.

1.4.1   Conventional Methods

Traditional and advanced ceramics can be synthesized by conventional methods such as precipitation from solution, powder mixing and fusion. These methods can be employed for both the laboratory and industrial scale. But conventional methods are not suitable for the synthesis of advanced ceramics. In the precipitation method, there is a limitation when two or more species are co-precipitated. This results in inhomogeneity which can adversely affect the electronic and mechanical properties of synthesized ceramic materials. Precipitation yields agglomerated powders. To reduce the particle size, grinding or milling process can be used but this leads to impurities. Conventional mixing methods are based on initial blending of starting materials, then grinding or milling, followed by calcination. This method is generally used for the synthesis of multicomponent oxide powders. Milling process can result in insufficient comminution of powder for occurrence of complete reaction during calcination. Alternative methods for the synthesis of advanced ceramics are preferred due to several reasons. Firstly, conventional methods show lack of versatility for the synthesis of broad range of materials. Secondly, chemical impurities are not suitable for properties of electro-ceramics (Segal 1997). Also, non-uniform compositions are not suitable for reproducible component fabrication. So, novel syntheses routes are desirable for the synthesis of highly pure advanced ceramics with improved properties.

1.4.2   Sol-Gel Method

Synthesis of advanced oxide and non-oxide materials has been widely explored with the use of aqueous and non-aqueous sols. This method has the advantage over conventional methods as low processing temperatures result in nano-meter sized powders. Sol-gel method can be utilized for colloidal and metal-organic compound processing (Prasad et al. 2018). Although colloidal processing results in homogeneous compositions, it is also suitable for processing of metal-organic compounds, i.e. alkoxides via hydrolysis. By avoiding sintering, alkoxides leads to preparation of aerogels, xerogels, and spherical oxide particles. Yet this method is not economical for mass production in comparison to conventional fusion methods.

1.4.3 Hydrothermal/Solvothermal Method

Nowadays, the hydrothermal method and solvothermal methods have gained immense attention due to the versatile synthesis availability of multifunctional advanced ceramics such as electronic ceramics, optical ceramics, catalysts, bio-ceramics, membranes, etc. Hydrothermal and solvothermal synthesis route comprises heating of reactants as solution/suspension in water or solvents, respectively. Reactants such as metal salts, metal powders, oxides, etc. via homogeneous nucleation results in the synthesis of advanced oxide and non-oxide ceramics with controlled growth of shape and size. This method provides a direct route for the synthesis of monodispersed powders. The main advantages of these methods are controlled size, shape, high chemical purity, and relatively low processing temperature. This method allows direct synthesis of crystallized powders without any further thermal treatments, thus resulting in nanostructured ceramics (Feng and Li 2017).

1.4.4 Non-aqueous Liquid Phase Reactions

This method is utilized mainly for the low temperature synthesis of non-oxide powders such as borides, carbides, nitrides, and sulphides. For example, silicon tetrachloride and ammonia are used for the synthesis of silicon nitride. Non-aqueous liquid phase reactions allow the synthesis of highly pure ceramic powders and can be utilized for commercial production (Segal 1997, Danforth 1992).

1.4.5 Gas Phase Reactions

This method is suitable for both laboratory as well as industrial scale production of advanced ceramics. Reactants can be solids, liquids or gaseous. A gas phase reaction can be carried out by using various techniques such as direct nitridation and carbothermic reduction, lasers, plasmas reactions and electron beam evaporation. Plasma synthesis can also be suitable for multi-component systems. Gas phase reactions can be desirable to obtain specific non oxide ceramics with high purity and narrow particle size distribution. These methods can be utilized for synthesis of powders and single crystal growth of advanced ceramics (Segal 1997, Cannon et al. 1982).

Various other synthesis techniques, in addition to above discussed, are utilized for the synthesis of advanced ceramics such as spray drying, pyrolysis, citrate gel, etc. Synthesis routes can be divided on the basis of end product and applications. For powder synthesis, methods such as gas phase reaction, sol gel method, hydrothermal, solvothermal, spray drying, chemical vapour deposition (CVD), vapour liquid synthesis etc. can be used. For the single crystal growth of advanced ceramics, gas phase processing, solid phase processing and liquid phase processing are best utilized. On the other hand, thin film growth of advanced ceramics can be achieved by physical vapour deposition (PVD), solid phase reactions and liquid phase reactions.

1.5 APPLICATIONS OF ADVANCED CERAMIC MATERIALS

Purity of basic powders used for preparation of advanced ceramic materials determine the properties and quality of processed materials. Properties can be altered depending upon the requirements such as narrow size distribution as well as small grain sizes by varying synthesis procedure. Different methods and processes have been developed to obtain the functional properties of a material. Researchers have attracted attention and engrossed in development of material components for functional applications depending upon chemical, electrical, magnetic, mechanical, optical, physical and thermal properties. Anti-corrosive applications, biomedical applications, energy transformation, information technology, sensor technology, solar cell technology, protective coatings, water remediation and waste water treatment etc. are examples of functional applications. The applications of some specific advanced ceramic materials according to their category and functions are listed in Table 1.1.

TABLE 1.1 Specific advanced ceramic materials useful in various applications on the basis of different functions (Otitoju et al. 2020)

Categories	Function	Applications	Specific Ceramic Materials
Electro-ceramics	Electrical	Conductors	B_4C, $LaCrO_3$, TiC, TiN
		Super conductors	$YBa_2Cu_3O_{7-\delta}$, $PrBa_2Cu_3O_{7-\delta}$, $YBa_2Cu_2O_7$
		Semi-conductors	ZnO, SnO_2, GaN, SnO_2, CdS, CoO, MoS_2, WO_3, BN, V_2O_5, ITO, $MgTi_2O_5$, Cu_2S, AlON, $ZnTiO_3$, $CaTiO_3$, CuO, CdS, SiC, $SrTiO_3$, Cu_2S, GaN, MnO, CoO, $BaTiO_3$, $BiFeO_3$, NiO, $LiNbO_3$, BeO, ZnO, TiO_2, TiC, HfN, ZrW_2O_8
		Insulators	$MgAl_2O_4$, Al_2O_3, MgO, $SrTiO_3$, SiO_2, SIALON, BeO
		Ferroelectric random memory	(Ba, Sr) TiO_3, PZT
		Non-linear conductors	SiC, CuO, PTC/NTC, V_2O_5, $LiNbO_3$, $LiTaO_3$
	Electronics	Electron emission	BaO, $SrTiO_3$, ZrW_2O_8, LaB_6
		Piezoelectric	$LiTaO_3$, $MgTiO_3$, ZnO, $LiNbO_3$, $LiTaO_3$, $BaTiO_3$, PZT, AlN, $LiCoO_3$
		Magnetic	Ferrites, NiO, Co_2O_3
	Mechanical	Heat resistant	Si_3N_4, SIALON, SiC, $Mg_2Al_4Si_5O_{18}$
		Lubrication	BN, MoS_2
		Machinable	Mica ceramics, BN composite
		High-strength	TiC, SiC, Si_3N_4, TiN, Diamond
		Water-resistant	B_4C, SiC
		Porous	SiO_2, Cordierite
	Optical	High refraction	ZrO_2, TiO_2, $MgAl_2O_4$, B_4C, BeO, TiC, Al_2TiO_5, SiC, mullite, Diamond, $ZrSiO_4$, $Mg_2Al_4Si_5O_{18}$
		Translucent	Al_2O_3, SIALON
		Photoconductive	CdS, GaN, Cu2S
		Photocatalysis	TiO_2, ZnO, NiO, CuO, CoO, $LiNbO_3$, CdS, MoS_2, $BiFeO_3$, $PbTiO_3$, Al_2O_3, Cu_2S, $SrTiO_3$, WO_3, $CaTiO_3$, ferrites, $ZnTiO_3$, CCTO, $ZnTiO_3$, $MgTi_2O_5$
		Transparency	Y_2O_3, ITO, SnO_2, ZnO
		Photochromic	WO_3
		Kerr effect	Lead lanthanum zirconate titanate
		Photoluminescence	Nd_2O_3, Er_2O_3
	Thermal	High thermal expansion	Al_2O_3, AlN, SiC, BeO, MgO, Si_3N_4
		Low thermal expansion	SiO_2, Cordierite, Al_2TiO_5
		Heat insulation	Si_3N_4, ZrO_2
Structural	Bio/chemical	Chemically resistant	SiC, Al_2O_3, TiB_2, BN
		Bio-active	HAP, $BaTiO_3$, $Ca_3ZrSi_2O_9$
		Bio-inert	Al_2O_3, ATZ, ZrO_2
		Membranes	SiC, $LaCrO_3$, SiO_2, Al_2O_3, TiO_2, ZnO, SiOC, mullite
		Catalyst	$LaCrO_3$, CuO, CoO, $MgAl_2O_4$, NiO, MnO, $ZnTiO_3$, SnO_2, $BaTiO_3$, Cu_2S, $MgTi_2O_5$
	Atomic energy		UO_2, SiC, B_4C, BeO, HfB_2, ZrB_2, ZrC, Ti_3SiC_2
	Cutting tools		Diamond, ZrC, HfB_2, Si_3N_4, TiB_2, ZTA, B_4C, ZrB_2, TiCN, Ti_3SiC_2, TiC, SIALON
	Aerospace		Al_2O_3, SiC, Si_3N_4, ITO, AlN, ZrB_2, HfB_2, TiCN, ZrO_2, TiC, Si_3N_4
	Tribological	Wear resistant	SiC, ZrC, HfN, B_4C, ZTA, TiCN, Ti_3SiC_2, TiC, ZrO_2
	Automotive	Sensors, heat engines	Y_2O_3, CuO, LaB_6, MgAlON, V_2O_5, NiO, $ZnTiO_3$, SnO_2, $SrTiO_3$, MoS_2, ZnO, ZrB_2, $LaCrO_3$, GaN, Ti_3SiC_2, $ZnTiO_3$, Si_3N_4, B_4C, SiC, HfB_2, PZT, CCTO, Al_2TiO_5, $Sc_2W_3O_{12}$, $Mg_2Al_4Si_5O_{18}$, SiOC, SIALON, Al_2O_3

Source: Data from Otitoju et al. "Advanced ceramic components: Materials, fabrication, and applications," *J. Ind. Eng. Chem.* 85, (2020): 34-65

Different type of advanced ceramic materials-based components have found their applications in different fields such as electronic applications comprising power electronics (Rödel et al. 2009, Tiwari et al. 2016), integrated circuitry (IC) systems, telecommunication, wave communication systems (Sebastian et al. 2015), phase shift systems (Sazegar et al. 2011), and automotive device (Frketic et al. 2017). Apart from these applications, ceramic-based components are also useful in different type of sensors (Kumar et al. 2015, 2017, 2020), resonators (Erhart et al. 2017), etc. Due to high thermal stability, these components can be used for high temperature and high frequency applications (Deepukumar et al. 2014).

Advanced ceramic materials are also useful in biomedical applications including biomedical sensors, boron neutron capture therapy (BNCT) and cancer therapy, drug carriers, bio-markers, bio-tags, dental prosthesis and artificial implantable processes. Different resin cements such as composite resins and zinc phosphate cements are useful for bonding dental applications. Due to biocompatibility and luminescence properties of advanced ceramic materials, these are useful for cancer detection, targeted drug delivery and biomedical sensors. Examples of ceramic materials used for biomedical applications include tetragonal boron-based compounds, Al_2O_3, nitrides, ceramic composites, ZrO_2 and hydroxy-appetite (Singh et al. 2016, Kaur et al. 2020a, Xu et al. 2007, Dziadek et al. 2017) etc. Figure 1.3 shows the different applications of advanced structural and functional ceramics.

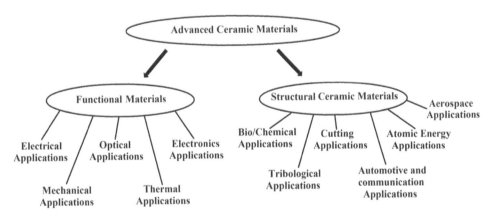

Fig. 1.3 Various applications of advanced ceramic materials

This book mainly focuses on the applications of advanced ceramics in the energy and environmental field. The increase in worldwide energy consumption, shortage of resources, and global warming immensely increased the demand for energy conversion, high efficiency with minimum resources, energy storage and with less or no emission of greenhouse gases and other pollutants. Also, the significance of environmental technologies is increasing day by day in correspondence to the need for energy due to the dramatic increase in air, water and soil pollution. There is a huge need of environmentally benign materials, reduction or elimination of the industrial pollutants. Advanced ceramics with additional functionality propose significant potential for more impact in the field of energy and environmental technologies. So, these applications are discussed below in detail:

1.5.1 Advanced Ceramics for Energy Applications

Chemical stability of advanced ceramic materials makes these materials useful in harsh environmental conditions (Gisele et al. 2013) and high-power load systems. As these materials possess magnetic properties along with electrical insulation, synergetic effects of both properties in ceramic materials are superior to metal systems and polymer-based compounds. Lower dielectric constant values with high thermal stability makes these materials useful in signal transmission process. Different types of losses like power dissipation loss and insertion loss can be minimized by varying different strategies for the improvement of quality factor of circuit-based systems. With the improvement in quality factor of devices, electrical noises suppression can be achieved and these ceramics are further useful for determining the feasibility of low-loss phase shifters, selective filters, non-reciprocal devices and dielectric resonator antenna (Zhou et al. 2017). As mentioned earlier, advanced ceramic materials are processed through various steps during synthesis procedure, which gives the opportunity to tune different properties, i.e. carrier mobility, band-gap energy, band structure of ceramics materials like borides, nitrides as well as oxides which makes these material useful in electronic applications. Different materials have also been used in electro-optic, optoelectronic, and semiconductor technology applications such as Ge_3N_4, SnO_2, ZnO, TiN, Si_3N_4, Hf_3N_4 etc. (Yu et al. 2016).

Microwave frequencies opened another area for exploration and applicability of advanced ceramic materials for microwave applications. Ceramic materials working in the millimeter frequency region are termed as microwave dielectric ceramics (Deepukumar et al. 2014). Some ceramic materials got attention of engineers and scientists due to their usefulness in communication systems including satellite systems, internet, defence systems, mobile communication and local area networks. Their usefulness leads to requirement of ceramic materials possessing specific properties related to a particular frequency region. Due to high thermal stability and medium permittivity of dielectric materials, these are useful in resonators in microwave communication applications. Advanced ceramic materials having complex or perovskite structure are used in resonators and receivers (Bi et al. 2017). Examples of useful ceramic materials for this application include barium, magnesium, zirconia-based titanites, complexes of tantalum as well as niobium etc.

Different ceramic materials possess properties which are useful for their use as electrical insulators. Basically, insulators ruptured the flow of current between different components by separating different electrical components in a circuitry boards supporting various devices. Due to extensive use of different technologies, applications of insulators-based ceramic materials in microelectronics increased abruptly. Single-fold to multi-fold insulator components are useful in daily use goods, automotive as well as aerospace technology. Due to high thermal conductivities and stability, ceramic materials-based insulators are useful for high voltage applications. Also, high dielectric strength, mechanical properties and high resistivity of some ceramic materials make them suitable candidates for insulator applications (Mukherjee 2011, Dou et al. 2019). Mica and SiC are well explored insulating materials due to their high dielectric strength.

High temperature applications open an area for the development of advanced structural ceramic materials. The main requirements of a structural ceramic material are high chemical stability, resistance to oxidation and creep deformation occurring at the interfaces, high thermal shock resistance as well as lower volatility. Advanced ceramics based on carbides and nitrides are promising candidates for high temperature gas turbine engines (Yang 2013).

Another field of application of advanced ceramics materials is components packaging in power electronics. Ceramic materials are useful as heat sink for heat dissipation process. Insulating properties of ceramic materials protects the connection from outside harmful chemical species and harsh environment. Generally, glass-based ceramic materials are used for packaging purpose of electronic components. Apart from glasses, alumina, silicon carbide and zirconia are useful in packaging applications due to high dielectric strength and high chemical resistance (Locatelli et al. 2014).

Ceramic materials are also used in energy conversion systems like solid oxide fuel cells. In these systems, electrical current is produced by the reaction occurring between air and fuels such as hydrogen or hydrocarbons. Zirconia-based materials are useful in membranes (Prakasam et al. 2018, Taroco et al. 2011). Some of the materials such as different structures of boron-based nitrides and titanium-based carbides are used for fuel cell preparation and avoiding leakage. Advanced ceramics have application in the energy storage such as batteries and supercapacitors. The supercapacitors arose as an efficient and alternative energy storage candidate to the batteries.

1.5.2 Advanced Ceramics for Environmental Applications

Environmental applications of advanced ceramics are mainly focused on two categories: (1) sensing applications, and (2) removal of pollutants. Sensor technology is a vast field which is affected by the development of different ceramic materials due to applicability and tenability of material properties. Basically, chemical sensing is a process where certain ions, molecules, complexes, movements are detected by analyte on sensing device or sensing materials in terms of different responses. Various types of sensors are used by consumers in their day to day routine from personal to industrial scale such as motion sensors, safety sensors, figure-print sensors, flame sensors, pressure sensors as well as humidity sensors. Due to tunable properties, advanced ceramic materials fulfilled the basic requirements of different type of sensors such as harsh working conditions, selectivity, sensitivity, stability as well as repeatability etc. Different types of gas sensors based on ceramic materials have been explored for detection of different gases such as CH_4, O_2, CO_2, NO_2, H_2, CO, NH_3 as well as volatile compounds etc. Oxides and nitrides-based advanced ceramic materials have been extensively explored for sensing applications (Kumar et al. 2015, 2017, 2020, Sajid and Feng 2014). Also, advanced oxide ceramics have been explored as a potential material for humidity sensors which measures the electrical resistance variation by water vapour absorption.

Water remediation and waste water treatment found the usefulness of advanced ceramic materials due to their tunable chemical, electrical and structural properties. Highly porous structures of ceramic materials can be prepared using different methods and these are useful in membranes, photocatalysis, treatment of coloured dyes used in different fields (He et al. 2019, García et al. 2012, Singh et al. 2018a, b) etc. High permeability, selectivity and high surface area makes ceramic material-based membranes useful in waste water treatment and water desalination. Various ceramic materials have been extensively explored for photocatalysis around the globe. Oxides, nitrides, and their composites-based ceramic materials are examples of materials used in water treatment and photocatalysis (Singh et al. 2017, 2018a, b, Krishnan et al. 2019, Singh et al. 2020). Oxides and nitride ceramic materials such as TiO_2, Al_2O_3, ZrO_2, BN, C_3N_4, SiO_2 etc. are preferred for membrane fabrication due to their high chemical stability, high mechanical strength, large surface area and more lifetime. The filter membranes of advanced ceramics can be utilized for clean drinking water. Apart from industrial dyes in water, the presence of heavy metal ions and pesticides also pose a huge threat to humans, animals and aquatic life. Metal oxide-based nanostructures and nanocomposites have gained tremendous research interest from researchers to be used for efficient removal of heavy metals from the contaminated water.

1.6 SUMMARY

From synthesis, properties and application perspective, advanced ceramics is elaborated in detail. Oxides, nitrides and carbides and their composites are promising candidates for fabrication of advanced ceramic materials. Properties of these materials can be modified as per functional requirement by varying synthesis conditions such as temperature, precursor amounts, solvent ratio, stabilizing gent, reducing agent, pH etc. Recent progress in advanced ceramics showed their usefulness towards various applications than traditional ceramics. Additionally, several application fields including environmental applications, biomedical, sensor technology, packaging, energy conversion devices, power electronic, electrical devices and waste water treatment etc. have found the usage of advanced ceramic materials.

1.7   REFERENCES

Aza, A.H.D., J. Chevalier, G. Fantozzi, M. Schehl and R.Torrecillas. 2002. Crack growth resistance of alumina, zirconia and zirconia toughened alumina ceramics for joint prostheses. Biomaterials 23: 937-945.

Bi, J.X., C.F. Xing, Y.H. Zhang, C.H. Yang and H.T. Wu. 2017. Correlation of crystal structure and microwave dielectric properties of $Zn_{1-x}Ni_xZrNb_2O_8$ ($0 \le x \le 0.1$) ceramics. J. Alloys Compd. 727: 123-134.

Boch, P. and J.-C. Niepce. 2001. Ceramic Materials: Processes, Properties and Applications. Hermès Science Publications, France.

Cannon, W.R., S.C. Danforth, J.H. Flint, J.S. Haggerty and R.A. Marra. 1982. Sinterable ceramic powders from laser-driven reactions: I, process description and modeling. J. Am. Ceram. Soc. 65: 324-330.

Danforth, S.C. 1992. Synthesis and processing of ultrafine powders for Si_3N_4 ceramics. Nanostruct. Mater. 1: 197-202.

Dou, L., X. Zhang, X. Cheng, Z. Ma, X. Wang, Y. Si, et al. 2019. Hierarchical cellular structured ceramic nanofibrous aerogels with temperature-invariant superelasticity for thermal insulation. ACS Appl. Mater. Interfaces. 11: 29056-29064.

Dziadek, M., E.S.-Zych and K.C.-Kowalska. 2017. Biodegradable ceramic-polymer composites for biomedical applications: A review. Mater. Sci. Eng. C. 71: 1175-1191.

Erhart, J., P. Půlpán and M. Pustka. 2017. Piezoelectric Ceramic Resonators. Springer International Publishing, Switzerland.

Feng, S.-H. and G.-H. Li. 2017. Hydrothermal and solvothermal syntheses. pp. 73-104. In: Ruren Xu and Yan Xu [eds.]. Modern Inorganic Synthetic Chemistry (Second Edition). Elsevier B.V. Amsterdam, Netherlands.

Frketic, J., T. Dickens and S. Ramakrishnan. 2017. Automated manufacturing and processing of fiber-reinforced polymer (FRP) composites: An additive review of contemporary and modern techniques for advanced materials manufacturing. Addit. Manuf. 14: 69-86.

García, C.M., D.E.-Quesada, L.P.-Villarejo, F.J.I.-Godino and F.A.C.-Iglesias. 2012. Sludge valorization from wastewater treatment plant to its application on the ceramic industry. J. Environ. Manage. 95: S343-S348.

Gisele Azimi, G., R. Dhiman, H.-M. Kwon, A.T. Paxson and K.K. Varanasi. 2013. Hydrophobicity of rare-earth oxide ceramics. Nature Mater. 12: 315-320.

Greil, P. 2002. Advanced engineering ceramics. Adv. Eng. Mater. 4: 247-254.

He, Z., Z. Lyu, Q. Gu, L. Zhang and J. Wang. 2019. Ceramic-based membranes for water and wastewater treatment. Colloids Surf. A: Physicochem. Eng. Asp. 578: 123513.

Heimann, R.B. 2010a. Non-oxide ceramics: Structure, technology, and applications. pp. 421-480. In: R.B. Heimann [ed.]. Classic and Advanced Ceramics: From Fundamentals to Applications. WILEY-VCH Verlag GmbH and Co. KGaA, Weinheim, Germany.

Heimann, R.B. 2010b. Introduction to advanced ceramic. pp. 157-174. In: R.B. Heimann [ed.]. Classic and Advanced Ceramics: From Fundamentals to Applications. WILEY-VCH Verlag GmbH and Co. KGaA, Weinheim, Germany.

Kaur, M., P. Singh, K. Singh, U.S. Gaharwar, R. Meena, M. Kumar, et al. 2020a. Boron nitride ([10]BN) a prospective material for treatment of cancer by boron neutron capture therapy (BNCT). Mater. Lett. 259: 126832.

Kaur, M., K. Singh, P. Singh, A. Kaur, R. Meena, G.P. Singh, et al. 2020b. Emerging aluminium nitride nanoparticles: Chemical synthesis and exploration of their biocompatibility and anticancer activity against cervical cancer cells. Nanomed. J. 7: 194-198.

Kormann, M., H. Ghanem, H. Gerhardand and N. Popovska. 2008. Processing of carbide-derived carbon (CDC) using biomorphic porous titanium carbide ceramics. J. Eur. Ceram. Soc. 28: 1297-1303.

Krishnan, U., M. Kaur, G. Kaur, K. Singh, A.R. Dogra, M. Kumar, et al. 2019. MoS_2/ZnO nanocomposites for efficient photocatalytic degradation of industrial pollutants. Mater. Res. Bull. 111: 212-221.

Kulik, O.P. 1999. Current state of development of new ceramic materials (review of foreign literature). Powder Metall. Met. Ceram. 38: 93-101.

Kumar, M., A. Kumar and A.C. Abhyankar. 2015. Influence of texture coefficient on surface morphology and sensing properties of w-doped nanocrystalline tin oxide thin films. ACS Appl. Mater. Interfaces. 7: 3571-3580.

Kumar, M., B. Singh, P. Yadav, M. Kumar, K. Singh, A.C. Abhyankar, et al. 2017. Effect of structural defects, surface roughness on sensing properties of Al doped ZnO thin films deposited by chemical spray pyrolysis technique. Ceram. Int. 43: 3562-3568.

Kumar, M., V. Bhatt and J.-H. Yun. 2020. Hierarchical 3D micro flower-like Co_3O_4 structures for NO_2 detection at room temperature. Phys. Lett. A. 384: 126477.

Locatelli, M.-L., R. Khazaka, S. Diaham, C.-D. Pham, M. Bechara, S. Dinculescu, et al. 2014. Evaluation of encapsulation materials for high-temperature power device packaging. IEEE Trans. Power Electron. 29: 2281-2288.

Mukherjee, M. 2011. Silicon Carbide: Materials, Processing and Applications in Electronic Devices. InTech, Janeza Trdine, Croatia.

Nair, D.M., K.M. Nair, M.F. McCombs, J.C. Malerbi and J.M. Parisi. 2014. Method of manufacturing high frequency receiving and/or transmitting devices from low temperature co-fired ceramic materials and devices made therefrom. U.S. Patent # 8,633,858 B2.

Otitoju, T.A., P.U. Okoye, G. Chen, Y. Li, M.O. Okoye and S. Li. 2020. Advanced ceramic components: Materials, fabrication, and applications. J. Ind. Eng. Chem. 85: 34-65.

Petzow, G. and M. Herrmann. 2002. Silicon nitride ceramics. pp. 47-167. In: M. Jansen [ed.]. High Performance Non-Oxide Ceramics II. Structure and Bonding, vol 102. Springer, Berlin, Heidelberg.

Prakasam, M., S. Valsan, Y. Lu, F. Balima, W. Lu, R. Piticescu, et al. 2018. Nanostructured pure and doped zirconia: synthesis and sintering for SOFC and optical applications. pp. 85-106. In: M. Liu [ed.]. Sintering Technology: Method and Application. InTech Open Limited, London, UK.

Prasad, S., V. Kumar, S. Kirubanandam and A. Barhoum. 2018. Engineered nanomaterials: nanofabrication and surface functionalization. pp. 305-340. In: Ahmed Barhoum and Abdel Salam Hamdy Makhlouf [eds.]. Emerging Applications of Nanoparticles and Architecture Nanostructures: Current Prospects and Future Trends. Elsevier Inc. Amsterdam, Netherlands.

Rödel, J., A.B.N. Kounga, M.W.-Eibl, D. Koch, A. Bierwisch, W. Rossner, et al. 2009. Development of a roadmap for advanced ceramics: 2010-2025. J. Eur. Ceram. Soc. 29: 1549-1560.

Sajjad, M. and P. Feng. 2014. Study the gas sensing properties of boron nitride nanosheets. Mater. Res. Bull. 49: 35-38.

Sazegar, M., Y. Zheng, H. Maune, C. Damm, X. Zhou, J. Binder, et al. 2011. Low-cost phased-array antenna using compact tunable phase shifters based on ferroelectric ceramics. IEEE Trans. Microw. Theory Techn. 59: 1265-1273.

Schwentenwein, M. and J. Homa. 2015. Additive manufacturing of dense alumina ceramics. Int. J. Appl. Ceram. Tec. 12: 1-7.

Sebastian, M.T., R. Ubic and H. Jantunen. 2015. Low-loss dielectric ceramic materials and their properties. Int. Mater. Rev. 60: 392-412.

Segal, D. 1997. Chemical synthesis of ceramic materials. J. Mater. Chem. 7: 1297-1305.

Shi, Y.F., Y. Meng, D.H. Chen, S.J. Cheng, P. Chen, H.F. Yang, et al. 2006. Highly ordered mesoporous silicon carbide ceramics with large surface areas and high stability. Adv. Funct. Mater. 16: 561-567.

Singh, B., G. Kaur, P. Singh, K. Singh, B. Kumar, A. Vij, et al. 2016. Nanostructured boron nitride with high water dispersibility for boron neutron, capture therapy. Sci. Rep. 6: 35535.

Singh, B., G. Kaur, P. Singh, K. Singh, J. Sharma, M. Kumar, et al. 2017. Nanostructured BN-TiO_2 composite with ultra-high photocatalytic activity. New J. Chem. 41: 11640-11646.

Singh, B., K. Singh, M. Kumar, S. Thakur and A. Kumar. 2020. Insights of preferred growth, elemental and morphological properties of BN/SnO_2 composite for photocatalytic applications towards organic pollutants. Chem. Phys. 531: 110659.

Singh, K., A. Thakur, A. Awasthi and A. Kumar. 2020a. Structural, morphological and temperature-dependent electrical properties of BN/NiO nanocomposites. J. Mater. Sci.: Mater. Electron. 31: 13158-13166.

Singh, K., M. Kaur, I. Chauhan, A. Awasthi, M. Kumar, A. Thakur, et al. 2020b. BN/NiO nanocomposites: Structural, defect chemistry and electrical properties in hydrogen gas atmosphere. Ceram. Int. 46: 26233-26237.

Singh, P., B. Singh, M. Kumar and A. Kumar. 2014. One step reduction of boric acid to boron carbide nanoparticles. Ceram. Int. 40: 15331-15334.

Singh, P., G. Kaur, K. Singh, B. Singh, M. Kaur, M. Kaur, et al. 2018a. Specially designed B_4C/SnO_2 nanocomposite for photocatalysis: Traditional ceramic with unique properties. Appl. Nanosci. 8: 1-9.

Singh, P., G. Kaur, K. Singh, M. Kaur, M. Kumar, R. Meena, et al. 2018b. Nanostructured boron carbide (B_4C): A biocompatible and recyclable photocatalyst for efficient waste water treatment. Materialia 1: 258-264.

Somia, S. 2013. Handbook of Advanced Ceramics: Materials, Applications, Processing, and properties. Academic Press, Waltham, MA, USA.

Taroco, H.A, J.A.F. Santos, R.Z. Domingues and T. Matencio. 2011. Ceramic Materials for Solid Oxide Fuel Cells. pp. 423-446. *In*: C. Sikalidis [ed.]. Advances in Ceramics – Synthesis and Characterization, Processing and Specific Applications. InTech, Janeza Trdine, Croatia.

Tiwari, A., R.A. Gerhardt and M. Szutkowska. 2016. Advanced Ceramic Materials. Scrivener Publishing, John Wiley and Sons, Inc. Hoboken, New Jersey, and Scrivener Publishing LLC, Beverly, Massachusetts.

Weimar, A.M. 1997. Carbide, Nitride and Boride Materials Synthesis and Processing. Chapman and Hall, London.

Xu, T., N. Zhang, H.L. Nichols, D. Shi and X. Wen. 2007. Modification of nanostructured materials for biomedical applications. Mater. Sci. Eng. C. 27: 579-594.

Yang, J. 2013. A silicon carbide wireless temperature sensing system for high temperature applications. Sensors 13: 1884-1901.

Yu, X., T.J. Marks and A. Facchetti. 2016. Metal oxides for optoelectronic applications. Nature Mater. 15: 383-396.

Zhou, D., L.-X. Pang, D.-W. Wang, C. Li, B.-B. Jin and I.M. Reaney. 2017. High permittivity and low loss microwave dielectrics suitable for 5G resonators and low temperature co-fired ceramic architecture. J. Mater. Chem. C. 5: 10094-10098.

2

Advanced Nanostructured Perovskite Oxides: Synthesis, Physical Properties, Structural Characterizations and Functional Applications

Kai Leng, Weiren Xia and Xinhua Zhu[*]

2.1 INTRODUCTION

The term of "perovskite" is coined from $CaTiO_3$ mineral crystal and named after the Russian mineralogist (Cheng and Lin 2010). Today, perovskite oxides have become a general name for functional oxides with an ABO_3 chemical formula. Due to the much flexibilities of A- and B-site cations (Pena and Fierro 2001), ABO_3 perovskite oxides exhibit much high chemical and structural adaptabilities, leading to large number of compounds. Perovskite oxides exhibit multifunctional properties such as dielectric, ferroelectric, piezoelectric, half-metallic ferromagnetic, antiferromagnetic, and so on, which have been widely investigated and used in microelectronic industry during the past century. As the geometry sizes of perovskite oxides are reduced to nanometer sizes, novel physical properties are discovered in nanostructured perovskite oxides, which are much different from that exhibited by the bulk and film counterparts. Therefore, nanoelectronic oxide devices based on nanostructured perovskite oxides are growing fast, which have an important impact on our daily life. Here, we address the synthesis methods, property and structural characterizations, and functional applications of nanostructured perovskite oxides.

2.2 DESCRIPTION OF ABO_3 PEROVSKITE STRUCTURE

2.2.1 Introduction

In the perovskite structure with an ABO_3 compositional formula, A- and B-sited cations with different ionic radii are bonded with oxygen anions in 12- and 6-coordinated numbers, respectively (Zhou et al. 2018). Ideally, perovskite structure can be described by a cube, where the A-sited ions enter the corners of the cube (0, 0, 0), and B-site ions occupy the body center of the cube, $\left(\frac{1}{2}, \frac{1}{2}, \frac{1}{2}\right)$, lying within a BO_6 octahedron. The face centered positions such as $\left(\frac{1}{2}, \frac{1}{2}, 0\right)$, $\left(\frac{1}{2}, 0, \frac{1}{2}\right)$, and

---

National Laboratory of Solid State Microstructures, School of Physics, Nanjing University, Nanjing 210093, China.
[*] Corresponding author: xhzhu@nju.edu.cn

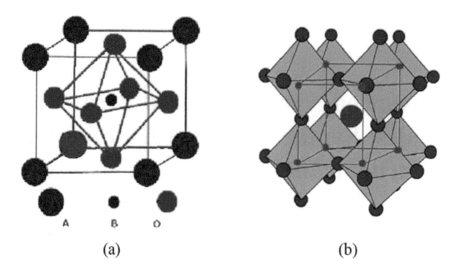

Fig. 2.1 (a) Three-dimensional view of conventional unit cell of a cubic ABO_3 perovskite structure, (b) ABO_3 perovskite structure formed by extending the corner-shared BO_6 octahedrons in three crystallographic directions

$\left(0, \dfrac{1}{2}, \dfrac{1}{2}\right)$, are occupied by oxygen anions, as illustrated in Fig. 2.1(a). The structure is also constructed

by extending the regular corner-shared BO_6 octahedrons in three crystallographic directions (Tan et al. 2014), as shown in Fig. 2.1(b). The multifunctional properties of perovskite oxides closely rely on the chemical selections of the A- and/or B-sited metal ions. However, due to the cationic substitution limit for the A- and/or B-sited ions, a deviation from cubic structure is generated, leading to the distorted structures with low symmetries such as tetragonal or orthorhombic symmetry. To express the distortion degree generated in the perovskite structure, Goldschmidt (1926) proposed the tolerance factor of perovskite oxides, t, which is given by Eq. (1)

$$t = \frac{r_A + r_O}{\sqrt{2}(r_B + r_O)} \tag{1}$$

Here, r_A, r_B, and r_O are the ionic radii of the A and B cations, and oxygen anions, respectively. It is well known that the phase structure stability of the formed perovskite structure by different cationic combinations can be predicted by the tolerance factor t (Galasso 1969). If the t value approaches 1.0, cubic perovskite is preferred to be formed (e.g. $t = 0.96$ for cubic $BaTiO_3$ (BTO) at temperature above 120°C; $t = 1.00$ for cubic $SrTiO_3$ (STO); and $t = 1.01$ for cubic $BaZrO_3$ (BZO)). While for $0.985 < t < 1.06$, at room temperature perovskites prefer to be formed without tilting; however, a tilted perovskite structure with antiphase boundary is formed under $0.964 < t < 0.985$. As t is smaller than 0.964, perovskites with phase and antiphase tilting are favored (Reaney et al. 1994). As t is further decreased, the stability of the perovskite phase will be further decreased, finally leading to the collapse of the perovskite matrix structure. However, for $0.70 < t < 0.75$, more stable bixbyte polymorph (α-Mn_2O_3) prefers to be formed. When t is over 1.05, a hexagonal structure is preferred to be formed, where AO_3 layers are hexagonally packed and the BO_6 octahedrons share faces along the hexagonal c-axis. Figure 2.2 illustrates the symmetry of ABO_3 oxides changing with the Goldschmidt tolerance factor, t (Dos Santos-García et al. 2015). Normally, perovskite structure formation requires t in the range of $0.75 < t < 1.0$ (Zhu et al. 2014a).

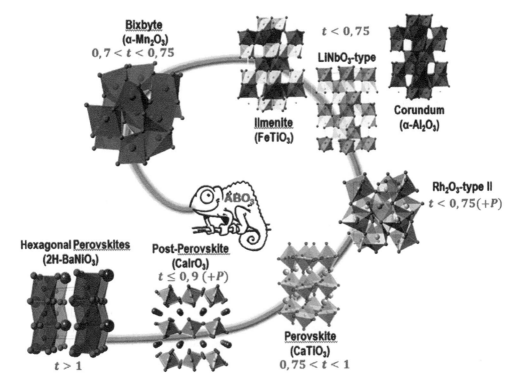

Fig. 2.2 Symmetries of the ABO_3 perovskite structure varying with the Goldschmidt tolerance factor (t). (Reprinted with permission from A.J. Dos Santos-Garcia, E. Solana-Madruga, C. Ritter, D. AvilaBrande, O. Fabelo and R. Saez-Puche, "Synthesis, structures and magnetic properties of the dimorphic Mn_2CrSbO_6 oxide," Dalton Trans. 44(2015): 10665-10672. Copyright 2015, The Royal Society of Chemistry.)

2.2.2 Classification of Perovskite Structures

Perovskite oxides can be crystallized in different phase structures with the symmetry ranging from high one (cubic system) to low one (triclinic system), which are dependent upon the compositional combinations of the A- and B-sited ions, and tilting and rotation of the BO_6 octahedrons. The most typical and important perovskite structures are introduced in this section.

2.2.2.1 Cubic Perovskite Structure

It is known that the t value for an ideal perovskite cubic structure is equal to 1.0, whereas for $t \neq 1.0$, distorted perovskite structures with lower symmetry will be formed. In the ideal cubic perovskite structure such as STO, its three-dimensional network of the crystal structure can be formed by extending the corner-shared (TiO_6) octahedrons along three crystallographic directions, where Sr^{2+} ions are located in the 12-fold cavities between the polyhedrons (seen in Fig. 2.1b). For the distorted ABO_3 perovskite structure with the t value smaller than 1.0, the (BO_6) octahedrons are tilted to fill the available space. Generally, for a cubic perovskite structure t should be in the range between 0.89 and 1.0.

2.2.2.2 Tetragonal Perovskite Structure

In the tetragonal perovskite structure such as tetragonal lead titanate ($PbTiO_3$, PTO), the tetragonal distortion (c/a) is as high as 1.064 due to the large radius of the Pb^{2+} ion with $6s^2$ lone-pair electrons (Damjanovic 1998). The high t value over 1.0 indicates that the Ti-O octahedral bond is elongated along the c direction as compared to the cubic case, leading to a larger lattice parameter c.

2.2.2.3 Rhombohedral Perovskite Structure

Some ABO_3 compounds exhibit rhombohedrally structural distortion. The representative sample is $BiFeO_3$ (BFO), which crystallizes in a rhombohedral perovskite structure with $R3c$ space group (R phase) and the tilted (FeO_6) octahedron (Damodaran et al. 2011). Due to the structural flexibility, BFO crystal structure is easily modulated by strain, compositions, and temperature. For example, BFO undergoes a phase transition from R phase to $Pbnm$ phase at high temperature of $\sim$ 1100 K (Arnold et al. 2010, Levin et al. 2011).

2.2.2.4 Orthorhombic Perovskite Structure

Among the ABO_3 perovskite-type compounds, the orthorhombic perovskite structure is very common. For example, many manganese perovskite $LnMnO_3$ compounds (Ln = trivalent rare-earth elements) crystallize in an orthorhombic perovskite structure (Muller and Roy 1974). The orthorhombic lattice parameters (a, b, and c) are bound with the pseudocubic lattice parameter, a_{ps}, which is described as follows:

$$a = b = a_{ps}/\sqrt{2} \tag{2}$$
$$c = a_{ps} \tag{3}$$

Due to the large misfits in the ionic sizes of orthorhombic perovskite compounds, the rotation of the BO_6 octahedrons about (110) axis can result in the O-type orthorhombic $GdFeO_3$ (GFO) structure, as shown in Fig. 2.3(a). In the orthorhombic GFO compound, the Fe-O-Fe bond angle is sensitive to the ionic radius of the A cation, reducing from 180° to 161° (Coey et al. 1999). In the O-type structure, the lattice parameters (a, b, and c) are satisfied with $a < c/\sqrt{2} < b$. However, in the O′-type orthorhombic structure (Fig. 2.3(b)), they are satisfied with $c/\sqrt{2} < a < b$. For example, in the $LaMnO_3$ crystal with an O′-type orthorhombic structure (space group $Pbnm$), the lattice constants were $a = 0.554$ nm, $b = 0.575$ nm, and $c = 0.770$ nm; and the Mn-O-Mn bond angle was $\sim$ 155°. Due to the large octahedral distortion caused by the Jahn-Teller effect of Mn^{3+} ion, the MnO_6 octahedrons are forced to be compressed along the c axis, leading to a shorter Mn-O bond length in c axis direction, whereas longer and shorter Mn-O bonds in the ab plane (Coey et al. 1999).

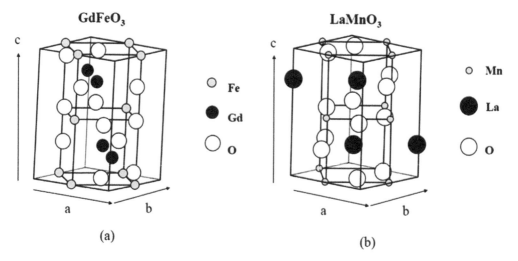

Fig. 2.3 (a) O-type orthorhombic $GdFeO_3$ (GFO) structure distorted from an ideal perovskite structure, (b) O′-type orthorhombic $LaMnO_3$ structure with severe octahedral distortion caused by the Jahn-Teller effect. (Reprinted with permission from J.M. Coey, D.M. Viret and S. von Molnaír, "Mixed-valence manganites," Adv. Phys. 48 (1999): 167-293. Copyright 1999, Taylor and Francis Ltd.)

2.2.2.5 Monoclinic Perovskite Structure

The monoclinic perovskite compounds have a $\sqrt{2}a_p \times \sqrt{2}a_p \times a_P$ unit cell with a space group of Cm, where a_P denotes the lattice parameter of a cubic perovskite. In the $BiCoO_3$-$BiFeO_3$ system, a monoclinic $\sqrt{2}a_p \times \sqrt{2}a_p \times a_P$ structure was found to exist across wide ranges of composition and temperature (Oka et al. 2012). A monoclinic phase was observed in the composition close to $BiCo_{0.3}Fe_{0.7}O_3$ at room temperature, as confirmed by small-angle XRD patterns (Fig. 2.4) and electron diffraction patterns (Fig. 2.5) (Oka et al. 2012). Furthermore, it is also found that the polarization of $BiCo_{0.3}Fe_{0.7}O_3$ can rotate between the $(001)_c$ and $(111)_c$ directions of a pseudocubic cell, which are closely related to composition and temperature.

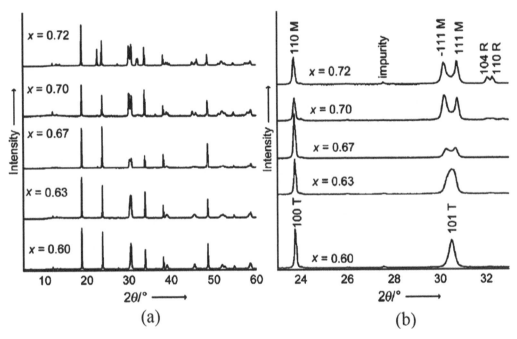

Fig. 2.4 (a) XRD patterns of the $BiCo_{1-x}Fe_xO_3$ (x = 0.60, 0.63, 0.67, 0.70, and 0.72) compounds measured at room temperature, (b) Local XRD patterns. The suffixes R, T, and M after the indices represent rhombohedral, tetragonal, and monoclinic phases, respectively. (Reprinted with permission from K. Oka, T. Koyama, T. Ozaaki, S. Mori, Y. Shimakawa and M. Azuma, "Polarization rotation in the monoclinic perovskite $BiCo_{1-x}Fe_xO_3$," Angew. Chem. Int. Ed. 51 (2012): 7977-7980. Copyright 2012, Wiley-VCH.)

2.2.2.6 Triclinic Perovskite Structure

Perovskite nickelate oxides, $MNiO_3$ with M = Bi or rare-earth (RE) elements, exhibit some interesting novel phenomena such as charge and orbital ordering, metal-insulator transition (MIT) (Alonso et al. 1999), negative thermal expansion (Azuma et al. 2011) and multiferroicity (Giovannetti et al. 2009). As one of perovskite nickelates, $BiNiO_3$ has a stable triclinic structure (with space group of P-1) across a temperature range of 5-420 K (Ishiwata et al. 2002, Carlsson et al. 2008). The unit cell of triclinic $BiNiO_3$ is schematically shown in Fig. 2.6(a), where four formula units are included (Pugaczowa-Michalska and Kaczkowski 2017). Figures 2.6(b)-(c) show the projections of the unit cell in the bc- and ab-planes, respectively. At room temperature, the triclinic $BiNiO_3$ has a G-type antiferromagnetic (AFM) insulating ground state with T_N = 300 K (Carlsson et al. 2008). The synchrotron radiation XRD patterns reveal that the valence state distribution in triclinic $BiNiO_3$ is $BiNiOBi_{0.5}^{3+}Bi_{0.5}^{5+}Ni^{2+}O_3$, and such charge distribution at Bi-sites (Bi^{3+} and Bi^{5+}) can be effectively restrained with external pressure (in the order of several GPa) or partial La-substituting the bismuth,

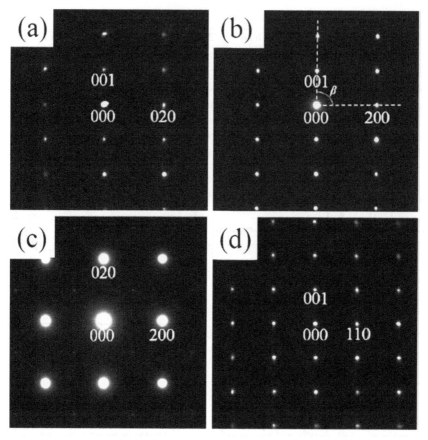

Fig. 2.5 Selected area electron diffraction patterns of the $BiFe_{0.30}Co_{0.70}O_3$ at 298 K taken along (a) (100), (b) (010), (c) (001), (d) (110) directions. The electron diffraction patterns are indexed by a cubic perovskite structure. The monoclinic angle b was estimated to be ~91.48° in the reciprocal lattice space. (Reprinted with permission from K. Oka, T. Koyama, T. Ozaaki, S. Mori, Y. Shimakawa and M. Azuma, "Polarization rotation in the monoclinic perovskite $BiCo_{1-x}Fe_xO_3$," Angew. Chem. Int. Ed. 51 (2012): 7977-7980. Copyright 2012, Wiley-VCH.)

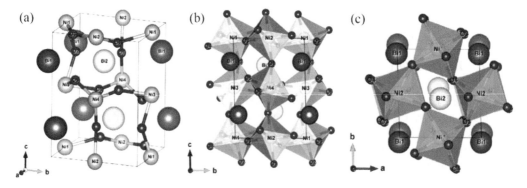

Fig. 2.6 Schematic diagrams of the triclinic $BiNiO_3$ with space group of P-1. (a) Unit cell composing of four $BiNiO_3$ formula units, (b) Unit cell projection in the bc plane with four different NiO_6 octahedrons, (c) Unit cell projection in the ab plane with two different NiO_6 octahedrons. (Reprinted with permission from M. Pugaczowa-Michalska and J. Kaczkowski, "DFT + U studies of triclinic phase of $BiNiO_3$ and La-substituted $BiNiO_3$," Comput. Mater. Sci. 126(2017): 407-417. Copyright 2017, Elsevier.)

driving the system into the orthorhombic metallic phase (Ishiwata et al. 2003, Oka et al. 2013). The triclinic phase was also reported in perovskite BFO epitaxial films, which were deposited on the STO substrates with (130)- and (120)-orientations. They have the tilting angles of 26° and 18° away from the (100) plane of the STO substrate. The lattice parameters of these films vs the tilted angle from the (100) towards the (110) are shown in Fig. 2.7, where almost linear relationship is observed (Yan et al. 2009). It is also found that the in-plane lattice parameters increase gradually as the tilting angle increased, but the out-of-plane parameter decrease gradually.

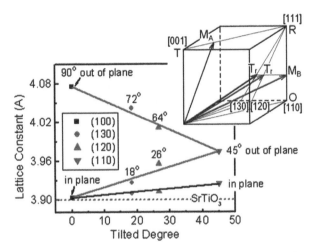

Fig. 2.7 Lattice constants of the (100)-, (130)-, (120)- and (110)-oriented BiFeO$_3$ (BFO) thin films vs the tilting angle between the film and substrate. Insert displays the M_A and M_B (triclinic phases), which are the results of the combination of the constraint stress from the (100)-, (130)-, (120)- and (110)-oriented substrates and the stable R phase of BFO, respectively. (Reprinted with permission from L. Yan, H. Cao, J.F. Li and D. Viehland, "Triclinic phase in tilted (001) oriented BiFeO$_3$ epitaxial thin films," Appl. Phys. Lett. 94 (2009) 132901 (1-4). Copyright 2005, American Institute of Physics.)

2.3   SYNTHESIS OF NANOSTRUCTURED PEROVSKITE OXIDES

2.3.1   Synthesis of Zero-Dimensional Nanostructured Perovskite Oxides

The wide applications of nanostructured perovskite oxides in different fields require their mass production in a uniform way. However, the traditional productions of perovskite oxides obtained from solid-state reactions do not meet the above requirements (Varma et al. 2016). To address these issues and to produce stoichiometric perovskite oxide powders with high quality, novel physical and chemical routes have been developed to synthesize perovskite ceramics in the nanopowder form, which are described as following.

2.3.1.1 Physical Routes

2.3.1.1.1 Mechanical Milling Method

Mechanical milling method is one of the up-to-date methods to synthesize fine ceramic powders by milling processing with high-energy balls, which is also called mechanosynthesis. This method not only allows for the synthesis of very fine powders but also synthesizes (at room temperature) those powders that are only synthesized under high temperature and/or high-pressure. In the mechanical milling process, many processing parameters can be controlled to achieve the required purpose (Suryanarayana et al. 2001). Some critical parameters include milling process (types, speed and time, and environment), operated temperature, and the external fields, electric or magnetic field, during milling process. Recently, perovskite oxide nanoparticles such as BTO (Stojanovic 2003, Phan et al. 2013, Neogi et al. 2015, Moghtada et al. 2018), PTO (Xue et al. 1999a, Leite et al. 2001, Yu et al. 2002, Stojanovic 2003), $Pb(Mg_{1/3}Nb_{2/3})O_3$ (PMN) (Wang et al. 1999, Xue et al. 1999b, 2001, Kong et al. 2001), $(Bi,Na)TiO_3$ (Lee et al. 2017), $(K, Na)NbO_3$ (Lee et al. 2017), $Pb(Zr,Ti)O_3$ (PZT) (Avvakumov et al. 1992, Xue et al. 1999c, Brankovic´ et al. 2003, Miclea et al. 2004), and LMO (Zhang and Saito 2000, Shannigrahi and Tan 2011, Blackmore et al. 2020) have been prepared by this method. Recently, at room temperature BFO multiferroic nanopowders are also prepared directly by mechanical milling method (Szafraniak et al. 2007, Perejon et al. 2013, Cristobal and Botta 2013). For more details about the mechanochemical synthesis of perovskite oxide powders and ceramics, the reader is referred to the previous reviews contributed by Kong et al. (2008) and Stojanovic (2003).

Despite the mechanical milling method being capable of synthesizing the nanometer-sized powders at room temperature, it faces a great problem in mass production for large-scale industrial applications due to the limited batch quantity and long processing time. The contamination produced from the milling media during the high-energy ball-milling process is also an issue to be solved.

2.3.1.1.2   Molten-Salt Method

Molten-salt method is one of the simple and effective methods for synthesizing nanoparticles at lower temperatures. In the molten-salt approach, inorganic molten salts not only serve as the medium but also have an impact on decreasing the system reaction temperature and enhancing the reaction rate and oxides. During the past years, many different perovskite oxide nanoparticles such as $LaAlO_3$ (LAO) (Maczka et al. 2012, Mendoza-Mendoza et al. 2012a, b), $YAlO_3$ (Lee 2013), $EuAlO_3$ (Lee 2017), BTO (Sahoo and Mazumder 2010, Zhang et al. 2012, Lee et al. 2016, Xue et al. 2017, Jiang et al. 2019), BFO (Chen et al. 2007, Zheng et al. 2012, Liu et al. 2013, Zhu et al. 2014c), PZT (Cho and Biggers 1983, Bortolani and Dorey 2010, Ahda et al. 2019), PMN (Yoon et al. 1998), $Pb(Fe_{1/2}Nb_{1/2})O_3$ (PFN) (Yoon et al. 1998), $Ba(M_{1/3}Nb_{2/3})O_3$ (BMN) (M = Mg, Zn) (Thirumal et al. 2001, Tian et al. 2008), perovskites $LaMO_3$ (M = Mn, Fe, Co, Ni) (Matei et al. 2007, Yang et al. 2010) have been obtained by this method. The important processing parameters such as the used molten salt types and quantities, annealing temperature and time, and heating or cooling rates have great impact on the quality of final products. Recent reviews relevant on this topic are available, which are contributed by Wu et al. (2017) and Xue et al. (2018).

2.3.1.1.3   Physical Vapor Deposition (PVD) Methods

PVD method is also used to synthesize perovskite oxide nanopowders, which is based on the formation of gaseous precursor molecules in aersol reactors via suitable physical methods (Swihart 2003). After the reaction between the vapor precursor molecules, tiny nuclei of the desired phase are formed, which is named as gas-to-particle conversion. Subsequently, those tiny nuclei will grow up through collision and coalescence process, and form the agglomerates (Wu et al. 1993). The key feature of the PVD methods is the manner of producing the vapor and/or plasma from the target material. The deposited temperature, time, and the produced particles per unit volume are the important processing parameters of physical vapor deposition methods, which have an important impact on the structural features of the final products (e.g. particle size and its distribution, grain boundaries, defect concentrations, and crystallinity). Recently, Seol et al. (2002) synthesized monodisperse PZT nanoparticles by gas phase vapor deposition with particle sizes across 4-20 nm. This process includes the following three steps: (i) producing amorphous and irregular PZT nanoparticles by laser ablation of PZT ceramics in O_2 atmosphere, (ii) crystallization of these nanoparticles by an on-line thermal treatment; (iii) separation of the crystalline nanoparticles by differential mobility analyzer, yielding monodisperse PZT nanoparticles with high quality. Perovskite nanoparticles prepared by this method exhibit the features of monodispersity, high purity and single-crystallinity, which will not only be suitable for investigation on the size effects of ferroelectrics, but also be useful for fabricating perovskite ferroelectric-based nanodevices.

2.3.1.1.4 Focus Ion Beam (FIB) Method

Recently, many perovskite oxide nanostructures are reported to be fabricated by FIB technology, which includes the following main steps: milling process, implantation, ion-induced deposition, and ion-assisted etching. FIB-based nanofabrication method has several advantages such as easy control of the morphology of the produced structures or even patterns, and facile operation. However, their slow milling speeds (especially for fabricating larger area nanostructures), the unexpected damage generated by the incident ions at the sample surface, limit their commercial applications. Morelli et al. (2013) grew BFO nanoisland arrays in large area by a mask-assisted FIB method. The diameters of epitaxial BFO nanoisland were ~250 nm. Nanoscale ferroelectric BTO dots were also prepared by direct FIB patterning the bulk BTO single crystal (Schilling et al. 2009). Figure 2.8 shows some STEM images of ferroelectric BTO nanodots. Ferroelectric domains with striped morphology were clearly observed within the well-defined quadrants, and the domain boundaries between these striped domains lie along the $<110>_{pseudocubic}$ orientation. This is typical of 90° domain walls, as observed in tetragonal BTO crystal.

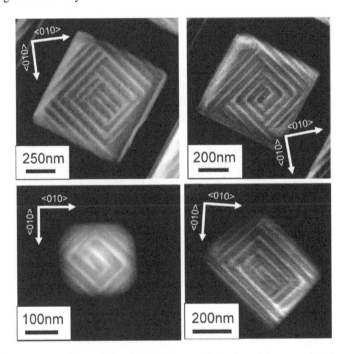

Fig. 2.8 Domain structures observed in the $BaTiO_3$ nanodots viewed by scanning transmission electron microscopy (STEM) along the $<100>_{pseudocubic}$ direction (which was almost vertical to the top surface of the dots). In many cases, packets of 90° stripe domains were found to form into quadrants. (Reprinted with permission from A. Schilling, D. Byrne, G. Catalan, K.G. Webber, Y.A. Genenko, G.S. Wu et al., "Domains in ferroelectric nanodots," Nano Lett. 9 (2009): 3359-3364. Copyright 2009, American Chemistry Society.)

2.3.1.2 Chemical Routes

2.3.1.2.1 Sol-Gel/Sol-Precipitation Synthesis

Sol-gel synthesis is widely applied to synthesize perovskite oxide nanoparticles; for example, BTO (Kavian and Saidi 2009, Panomsuwan and Manuspiya 2019), PTO (Lee et al. 2006, Bhatti et al. 2016, Sagadevan et al. 2016) as well as rare earth-doped BTO nanoparticles (Gomes et al. 2016) have been obtained. This route involves the hydrolysis and condensation of alkoxide-based precursor solution, and drying and pyrolysis of gel to remove the organic matter. Several distinct steps are involved in this route (Cushing et al. 2004): (i) the sol formation, (ii) gel formation via polycondensation reaction, (iii) drying, (iv) dehydration, and (v) annealing the gel at high temperatures. To modulate the morphology, particle sizes and their distribution of the final products, the processing parameters such as annealing temperature and time, atmosphere and heating speed, the used reactants and their concentrations, and pH value should be optimized. The perovskite oxide nanostructure synthesized via sol-gel process usually exhibits different morphologies such as irregular shapes (Hwang et al. 2004, Kavian and Saidi 2009, Li et al. 2009b, Ashiri et al. 2011), spheres (O'Brien et al. 2001, Brutchey and Morse 2006), or cubes (Adireddy et al. 2010, Caruntu et al. 2015). Examples of the nanoparticles synthesized by sol-gel process with different morphologies are illustrated in Figure 2.9 (Kostopoulou et al. 2018, 2019).

The sol-precipitation process is another method for synthesizing perovskite oxide nanoparticles, where strong alkaline conditions are often utilized to yield crystalline perovskite oxide nanoparticles at low temperatures. For example, at low temperature of 80°C single-crystalline and cubic 20-nm BTO nanoparticles were obtained by one-step sol-precipitation route under strong alkaline conditions (Fan et al. 2005). Similarly, STO nanoparticles with particle sizes across 40-60 nm were also synthesized at 160°C (Hao et al. 2014).

2.3.1.2.2   Co-precipitation Method

The co-precipitation method is a simple approach to yield fine perovskite oxide powders at low temperatures, where various ionic species are co-precipitated from a solution phase to form solid particles. In this method, several types of precursors (e.g. oxides, alkoxides, inorganic salts and nitrates) can be employed. The properties of final precipitated products are closely related to the processing variables such as the solution pH value, co-precipitation rate, washing, drying, and the synthesis temperature. By carefully controlling the above factors, one can obtain homogeneous perovskite oxide nanopowders. For instance, Fox et al. (1991) synthesized the PTO nanopowders by using the co-precipitated method. The particle sizes of the obtained cubic PTO powders varied from 20 nm to 200 nm, whereas the particle sizes of the obtained tetragonal PTO powders were across 100-400 nm. Similarly, Dyakonov et al. (2009) prepared the $(La_{0.7}Sr_{0.3})_{0.9}Mn_{1.1}O_3$ manganite nanoparticles with different average particle sizes. All these nanosized particles show ferromagnetic-like ordering with close phase transition temperatures. Their magnetizations are decreased by increasing the particle sizes, which is ascribed to the high contribution to the magnetization from the surface within smaller particles.

2.3.1.2.3   Pechini Method

The Pechini method is also named as the polymeric precursor method, which involves the following major steps: (i) intensive mixing of the positive ions in a solution, (ii) conversion of the solution into a polymer gel, (iii) polymer matrix removal, and (iv) formation of homogeneous oxide precursors (Shandilya et al. 2016). Today, the Pechini method is widely applied to synthesize perovskite oxide nanomaterials. Jinga et al. (2010) synthesized the $BaMg_{1/3}(Ta_{1-x}Nb_x)_{2/3}O_3$ ($x = 0$-0.2) nanoparticles by a modified Pechini method at 750°C. Similarly, Ianculescu et al. (2007) also prepared the $Ba_{1-x}Sr_xTiO_3$ ($x = 0$-0.35) nanopowders at 850°.

2.3.1.2.4   Hydrothermal Synthesis

Hydrothermal synthesis is a powerful tool for synthesizing perovskite nanoparticles with controllable size and morphology. In this process, starting materials are placed in a closed autoclave under a certain temperature and pressure, where the combination of this and increases in pressure drive the formation of the desired products. The advantages of this method are controllable particle sizes and their distribution, and the morphology of the final products, which are strongly dependent on the starting precursors, reaction temperature and time, pH value, the types and concentrations of mineralizers (Zhu et al. 2005, Miron 2008). Typical examples such as BTO, STO, PTO and BFO nanoparticles synthesized by this method with different morphologies are shown in Fig. 2.9. As compared with the sol-gel process, hydrothermal process can be performed at much lower temperature to yield nanocrystalline materials. For example, single-phase BFO nanocrystallites with controllable morphology and particle size were synthesized at 200°C by hydrothermal process (Niu et al. 2015, Gao et al. 2015, Zhang et al. 2016). Such synthesized temperature is much lower than that required in the sol-gel process (e.g. 650-750°C). However, there are some disadvantages for hydrothermal synthesis process, which includes its high cost of the equipment (the use of an expensive autoclave) and the impossibility of *in situ* monitoring the hydrothermal reaction process ("black box") due to the extremely harsh reaction environment.

2.3.1.2.5 Solvothermal Synthesis

Solvothermal synthesis is a modified hydrothermal process, where the reaction takes place in non-aqueous solutions such as NH_3, methanol, ethanol, and *n*-propanol. The representatives of the perovskite family such as BTO, $BaZrO_3$ (BZO) and $LiNbO_3$ nanoparticles were prepared by this method (Niederberger et al. 2004). Nowadays, perovskite oxide $MTiO_3$ (M = Ba, Sr, Ca, Pb) nanoparticles were systematically synthesized by solvothermal method (Kimijima et al. 2014, Caruntu et al. 2015). The morphology of these perovskite oxides can be spherical, cubic or hollow shapes, and most of them are without capping agents (Kostopoulou et al. 2018). Some typical examples are shown in Fig. 2.9.

2.3.1.2.6 Microwave-Hydrothermal Synthesis

In the microwave-hydrothermal process microwave is introduced, which offers an enhanced crystallization kinetics of the hydrothermal process. Recently, numerous perovskite oxide nanopowders have been synthesized by this method. The typical examples are BTO (Sun et al. 2006, Zhu et al. 2008, Chen et al. 2016), BST (Pązik et al. 2007, Chen et al. 2015), BFO (Joshi et al. 2008, Zhu et al. 2011, Prado-Gonjal et al. 2011, Ponzoni et al. 2013), and BMN (Dias et al. 2007, 2009). For more details about the reaction mechanism of microwave-hydrothermal process and its applications to functional materials, we suggest the readers to read the recent review papers contributed by Yang and Park (2019), Xia et al. (2020), and Zhu and Hang (2013).

2.3.1.2.7 Sonochemical Method

Sonochemical method is one kind of advanced wet chemical processing method, which leads to the formation of the final products based on the ultrasound irradiation effect and acoustic cavitation (Moghtada and Ashiri 2016). The advantages of this method involve high purity, controllable particle size and morphology, and short synthesis time at lower temperatures as compared with other methods (Moghtada and Ashiri 2016). Yu et al. (2003) performed the pioneering works on synthesizing the perovskite nanoparticles by this method. Later, BTO (Utara and Hunpratub 2018, Dang et al. 2010), STO (Moghtada and Ashiri 2015), and BFO (Dutta et al. 2010) nanostructures with different morphologies (e.g. irregular, spherical, rod-like and polygonal shapes) were synthesized by the same method.

2.3.1.2.8 Microemulsion Method

In a microemulsion process system, micrometer-sized droplets (micelle) are distributed in the immiscible solvent, and the micelle is surrounded by an amphiphilic surfactant species at the surface (Lopez-Quitela and Rivas 1993, Kumar and Mittal 1999, Paul and Moulik 2001). The microemulsion technique has attracted much attention due to the nanosized aqueous droplets, which play the role of nanoreactors for synthesizing the perovskite oxide nanoparticles (Pileni 2003). In addition, the sizes of nanoreactors can be easily controlled by the parameters of the microemulsion process system. For example, in a water/oil microemulsion system, the nanosized spherical aqueous water micelles are distributed in an oil matrix, which act as nanoreactors to synthesize perovskite oxide nanoparticles such as BTO, STO, and BST (Su et al. 2007, Chen and Zhu 2007). Their particle sizes were controlled by adjusting the volume ratio of water to the surfactant. The BFO nanoparticles (average particle size 21 nm) were synthesized by this method (Das et al. (2007) at 400°C. The used microemulsion system consisted of CTAB/water/isooctane/butanol.

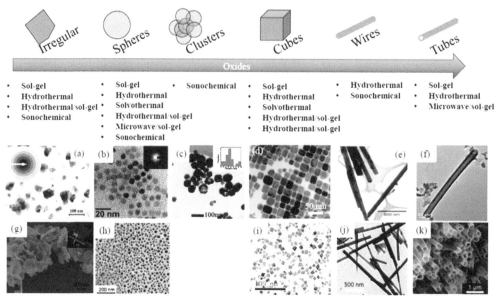

Fig. 2.9 Summary of various perovskite oxide nanostructures synthesized via chemical routes. (a) BaTiO$_3$ nanocrystals with irregular morphology synthesized by sol-gel method (Reprinted with permission from R. Ashiri, A. Nemati, M. S. Ghamsari, S. Sanjabi and M. A. Aalipour, "Modified method for barium titanate nanoparticles synthesis," Mater. Res. Bull. 46(2011): 2291-2295. Copyright 2011, Elsevier), (b) BaTiO$_3$ nanoparticles with spherical morphology synthesized by sol-gel method (Reprinted with permission from S. O'Brien, L. Brus and C.B. Murray, "Synthesis of monodisperse nanoparticles of barium titanate: Toward a generalized strategy of oxide nanoparticle synthesis," J. Am. Chem. Soc. 123(2001): 12085-12086. Copyright 2001, American Chemistry Society), (c) BaTiO$_3$ clusters synthesized by the sonochemical method (Reprinted with permission from S. Utara and S. Hunpratub, "Ultrasonic assisted synthesis of BaTiO$_3$ nanoparticles at 25°C and atmospheric pressure," Ultrason. Sonochem. 41(2018): 441-448. Copyright 2018, Elsevier), (d) BaTiO$_3$ nanocubes synthesized by solvothermal method (Reprinted with permission from D. Caruntu, T. Rostamzadeh, T. Costanzo, S.S. Parizi and G. Carunt, "Solvothermal synthesis and controlled self-assembly of monodisperse titanium-based perovskite colloidal nanocrystals," Nanoscale. 7(2015): 12955-12969. Copyright 2015, The Royal Society of Chemistry), (e) BaTiO$_3$ nanorods synthesized by hydrothermal method (Reprinted with permission from U.A. Joshi and J.S. Lee, "Template-free hydrothermal synthesis of single-crystalline barium titanate and strontium titanate nanowires," Small. 1 (2005): 1172-1176. Copyright 2005, Wiley-VCH), (f) an isolated BaTiO$_3$ nanotube synthesized by using anodic aluminium oxide template (Reprinted with permission from Y.Y. Chen, B.Y. Yu, J.H. Wang, R.E. Cochran and J.J. Shyue, "Template-based fabrication of SrTiO$_3$ and BaTiO$_3$ nanotubes," Inorg. Chem. 48(2009): 681-686. Copyright 2000, American Chemistry Society), (g) PbTiO$_3$ nanosheets synthesized by hydrothermal method (Reprinted with permission from S.Q. Deng, G. Xu, H.W. Bai, L.L. Li, S. Jiang, G. Shen et al., "Hydrothermal synthesis of single-crystalline perovskite PbTiO$_3$ nanosheets with dominant (001) facets," Inorg. Chem. 53(2014): 10937-10943. Copyright 2014, American Chemistry Society), (h) spherical BiFeO$_3$ nanoparticles synthesized by microwave-hydrothermal method (Reprinted with permission from X.H. Zhu, Q.M. Hang, Z.B. Xing, Y. Yang, J.M. Zhu, Z.G. Liu et al. "Microwave hydrothermal synthesis, structural characterization, and visible-light photocatalytic activities of single-crystalline bismuth ferric nanocrystals," J. Am. Ceram. Soc. 94(2011): 2688-2693. Copyright 2011, The American Ceramic Society), (i) SrTiO$_3$ nanocubes synthesized from a rapid sol-precipitation method (Reprinted with permission from Y.N. Hao, X.H. Wang and L.T. Li, "Highly dispersed SrTiO$_3$ nanocubes from a rapid sol-precipitation method, Nanoscale 6(2014): 7940-7946. Copyright 2014, The Royal Society of Chemistry), (j) KNbO$_3$ nanowires synthesized by hydrothermal method (Reprinted with permission from A. Magrez, E. Vasco, J.W. Seo, C. Dieker, N. Setter and L. Forró, "Growth of single-crystalline KNbO$_3$ nanostructures," J. Phys. Chem. B. 110(2006): 58-61. Copyright 2006, The American Chemistry Society), (k) PbTiO$_3$ nanotubes synthesized by AAO membrane template-assisted method (Reprinted with permission from P.M. Rørvik, K. Tadanaga, M. Tatsumisago, T. Grande and M.A. Einarsrud, "Template-assisted synthesis of PbTiO$_3$ nanotubes," J. Eur. Ceram. Soc. 29(2009): 2575-2579. Copyright 2009, Elsevier.)

2.3.2   Synthesis of One-Dimensional Nanostructured Perovskite Oxides

Today, 1D perovskite oxide nanostructures have been fabricated via two kinds of technical routes, which are named as top-down and bottom-up approaches, respectively. Among top-down approaches, FIB-based technique is the most commonly used one for the synthesis of 1D nanostructured perovskite oxides. Although the top-down approaches are simple but time-consuming, the later sizes of the fabricated nanostructures are limited, which necessitates the application of bottom-up approaches, where 1D nanostructured perovskite oxide nanostructures are constructed by using the basic building blocks (e.g. atoms or molecules). This process is very similar to the nature where complex biological systems are constructed by using proteins and other macromolecules. In this section, various techniques used for fabricating 1D nanostructured perovskite oxides are briefly introduced.

2.3.2.1   Top-Down Approaches

Recently, FIB method was used to fabricate large aspect-ratio (thickness = 150 nm, aspect ratio >300) single crystal perovskite $La_{2/3}Ca_{1/3}MnO_3$ nanowires, which preserved their magnetic properties down to lateral widths of 150 nm (Lorena et al. 2014). Details are shown in Fig. 2.10. It is observed that the narrowest 150-nm nanowire has magnetoresistance value as high as 34% under an applied magnetic field of 0.1 T. That can be attributed to the strain release at the edges as well as the destabilization of the insulating regions. Schilling et al. (2007) also used the FIB technique to fabricate perovskite BTO nanowires cut directly from single-crystal BTO. Such FIB process are schematically shown in Fig. 2.11, where the STEM image of the nanowires exhibits stripe contrast oriented at 45° to the column axes, indicating that the polar reorientations are morphologically controlled. Thus, the FIB process provides an advantage of artificial morphology control of the produced perovskite oxide nanostructures without considering the time consumed and low-throughput.

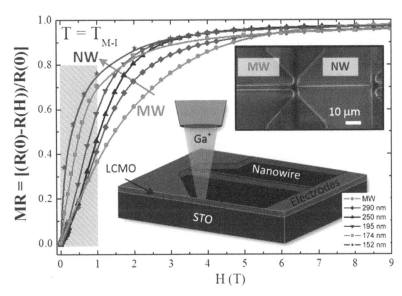

Fig. 2.10   Magnetoresistance (MR = 1- R(H))/R(0)) of the $La_{2/3}Ca_{1/3}MnO_3$ nanowires with different widths (*w* = 290, 250, 195, 174 and 152 nm) compared with that of the micro-wire (width = 5 μm) at $T = T_{MI}$. Insets are schematic diagrams illustrating the fabrication of $La_{2/3}Ca_{1/3}MnO_3$ nanowires by FIB and SEM image of the $La_{2/3}Ca_{1/3}MnO_3$ nanowire and micro-wire. (Reprinted with permission from M. Lorena, M. Luis, P.A. Algarabel, A.R. Luis, C.M. Rodríguez, D.T. José M. et al., "Enhanced magnetotransport in nanopatterned manganite nanowires," Nano Lett. 14(2014): 423-428. Copyright 2014, American Chemistry Society.)

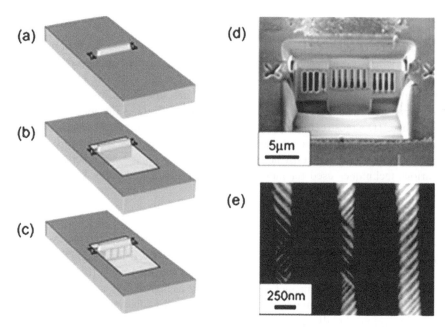

Fig. 2.11 (a)-(c) Schematic diagrams illustrating the process for fabricating the nanoscale columns cut from BaTiO$_3$ single crystals by FIB technique. (a) First, a protective Pt bar was deposited by local ion-beam induced breakdown of Pt-based organic precursor gases between two alignment crosses, (b) Automated software then allowed the milling of trenches on either side of a BaTiO$_3$ lamella, (c) The sample was then reoriented to allow milling into the lamellar face to leave columns of material with schematic illustration, (d) secondary electron image. (e) STEM image obtained with a high-angle annular dark field detector reveals distinct contrast from stripe domains. (Reprinted with permission from A. Schilling, R.M. Bowman, G. Catalan, J.F. Scott and J.M. Gregg, "Morphological control of polar orientation in single-crystal ferroelectric nanowires," Nano Lett. 7(2007): 3787-3791. Copyright 2007, American Chemistry Society.)

2.3.2.2 Bottom-Up Approaches

2.3.2.2.1 Template-Free Synthesis

Template-free methods (e.g. hydrothermal process, solvothermal process, electrophoretic deposition method) have been utilized to fabricate 1D perovskite oxide nanostructures (e.g. nanowires, nanorods, nanotubes) (Buscaglia and Buscaglia 2016). For example, Joshi and Lee (2005) synthesized single crystalline perovskite BTO and STO nanowires (diameters across 50-100 nm and lengths up to several micrometres) by solution-based template free method. Urban et al. (2002) synthesized the BTO and STO nanowires (diameters across 5-60 nm) by solvothermal synthesis. By template-free hydrothermal method, single-crystalline BTO nanowires (Joshi et al. 2006, Liu et al. 2011), PTO nanowires (Hu et al. 2006b, Gu et al. 2007), and PZT nanowires (Wang et al. 2012) were also synthesized. Due to organic templates not used in this process, it does not require the removal of the organic templates after the finishing synthesis. Single-crystalline perovskite nanowires/nanorods such as BTO (Mao et al. 2003, 2007, Li et al. 2014, Xue et al. 2019), PTO (Deng et al. 2005), (K, Na)NbO$_3$ (KNN) (Cheng et al. 2013), BaMnO$_3$ and Ba(Ti$_{0.5}$Mn$_{0.5}$)O$_3$ (Hu et al. 2006a, 2009) are also synthesized by MSS method. Bao et al. (2009) used a modified hydrothermal process to synthesize nanocrystalline BTO nanowires. Tetragonal PTO nanowires with necklace-like morphology were also synthesized by electro-spinning method (Lu et al. 2006). Similarly, BFO nanofibers were synthesized via a modified electrospinning process (Fei et al. 2015). In addition to the ferroelectric oxide nanowires, perovskite manganite nanowires such as La$_{0.5}$Ca$_{0.5}$MnO$_3$ (Zhang et al. 2004a), La$_{0.5}$Sr$_{0.5}$MnO$_3$ (Zhu et al. 2002a), La$_{0.5}$Ba$_{0.5}$MnO$_3$ (Zhu et al. 2002b) and Pr$_{0.5}$Ca$_{0.5}$MnO$_3$

(Rao et al. 2005) were also synthesized by hydrothermal method. Recently, Datta et al. (2016) performed detailed studies on the hydrothermal growth mechanism of perovskite manganite $La_{1-x}A_xMnO_3$ (A = Sr, Ca; x = 0.3 and 0.5) nanowires. They found that the shape of the nanostructures were controlled by the amount of KOH. They proposed a possible phase diagram (Fig. 2.12) for the hydrothermal synthesis of perovskite manganite $La_{1-x}A_xMnO_3$ nanostructures. Such phase diagram is closely dependent on the amount of KOH (mineralizer) and the synthesized temperature, as shown in Fig. 2.12.

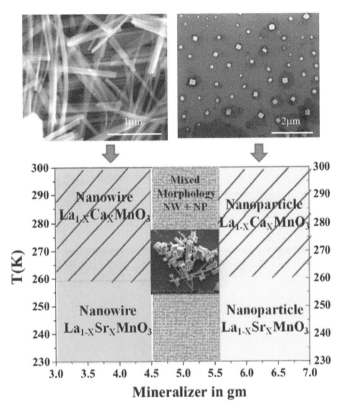

Fig. 2.12 Phase diagram of the hydrothermal synthesis of perovskite oxide manganites $La_{1-x}A_xMnO_3$ (A = Sr, Ca; x = 0.3 and 0.5) nanostructures as a function of the amount of mineralizer (KOH) and the synthesized temperature. The nanowires are formed at lower amount of KOH (3.0-4.5 g) and the nanocrystallites are formed at the higher amount of KOH (5.5-7.0 g). The $La_{1-x}Sr_xMnO_3$ samples are formed at 230-300°C whereas $La_{1-x}Ca_xMnO_3$ samples are grown at 260-300°C marked in the shaded region with diagonal stripes. The formation of mixed shape/size of nanowire and faceted nanoparticle is high in the region where the amount of KOH varies in the range (4.5-5.5 g). (Reprinted with permission from S. Datta, A. Ghatak and B. Ghosh, "Manganite ($La_{1-x}A_xMnO_3$; A = Sr, Ca) nanowires with adaptable stoichiometry grown by hydrothermal method: Understanding of growth mechanism using spatially resolved techniques," J. Mater. Sci. 51(2016): 9679-9695. Copyright 2016, Springer.)

Electrospinning is also an effective technique for fabricating 1D ceramic nanofibers with controllable dimensions and morphologies. This process involves three steps: (i) electrospinning solution preparation (e.g. selection of solvents and polymers, solution concentration optimization, solution viscosity, etc.), (ii) controlling the electrospinning parameters (e.g. flow rate, applied voltage, collector type), and (iii) annealing condition optimization (e.g. annealing temperature, ramping rate, annealing time, etc.). The electrospinning process has advantages of low cost, mass production fibers with uniform diameters, and considerable fiber length. Recently, by the electrospinning technique

Bharathkumar et al. (2015) obtained the BFO fibrous mat and mesh nanostructures, where the plate and drum collectors are used, respectively, as schematically shown in Fig. 2.13. The average sizes of the fibers in the mat and mesh were 200 nm and 150 nm, respectively. The BFO mesh and mat nanostructures exhibited a weak ferromagnetic properties due to the reduction in dimension and suppression in the cycloidal spin structures. The BFO nanofibers with diameter of ~ 170 nm were also synthesized via a sol-gel-modified electrospinning process, and then thermally annealed (Fei et al. 2015). The BFO nanofibers synthesized under optimized conditions had good crystalline quality without impurity phase. They exhibited excellent ferroelectric and photovoltaic properties. Li et al. (2009a) also synthesized BTO nanofibers (diameters across 92-182 nm) by electrospinning. Besides the BFO and BTO nanofiber, electrospun nanofibers of PTO (Selvaraj et al. 1992, Toyoda et al. 1997), PZT (Selvaraj et al. 1992), and NaTaO$_3$ (Yi and Li 2010) were also prepared.

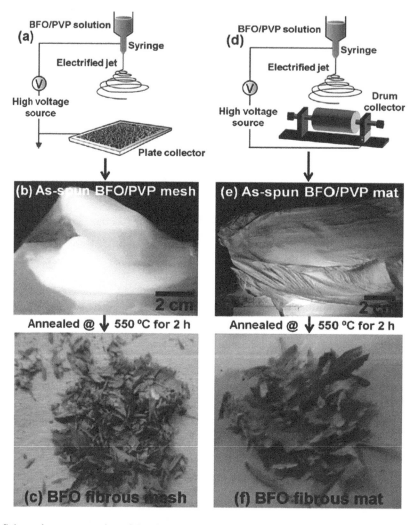

Fig. 2.13   Schematic representation of the electrospinning process and the photographic images of BiFeO$_3$. (a) Fibrous mesh in a plate collector, (b) as-spun mesh, (c) annealed mesh, (d) fiber mat in a drum collector, (e) as-spun mat, (f) annealed mat. (Reprinted with permission from S. Bharathkumar, M. Sakar, V.K. Rohith and S. Balakumar, "Versatility of electrospinning in the fabrication of fibrous mat and mesh nanostructures of bismuth ferrite (BiFeO$_3$) and their magnetic and photocatalytic activities," Phys. Chem. Chem. Phys. 17 (2015): 17745-17754. Copyright 2015, The Royal Society of Chemistry

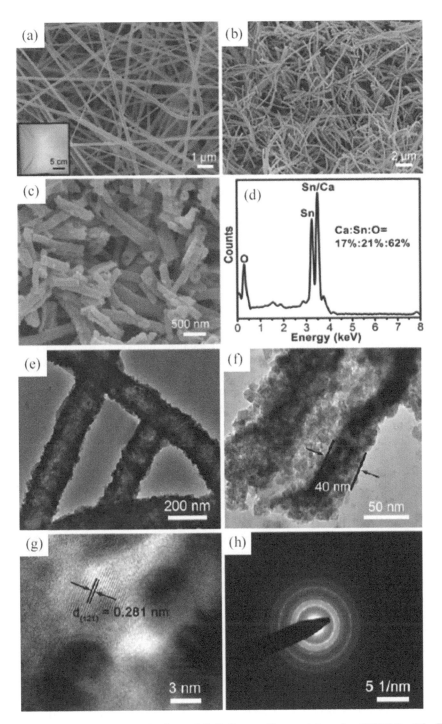

Fig. 2.14 Microstructural characterizations of CaSnO$_3$ nanofibers calcinated at 700°C for 5 h. FE-SEM images of (a) as-spun nanofibers and (b, c) calcinated nanofibers and nanotubes, (d) EDX spectra of the calcinated nanotubes, (e)-(f) TEM images of calcinated nanotubes, (g) HRTEM image of calcinated nanotubes, (h) SAED pattern of calcinated nanotubes. Inset is the photo of the electrospun precursor film. (Reprinted with permission from L.L. Li, S.J. Peng, J. Wang, Y.L. Cheah, P.F. Teh, Y. Ko et al., "Facile approach to prepare porous CaSnO$_3$ nanotubes via a single spinneret electrospinning technique as anodes for lithium ion batteries," ACS Appl. Mater. Interfaces. 4(2012): 6005-6012. Copyright 2012, American Chemistry Society.)

Template-free methods are also used to synthesize perovskite oxide nanotubes. For example, by single spinneret electrospinning technique, Li et al. (2012) synthesized perovskite $CaSnO_3$ nanotubes. Figure 2.14(a) is the SEM image of the as-spun nanofibers. Figures 2.14(b)-(c) are the SEM image of the calcinated nanofibers. The EDX spectra (Fig. 2.14d) of the calcinated nanofibers reveal the elemental composition ratio of Ca:Sn:O = 17%:21%:62%, indicating the formation of $CaSnO_3$. The microstructures of $CaSnO_3$ nanotubes were examined by TEM and HRTEM, as shown in Fig. 2.14(e) and Fig. 2.14(f), respectively. As it can be seen, the $CaSnO_3$ nanotubes consist of individual nanocrystallites with size ~ 10 nm. Different TEM contrast is observed between the edge and middle part of the nanotubes, indicating the formation of nanotubes. The wall thickness of the nanotubes was ~ 40 nm and their interior diameters were ~ 90 nm. Figure 2.14(g) shows the HRTEM image of $CaSnO_3$ nanotubes, where the lattice fringes with spacing of ~ 0.281 nm are well-resolved, corresponding to the (121) interplane distance of $CaSnO_3$. That matches well in with the data JCPDS card (31-0312) (d_{121} = 0.2789 nm). Figure 2.14(h) displays the SAED pattern obtained from the $CaSnO_3$ nanotubes, which consists of polycrystalline diffraction rings, confirming the polycrystalline nature of the $CaSnO_3$ nanotubes.

2.3.2.2.2 Template-Based Synthesis

The template-assisted method is a much effective method for mass production of regular arrays of perovskite oxide nanostructures. Nowadays, the most used templates are anodic aluminum oxide (AAO), silicon, and polycarbonate membranes (Limmer et al. 2002, Huczko 2000). Among the developed template-based methods, sol-gel based template method is widely used to fabricate ordered perovskite oxide nanostructures, e.g. BTO nanowires (Boucher et al. 2011, Anuradha 2014), PTO nanotubes (Zhao et al. 2006, Liu et al. 2008, Rørvik et al. 2009), and PZT nanowires (Zhang et al. 2004b, Wen et al. 2005, Shen et al. 2011). Furthermore, perovskite manganite nanowires and ordered arrays such as $La_{1-x}Ca_xMnO_3$ (LCMO, x = 0.20, 0.67) were also obtained by sol-gel based template (AAO membrane) method (Shankar and Raychaudhuri 2004). By the same method, perovskite oxide $La_{0.825}Sr_{0.175}MnO_3$ nanowires were also synthesized (Chen et al. 2005). Hernandez et al. (2002) also prepared perovskite BTO and PTO nanotubes by using AAO templates combined with the sol-gel process. PZT nanotubes (Luo et al. 2003, Kim et al. 2008, Nourmohammadi et al. 2009) and multiferroic BFO nanotubes (Park et al. 2004, Zhang et al. 2005, Xu et al. 2007) were also synthesized. In general, the perovskite oxide nanotubes produced by sol-gel based template method exhibit polycrystalline structure, and few ones have single-crystalline nature. That is attributed to the heterogeneous nucleation taking place on the interpore walls. Figure 2.15 shows the typical process for synthesizing the PTO and PZT nanotubes by sol-gel based template method, and their structural characterizations. To improve the filling effect, sol-gel process was replaced by the electrodeposition, where a direct electrophoretic current drives the PZT sol into the channels of the AAO template (Nourmohammadi et al. 2008). Similarly, $BiScO_3$-$PbTiO_3$ nanotube arrays were also obtained by sol-gel based template (AAO membranes) method (Zhong et al. 2016). The as-prepared $BiScO_3$-$PbTiO_3$ nanotubes were polycrystalline with a perovskite structure. They exhibit good piezoelectricity and the piezoelectric coefficient was measured to be 60 pm/V from the nanotubes with wall thickness of 30 nm. Alexe et al. (2006) also fabricated ferroelectric PZT nanotubes by using positive templates (e.g. silicon and ZnO nanowires), and further constructed the Pt/PZT/Pt nanotubes by rf magnetron sputtering and/or PLD method. $La_{2/3}Ca_{1/3}MnO_3$ nanotube ordered arrays were also prepared by using AAO templates under irradiation of microwave (Sousa et al. 2009). This method offers a new approach to fabricate nanotube arrays at relatively low temperatures. Porous polycarbonate membranes were also used to prepare rare-earth manganese oxide nanotubes such as $La_{0.325}Pr_{0.30}Ca_{0.375}MnO_3$, where the porous pores are wetted by the liquid precursor, and then heat-treated by microwave irradiation and further annealed at 800°C (Levy et al. 2003). Zhang and Chen (2005) prepared the perovskite $La_{0.59}Ca_{0.41}CoO_3$ nanotubes by a sol-gel template method. Tagliazucchi et al. (2006) synthesized the perovskite $LaNiO_3$ nanotubes by using a template-inorganic precursor at relatively low annealing temperature. The polycrystalline structure was revealed for the LaNiO$_3$ nanotubes, and the nanotube wall consisted of nanocrystallites with sizes across 3-5 nm.

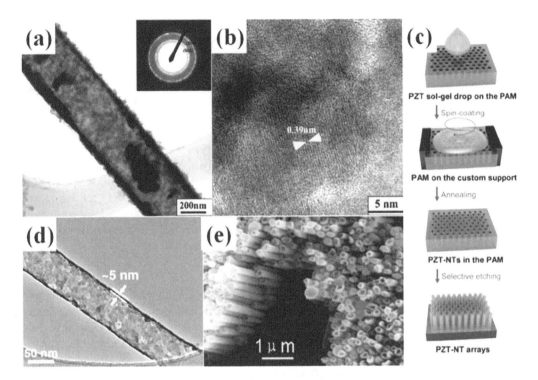

Fig. 2.15 Structural characterizations of the PTO and PZT nanotubes synthesized by sol-gel template technique. (a, b) TEM and HRTEM images of the prepared PTO nanotubes. Inset is selected-area electron diffraction pattern of the PTO nanotubes, (c) Schematic illustration of the synthesis procedure of PZT nanotube arrays, (d) Single PZT nanotube with a wall thickness of about 5 nm, (e) SEM images of the BFO nanotubes. (a, b) (Reproduced with permission from L.F. Liu, T.Y. Ning, Y. Ren, Z.H. Sun, F.F. Wang, W.Y. Zhou et al., "Synthesis, characterization, photoluminescence and ferroelectric properties of PbTiO$_3$ nanotube arrays," Mater. Sci. Eng. B. 149(2008): 41-46. Copyright 2008, Elsevier). (c, d) (Reproduced with permission from J. Kim, S.A. Yang, Y.C. Choi, J.K. Han, K.O. Jeong, Y.J. Yun et al., "Ferroelectricity in highly ordered arrays of ultra-thin-walled Pb(Zr,Ti)O$_3$ nanotubes composed of nanometer-sized perovskite crystallites," Nano Lett. 8(2008): 1813-1818. Copyright 2008, American Chemistry Society). (e) (Reproduced with permission from X.Y. Zhang, C.W. Lai, X. Zhao, D.Y. Wang and J.Y. Dai, "Synthesis and ferroelectric properties of multiferroic BiFeO$_3$ nanotube arrays," Appl. Phys. Lett. 87(2005): 143102 (1-3). Copyright 2005, American Institute of Physics.)

Zhu et al. (2006) reported on the PZT nanorings with ring thicknesses across 5-10 nm and internal diameters close to 5 nm, which were fabricated by first conformal wetting the channels of AAO template with PZT sol solution and then cross-sectioning the coated pores normal to the cylindrical axes. Byrne et al. (2008) reported the fabrication of periodic arrays of PZT nanorings with a composition close to MPB (morphotropic phase boundary) by a self-assembly technique. The PZT nanoring had a diameter of 100 nm and wall thickness ~ 10 nm. Recently, ordered arrays of Pb(Zr$_{0.2}$Ti$_{0.8}$)O$_3$ nanodiscs and nanorings in wafer-size scale were also fabricated by Han et al. (2009). The epitixal growth of the Pb(Zr$_{0.2}$Ti$_{0.8}$)O$_3$ nanostructures was confirmed by XRD patterns and selected area electron diffraction patterns. The Pb(Zr$_{0.2}$Ti$_{0.8}$)O$_3$ nanodiscs and nanorings still retained ferroelectricity, as confirmed by the PFM measurements. Ordered BFO nanoring arrays were fabricated by AAO template-assisted PLD and ion beam etching methods (Tian et al. 2017). The BFO nanorings had an internal diameter of 12 nm and 30 nm wall thickness, and heights of around 10 nm. The isolated BFO nanorings exhibit polarization reversal behavior, indicating the ordered BFO nanorings have a huge potential for the high-density ferroelectric memory devices.

2.3.3 Synthesis of Two-Dimensional Perovskite Oxide Nanostructures

With stimulation of the experimental discovery of 2D graphene and transition-metal dichalcogenides, new interest in 2D materials has awakened. Due to the high tunabilities of the composition, structure, and functionality, much attention has been given to 2D perovskite oxide nanostructures, which have much potential applications in nonvolatile logic devices, spintronic devices, and photovoltaic devices. In general, two main strategies are developed for fabricating 2D perovskite oxide nanostructures: top-down and bottom-up methods (Sun et al. 2017). In the top-down method, FIB milling as well as electron beam direct writing (EBDW) are often used to fabricate perovskite oxide nanostructures, while bottom-up method makes use of 2D materials at a molecular level in a specific medium or on a substrate, which is applicable for the synthesis of 2D perovskite oxide nanostructures. However, this approach has difficulties in controlling the anisotropic growth into atomically thin perovskite oxide nanosheets. Here, we summarize the methods for fabricating 2D perovskite oxide nanostructures from the recent published references.

2.3.3.1 *Top-Down Methods*

In the past decade, many top-down methods (e.g. FIB, electron beam lithography (EBL), nanoimprint lithography (NIL)) have been developed to synthesize 2D perovskite oxide nanostructures. Ganpule et al. (1999) fabricated ferroelectric capacitors based on the $Pb(Nb_{0.04}Zr_{0.28}Ti_{0.68})O_3$ and $SrBi_2Ta_2O_9$ thin films by FIB milling method, the capacitor sizes were reduced from 1 mm^2 to 0.01 mm^2, and the lateral sizes are as small as 70 nm, as shown in Fig. 2.16(a). In addition, Schilling et al. (2009) also fabricated the BTO nanodots by FIB method, which were cut directly from bulk BTO crystal. The domain structures in the BTO nanodots were investigated by STEM (Schilling et al. 2009, Ahluwalia et al. 2013). In contrast with the FIB technique, there is no need of a mask in EBDW for fabricating ferroelectric nanostructures. Alexe et al. (1999) performed the pioneering works on fabricating periodic arrays of perovskite ferroelectric oxide nanostructure, which were used to develop high density ferroelectric memory devices. They fabricated periodic arrays of $SrBi_2Ta_2O_9$ and PZT nanocapacitors with lateral sizes down to ~ 100 nm by EBL method, as shown in Fig. 2.16(b). NIL method is one kind of reliable technique for replicating nanoscale features. This technology can faithfully reproduce patterns with better resolution and uniformity as compared with those generated by photolithography technique. There are two main NIL methods, which are schematically shown in Fig. 2.17. Harnagea et al. (2003) fabricated ferroelectric PZT cell arrays by imprinting precursor gel deposited by CSD method, where the lateral sizes of the PZT cells were below 300 nm. Figure 2.18(a) dispalys the SEM image of the mould made from microporous silicon with 500 nm pitch and 350 nm pore size. Figure 2.18(b) presents the AFM topography image of the crystallized cells. Figure 2.18(c) shows the AFM cross-section profile obtained along the line marked in Fig. 2.18(b), which illustrates that the PZT cell has a lateral size of ~ 350 nm and an average height of 50 nm. It is noticed that the mold shape is well kept in the sol-gel deposited PZT film. Piezoresponse measurements revealed that the local hysteresis loop of the PZT cell with lateral size of 300 nm is almost identical to that from a continuous film. Lan et al. (2010) patterned a variety of ferroelectric PZT nanostructures with feature size of 180 nm by using sol-gel based soft lithography technique. Figure 2.19(a) displays the patterning process of PZT nanostructure by soft lithography technique, and Fig. 2.19(b) depicts the SEM images of the PZT microstructures with different morphologies (e.g. rectangles, panes, homocentric cirques, and dots). It is found that the pattern of master mold is well transferred to the patterned films.

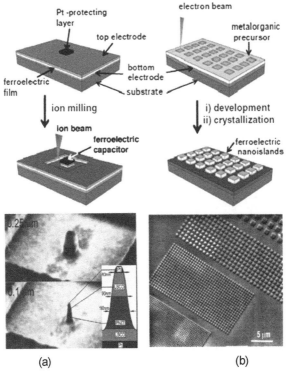

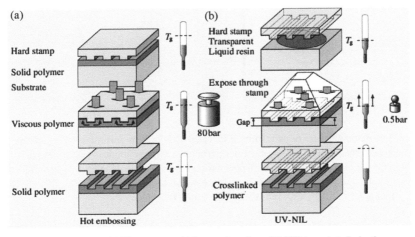

Fig. 2.16 Two typical examples of top-down methods for fabricating perovskite oxide ferroelectric nanostructures. (a) Test structures fabricated by FIB milling in the size range from 1 μm² to 0.01 μm². The contrast from the various layers in the heterostructure is also indicated. Inset is a schematic of the shape and dimensions of the 0.1 μm test structure. (Reproduced with permission from C.S. Ganpule, A. Stanishevsky, Q. Su, S. Aggarwal, J. Melngailis, E. Williams et al., "Scaling of ferroelectric properties in thin films," Appl. Phys. Lett. 75(1999): 409-411. Copyright 1999, American Institute of Physics), (b) EB direct writing of test patterns of SBT and PZT cells with lateral dimension between 1 μm and 125 nm. (Reprinted with permission from M. Alexe, C. Harnagea, D. Hesse and U. Gosele, "Patterning and switching of nanosize ferroelectric memory cells," Appl. Phys. Lett. 75(1999): 1793-1795. Copyright 1999, American Institute of Physics.)

Fig. 2.17 Schematic diagram of NIL process: (a) hot embossing, (b) UV-imprint. In both cases, a thickness profile is generated in the thin polymer layer. After removing the residual layer, the remaining polymer can serve as a masking resist which can be used for pattern transfer. (Reprinted with permission from H. Schift and A. Kristensen, Nanoimprint Lithography, pp. 239-278. In: B. Bhushan [ed.]. Handbook of Nanotechnology. Copyright 2007, Springer.)

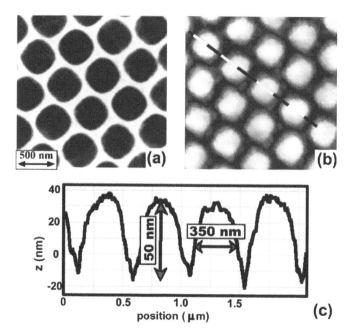

Fig. 2.18 (a) SEM image of a part mold made from microporous silicon, (b) AFM topography image of a part of crystallized PZT cell arrays obtained by imprinting a sol-gel film using the mold shown in (a), (c) AFM cross-section profile taken along the line shown in (b). (Reprinted with permission from C. Harnagea, M. Alexe, J. Schilling, J. Choi, R.B. Wehrspohn, D. Hesse et al., "Mesoscopic ferroelectric cell arrays prepared by imprint lithography," Appl. Phys. Lett. 83(2003): 1827-1829. Copyright 2003, American Institute of Physics.)

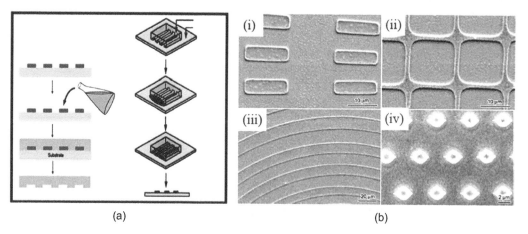

Fig. 2.19 (a) Schematic diagram of patterning PZT microstructure by soft lithography technique, (b) SEM images of some patterned $Pb(Zr_{0.52}Ti_{0.48})O_3$ microstructures: (i) rectangles, (ii) panes, (iii) homocentric cirques, and (iv) circular dots. (Reprinted with permission from J. Lan, S.H. Xie, L. Tan and J.Y. Li, "Sol-gel based soft lithography and piezoresponse force microscopy of patterned $Pb(Zr_{0.52}Ti_{0.48})O_3$ microstructures," J. Mater. Sci. Technol. 26(2010): 439-444. Copyright 2010, Elsevier.)

2.3.3.2 Bottom-Up Methods

2D perovskite oxide nanostructures are also fabricated by bottom-up methods, in which atoms, molecules, or nanoparticles are used as building blocks that are self-assembled together. In contrast to the top-down approach, this approach exhibits some advantages. For example, 2D perovskite oxide nanostructures can be directly prepared at nanoscale. In addition, they exhibit single-crystalline nature with free of lattice defects at the surface. However, the bottom-up method has difficulties in controlling the shape, size, and precise localization of individual perovskite oxide nanostructure since the bottom-up approach principally relies on either self-assembly processes or thermodynamically driven reorganization. In recent years, some bottom-up methods have been developed. Among them, chemical solution deposition (CSD) and metal-organic chemical vapor deposition (MOCVD) are representative ones, which work on the structural instability of very thin films (several nanometers in thickness) at high temperatures. Such structural instability will drive the thin film to separate into nanoislands in order to reduce the interfacial energy between the film and the substrate. The structural characteristics of the resulted 2D perovskite oxide nanostructures (e.g. shape and size of 2D ferroelectric PTO and PZT nanoisland arrays) are dependent on the film thickness, annealing temperature, and the substrate orientations (Szafraniak et al. 2003, Nonomura et al. 2003). Recently, perovskite oxide nanosheets (e.g. $ZnSnO_3$ nanoplates with the (111) facets as exposing surface plates) are grown by one-step hydrothermal process (Guo et al. 2017). Meanwhile, 2D perovskite $LaNiO_3$ nanosheets with a mesoporous structure are also synthesized by sol-gel method and subsequent annealing process (Li et al. 2017). Hong et al. (2017) first reported the epitaxial growth of freestanding ultrathin STO membranes (with a few unit cell thicknesses) on a STO (001) substrate by PLD method, where a water-soluble sacrificial layer ($Sr_3Al_2O_6$) acts as a buffer layer (Fig. 2.20a), which can be removed by immersing into deionized water at room temperature. Thus, the top STO layer is freely released. By this way, the grown film thickness was controlled in single unit cell accuracy (Fig. 2.20c). Similarly, freestanding STO and BFO ultrathin films were also grown by reactive molecular beam epitaxy, which were then transferred onto diverse substrates (Ji et al. 2019a). It is found that the freestanding BFO films exhibit giant tetragonality

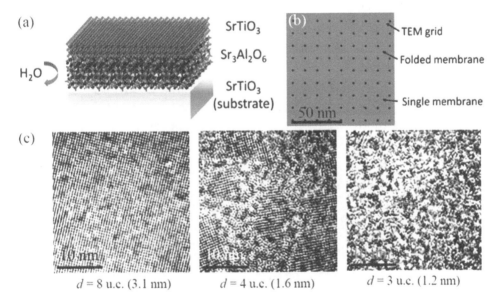

Fig. 2.20 Freestanding $SrTiO_3$ membranes with a few unit cell thicknesses grown by PLD method. (a) A schematic $SrTiO_3/Sr_3Al_2O_6$ film heterostructure grown on a $SrTiO_3$ (001) substrate. The $Sr_3Al_2O_6$ layer is dissolved in water to release the top $SrTiO_3$ layer at room temperature, (b) Optical image of a suspended $SrTiO_3$ membrane (6 unit cells (u.c.) in thickness) on a silicon nitride transmission electron microscopy (TEM) grid with 2-mm diameter holes, (c) High-resolution TEM (HRTEM) images of $SrTiO_3$ membranes with different thicknesses. (Reprinted with permission from S.S. Hong, J.H. Yu, D. Lu, A.F. Marshall, Y. Hikita, Y. Cui et al., "Two-dimensional limit of crystalline order in perovskite membrane films," Sci. Adv. 3(2017): eaao5173 (1-5). Copyright 2017, American Association for the Advancement Science.)

and polarization as the film approaches the 2D limit. Figure 2.21 shows the growth, release, and transfer process of ultrathin freestanding STO films, and their microstructural characterizations are demonstrated in Fig. 2.22. The present results reveal that critical thickness does not exist for a freestanding ultrathin perovskite oxide films. In addition, freestanding single-crystalline ferroelectric BTO membranes were also deposited on STO (001) substrates by PLD method, where water-soluble $Sr_3Al_2O_6$ was used as the sacrificial buffer layer in the $BTO/Sr_3Al_2O_6$ heterostructures (Dong et al. 2019). Freestanding BTO membranes were obtained by immersing the $BTO/Sr_3Al_2O_6$ heterostructures in deionized water to remove the sacrificial $Sr_3Al_2O_6$. Figure 2.23 shows the growth and characterizations of the freestanding BTO membranes. The heteroepitaxial growth of the BTO thin film on the $Sr_3Al_2O_6$ confirmed the reciprocal space mapping around the pseudocubic (002) and (103) reflections of BTO film (Fig. 2.23b). Planar HR-STEM images reveal the single-crystalline nature of the BTO membrane (Fig. 2.23g). Vertical PFM phase image of the BTO membranes demonstrates explicit ferroelectric domains switching after both positive- and negative-bias poling (Fig. 2.23h), which is also reflected by the piezoelectric response loops (Fig. 2.23i). It is found that the BTO membranes have super-elasticity and ultraflexibility, as confirmed by an *in-situ* bending test (Fig. 2.24). Such super-elasticity is due to the dynamic evolution of ferroelectric nanodomains. The ultraflexible epitaxial ferroelectric BTO membranes would enable them to be used in flexible sensors, memories, and electronic skins. At present, the perovskite oxide 2D nanostructures are still at their initial stage, and their applications in the new generation of multifunctional electronic devices have a bright future.

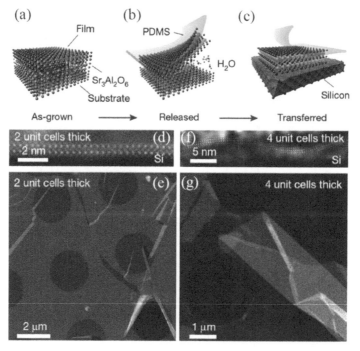

Fig. 2.21 Growth and transfer of ultrathin freestanding SrTiO₃ films. (a) Schematic of a film with an Sr₃Al₂O₆ buffer layer, (b) The sacrificial Sr₃Al₂O₆ layer is dissolved in water to release the top oxide films with the mechanical support of PDMS, (c) New heterostructures and interfaces are formed when the freestanding film is transferred onto the desired substrate, (d) Atomically resolved cross-sectional and (e) low-magnification plan-view HAADF images of a two-unit-cell freestanding SrTiO₃ film transferred to a silicon wafer and a holey carbon TEM grid, respectively, (f) Atomically resolved cross-sectional and (g) low-magnification plan view HAADF images of a representative four unit cell freestanding SrTiO₃ film, showing the excellent flexibility of ultrathin freestanding films. (Reprinted with permission from D.X. Ji, S.H. Cai1, T.R. Paudel, H.Y. Sun, C.C. Zhang, L. Han et al., "Freestanding crystalline oxide perovskites down to the monolayer limit," Nature 570(2019): 87-90. Copyright 2019, Nature Publishing Group.)

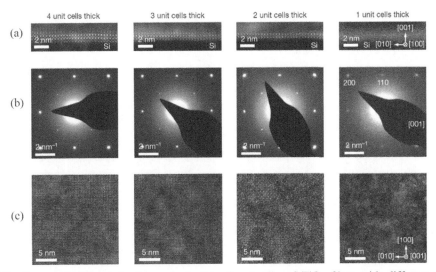

Fig. 2.22 Structural characterizations of ultrathin freestanding SrTiO₃ films with different unit-cell thicknesses. (a) Cross-sectional HAADF images, (b) selected area electron diffraction patterns, (c) plan-view HAADF images of ultrathin freestanding SrTiO₃ films with different unit-cell thicknesses, showing no critical thickness limitation for the freestanding crystalline perovskite oxide films. Since the one unit-cell freestanding film is extremely sensitive to the electron beam, it can survive only at low-dose SAED measurements. (Reprinted with permission from D.X. Ji, S.H. Cai1, T.R. Paudel, H.Y. Sun, C.C. Zhang, L. Han et al., "Freestanding crystalline oxide perovskites down to the monolayer limit," Nature 570 (2019): 87-90. Copyright 2019, Nature Publishing Group.)

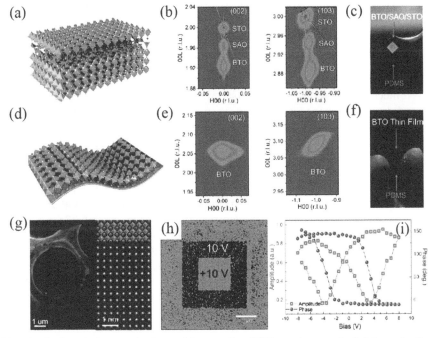

Fig. 2.23 Synthesis and characterizations of freestanding BTO membranes. (a,d) Schematic diagrams, (b, e) reciprocal space mappings around the (002)- and (103)-diffraction reflections of BTO film, and (c,f) optical photographs of the BTO/Sr₃Al₂O₆/STO heterostructure and flexible BTO membrane, respectively. r.l.u., relative light units, (g) Planar HAADF-STEM image of freestanding BTO membrane, (h) Vertical PFM phase image of BTO transferred on Pt/Si, and (i) Piezoelectric response. a.u., arbitrary units. (Reprinted with permission from G.H. Dong, S.Z. Li, M.T. Yao, Z.Y. Zhou, Y.Q. Zhang, X. Han et al., "Super-elastic ferroelectric single-crystal membrane with continuous electric dipole rotation," Science 366(2019): 475-479. Copyright 2019, American Association for the Advancement Science.)

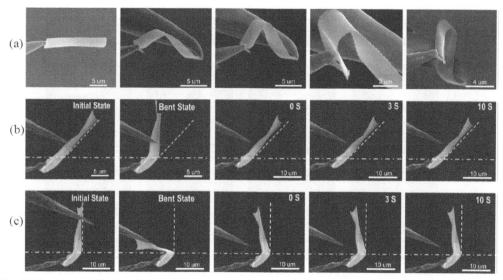

Fig. 2.24 *In-situ* SEM bending test of BTO nanobelts. (a) Series of SEM images with the bending process of a BTO nanobelt (20 μm × 4 μm × 120 nm), (b, c) SEM images of the recovery process after different bending states for a BTO nanobelt (20 μm × 4 μm × 60 nm). (Reprinted with permission from G.H. Dong, S.Z. Li, M.T. Yao, Z.Y. Zhou, Y.Q. Zhang, X. Han et al., "Super-elastic ferroelectric single-crystal membrane with continuous electric dipole rotation," Science 366(2019): 475-479. Copyright 2019, American Association for the Advancement Science.)

2.3.4   Synthesis of Three-Dimensional Perovskite Oxide Nanostructures

Complex 3D architectures that are constructed with 1D nanostructures (e.g. nanowires, nanorods, nanotubes) exhibit some unique properties, which are much different from the mono-morphological structures because they have the features of all involved nanoscale building blocks (Li and Wang 2010). In recent years, coherent 3D perovskite oxide nanostructures built by self-organization have received much attention since the self-organized 3D nanostructures exhibit novel physical properties and have promising applications in nanotechnology. These 3D perovskite oxide nanostructures provide the possibility of constructing oxide nanodevices. Recently, well defined 3D epitaxial (La,Sr)MnO$_3$ (LSMO) nanorods were fabricated by PLD method (Jiang et al. 2004, Jiang and Meletis 2006). Figure 2.25(a) shows a planar TEM image of the LSMO nanorods, where well-defined and ordered array of 2D orthogonal blocks are clearly observed. Figure 2.25(b) demonstrates the cross-sectional TEM image of the LSMO/LAO heterostructure, in which two-layered structure are observed. One is a continuous layer with thickness of 70 nm connected with the substrate and another one is the nanocolumnar layer with the nanocolumn's lateral size of ~ 25 nm and height of ~ 70 nm. These nanocolumns are nearly periodically distributed along the interface, and they are separated by straight boundaries with a thickness of 2-4 nm and bright TEM contrast. The SAED patterns taken from the LSMO nanorod layer and the LSMO/LAO interface along the direction are displayed in Figs. 2.25(c)-(d), respectively. Based on the SAED patterns, it can be concluded that all nanocolumns exhibit single-crystalline nature and they have in-plane orientation relationship with the substrate: (100)°$_{LSMO}$//(100)$_{LAO}$. Recently, Kang et al. (2017) reported on the vertical growth and ordered ferroelectric PZT single-crystalline nanorod arrays in Pt-covered silicon wafers by sputtering method. The synthesis procedure of the close-packed PZT nanorod arrays on silicon substrate is schematically shown in Fig. 2.26. The growth details can be found in the reference (Kang et al. 2017). Structural characterizations of these close-packed PZT nanorod arrays are shown in Figs. 2.27(a)-(c). It was observed that the diameter of PZT nanorods was ~ 200 nm and height was ~ 600 nm (the aspect ratio was 1:3), and inter-nanorod spacing was several nanometers. The single-crystalline of the nanorod was confirmed by the SAED pattern, and neither interfacial defects nor dislocations were observed. With increasing deposition time, the PZT nanorods vertically grew and their lateral sizes also increased simultaneously. As demonstrated in Figs. 2.27(d)-(e), the deposition temperature was the critical factor for controlling the vertical growth of PZT nanorods. By using AAO template as a mask, Tian et al. (2016) fabricated the $BFO/CoFe_2O_4/SrRuO_3$ heterostructured nanodot arrays by PLD method. The lateral sizes of the nanodots were ~ 70 nm and the spacing between the two dots was ~ 110 nm. An enhancement of piezoelectric effect was observed in a single nanodot due to a strong magnetoelectric coupling.

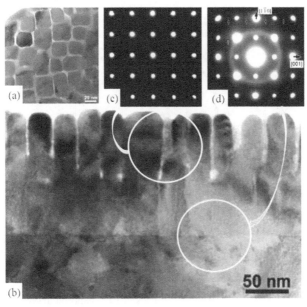

Fig. 2.25 (a) Plan-view, (b) cross-sectional TEM of $(La,Sr)MnO_3/LaAlO_3$; (c, d) SEAD patterns taken from the $(La,Sr)MnO_3$ continuous layer/$LaAlO_3$ substrate interface and the $(La,Sr)MnO_3$ nanorod layer, respectively. (Reprinted with permission from J.C. Jiang, L.L. Henry, K.I. Gnanasekar, C.L. Chen and E.I. Meletis, "Self-assembly of highly epitaxial $(La,Sr)MnO_3$ nanorods on (001) $LaAlO_3$," Nano Lett 4(2004): 741-745. Copyright 2004, American Chemistry Society.)

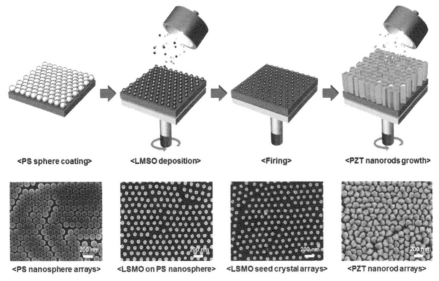

Fig. 2.26 Schematic diagrams illustrating the synthesis procedure for fabricating close-packed PZT nanorod arrays on 3 in. Si wafer. (Reprinted with permission from M.G. Kang, S.Y. Lee, D. Maurya, C. Winkler, H.C. Song, R.B. Moore et al. 2017. Wafer-scale single-crystalline ferroelectric perovskite nanorod arrays. Adv. Funct. Mater. 27(2017): 1701542 (1-9). Copyright 2017, Wiley-VCH.)

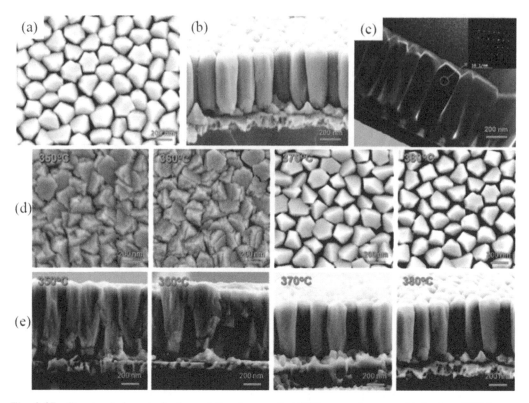

Fig. 2.27 Structural characterizations of the close-packed PZT nanorod arrays. (a) Surface SEM image, (b) Cross-sectional SEM image, (c) TEM image. Inset is the selected area electron diffraction pattern. (d,e) Surface and cross-sectional SEM images of the PZT nanorod arrays with different deposition temperatures. (Reprinted with permission from M.G. Kang, S.Y. Lee, D. Maurya, C. Winkler, H.C. Song, R.B. Moore et al., "Wafer-scale single-crystalline ferroelectric perovskite nanorod arrays," Adv. Funct. Mater. 27 (2017): 1701542(1-9). Copyright 2017, Wiley-VCH.)

Recently, 3D perovskite ferroelectric nanostructures have been fabricated by combining PLD technique with AAO templates. For example, ferroelectric Pt/PZT/Pt nanocapacitor arrays were fabricated with ultra-high density of ~ Tb/inch^2 density (Lee et al. 2008). 3D nano-templated PLD method is newly developed, where an inclined substrate was deposited onto both side surfaces of a 3D nano-patterned substrate. By this method, Nguyen et al. (2013) prepared 3D perovskite $(La_{0.275}Pr_{0.35}Ca_{0.375})MnO_3$ (LPCMO) nanobox on MgO(001) substrate. The wall-widths of the LPCMO nanoboxes varied across 30-160 nm, which was controlled via the deposition time. An insulator-metal transition was observed in these LPCMO nanoboxes, and this phase transition temperature was higher than that observed in the corresponding film counterpart. That inplies the LPCMO nanoboxes have huge potentials in oxide spintronic devices. 3D-nanotemplated PLD technique offers a new approach to fabricate 3D perovskite oxide nanostructures.

2.4 PHYSICAL PROPERTIES OF NANOSTRUCTURED PEROVSKITE OXIDES

2.4.1 Ferroelectric Properties

Ferroelectricity is defined as a property that crystalline substance has a spontaneous electric polarization that can be reversed by an electric field (Retot et al. 2008, Cross 2011). It was first discovered in the Rochelle salt in 1920. However, in the mid-1940s ferroelectricity was discovered in BTO, PZT, and other perovskite-based materials. This discovery has opened a door for commercial applications of ferroelectric materials (Haertling 1999, Rigoberto et al. 2011). Since the electronic devices based on perovskite ferroelectric materials are physically operated on the spontaneous polarization, the spontaneous polarization is reduced by decreasing the physical sizes of perovskite oxide nanostructures. Thus, the size effect becomes an important issue for ferroelectric nanodevices based on the reduced perovskite ferroelectric nanostructures. Ishikawa et al. (1988) performed the pioneering work on the size effect of PTO nanoparticles by Raman spectroscopy. They found that as the particle size was reduced, ferroelectric phase transition temperature (T_C) was also decreased, and the critical particle size was estimated to be 12.6 nm, as demonstrated in Fig. 2.28(a). Similarly, the size effects in BFO (Selbach et al. 2007) and BTO (Yashima et al. 2005) nanoparticles were also studied, and the results are shown in Figs. 2.28(b)-(d). The particle size effect on the infrared phonon modes and the ferroelectric phase transitions of the BFO nanoparticles were also studied by Chen et al. (Chen et al. 2010a). Figure 2.29 shows some selected infrared phonon modes such as E(TO1),

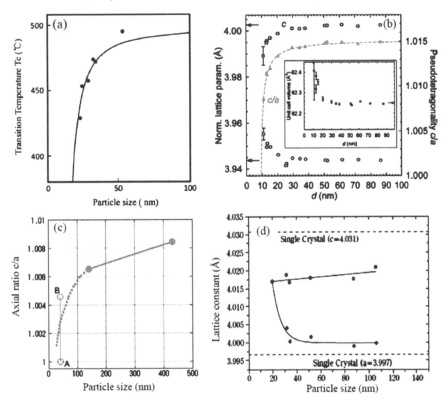

Fig. 2.28 Size effects in PbTiO₃, BiFeO₃, and BaTiO₃ nanoparticles. (a) Ferroelectric phase transition temperature of PbTiO₃ nanoparticles vs. the particle diameter. (Reprinted with permission from K. Ishikawa, K. Yoshikawa and N. Okada, "Size effect on the ferroelectric phase transition in PbTiO₃ ultrafine particles," Phys. Rev. B. 37(1988): 5852-5855. Copyright 1988, American Physics Society), (b) Normalized lattice parameters of nanocrystalline BiFeO₃ vs. the crystallite size (lattice parameters *a* (open circle), *c* (open square), and pseudo-tetragonality *c*/*a* (dotted line). Inset shows the unit-cell volume vs. the crystallite size. (Reprinted with permission from S.M. Selbach, T. Tybell, M.-A. Einarsrud and T. Grande, "Size-dependent properties of multiferroic BiFeO₃ nanoparticles," Chem. Mater. 19(2007): 6478-6484. Copyright 2007, American Chemistry Society), (c)-(d) Size effects on the axial ratio *c*/*a* and lattice constants of the tetragonal BaTiO₃ nanoparticles. (Reprinted with permission from M. Yashima, T. Hoshina, D. Ishimura, S. Kobayashi, W. Nakamura, T. Tsurumi et al. "Size effect on the crystal structure of barium titanate nanoparticles," J. Appl. Phys. 98(2005): 014313(1-8). Copyright 2005, American Institute of Physics.)

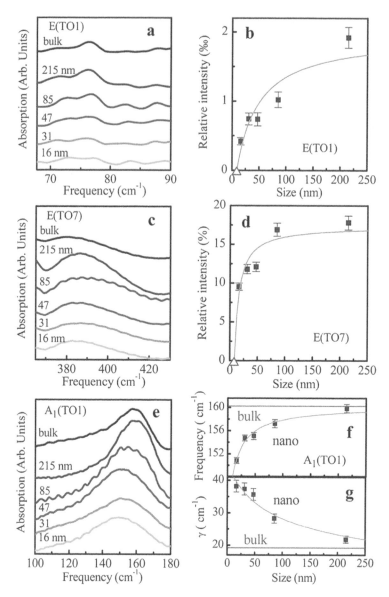

Fig. 2.29 Selected infrared phonon modes of BiFeO₃ nanoparticles vs. the particle size at 300 K. The selected infrared modes include (a) E(TO1), (c) E(TO7), and (e) A1(TO1) symmetries. Important trends quantified include, (b) the relative intensity of the E(TO1) mode vs. size, (d) the relative intensity of the E(TO7) mode vs. size, (f) the frequency of the A1(TO1) mode vs. size, and (g) the damping of the A1(TO1) mode vs. size. The green solid lines are the calculated curves fit to the data, described in the text. Open triangles mark the critical sizes for the E (TO1) mode (13.5 nm) and E (TO7) mode (8.1 nm), respectively. (Reprinted with permission from P. Chen, X. S. Xu, C. Koenigsmann, A. C. Santulli, S. S. Wong and J. L. Musfeldt, "Size-dependent infrared phonon modes and ferroelectric phase transition in BiFeO₃ nanoparticles," Nano Lett. 10(2010): 4526-4532. Copyright 2010, American Chemistry Society.)

E(TO7), and A1 (TO1) of BFO vs the particle size (diameter) at 300 K. The results reveal that the paraelectric phase is cubic, and the A1(TO1) mode is softened over 6% and dampened considerably as the crystalline size decreases. However, more recent experimental works (Ghosez and Rabe 2000, Fong et al. 2004), in conjunction with the theoretical calculations (Junquera and Ghosez 2003), demonstrate that the critical size of ferroelectricity in PTO thin film is only 1.20 nm, just 3 unit cells of PTO. For the tetragonal BTO particles, they still exhibit nanoscale ferroelectric property as the particle sizes are across 6-12 nm, as confirmed by electrostatic force microscopy (EFM) images (Fig. 2.30) (Nuraje et al. 2006). Details are described in the figure caption. In addition, ferroelectric properties of nanowires and nanotubes are investigated by PFM and EFM. Yun et al. (2002) measured the nanoscale ferroelectricity of a single BTO nanowire by EFM. They found that ferroelectricity still retained in a single BTO nanowire with diameter as small as 12 nm, and nonvolatile electric polarization could be switched. Figure 2.31 shows the EFM images of the BTO nanowires where four distinct polarization domains can be switched independently by an external field. That indicates there exists a remnant polarization vertical to the wire axis even in the nanowires with diameters across 10-20 nm (Yun et al. 2002). The data storage density can be reached as high as 1 Terabit/cm^2 in real memory devices which work on the nonvolatile polarization domains with area of 100 nm^2. Besides the BTO nanowires, the polarization switching of PZT nanotubes was also reported by Luo et al. (2003). Size effect is observed in perovskite oxide nanowires. Figure 2.32(a)-(b) shows the Curie temperature (T_c) and polarization (P) of the PZT nanowire at 64 K vs the nanowire diameter, which are theoretically calculated by *ab initio* first-principles calculation (Naumov and Fu 2005). They predict a sharp paraelectric to ferroelectric transition, a polarization appearing along the major axis of the nanowire. In experiment, Spanier et al. (2006) reported the T_c of BTO nanowire vs the nanowire diameter, as displayed in Fig. 2.32(c). It is noticed that T_C reduces linearly with $1/d_{nw}$ and falls down to 300 K when d_{nw} is equal to 3 nm. First-principles calculations of BTO nanowire indicate a critical diameter of 1.2 nm for the ferroelectric instability (Geneste et al. 2006). Kim et al. (2008) reported on the ferroelectric switching behavior of highly-ordered arrays of PZT nanotubes, as depicted in Fig. 2.33. The remnant polarization (P_r) of the PZT nanotubes was measured to be 1.5 μC/cm^2, and the coercive field (E_c) was 86 kV/cm.

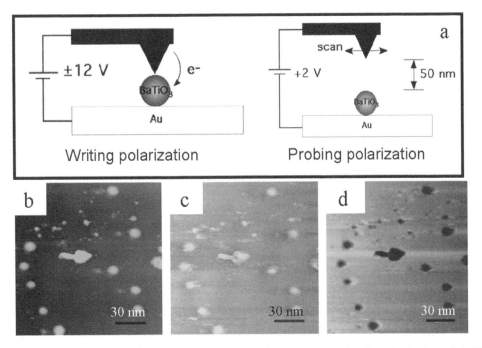

Fig. 2.30 (a) Schematic diagram of manipulating and probing the electric polarization of BaTiO$_3$ nanoparticles with EFM, (b) Topological AFM image of BaTiO$_3$ nanoparticles, (c)-(d) EFM images of BaTiO$_3$ nanoparticles with V_{probe} = +2 V after V_{write} = ± 12 V applied on the nanoparticles across a conductive AFM tip and a gold substrate. (Reprinted with permission from N. Nuraje, K. Su, A. Haboosheh, J. Samson, E.P. Manning, N. Yang et al., "Room temperature synthesis of ferroelectric barium titanate nanoparticles using peptide nanorings as templates," Adv. Mater. 18(2006): 807-811. Copyright 2006, Wiley-VCH.)

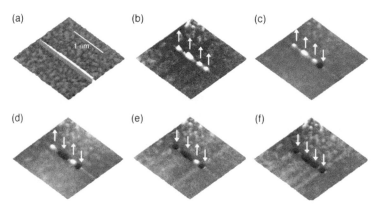

Fig. 2.31 (a) Three-dimensional topographic image of a BaTiO$_3$ nanowire with a diameter of 12 nm, (b)-(f) Successive EFM images showing that four distinct polarization domains can be independently manipulated by an external electric field. In these EFM images, the bright and dark colors correspond to a resonance frequency shift of +10 Hz and –10 Hz, respectively, and the white arrows indicate the polarization directions. The upward and downward polarization spots were written with $V_{tip} = -10$ V and $V_{tip} = +10$ V, respectively. The distance between the tip and the top surface of the nanowire was 10 nm during the writing procedure and 35 nm during the reading procedure. (Reprinted with permission from W.S. Yun, J.J. Urban, Q. Gu and H.K. Park, "Ferroelectric properties of individual barium titanate nanowires investigated by scanned probe microscopy," Nano Lett. 2(2002): 447-450. Copyright 2002, American Chemical Society.)

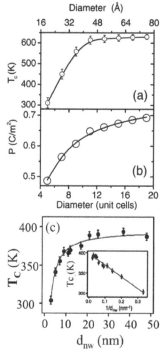

Fig. 2.32 (a) Critical T_c temperature, and (b) magnitude of polarization at 64 K, as a function of the diameter of PZT nanowire. The diameter is in units of Å (bulk lattice constant) for the upper (lower) horizontal axis. The fitted result using the $1/d_{nw}$ scaling law is shown as a solid line in (b), (a,b) (Reprinted with permission from I.I. Naumov and H.X. Fu, 2005. Spontaneous polarization in one-dimensional Pb(ZrTi)O$_3$ nanowires. Phys. Rev. Lett. 95(2005): 247602 (1-4). Copyright 2005, American Physical Society.), (c) T_c of BTO nanowire varies as a function of nanowire diameter. Inset is T_c decreasing linearly with $1/d_{nw}$. (Reprinted with permission from J.E. Spanier, A.M. Kolpak, J.J. Urban, I. Grinberg, O.Y. Lian, W.S. Yun et al., "Ferroelectric phase transition in individual single-crystalline BaTiO$_3$ nanowires," Nano Lett. 6(2006): 735-739. Copyright 2006, American Chemical Society.)

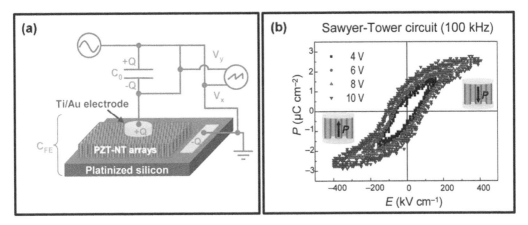

Fig. 2.33 (a) A schematic diagram of the Sawyer-Tower circuit used to minimize the contribution of leakage current to the *P-E* hysteresis measurements, (b) Unambiguous *P-E* hysteresis loops obtained at 100 kHz with a Sawyer-Tower circuit with applied voltages between 4 and 10 V. The measured P_r and E_c are about 1.5 μC/cm^2 and 86 kV/cm, respectively. (Reprinted with permission from J. Kim, S.A. Yang, Y.C. Choi, J.K. Han, K.O. Jeong, Y.J. Yun et al., "Ferroelectricity in highly ordered arrays of ultra-thin-walled Pb(Zr,Ti)O$_3$ nanotubes composed of nanometer-sized perovskite crystallites," Nano Lett. 8(2008): 1813-1818. Copyright 2008, American Chemical Society.)

2.4.2 Dielectric Properties

Perovskite oxide nanoparticles exhibit excellent dielectric properties (e.g. high dielectric constant and low dielectric loss). As representative samples, BTO and STO nanoparticles are widely investigated and used in the modern microelectronic devices (O'Brien et al. 2001, Bansal et al. 2006, Bassano et al. 2009, Dang et al. 2011). For example, Huang et al. (2006) synthesized well-isolated BTO nanocrystals on different substrates (e.g. Si, Si/SiO$_2$, Si$_3$N$_4$/Si, and Pt-coated Si substrate), from which nanocrystallined BTO films were fabricated. The nanocrystallined BTO films exhibit ferroelectric behavior. Their dielectric constants were measured to be 85-90 across 1 KHz-100 KHz, while the dielectric losses were of 0.03-0.04. That enables nanocrystallined BTO films to be used in thin film capacitances. Recently, perovskite-type strontium hafniate (SrHfO$_3$) nanoparticles were also reported by Karmaoui et al. (2016). They synthesized monodisperse, ultrasmall SrHfO$_3$ nanoparticles via a non-aqueous sol-gel process. Figure 2.34(a) shows the frequency dependence of dielectric properties of the SrHfO$_3$ nanoparticles at room temperature. The data demonstrate that the dielectric constant is almost independent of the frequency within the investigated frequency range, similar to that reported for SrHfO$_3$ thin films. The dielectric constant of the SrHfO$_3$ nanoparticles (with average diameter of 2.5 nm) was 17 at 100 kHz, which was comparable to the value of 21 for the SrHfO$_3$ films with thickness of 59 nm (Black et al. 2011), but smaller than the value of 31 (at 1 MHz) of bulk SrHfO$_3$ ceramics made from nanocrystalline powders (Thomas et al. 2011). The room temperature dielectric loss (tan δ) decreases with increasing the frequency; as expected, they are in the range of 0.02-0.07 within the measured frequency range from 10^2-10^6 Hz. Recently, STO nanocubes with sizes below 10 nm were prepared by a rapid sol-based precipitation method. These STO nanocubes have good crystallinity, high dispersibility, and narrow size distribution (Hao et al. 2014). High-quality STO nanostructured films that were made from nanocubes, were fabricated on Pt-coated Si substrate by spin-coating method, and their dielectric properties are shown in Fig. 2.34(b). It is noticed that both dielectric constant (ε_r) and dielectric loss (tan δ) remain constant values across frequency of 10^2 Hz-10^6 Hz, only slight reduction of dielectric constant appears in a higher frequency. The value of ε_r is in the range of 13-19, which is comparable to that reported by Huang et al. (2013b). Frequency dependence of the dielectric properties of BFO nanotube arrays was also investigated by Zhang et al. (2005), as depicted in Fig. 2.34(c). They found that below 10^3 Hz, dielectric constant and dielectric loss decreased fast, and then they remained fairly constant afterwards. The values of dielectric constant and dielectric loss were measured to be 90 and 0.04, respectively at 10^5 Hz, matching well with that reported previously (Teowee et al. 1997, Yun et al. 2004).

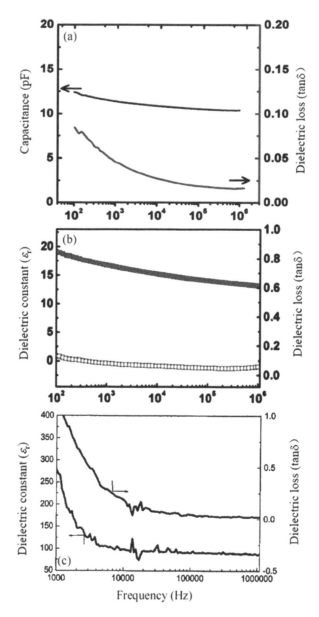

Fig. 2.34 Dielectric properties of (a) SrHfO$_3$, (b) SrTiO$_3$ nanoparticles, and (c) BFO nanotube arrays measured as a function of the frequency. (a) (Reprinted with permission from M. Karmaoui, E.V. Ramana, D.M. Tobaldi, L. Lajaunie, M.P. Graca, R. Arenal et al., "High dielectric constant and capacitance in ultrasmall (2.5 nm) SrHfO$_3$ perovskite nanoparticles produced in a low temperature non-aqueous sol-gel route," RSC Adv. 6(2016): 51493-51502. Copyright 2016, The Royal Society of Chemistry.), (b) (Reprinted with permission from Y.N. Hao, X.H. Wang and L.T. Li, "Highly dispersed SrTiO$_3$ nanocubes from a rapid sol-precipitation method," Nanoscale. 6(2014): 7940-7946. Copyright 2014, The Royal Society of Chemistry.), (c) (Reprinted with permission from X.Y. Zhang, C.W. Lai, X. Zhao, D.Y. Wang and J.Y. Dai, "Synthesis and ferroelectric properties of multiferroic BiFeO$_3$ nanotube arrays," Appl. Phys. Lett. 87 (2005): 143102 (1-3). Copyright 2005, American Institute of Physics.)

2.4.3 Piezoelectric Properties

Piezoelectricity is a form of electricity generated as certain crystals are subjected to a stress. Curie brothers first discovered the piezoelectricity in 1880. However, it was in the 1950s that researches on piezoelectricity gained a breakthrough after the discovery of BTO, PZT, and the family of perovskite oxide materials (Ortega-San-Martin 2020). Recently, in piezoelectric nanostructures an enhanced piezoelectric effect is observed (Fang et al. 2013). These distinct features make them attractive in the nanoelectromechanical systems (Park et al. 2010, Lew et al. 2011, Sen et al. 2012). To date, the piezoelectric properties of 0D and 1D piezoelectric nanostructures are characterized by PFM technique. For example, Vasudevan et al. (2011) used PFM to characterize the piezoresponse characters of BFO nanoparticles. Ferroelectric nature of the BTO nanoparticles was confirmed by the typical symmetric piezoresponse and phase loops obtained from a single nanoparticle, as shown in Figs. 2.35(a)-(b). In addition, the ferroelectric domain structures within the nanoparticles are similar to that observed in BFO thin films (seen in Figs. 2.35(c)-(d)). The piezoelectric properties of monocrystalline PZT nanowires (Wang et al. 2007), (K,Na)NbO$_3$ (KNN) nanorods (Cheng et al. 2013), and polycrystalline BFO nanofibers (Xie et al. 2012) were also investigated, respectively. Figures 2.36(a)-(c) show the AFM/PFM characterizations of 1D PZT nanorods (Wang et al. 2007). The PFM characterizations of the BTO nanowires were performed in both vertical and lateral modes by Wang et al. (2006a, b). Figures 2.36(d)-(f) depict the BTO nanowire morphology, lateral amplitude curves, and phase hysteresis loops, respectively. Figures 2.36(g) shows the AFM image of a single KNN nanorod with length of ~ 25 μm and width of 500 -700 nm. The amplitude butterfly

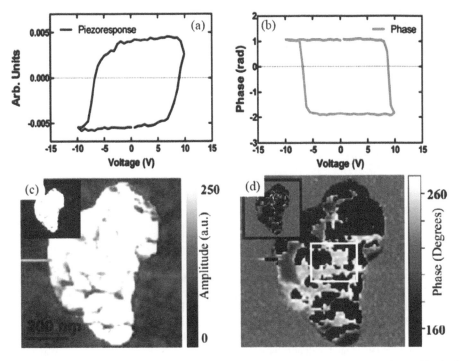

Fig. 2.35 (a) Piezoresponse, (b) phase hysteresis loops of a single BiFeO$_3$ nanoparticle, (c) Vertical PFM amplitude, (d) phase images of a nanoparticle cluster, before and after applying +10 V, 5 s pulse to the center of the cluster. Insets in (c)-(d) show the PFM amplitude and phase before applying the bias, respectively. The above measurement was carried out on the LSMO surface. (Reprinted with permission from R.K. Vasudevan, K.A. Bogle, A. Kumar, S. Jesse, R. Magaraggia, R. Stamps et al., "Ferroelectric and electrical characterization of multiferroic BiFeO$_3$ at the single nanoparticle level," Appl. Phys. Lett. 99 (2011) 252905 (1-4). Copyright 2011, American Institute of Physics.)

curves and phase hysteresis loops are observed, as shown in Fig. 2.36(h). Such typical amplitude-voltage and phase-voltage hysteresis loops indicate a homogeneity of the nanorods. Figures 2.37(a) is the piezoelectric hysteresis loop obtained from a PZT nanotube whose outer diameter is 700 nm and wall thickness is 90 nm (Luo et al. 2003). This is an experimental proof of the piezoelectricity in the PZT nanotubes. The effective piezoelectric coefficient was estimated to be about 90 pm/V, comparable to those reported for PZT thin films. Figure 2.37(b) shows the piezoelectric properties of BFO nanotubes synthesized by sol-gel template method (Zhang et al. 2005). A clear piezoresponse d_{33} hysteresis loop was observed, indicating the ferroelectric behavior of the BFO nanotubes. It is also noticed that the d_{33} value decreases at high electric field, which is ascribed to the electric field-induced lattice hardening (Zeng et al. 2004).

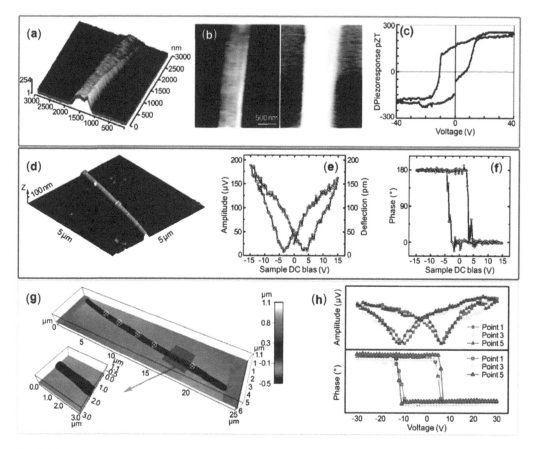

Fig. 2.36 AFM/PFM measurements of (a) PZT, (b) BTO, and (c) KNN nanorods. (a) 3D drawing of the amplitude of PZT nanowire, (b) The amplitude and phase information of PZT nanowire poled with voltage of –40 V applied on the bottom electrode, (c) Local hysteresis loop of piezoresponse of PZT nanowire, (d) AFM image of BTO nanowire, (e) Lateral PFM phase loop, (f) amplitude butterfly curve of BTO nanowire, (g) Morphology of the KNN nanorod, (h) Piezoresponse and phase information obtained from three points as marked in (g). (a-c) (Reprinted with permission from J. Wang, C.S. Sandu, E. Colla, Y. Wang, W. Ma, R. Gysel et al., "Ferroelectric domains and piezoelectricity in monocrystalline Pb(Zr,Ti)O₃ nanowires," Appl. Phys. Lett. 90(2007): 133107 (1-3). Copyright 2007, American Institute of Physics.). (d-f) (Reprinted with permission from Z. Wang, J. Hu and M.F. Yu, "One-dimensional ferroelectric monodomain formation in single crystalline BaTiO₃ nanowire," Appl. Phys. Lett. 89(2006): 263119 (1-3). Copyright 2006, American Institute of Physics.). (g-h) (Reprinted with permission from L.Q. Cheng, K. Wang, J.F. Li, Y. Liu, J. Li, "Piezoelectricity of lead-free (K,Na)NbO₃ nanoscale single crystals," J. Mater. Chem. C. 2 (2014): 9091-9098. Copyright 2014, The Royal Society of Chemistry.)

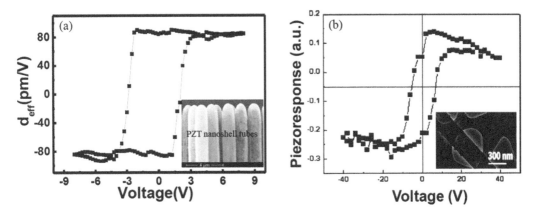

Fig. 2.37 (a) Piezoelectric hysteresis loop obtained from a PZT nanotube. Inset is SEM image of the capped tips of PZT nanotubes. (Reprinted with permission from Y. Luo, I. Szafraniak, N.D. Zakharov, V. Nagarajan, M. Steinhart, R.B. Wehrspohn et al., "Nanoshell tubes of ferroelectric lead zirconate titanate and barium titanate," Appl. Phys. Lett. 83(2003): 440-442. Copyright 2003, American Institute of Physics), (b) Piezoelectric hysteresis loop obtained from the multiferroic BFO nanotubes prepared by sol-gel template method. Inset is SEM image of the single PZT nanotube. (Reprinted with permission from X.Y. Zhang, C.W. Lai, X. Zhao, D.Y. Wang and J.Y. Dai, "Synthesis and ferroelectric properties of multiferroic BiFeO₃ nanotube arrays," Appl. Phys. Lett. 87(2005): 143102 (1-3). Copyright 2005, American Institute of Physics.)

2.4.4 Magnetic Properties

Perovskite oxide manganites with the formula of $R_{1-x}A_xMnO_3$ (e.g., R = La, Pr, Sm, A = Ca, Sr, Ba, or Pb) exhibit colossal magneto-resistance, magnetocaloric effect, and multiferroic properties, which enable them to use spintronic devices, MERAM devices. Similar to the ferroelectric properties, magnetic properties of perovskite oxide nanoparticles also display the size effect. Up to date, several research groups have explored the size-dependent magnetic properties of the BFO nanostructures (Park et al. 2007, Zavaliche et al. 2007, Goswami et al. 2011, Huang et al. 2013a); details are presented in Fig. 2.38. It is noticed that the saturation magnetization (M_s) increases as the particle size decreases (see Figs. 2.38(a), (b) and (d)). Such enhanced magnetic properties are ascribed to the broken spiral of magnetic order in BFO particles below 62 nm (Park et al. 2007). While the particle sizes of BFO nanoparticles approach 62 nm (the spiral spin period), they display much enhanced magnetization, as shown in Fig. 2.38(b). That can be attributed to an enhanced rotation distortion of the FeO_6 octahedron (Huang et al. 2013a). Figure 2.38(c) presents the theoretical results of the critical temperatures $(T_C$ and $T_N)$ varying with the particle size in nanometer scale (Zavaliche et al. 2007). They are confirmed by the experimental data (Fig. 2.38(d)). Clearly, the size effect has important impact on the magnetic properties of BFO nanoparticles (Dutta et al. 2013). Liu et al. (2011) measured the magnetic properties of single-crystalline BFO nanowires synthesized by hydrothermal method. Figure 2.39(a) shows the magnetizations of the BFO nanowires vs temperature, which were measured in a field of 100 Oe under field-cooling (FC) and zero-field-cooling (ZFC) modes. A sharp cusp was observed in the *M-T* curve in the ZFC mode, corresponding to the blocking temperature, or freezing temperature (T_f). As the temperature is decreased close to T_f (= 55 K), the ZFC and FC curves of BFO nanowires become separated. The irreversibility between the ZFC and FC magnetizations at low temperature and the observed sharp cusp (T_f) are the typical characteristics of spin-glass behavior. The *M-H* loops of the BFO nanowires measured at 5 K and 300 K with an external magnetic field varying from –6000 to 6000 Oe are shown in Fig. 2.39(b). At 5 K, a clear *M-H* hysteresis loop was observed, whereas at 300 K the magnetic properties of the BFO nanowires become much weaker.

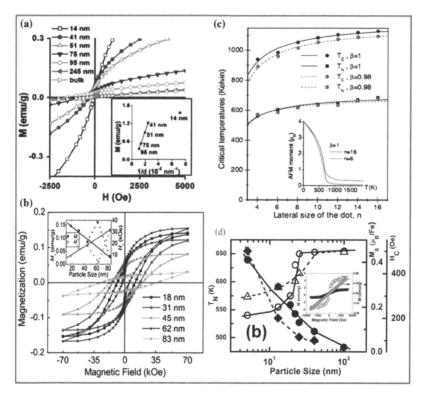

Fig. 2.38 (a) Magnetization of BFO nanoparticles with different sizes, (b) M-H loops of BFO nanoparticles with different sizes, and the inset in (b) shows size-dependent M_s and H_c, (c) Size dependency of critical temperatures in BFO nanodots, (d) Variations in T_N, Ms, and H_c with particle sizes. (a) (Reprinted with permission from T.J. Park, G.C. Papaefthymiou, A.J. Viescas, A.R. Moodenbaugh, and S.S. Wong, "Size-dependent magnetic properties of single-crystalline multiferroic BiFeO$_3$ nanoparticles," Nano Lett. 7(2007): 766-772. Copyright 2007, American Chemical Society.). (b) (Reprinted with permission from F. Huang, Z. Wang, X. Lu, J. Zhang, K. Min, W. Lin et al., Peculiar magnetism of BiFeO$_3$ nanoparticles with size approaching the period of the spiral spin structure," Sci. Rep. 3(2013) 2907 (1-7). Copyright 2013, Nature Publishing Group.). (c) (Reprinted with permission from F. Zavaliche, T. Zhao, H. Zheng, F. Straub, M.P. Cruz, P.L. Yang et al., "Electrically assisted magnetic recording in multiferroic nanostructures," Nano Lett. 7(2007): 1586-1590. Copyright 2007, American Chemical Society). (d) (Reprinted with permission from S. Goswami, D. Bhattacharya and P. Choudhury, "Particle size dependence of magnetization and noncentrosymmetry in nanoscale BiFeO$_3$," J. Appl. Phys. 109 (2011): 07D737 (1-3). Copyright 2011, American Institute of Physics.)

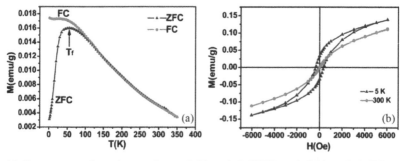

Fig. 2.39 (a) Temperature dependence of zero-field-cooled (ZFC) and field-cooled (FC) susceptibility measured in a field of 100 Oe for BiFeO$_3$ nanowires, (b) M-H hysteresis loops for BiFeO$_3$ nanowires measured at 5 and 300 K. (Reprinted with permission from B. Liu, B. Hu and Z. Du, "Hydrothermal synthesis and magnetic properties of single-crystalline BiFeO$_3$ nanowires," Chem. Commun. 47(2011): 8166-8168. Copyright 2011, American Chemical Society.)

2.4.5   Multiferroic Properties

Multiferroics materials possess strong couplings among the ferroic order parameters (e.g. ferromagnetic, ferroelectric, or ferroelastic ordering), leading to simultaneous ferroelectricity, ferromagnetism, and ferroelasticity (Schmid 1994). Particularly, in a magnetoelectric (ME) multiferroic material, the ferroelectric polarization controlled by magnetic field or magnetization controlled by electrical field opens up an entirely new perspective of the next generation of revolutionary memory devices and sensors (Fiebig et al. 2002, Wang et al. 2003, Hur et al. 2004). However, there are few materials exhibiting multiferroic behavior at room temperature (Hill 2000, Martin et al. 2008). Bismuth ferrite ternary oxides (BFO) and its derivative compounds with a perovskite structure are promising multiferroic compounds at room temperature (Catalan and Scott 2009, Kan et al. 2012), but their ME coupling coefficients are rather small (Suryanarayana 1994). To increase the ME coupling of the BFO at RT, several strategies have been developed (Wu et al. 2016). Figure 2.40 displays the ME coefficients from the BFO-based materials in comparison with other materials, where the largest ME coupling coefficient (315 mV cm^{-1} Oe^{-1} at 50 kHz) was achieved in the $0.3Co_{0.7}Zn_{0.3}Fe_2O_4$-$0.7Bi_{0.9}La_{0.1}FeO_3$ nanocomposite. That is ascribed to the strong coupling between high piezomagnetic and piezoelectric effects, which are induced by small grain sizes (Kumar and Yadav 2013). Obviously, the ME coefficients of the BFO-based materials are still smaller than that of other materials (Wu et al. 2016). There exists a large space for improving the ME coefficients of BFO-based materials. It is expected that the formation of nanocomposites composed of strong magnetic phase and BFO is an effective approach to achieve higher ME coefficient. Figure 2.41 shows the ME properties of epitaxial BFO-$CoFe_2O_4$ (CFO) (001) vertical nanocomposites, where the transverse (α_{31}) and longitudinal (α_{33}) magnetoelectric susceptibilities were quantitatively determined (Oh et al. 2010). It was noted that a high value of α_{31} (~ 60 mV cm^{-1} Oe^{-1} at a magnetic field of ~ 6 kOe) was achieved in the BFO-CFO (001) vertical nanocomposite (Fig. 2.41(a)), which was larger than the reported value (α_{33} ~16 mV cm^{-1} Oe^{-1}) for the 0-3 structured $NiFe_2O_4$-$PbZr_{0.52}Ti_{0.48}O_3$ composites (Ryu et al. 2006). Therefore, enhanced ME properties were achieved in the CFO-BFO nanocomposites. In the heterostructures, however, some important issues such as switching reliability and ME coupling mechanism still remain a challenge (Heron et al. 2014). Much work is needed to be done to satisfy the requirements of practical applications of BFO-based multiferroic materials.

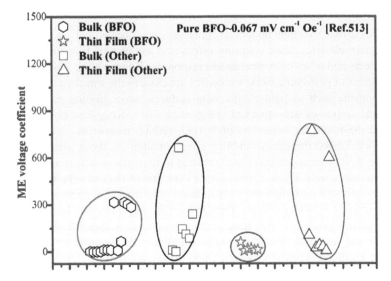

Fig. 2.40 Magnetoelectric properties of selected BFO materials (unit of ME coupling: mV cm^{-1} Oe^{-1}). (Reprinted with permission from J.A. Wu, Z. Fan, D.Q. Xiao, J.G. Zhu and J. Wang, "Multiferroic bismuth ferrite-based materials for multifunctional applications: Ceramic bulks, thin films and nanostructures," Prog. Mater. Sci. 84(2016): 335-402. Copyright 2016, Elsevier.)

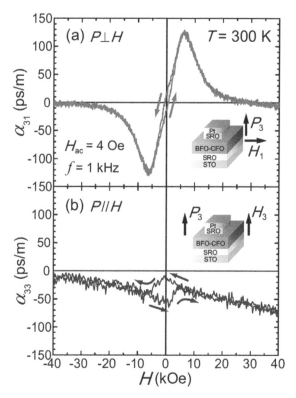

Fig. 2.41 (a) Transverse (α_{31}), (b) longitudinal (α_{33}) magnetoelectric susceptibilities of the BFO-CoFe$_2$O$_4$ (CFO) (001) vertical nanocomposites at 300 K. (Reprinted with permission from Y.S. Oh, S. Crane, H. Zheng, Y.H. Chu, R. Ramesh and K.H. Kim, "Quantitative determination of anisotropic magnetoelectric coupling in BiFeO$_3$-CoFe$_2$O$_4$ nanostructures," Appl. Phys. Lett. 97(2010): 052902(1-3). Copyright 2011, American Institute of Physics.)

2.4.6 Optical Properties

Perovskite oxides exhibit excellent optical and photoluminescence properties. As light is introduced into nanoscale ferroelectrics, their coupling offers new avenues of functionality such as photo-ferroelectric effects and other opto-mechanical responses (Kundys et al. 2010, Schick et al. 2014). When the diameters of perovskite oxide nanowires are close to the wavelength of the incident light, the novel phenomena such as light-confinement features, wave guiding patterns, and enhanced nonlinear optical responses are observed (Jung et al. 2011, Krogstrup et al. 2013). Kim et al. (2013) reported the nonlinear behaviors of KNbO$_3$ (KNO) nanowires, and they found that the KNO nanowires exhibited frequency-doubling effect. In addition, the second harmonic generation efficiency of the KNO nanowires was closely related to their phase structures, e.g. the monoclinic phase KNO nanowires were about 3 times larger than that of the orthorhombic phase nanowires. Moreover, Nah et al. (2014) studied the light-induced polarization dynamics at femtosecond scale in an optically trapped KNO nanowires. They found that a spontaneous ferroelectric polarization inclined to the optical axis at 45° was generated in an optically trapped KNO nanowire, as shown in Figs. 2.42(a)-(b), respectively. Meanwhile, the large amplitudes, reversible photoinduced modulations in nonlinear optical properties and polarization were also observed in a single nanowire; details are shown in Figs. 2.42(e)-(h). Recently, systematic studies on the second harmonic generation properties of XNbO$_3$ (X = Li, Na, K) nanowires were carried out by Dutto et al. (2011). Among them, the LiNbO$_3$ (LNO) nanowires exhibited the strongest second harmonic generation response. In addition, the intensity of the second harmonic generation signal was closely related

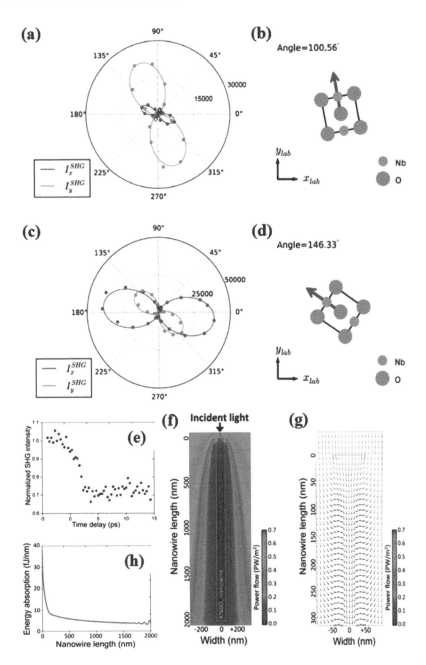

Fig. 2.42 (a) Polarization dependence of second harmonic generations observed in a trapped nanowire with a polarizer parallel to the x-axis (dark) and y-axis (light), (b) Schematic unit cell of the nanowire used at an azimuthal angle of 100.56° for the spontaneous polarization direction (arrow), (c) Polar plots from another trapped nanowire, (d) Schematic unit cell of the nanowire used in (c) at an azimuthal angle of 146.33° (arrow), (e) Time-resolved second harmonic generation signals of a trapped nanowire as a function of pump-probe time delays upon photoexcitation, (f) 3D finite difference time domain (FDTD) simulation showing the amplitude of the power flow over a free-standing nanowire (100 nm in diameter and 2 μm in length) in water. Incident plane wave ($E = 0.53$ GV/m, 50 fs) propagating along the long axis of the nanowire, (g) Calculated directions of the power flow shown in (f) and (h) energy absorption distribution per pulse within successive 1 nm slices as a function of the nanowire length. (Reprinted with permission from S. Nah, Y.H. Kuo, F. Chen, J. Park, R. Sinclair, A.M. Lindenberg, "Ultrafast polarization response of an optically trapped single ferroelectric nanowire," Nano Lett. 14(2014): 4322-4327. Copyright 2014, American Chemical Society.)

to the generated polarization and the crystallinity degree of the $XNbO_3$ nanowires. Despite all the three $XNbO_3$ nanowires having a similar coupling waveguiding efficiency as potential waveguides, $NaNbO_3$ nanowiress are regarded as the superior choice due to their highest aspect ratio. For 1D perovskite nanostructures, their optical properties have promising applications in photonic crystals and sub-wavelength optics.

In the last decade, terahertz (THz) technology has been used as powerful tool for non-destructive imaging (Davies et al. 2008). Recently, Scott et al. (2008) reported on a strong THz emission from PZT nanotubes; however, such strong emission was not observed in bulk PZT and films. The THz radiation exhibits short emitting time (0.2 ps) and broader spectrum peak (from 2 to 8 THz), which supplements the conventional semiconductor THz emitters. It is believed that after optical rectification, the THz emitters based on PZT nanotubes will find promising applications in the fields of THz technology.

2.4.7 (Photo-) Catalytic Properties

Perovskite oxide materials exhibit high photocatalytic activities and stabilities in various environments, which are used as efficient photocatalysts to degrade organic contaminants under visible light irradiation (Li et al. 2010, Thirumalairajan et al. 2012, Lam et al. 2017). For example, under visible light irradiation perovskite $LaFeO_3$ nanoparticles can produce the electron-hole pairs to degrade contaminant in water and split water into H_2 and O_2 (Parida et al. 2010). Therefore, much attention has been given to perovskite $LaFeO_3$ nanoparticles due to their abundance, non-toxicity, narrow optical bandgap (2.1 eV), and structural stability (Ding et al. 2010). Since the photocatalytic properties of $LaFeO_3$ nanoparticles are dependent upon their stoichiometric compositions and microstructures, the experimental parameters of synthesis methodologies must be well controlled (Ansari et al. 2019, Kucharczyk et al. 2019). Figure 2.43 shows the photocatalytic activities of the perovskite $LaFeO_3$ nanoparticles (annealed at different temperatures), which are valuated by degrading the typical organic dyes, methyl orange (MO) and methyl blue (MB) under a visible light illumination (Shen et al. 2016). It was found that annealing temperature and annealing methods (conventional calcination vs vacuum microwave calcination) have an important impact on controlling the photocatalytic activities of $LaFeO_3$ nanoparticles. Recently, Bi-based oxides such as BFO have been developed as a new visible light photocatalyst (McDonnell et al. 2013, Wu et al. 2016, Lam et al. 2017). Lam et al. (2017) summarized the photocatalytic activities of BFO with respect to different pollutants. It is noticed that the BFO exhibits different activities when operated in different process systems. However, the photo-degradation efficiency of organic pollutants is still relatively low for pristine BFO. To solve this problem, many modification approaches (e.g. morphology control, chemical doping, heterojunction structure construction) have been developed (Peng et al. 2015, Wu et al. 2016, Zhang et al. 2016, Lam et al. 2017). It is expected that the photocatalytic performance of BFO nanoparticles will be improved after using the above methods.

Besides the perovskite oxide nanoparticles, good photocatalytic activity is also reported in 1D perovskite oxide nanowires such as $KNbO_3$ (Ding et al. 2008), $NaNbO_3$ (Lv et al. 2010, Saito and Kudo 2010), $NaTaO_3$ (Yi and Li 2010), STO (Miyauchi 2007), and BFO nanowires (Gao et al. 2006). It is reported that BFO nanowires exhibit photoinduced oxidization ability, producing oxygen in the $AgNO_3/H_2O$ system with an initial efficiency of 1876 mol $h^{-1}g^{-1}$ (Gao et al. 2006). Further enhanced photoactivity was achieved by decorating BFO nanowires with Au nanoparticles, as shown in Fig. 2.44(a) (Li et al. 2013a). Figure 2.44(b) demonstrates the oxygen evolution as a function of time for the pristine BFO nanowires and a series of Au-BFO hybrid nanowires. It is found that the Au-BFO hybrid nanowires have better visible-light photocatalytic properties than the pure BFO nanowires. Besides the Au-BFO hybrid nanowires, BFO-based nanocomposites such as Ag-BFO (Lu et al. 2015), BFO-grapheme (Li et al. 2013b, c), BFO-γ-Fe_2O_3 (Guo et al. 2011), BFO-TiO_2 (Sarkar et al. 2015), and STO-BFO core/shell (Luo and Maggard 2006) also exhibit enhanced photocatalytic properties. Up to date, there are many investigations on modifying morphology, chemical compositions and interfaces within the nanocomposite structures, in order to improve the photocatalytic properties of BFO-based nanocomposites.

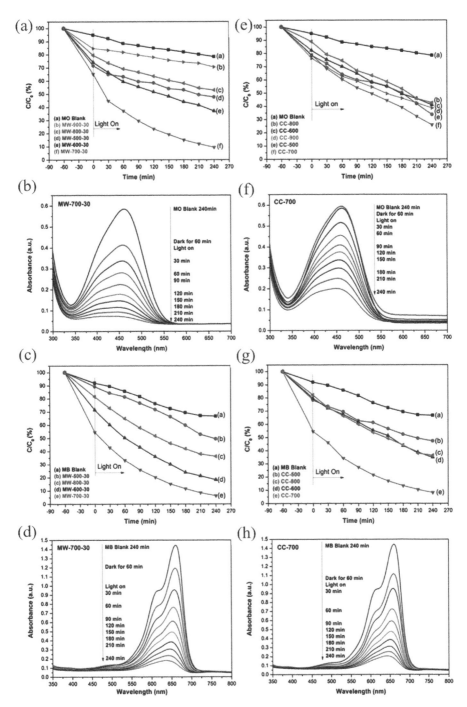

Fig. 2.43 Photocatalytic activities of the perovskite LaFeO₃ nanoparticles (annealed at different temperatures) evaluated by the degradation of the typical organic dyes, methyl orange (MO) and methyl blue (MB), under a visible light illumination. Effect of vacuum microwave annealing temperature of LaFeO₃ on the degradation efficiency of MO(A), MB(C), the temporal evolution of the MO(B) and MB(D) absorption spectrum of the sample MW-700-30 as a photocatalyst. Effect of conventional annealing temperature on the degradation efficiency of MO(E), MB(G), the temporal evolution of the MO(F) and MB(G) absorption spectrum of the sample CC-700 as a photocatalyst. (Reprinted with permission from H.F. Shen, T. Xue, Y.M. Wang, G.Z. Cao, Y.J. Lu, G.L. Fang, "Photocatalytic property of perovskite LaFeO₃ synthesized by sol-gel process and vacuum microwave calcination," Mater. Res. Bull. 84 (2016): 15–24. Copyright 2016, Elsevier.)

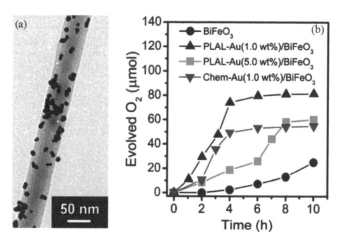

Fig. 2.44 (a) TEM image of BiFeO$_3$ nanowires decorated with Au nanoparticles, (b) Oxygen evolved upon visible light ($\lambda > 380$ nm) illumination of the FeCl$_3$ suspension (4 mmol L^{-1}, 50 mL) containing the photocatalysts (50 mg). (Reprinted with permission from S. Li, J. Zhang, M.G. Kibria, Z. Mi, M. Chaker, D. Ma et al., "Remarkably enhanced photocatalytic activity of laser ablated Au nanoparticle decorated BiFeO$_3$ nanowires under visible-light," Chem. Commun. 49(2013): 5856-5858. Copyright 2013, American Chemical Society.)

2.5 STRUCTURAL CHARACTERIZATIONS OF NANOSTRUCTURED PEROVSKITE OXIDES

Along with the development of the fabricated methods for nanostructured perovskite oxides, the characterization techniques of nanostructured perovskite oxides have also been developed, which can provide invaluable structural information on the microstructural, crystallographic, and atomic features. Up to date, much advanced technical approaches including x-ray diffraction (XRD), electron diffraction, convergent beam electron diffraction, transmission electron microscopy (TEM), high-resolution transmission electron microscopy (HRTEM) and related spectroscopies such as x-ray energy dispersive spectroscopy (EDS) and electron energy loss spectroscopy (EELS) have been widely used to characterize the macroscopic and microscopic details of nanostructured perovskite oxides. STEM, in conjunction with high-angle annular dark field (HAADF) imaging (or Z-contrast incoherent imaging), has the ability to determine the chemical compositions and structures of materials at the atomic scale. By combination of Z-contrast imaging and high-resolved EELS spectroscopy, the interfacial structures and the elemental/chemical environment at/around interfaces within nanostructured perovskite oxides can be determined at atomic scale. In recent years, the Cs-corrected HRTEM/STEM microscopy allows one to reach a spatial sub-Å resolution and a sub-eV energy resolution, characterizing individual nanoscale structure at sub-Å available. SPM-based techniques, in particular PFM, can measure the local ferroelectric/piezoelectric properties of nanostructured perovskite oxides. Besides the PFM technique, atomic force microscopy (AFM), conducting AFM (c-AFM), scanning tunneling microscopy, and EFM techniques have been developed to examine the ferroelectric domain structures within the nanostructured perovskite oxides. Here, we shortly introduce the working principles of these techniques and provide some of their typical applications.

2.5.1 Microstructural Characterizations

2.5.1.1 X-Ray Diffraction (XRD)

X-ray diffraction (XRD) is a very common method to determine the phase structures of all kinds of matter ranging from fluids, to powders, and crystals. In XRD experiment, the incident x-rays are diffracted by the crystalline phase in the specimen, following the Bragg's law $2d \sin \theta = \lambda$, where d is the spacing between atomic planes in the crystalline phase, λ is the wavelength of the x-ray, and θ is the diffraction angle. The intensity of the diffracted x-rays is measured vs the diffraction angle 2θ and the specimen's orientation. By analyzing the XRD data, the critical features such as crystal structure, crystallite size, lattice parameters (a, b, and c), lattice volume, and strain can be determined.

2.5.1.2 Scanning Electron Microscopy (SEM)

SEM makes use of a fine electron beam to reveal the microstructures of the investigated materials. Its operation is based on various signals (e.g. secondary electrons, internal currents, photon emission, etc.) that are generated by the interaction between the electron beam and the sample. SEM images can not only provide the topographical information, but also the chemical compositional information of the material. Therefore, the morphology, grain size and its distribution, and surface compositional information of nanostructured perovskite oxides can be easily obtained from their SEM images.

2.5.1.3 Transmission Electron Microscopy (TEM)

TEM is a microscope working in transmission mode. By making use of the transmitted electrons and the forward diffracted electrons, both bright-field and dark-field TEM images can be formed, respectively, which reveal the atomic structures and defects present in the material. Currently, TEM is widely used to characterize the microstructures of nanostructured perovskite oxides synthesized by different methods, which offers a variety of other signals much beyond the imaging. For example, electron diffraction is a much efficient technique, which provides the crystallographic data in reciprocal space. Nowadays, TEM is widely used to reveal microstructures of nanostructured perovskite oxides synthesized via chemical routes.

2.5.1.4 Scanning Transmission Electron Microscopy (STEM)

STEM makes use of high-angle elastically-scattered electrons by an annular dark-field detector, forming the Z-contrast incoherent images, or HAADF images, as schematically shown in Fig. 2.45(a). In a crystal, atoms with higher atomic number (Z) produce more scattering, exhibiting the bright spots in the image. Therefore, STEM images can be explained directly in terms of atomic types and positions. As the investigated samples involve nanoparticles with heavy elements as compared with the matrix material, their different contrast in the HAADF-STEM image is closely correlated with their atomic numbers and specimen thickness. Thus, HAADF-STEM image can be used as an effective method to distinguish the nanoparticles of interest from the matrix (Utsunomiya and Ewing 2003). In the perovskite oxide nanoparticles, the dopant atom positions in the host nanoparticles can be identified by using high-resolution STEM image combined with multivariate statistical analysis technique. Details are shown in Fig. 2.46 (Rossell et al. 2012). The present results offer a direct method in searching for the Ba dopants in the STO nanoparticles, which can be extended into determining the spatial distribution of impurity atoms in perovskite nanocrystals.

2.5.1.5 High-Resolution Transmission Electron Microscopy (HRTEM)

When nearly parallel electron beam travels through the sample, direct (transmitted) beam and the diffracted beams will be produced, and they interfere with each other to form a "lattice" image, as schematically shown in Fig. 2.45(b). Therefore, in principle, HRTEM images are interference patterns produced by the forward-scattered and diffracted electron waves from the specimen. Due to the ability for offering the structural information about local atomic structures, HRTEM images are widely used to identify structural defects in nanocrystals, and reveal the atomic distributions at interface between heterostructures. Up to date, as a powerful tool HRTEM image is widely used to identify structural defects in nanostructured perovskite oxides. For example, Fig. 2.47(a) shows a cross-sectional HRTEM image of the (001)-oriented $Pb(Zr_{0.52}Ti_{0.48})O_3$ (PZT) nanoisland deposited on cubic Nb-doped STO (001) substrates, where an edge dislocation (indicated by T) with a Burgers vector $b = a(100)$ is revealed (Chu et al. 2004). The schematic diagram of the interface between PZT and STO perovskite unit cells is shown in Fig. 2.47(b), and Fig. 2.47(c) is bright-field planar TEM image recorded under $g = (220)$, the dark contrast illustrates the network of misfit dislocations. Such misfit dislocations could induce the polarization instability of (001)-oriented epitaxial PZT nanoislands with a height of ~ 10 nm.

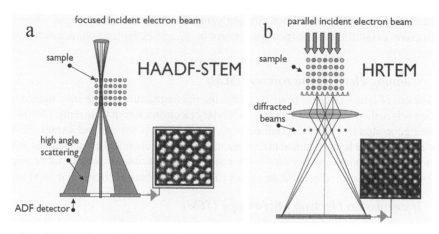

Fig. 2.45 Schematic diagrams for the formation of (a) STEM, (b) HRTEM images

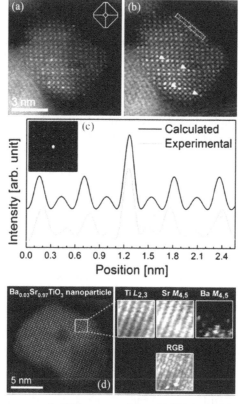

Fig. 2.46 Imaging Ba dopant atoms in STO nanoparticles. (a) HAADF-STEM image of a Ba-doped STO nanoparticle oriented along (001) with an inset showing good agreement with the truncated octahedral shape, (b) Processed image using a smooth low-pass filter set at a spatial frequency corresponding to 0.16 nm to reduce the high-frequency noise. Arrowheads indicate the presence of substitutional heavy Ba atoms in Sr positions, (c) Intensity profile taken along the region outlined with a box in panel *b* and calculated HAADF-STEM profile of a 0.4 nm thick STO crystal containing a substitutional Ba atom. Inset: corresponding simulated HAADF image, (d) HAADF-STEM image of a Ba-doped STO nanoparticle along the (011) direction. The Sr-$M_{4,5}$, Ti-$L_{2,3}$, and Ba-$M_{4,5}$ edges collected from the area outlined with white square in panel *d* are used to generate the elemental atomic-resolution maps. (Reprinted with permission from M.D. Rossell, Q.M. Ramasse, S.D. Findlay, F. Rechberger, R. Erni and M. Niederberger, "Direct imaging of dopant clustering in metal-oxide nanoparticles," ACS NANO. 6(2012): 7077-7083. Copyright 2012, American Chemical Society.)

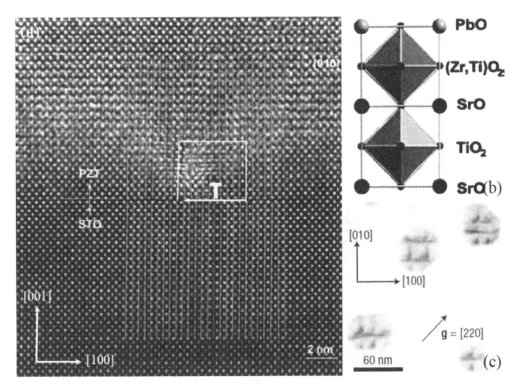

Fig. 2.47 Interfacial HRTEM image between a $Pb(Zr_{0.52}Ti_{0.48})O_3$ (PZT) nanoisland and Nb-doped STO (001) substrate, where an edge dislocation with a Burgers vector $b = a(100)$ is highlighted, (b) The schematic diagram of the interface between PZT and STO perovskite unit cells, (c) Bright-field planar view image recorded under $g = (220)$, the dark contrasts showing the network of misfit dislocations. (Reprinted with permission from M.W. Chu, I. Szafraniak, R. Scholz, C. Harnagea, D. Hesse, M. Alexe et al. "Impact of misfit dislocations on the polarization instability of epitaxial nanostructured ferroelectric perovskites," Nat. Mater. 3(2004): 87-90. Copyright 2004, Nature Publishing Group.)

2.5.1.6   Spherical-Corrected HRTEM/STEM

Traditionally, the spacial resolutions of HRTEM and STEM images are limited by their spherical aberration (Cs) of magnetic lenses. To solve this problem, spherical aberration correctors proposed by Rose (1990) have been developed to reduce Cs value of the objective lens. At present, the Cs-corrected HRTEM provides atomic scale resolution at 0.5 Å, light elements such as oxygen atoms and even their vacancies can be imaged (Varela et al. 2005, Urban 2008). With the help of the Cs-corrected imaging technique, Jia et al. (2008) carried out the pioneering works on atomic-scale revealing the electric dipoles near (charged and uncharged) 180° domain walls in epitaxial $PbZr_{0.2}Ti_{0.8}O_3$ thin films. Figure 2.48(a) is atomic-scale image of the electric dipoles formed in the 10-nm thick $PbZr_{0.2}Ti_{0.8}O_3$ film. The image was viewed along the direction and recorded in negative spherical-aberration imaging conditions. A large difference between the charged and uncharged domain walls can be revealed at atomic scale. Recently, Jia et al. (2011) reported a vortex ferroelectric domain structure in the ferroelectric $Pb(Zr,Ti)O_3$ thin film, where the electric dipoles rotate in the flux-closure pattern, as shown in Fig. 2.48(b). Figure 2.48(c) shows the map of the atomic diaplacement vectors, which offers the direct evidence of a continuous rotation of the dipole vector, forming a vortex domain structure. Similarly, Nelson et al. (2011) also reported vortex-like nanodomain arrays at the interface between the epitaxial BFO films and $TbScO_3$ substrates. Such a flux-closure domain texture was observed in the BFO film by mapping the electric polarization vector with atomic resolution via HRTEM images. More recently, in the PTO/STO ($PbTiO_3/SrTiO_3$)

multilayers/superlattices, periodic arrays of polar closure domains and vortices were also observed (Tang et al. 2015, Yadav et al. 2016). Besides the polarization vortice domain states, topological domain states such as polar-skyrmion bubble domains are also revealed by atomic-resolution STEM image in the compressively strained PTO/STO superlattices (Das et al. 2019).

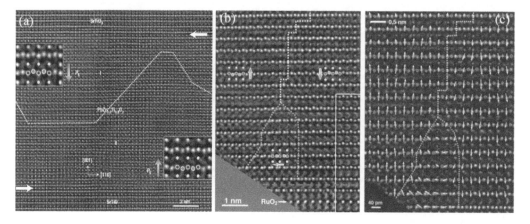

Fig. 2.48 (a) Atomic-scale imaging of the electric dipoles formed by the relative displacements of the Zr/Ti cation columns and the O anion columns in the approximately 10-nm-thick $PbZr_{0.2}Ti_{0.8}O_3$ (PZT) layer sandwiched between two $SrTiO_3$ (STO) layers. The image was viewed along the $(\bar{1}00)$ direction and recorded under negative spherical-aberration imaging conditions. (Reprinted with permission from C.L. Jia, S.B. Mi, K. Urban, I. Vrejoiu, M. Alexe and D. Hesse, "Atomic-scale study of electric dipoles near charged and uncharged domain walls in ferroelectric films," Nat. Mater. 7(2008): 57-61. Copyright 2008, Nature Publishing Group), (b) Atomic-resolution image of a vortex domain structure with continuous dipole rotation in PZT close to the interface to the STO substrate. In the image, two larger domains with 180° orientation can be identified. The inset at the lower right shows a calculated image demonstrating the excellent match between the atomic model and the specimen structure, (c) Map of the atomic displacement vectors. The length of the arrows represents the modulus of the displacements with respect to the scale bar in the lower left corner. The arrowheads point into the displacement directions. (b, c) (Reprinted with permission from C.L. Jia, K.W. Urban, M. Alexe, D. Hesse and I. Vrejoiu, "Direct observation of continuous electric dipole rotation in flux-closure domains in ferroelectric Pb(Zr,Ti)O$_3$, Science 331(2011): 1420-1423. Copyright 2011, American Association for the Advancement of Science.)

2.5.2 Spectroscopic Characterization

2.5.2.1 Energy Dispersive X-ray Spectroscopy (EDS)

EDS is one kind of spectroscopic technique used for chemical characterization of the samples. It works on the interactions between electromagnetic radiation and matter. By analyzing the characteristic *x*-rays emitted by the matter in response to being hit with charged particles, the elements in the samples can be uniquely identified. While the TEM images provide real-time pictures about the sizes and morphologies of the investigated samples, the supplementary EDS data offer the exact compositions of the samples. Several examples have demonstrated the use of EDS spectra to analyze the chemical compositions of oxide perovskite nanocrystals, particularly in the determination of the composition of substituted or nanoparticle composite materials (Lee et al. 2006, Zhu et al. 2014a, Ji et al. 2019b).

2.5.2.2 Electron Energy Loss Spectroscopy (EELS)

EELS is a powerful technology for obtaining electronic structures of nanometer-scale materials, making use of the energy loss of the inelastic scattering electrons. The energy losses are processed by an electron spectrometer and counted by a suitable detector system. The local specimen compositional concentrations, electronic and valence structures, and the nearest neighbor atomic spacings can be determined quantitatively from the measured signals after background subtraction. The signals in EELS appear in the form of ionization edges on a large background. Datta et al. (2013) have performed chemical analyses of perovskite manganite $La_{0.5}Sr_{0.5}MnO_3$ (LSMO) nanowires by EELS spectra, and quantitatively determined the Mn chemical valence (Wang et al. 2000). Figure 2.49(a) shows the element mapping of the LSMO nanowire obtained by L map of La, Sr, and Mn elements, and K map of O element. Obviously, the constituent elements are uniformly distributed within the nanowire. Figure 2.49(b) shows the intensity ratios of L_3 to L_2 lines determined from Mn oxide compounds with different chemical valences; based on such calibration curve, the chemical valence of Mn element in the LSMO nanowire was determined to be 3.50.

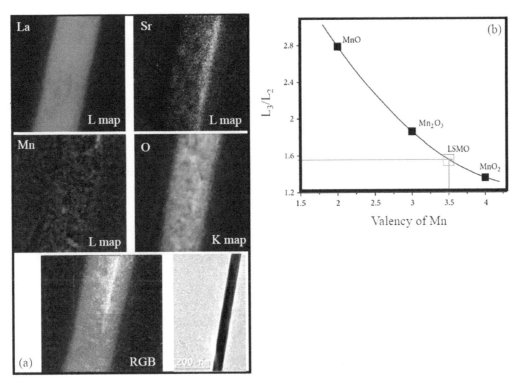

Fig. 2.49　(a) Chemical element mapping obtained by energy-filtered transmission electron microscope (EF-TEM) image of each constituent element: L map of La, Sr, and Mn and K map of O in LSMO nanowire. Inset is the TEM image of a single LSMO nanowire, (b) Intensity ratios of L_3 to L_2 lines of Mn-based oxides with different chemical valences of Mn, where the valence of Mn in the LSMO nanowire was estimated from the calibration curve. (Reprinted with permission from S. Datta, S. Chandra, S. Samanta, K. Das, H. Srikanth and B. Ghosh, "Growth and physical property study of single nanowire (diameter ~ 45 nm) of half doped manganite," J. Nanomater. 2013 (2013): 162315 (1-6). Copyright 2013, Hindawi Publishing Corporation.)

2.5.2.3　X-ray Photoelectron Spectroscopy (XPS)

In XPS spectra, the elements present in the sample are determined by measuring the kinetic energies of the ejected photoelectrons, and the element concentrations are also determined from the measured intensities of photoelectrons without standard samples. In a solid, XPS normally can probe a layer with depth of 2-20 atomic layers, which is dependent upon the following factors such as the examined sample, the energy of the photoelectrons, and the incident angle with respect to the sample surface. The chemical valences of the elements within the surface layer can be also determined by the XPS spectra. Dudrica et al. (2019) investigated the electronic properties of the $Nd_{0.6-x}Bi_xSr_{0.4}MnO_3$ (x = 0, 0.05 and 0.1) nanoparticles by XPS measurements, the results are shown in Fig. 2.50. It is noticed that a chemical shifting to lower binding energies appears for Mn 3s, Mn 2p, Nd 3d, and O 1s core levels as the Bi doping content increases. The existence of dual chemical valences (Mn^{4+} and Mn^{3+} ions) in all samples was confirmed by the XPS spectra, and the bismuth was present as Bi^{3+} ions in the perovskite lattice.

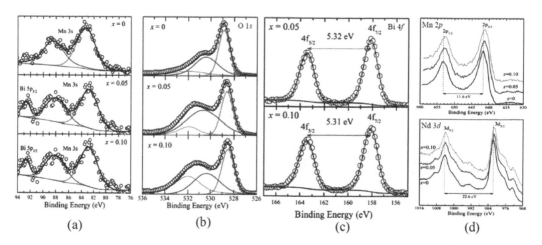

Fig. 2.50 XPS spectra of (a) Mn 3s and Bi 5p3/2, (b) O 1s, (c) Bi 4f, (d) local Mn 2p and Nd 3d core levels in the $Nd_{0.6-x}Bi_xSr_{0.4}MnO_3$ (x = 0, 0.05, 0.1) nanoparticles. (Reprinted with permission from R. Dudric, R. Bortnic, G. Souca, R. Ciceo-Lucacel, R. Stiufiuc and R. Tetean, "XPS on $Nd_{0.6-x}Bi_xSr_{0.4}MnO_3$ nanopowders," Appl. Surf. Sci. 487 (2019): 17-21. Copyright 2019, Elsevier.)

2.6 FUNCTIONAL APPLICATIONS OF NANOSTRUCTURED PEROVSKITE OXIDES

2.6.1 Ferroelectric Memory Devices

In ferroelectric materials, the spontaneous polarization results from the electric dipoles that are formed by the relative displacements of constituent ions in the crystal and its direction can be switched by external electric field. In a ferroelectric random access memory (FeRAM), the information data are stored in two states with spontaneous polarization up or down direction, which represents the two states of "0" and "1" in the binary logic (Fig. 2.51(a)). Up to date, two types of FeRAM cells have been developed, named as 1T-1C and 2T-2C configurations, respectively, as shown in Figs. 2.51(b)-(c) (Setter et al. 2006). Due to the larger space of the 2T-2C cell, 1T-1C configuration becomes the popular cell because of its smaller cell size. However, the ferroelectric capacitor suffers from the endurance problem, and the lifetime of FeRAM cell is much reduced due to the fatigue question in ferroelectric material. To resolve the above issues involved in FeRAM cell, much effort has been made to develop the ferroelectric gate field effect transistor (FeFET) (Kohlstedt and Ishiwara 2003), as shown in Fig. 2.51(d). In principle, FeFET is similar to a conventional Si MOSFET but with a ferroelectric material as gate dielectrics. This kind of FET-type FeRAMs have several advantages such as non-volatile data storage, non-destructive data readout, and single-transistor-type cell structure. Despite the FET-type FeRAMs being first studied in 1950s, up to date, it is not commercialized. The main reason is the difficulty in controlling the interfacial structures (e.g. interdiffusion, defects, undesired phase, electronic trapping states) formed between the perovskite ferroelectric gate and semiconductor since any imperfection formed at the interface will deteriorate the performance of the device (Ishiwara 2012). In order to suppress charge traps formed at the Si/ferroelectric interface and to improve the retention time, several kinds of buffer layers such as SiO_2, CeO_x, Si_3N_4, HfO_2 or HfAlO have been inserted between the Si and the ferroelectrics (Ishiwara 2012). The most encouraging result is using gate oxides such as HfO_2 or

HfAlO. Mathews et al. (1997) reported on the retention time of 30 days for the gate layer sequence of Pt/SBT/HfO$_2$/Si. However, the buffer layer insertion also leads to other problems such as current injection and leakage of current problem. Under such background, all perovskite-based FeFET was developed, where Ca-doped La$_{0.7}$Ca$_{0.3}$MnO$_3$ was used as the semiconductor channel material, and Pb(Z$_{r0.2}$Ti$_{0.8}$)O$_3$ was used as the ferroelectric gate in the prototypical epitaxial field effect device (Wu et al. 2001). The carrier concentration of the semiconductor channel was modulated by changing the manganate stoichiometry. An enhanced interface characteristic was achieved by epitaxial growth of ferroelectrics onto the manganates, and the all perovskite field-effect devices had retention time in the order of hours. Recently, 1D ferroelectric BTO nanowires also provide the possibility to store high-density information by polarizing the ferroelectric domains via an SPM tip, and data storage density can be reached as high as ~ 1 Terabit/cm^2 in memory device (Yun et al. 2002).

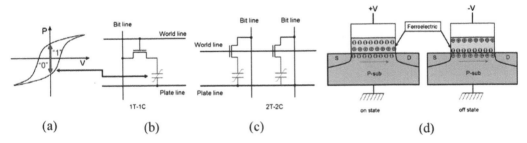

Fig. 2.51 Schematic diagrams of (a) the *P-E* hysteresis loop in ferroelectrics, where two states of "0" and "1" in the binary logic are represented by the direction of the spontaneous polarization, (b) 1T-1C, (c) 2T-2C FRAM bit cells, (d) Charge diagrams for the on-state (left) and off-state (right) of a FeFET after applying a positive and subsequently a negative voltage pulse at the gate, respectively. (b, c) (Reprinted with permission from N. Setter, D. Damjanovic, L. Eng, G. Fox, S. Gevorgian, S. Hong et al., "Ferroelectric thin films: Review of materials, properties, and applications," J. Appl. Phys. 100 (2006): 051606 (1-46). Copyright 2006, American Institute of Physics). (d) (Reprinted with permission from Kohlstedt, H. and H. Ishiwara, Ferroelectric field effect transistors, pp. 389-399. In: R. Waser [ed.]. Nanoelectronics and Information Technology – Advanced Electronic Materials and Novel Devices. Copyright 2003, WILEY-VCH Verlag

2.6.2 Multiferroic Devices

Applications of the perovskite multiferroic nanostructures in the field of spintronics have received much attention, and some possible devices such as four-state logic devices have been proposed and fabricated (Binek and Doudin 2005, Gajek et al. 2007, Bibes and Barthelemy 2008). A schematic diagram for such a multiferroic device based on BFO is shown in Fig. 2.52. Such MERAM device makes use of the ME coupling at the interface between a multiferroic material and a ferromagnet, which allows the magnetization of the ferromagnetic layer to be switched by a voltage (Zheng et al. 2006). Another example is multiferroic tunnel junction, which can be used as a spin filter device controlled both electrically and magnetically (Tsymbal and Kohlstedt 2006, Bea et al. 2008). In such a spin filter device, the BFO ultrathin layer with thickness of 1-2 nm acts as a tunnel barrier in LSMO/BFO/Co magnetic tunnel junctions, where the magnetic state of ferromagnetic electrode is controlled by the ferroelectric state, and thus the tunneling magnetoresistance is modified. A positive tunnel magnetoresistance (TMR) of 25% was achieved at low temperature and it decreased symmetrically with the bias voltage. Other potential applications of multiferroic thin films in spintronics include the spin wave devices controlled by electric field. Due to the hysteretic nature of the ME effect, ME nanostructures are also used for memory devices. In principle, the linear ME effect could permit data to be written electrically and read magnetically in memory devices. Zavaliche et al. (2007) also proposed a concept and experimental setup for proofing such memory device.

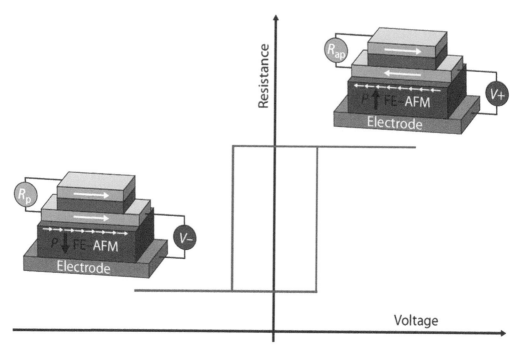

Fig. 2.52 Schematic diagram of an MERAM device, whose binary information is stored by the magnetization direction (thick white arrows) of the bottom ferromagnetic layer read by the resistance of the magnetic tri-layer (Rp: when the magnetizations of the two ferromagnetic layers are parallel, and Rap: antiparallel), and written by applied voltage (V) across the multiferroic ferroelectric - antiferromagnetic layer (FE-AFM). (Reprinted with permission from M. Bibes and A. Barthelemy, "Towards a magnetoelectric memory," Nat. Mater. 7(2008): 425-426. Copyright 2008, Nature Publishing Group.)

2.6.3 Energy Storage Devices

Perovskite oxide materials exhibit excellent catalytic activity, electrical conductivity, and durability, which find applications in energy storage. Thanks to their ion migration through perovskite lattices, perovskite oxide materials can be applied to batteries as good electrodes. Electrochemical measurements reveal that perovskite oxide nanoparticles display excellent catalytic activity for oxygen reduction, as well as a higher discharge plateau and specific capacity in contrast with the bulk counterpart (Fu et al. 2012). In recent years, perovskite oxide nanocrystals are widely studied for such applications. Despite perovskite oxide nanocrystals such as $LaNi_{1-x}Mn_xO_3$ (Du et al. 2014), $La_{0.6}Sr_{0.4}CoO_3$ (Sun et al. 2014) and $Ba_{0.9}Co_{0.5}Fe_{0.4}Nb_{0.1}O_{3-\delta}$ (Jin et al. 2014) being used in batteries since 2014, they exhibit less than 50 battery cycles. Xu et al. (2016) reported the cycling ability up to 155 cycles for $LaNiO_3$ nanoparticles. Recently, perovskite oxide nanoparticles with morphologies ranging from spherical or random shapes to nanocubes and anisotropic ones or double functional structures such as core-shell structure, decorated structure with a second material (e.g. metal, carbon, or oxides), or composites have been investigated to improve their catalytic performance in batteries (Kostopoulou et al. 2018, 2019). Besides the size and the morphology of perovskite oxide nanoparticles, the structural quality of nanocrystals and defects in the lattice, chemical doping at A- and/or B-site in perovskite lattice, and the synergetic effects in the double functional morphologies also influence their electrochemical performance. Details are referred to in the review paper contributed by Kostopoulou et al. (2018).

Supercapacitors have important applications in energy storage because of their high power density, moderate energy density, long cycle life, and environment-friendly nature. As an essential electro-chemical double layer capacitor, supercapacitor is composed of two electrodes separated by an ion-permeable membrane and an electrolyte ionically connected to both electrodes. It can be classified into three types (Poonam et al. 2019). In supercapacitors, perovskite oxide nanoparticles are used as electrode materials for energy storage. As a representative pseudocapacitance material, lanthanum-based perovskite nanocrystals have received much attention because of their fully reversible redox reactions and high intrinsic capacity, as well as good thermal stability and low cost (Mefford et al. 2014). The first reported nanostructured perovskite oxide used for pseudocapacitive electrode material was cubic crystalline $LaMnO_{3-\delta}$ (where $\delta = 0.09 \pm 0.02$) nanoparticles with irregular-shape, and the reported capacitance was 586.7-609.8 F/g (Mefford et al. 2014). To improve the capacitance of supercapacitor, introducing a secondary phase (La_2O_3 attached on the nanocrystals) or doping at B site of $LaMnO_3$ with Fe, Cr, and Ni elements have been performed. It was found that the perovskites $LaMnO_{3-\delta}$ doped with Ni at the B site had the best electrical conductivity and a capacitance as high as a few hundred Farah. Specifically, $LaNiO_3$ nanosheets (Li et al. 2017), hollow nanospheres (Shao et al. 2017), and randomly-shaped nanocrystals (Che et al. 2018) exhibit different effects on the capacitance and cycling stability. Figure 2.53 demonstrates the effects of morphologies of perovskite nanocrystals on the cyclic voltammetry curves measured under different scan rates. Other perovskite oxide nanocrystals have also been introduced for supercapacitor applications. For example, BFO nanoplates display a capacitance of 254.6 F/g (Yin et al. 2018), $LaFeO_3$ nanocrystals with Na and Mn substitutions 56.4 F/g (Rai and Thakur 2017), Ba/Ca-doped at Sr-site and Co/Fe/Ni-doped at Mn-site $SrMnO_3$ nanofibers 321.7- 446.8 F/g (George et al. 2018), and Co-doped $SrTiO_3$ nanocubes 75.28 F/g (Songwattanasin et al. 2019).

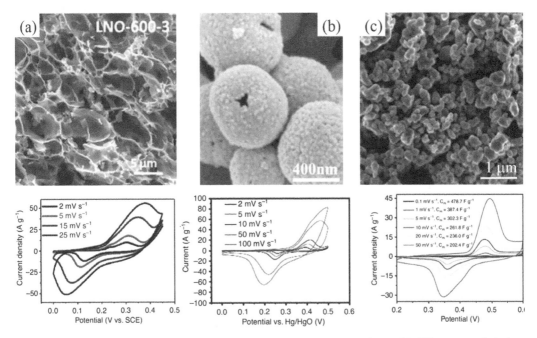

Fig. 2.53 $LaNiO_3$ perovskite nanocrystals used as electrodes in supercapacitors with different morphologies. (a) Nanosheets, (b) hollow nanospheres, (c) irregular-shaped nanocrystals (above figures), and their cyclic voltammetry measurements at different scan rates (below figures). (a) (Reproduced with permission from Z.J. Li, W.Y. Zhang, H.Y. Wang and B.C. Yang, "Two-dimensional perovskite $LaNiO_3$ nanosheets with hierarchical porous structure for high-rate capacitive energy storage. Electrochim. Acta. 258 (2017): 561-570. Copyright 2017, Elsevier). (b) (Reproduced with permission from T.Y. Shao, H.H. You, Z.J. Zhai, T.H. Liu, M. Li and L. Zhang, "Hollow spherical $LaNiO_3$ supercapacitor electrode synthesized by a facile template-free method," Mater. Lett. 201 (2017): 122-124. Copyright 2017, Elsevier). (c) (Reproduced with permission from W. Che, M. Wei, Z. Sang, Y. Ou, Y. Liu and J. Liu, "Perovskite $LaNiO_{3-\delta}$ oxide as an anion-intercalated pseudocapacitor electrode," J. Alloy Compd. 731(2018): 381-388. Copyright 2018, Elsevier.)

2.6.4 Solar Energy Conversion

Perovskite oxides have been widely investigated for solar energy conversion because of their suitable bandgap values, high stability, and solar absorption (Kudo and Miseki 2009, Chen et al. 2010b, Nuraje 2012). For solar energy conversion, perovskite materials such as BTO, PTO, STO, $KTaO_3$, $NaTaO_3$, $CaTiO_3$, and BFO have been widely investigated. It was reported that under visible-light irradiation, BFO nanoparticles coated with STO could produce H_2 while pure BFO failed to do so (Luo and Maggard 2006). In this case, BFO with a small bandgap can be used as a visible-light sensitizer in a STO photocatalytic material. In addition, Joshi et al. (2008) also demonstrated the O_2 production ability of BFO nanocubes. Recently, Deng et al. (2011) found that the photocatalytic activities of BFO nanoparticles were closely related with their particle sizes and morphologies, e.g. BFO nanoparticles with $\{100\}_c$ dominant cube exhibit good photocatalytic performance. In addition, the visible-light response of BFO nanoparticles can be improved by suitable elements doping or forming heterojunction (Gao et al. 2015). In addition to BFO nanoparticles, perovskite BFO nanowires also display the photovoltaic behavior, which can be used for photovoltaic device such as solar cell. Figure 2.54(a) is schematic diagram for measuring I-V curve of the photovoltaic devices based on random BFO nanofibers (Fei et al. 2015), and the measured results are shown in

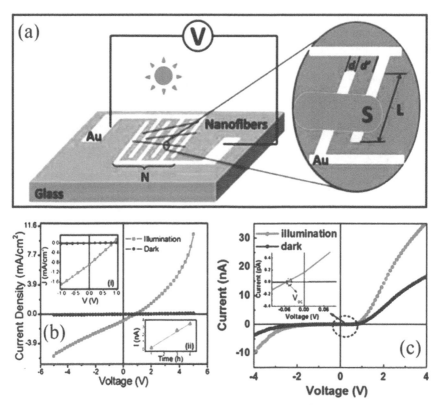

Fig. 2.54 (a) Schematic illustration for measuring the I-V characteristics of photovoltaic devices based random-oriented $BiFeO_3$ nanofibers, (b) I - V curves for $BiFeO_3$ nanofibers under dark and illumination conditions. Inset (i) shows expanded view of current density behavior around zero-bias. Inset (ii) shows averaged photocurrent after several measurements for different deposition time from 1 to 4 hr. (a,b) (Reproduced with permission from L. Fei, Y. Hu, X. Li, R. Song, L. Sun, H. Huang et al., Electrospun bismuth ferrite nanofibers for potential applications in ferroelectric photovoltaic devices," ACS Appl. Mater. Interfaces. 7 (2015): 3665-3670. Copyright 2015, American Chemical Society), (c) I - V curve for $BiFeO_3$ nanowires in dark and under illumination. Inset shows enlarged I - V curve in the portion of dotted circle. (Reproduced with permission from K. Prashanthi, P. Dhandharia, N. Miriyala, R. Gaikwad, D. Barlage and T. Thundat, "Enhanced photo-collection in single $BiFeO_3$ nanowire due to carrier separation from radial surface field," Nano Energy. 13 (2015): 240-248. Copyright 2015, American Chemical Society.)

Figs. 2.54(b)-(c) for the BFO nanofibers and nanowires, respectively. The current density of the BFO nanofibers with a diameter of ~ 200 nm was measured to be about 1 mA/cm^2, 2-10 times higher than that of BFO thin films (Fei et al. 2015). The enhanced photovoltaic properties observed in BFO nanofibers are ascribed to that the ferroelectric domains in the free-standing nanofibers are easily switched, and the nanofibers trap more photons due to their geometric confinements. Furthermore, the electrons and holes are driven by depolarization field in opposite direction, preventing the electrons and holes from their combinations. Thus, the photovoltaic performance is enhanced (Prashanthi et al. 2015). In addition, perovskite oxide photocatalysts also have promising applications in environmental remediation and solar fuel fields. A bright future for the applications of nanostructured perovskite oxides in photocatalytic and photovoltaic devices is envisioned.

2.7 CONCLUSIONS AND PERSPECTIVES

We review the recent progress in the synthesis, physical properties, characterizations, and functional applications of nanostructured perovskite oxides. By using physical methods such as mechanical milling method, molten-salt method, physical vapor deposition methods, and focus ion beam method, low-dimensional perovskite oxide nanostructures such as nanoparticles/nanoislands/nanodots can be synthesized. Similarly, the chemical routes including sol-gel/sol-precipitation synthesis, co-precipitation method, Pechini method, hydrothermal process, solvothermal process, microwave-hydrothermal process, sonochemical method, and microemulsion method, can be used to synthesize perovskite oxide nanoparticles/nanoislands/nanodots. Nowadays, top-down and bottom-up approaches are widely applied to fabricate 1D perovskite oxide nanostructures, and 2D perovskite oxide nanostructures (e.g. nanosheets/free standing perovskite oxide membranes). In recent years, 3D perovskite oxide nanostructures are also fabricated by using 3D-nanotemplate PLD technique (PLD process combined with the AAO template). Perovskite oxides nanostructures are structurally characterized by microscopic analytic techniques, such as XRD, SEM, TEM, HRTEM, and spherical-corrected HRTEM/STEM as well as related spectroscopies such as EDS, EELS and XPS techniques. The characterizations of physical properties reveal the unique properties of perovskite oxide nanostructures, which are much different from their bulk counterparts. One-dimensional perovskite oxide nanostructures such as nanowires and nanotubes are widely investigated and they are used as building blocks for constructing nanostructured devices, which have promising applications in various technical fields such as oxide electronics, spinelectronics, optoelectronics, solar energy storage and photovoltaic devices. It is envisioned that the future of perovskite oxides nanostructures is very bright.

2.8 ACKNOWLEDGEMENTS

The authors would like to acknowledge the financial supports from National Natural Science Foundation of China (Grant No. 11674161, 11174122), and Natural Science Foundation of Jiangsu Province (Grant No. BK20181250), and undergraduate teaching reform projects from Nanjing University (grant nos. X20191028402 and 202010284036X).

2.9 REFERENCES

Adireddy, S., C.K. Lin, B.B. Cao, W.L. Zhou and G. Caruntu. 2010. Solution based growth of monodisperse cube-like BaTiO$_3$ colloidal nanocrystals. Chem. Mater. 22: 1946-1948.

Ahda, S., L. Zulhijah, N. Darmawan and A. Dymiati. 2019. Characterization of phase transitions on PbZr$_x$Ti$_{1-x}$O$_3$ nanocrystal ceramic materials synthesized using the molten salt method. Malaysian J. Fundam. Appl. Sci. 15: 885-889.

Ahluwalia, R., R.N. Ng, A. Schilling, R.G. McQuaid, D.M. Evans, J.M. Gregg, et al. 2013. Manipulating ferroelectric domains in nanostructures under electron beams. Phys. Rev. Lett. 111: 165702(1-5).

Alexe, M., C. Harnagea, D. Hesse and U. Gösele. 1999. Patterning and switching of nanosized ferroelectric memory cells. Appl. Phys. Lett. 75: 1793-1795.

Alexe, M., D. Hesse, V. Schmidt, S. Senz, H.J. Fan, M. Zacharias, et al. 2006. Ferroelectric nanotubes fabricated using nanowires as positive templates. Appl. Phys. Lett. 89: 172907(1-3).

Alonso, J.A., J.L. García-Muñoz, M.T. Fernández-Díaz, M.A.G. Aranda, M.J. Martínez-Lope and M.T. Casais. 1999. Charge disproportionation in RNiO$_3$ perovskites: Simultaneous metal-insulator and structural transition in YNiO$_3$. Phys. Rev. Lett. 82: 3871-3874.

Ansari, A.A., N. Ahmad, M. Alam, S.F. Adil, M.E. Assal, A. Albadri, et al. 2019. Optimization of redox and catalytic performance of LaFeO$_3$ perovskites: Synthesis and physicochemical properties. J. Electron. Mater. 48: 4351-4361.

Anuradha, T.V. 2014. Template-assisted sol-gel synthesis of nanocrystalline BaTiO$_3$. E-J. Chem. 7: 894-898.

Arnold, D.C., K.S. Knight, G. Catalan, S.A.T. Redfern, J.F. Scott, P. Lightfoot, et al. 2010. The b to c transition in BiFeO$_3$: A powder neutron diffraction study. Adv. Funct. Mater. 20: 2116-2123.

Ashiri, R., A. Nemati, M.S. Ghamsari, S. Sanjabi and M.A. Aalipour. 2011. Modified method for barium titanate nanoparticles synthesis. Mater. Res. Bull. 46: 2291-2295.

Avvakumov, E.G., N.V. Kosova, I.P. Bykov, V.V. Melikhov, B.G. Polkovnichenko and E.T. Devyatkina. 1992. Mechanochemical synthesis of lead-zirconate-titanate from hydrated zirconium and titanium-oxides. Inorg. Mater. 28: 1771-1775.

Azuma, M., W.-T. Chen, H. Seki, M. Czapski, O. Smirnova, K. Oka, et al. 2011. Colossal negative thermal expansion in BiNiO$_3$ induced by intermetallic charge transfer. Nat. Commun. 2: 347(1-5).

Bansal, V., P. Poddar, A. Ahmad and M. Sastry. 2006. Room-temperature biosynthesis of ferroelectric barium titanate nanoparticles. J. Am. Chem. Soc. 128: 11958-11963.

Bao, N., L. Shen, A. Gupta, A. Tatarenko, G. Srinivasan and K. Yanagisawa. 2009. Size-controlled one-dimensional monocrystalline BaTiO$_3$ nanostructures. Appl. Phys. Lett. 94: 253109(1-3).

Bassano, A., M.T. Buscaglia, V. Buscaglia and P. Nanni. 2009. Particle size and morphology control of perovskite oxide nanopowders for nanostructured materials. Integr. Ferroelectr. 109: 1-17.

Béa, H., M. Gajek, M. Bibes and A. Barthélémy. 2008. Spintronics with multiferroics. J. Phys.: Condensed Matt. 20: 434221(1-11).

Bharathkumar, S., M. Sakar, V.K. Rohith and S. Balakumar. 2015. Versatility of electrospinning in the fabrication of fibrous mat and mesh nanostructures of bismuth ferrite (BiFeO$_3$) and their magnetic and photocatalytic activities. Phys. Chem. Chem. Phys. 17: 17745-17754.

Bhatti, H.S., S.T. Hussain, F.A. Khan and S. Hussain. 2016. Synthesis and induced multiferroicity of perovskite PbTiO$_3$: A review. Appl. Surf. Sci. 367: 291-306.

Bibes, M. and A. Barthelemy. 2008. Towards a magnetoelectric memory. Nat. Mater. 7: 425-426.

Binek, C. and B. Doudin. 2005. Magnetoelectronics with magnetoelectrics. J. Phys.: Condensed Matter. 17: L39-L44.

Black, K., M. Werner, R. Rowlands-Jones, P.R. Chalker and M.J. Rosseinsky. 2011. SrHfO$_3$ films grown on Si(100) by plasma-assisted atomic layer deposition. Chem. Mater. 23: 2518-2520.

Blackmore, R.H., M.E. Rivas, T.E. Erden, T.D. Tran, H.R. Marchbank, D. Ozkaya, et al. 2020. Understanding the mechanochemical synthesis of the perovskite LaMnO$_3$ and its catalytic behaviour. Dalton Trans. 49: 232-240.

Bortolani, F. and R.A. Dorey. 2010. Molten salt synthesis of PZT powder for direct write inks. J. Eur. Ceram. Soc. 30: 2073-2079.

Boucher, R., P. Renz, C. Li, T. Fuhrlich, J. Bauch, K.H. Yoon, et al. 2011. Large coercivity and polarization of sol-gel derived BaTiO$_3$ nanowires. J. Appl. Phys. 110: 064112(1-5).

Brankovic, Z., G. Brankovic, C. Jovalekic, Y. Maniette, M. Cilense and J.A. Varela. 2003. Mechanochemical synthesis of PZT powders. Mater. Sci. Eng. A. 345: 243-248.

Brutchey, R.L. and D.E. Morse. 2006. Template-free, low-temperature synthesis of crystalline barium titanate nanoparticles under bio-inspired conditions. Angew. Chem. Int. Edit. 45: 6564-6566.

Buscaglia, V. and M.T. Buscaglia. 2016. Synthesis and properties of ferroelectric nanotubes and nanowires: A review. pp. 200-231. *In*: M. Algueró, J.M. Gregg and L. Mitoseriu [eds.]. Nanoscale Ferroelectric and Multiferroics: Key Processing and Characterization Issues, and Nanoscale Effects. John Wiley and Sons, Ltd., New Delhi, India.

Byrne, D., A. Schilling, J.F. Scott and J.M. Gregg. 2008. Ordered arrays of lead zirconium titanate nanorings. Nanotechnology 19: 165608(1-5).

Carlsson, S.J.E., M. Azuma, Y. Shimakawa, M. Takano, A. Hewat and J.P. Attfield. 2008. Neutron powder diffraction study of the crystal and magnetic structures of $BiNiO_3$ at low temperature. J. Solid State Chem. 181: 611-615.

Caruntu, D., T. Rostamzadeh, T. Costanzo, S.S. Parizi and G. Carunt. 2015. Solvothermal synthesis and controlled self-assembly of monodisperse titanium-based perovskite colloidal nanocrystals. Nanoscale 7: 12955-12969.

Catalan, G. and J.F. Scott. 2009. Physics and applications of bismuth ferrite. Adv. Mater. 21: 2463-2485.

Che, W., M. Wei, Z. Sang, Y. Ou, Y. Liu and J. Liu. 2018. Perovskite $LaNiO_{3-\delta}$ oxide as an anion-intercalated pseudocapacitor electrode. J. Alloy Compd. 731: 381-388.

Chen, F., H.W. Liu, K.F. Wang, H. Yu, S. Dong, X.Y. Chen, et al. 2005. Synthesis and characterization of $La_{0.825}Sr_{0.175}MnO_3$ nanowires. J. Phys. Condens. Matter. 17: L467-L475.

Chen, J., M.C. Che and F. Yan. 2015. Synthesis of barium strontium titanate nanopowders by microwave hydrothermal method. Adv. Appl. Ceram. 114: 344-349.

Chen, J., X.R. Xing, A. Watson, W. Wang, R.B. Yu, J.X. Deng, et al. 2007. Rapid synthesis of multiferroic $BiFeO_3$ single crystalline nanostructures. Chem. Mater. 19: 3598-3600.

Chen, P., X.S. Xu, C. Koenigsmann, A.C. Santulli, S.S. Wong and J.L. Musfeldt. 2010a. Size-dependent infrared phonon modes and ferroelectric phase transition in $BiFeO_3$ nanoparticles. Nano Lett. 10: 4526-4532.

Chen, P., Y.T. Zhang, F.Q. Zhao, H.X. Gao, X.B. Chen and Z.W. An. 2016. Facile microwave synthesis and photocatalytic activity of monodispersed $BaTiO_3$ nanocuboids. Mater. Charact. 114: 243-253.

Chen, W. and Q.A. Zhu. 2007. Synthesis of barium strontium titanate nanorods in reverse microemulsion. Mater. Lett. 61: 3378-3380.

Chen, X., S. Shen, L. Guo and S.S. Mao. 2010b. Semiconductor-based photocatalytic hydrogen generation. Chem. Rev. 110: 6503-6570.

Cheng, L.Q., K. Wang and J.F. Li. 2013. Synthesis of highly piezoelectric lead-free $(K,Na)NbO_3$ one-dimensional perovskite nanostructures. Chem. Commun. 49: 4003-4005.

Cheng, Z. and J. Lin. 2010. Layered organic-inorganic hybrid perovskites: Structure, optical properties, film preparation, patterning and templating engineering. Cryst. Eng. Comm. 12: 2646-2662.

Cho, S.H. and J.V. Biggers. 1983. Characteization and sintering of lead zirconate-titanate powders. J. Am. Ceram. Soc. 66: 743-746.

Chu, M.W., I. Szafraniak, R. Scholz, C. Harnagea, D. Hesse, M. Alexe, et al. 2004. Impact of misfit dislocations on the polarization instability of epitaxial nanostructured ferroelectric perovskites. Nat. Mater. 3: 87-90.

Coey, J.M., D.M. Viret and S. von Molnaír. 1999. Mixed-valence manganites. Adv. Phys. 48: 167-293.

Cristobal, A.A. and P.M. Botta. 2013. Mechanochemically assisted synthesis of nanocrystalline $BiFeO_3$. Mater. Chem. Phys. 139: 931-935.

Cross, L.E. 2011. Relaxor ferroelectrics. Ferroelectrics. 76: 241-267.

Cushing, B.L., V.L. Kolesnichenko and C.J. O'Connor. 2004. Recent advances in the liquid-phase synthesis of inorganic nanoparticles. Chem. Rev. 104: 3893-3946.

Damjanovic, D. 1998. Ferroelectric, dielectric and piezoelectric properties of ferroelectric thin films and ceramics. Rep. Prog. Phys. 61: 1267-1324.

Damodaran, A.R., C.W. Liang, Q. He, C.Y. Peng, L. Chang, Y.H. Chu, et al. 2011. Nanoscale structure and mechanism for enhanced electromechanical response of highly strained $BiFeO_3$ thin films. Adv. Mater. 23: 3170-3175.

Dang, F., K. Kato, H. Imai, S. Wada, H. Haneda and M. Kuwabara. 2010. A new effect of ultrasonication on the formation of $BaTiO_3$ nanoparticles. Ultrason. Sonochem. 17: 310-314.

Dang, F., K. Mimura, K. Kato, H. Imai, S. Wada, H. Haneda, et al. 2011. Growth of monodispersed $SrTiO_3$ nanocubes by thermohydrolysis method. Cryst. Eng. Commun. 13: 3878-3883.

Das, N., R. Majumdar, A. Sen and H.S. Maiti. 2007. Nanosized bismuth ferrite powder prepared through sonochemical and microemulsion techniques. Mater. Lett. 61: 2100-2104.

Das, S., Y.L. Tang, Z. Hong, M.A.P. Gonçalves, M.R. McCarter, C. Klewe, et al. 2019. Observation of room-temperature polar skyrmions. Nature 568: 368-372.

Datta, S., S. Chandra, S. Samanta, K. Das, H. Srikanth and B. Ghosh. 2013. Growth and physical property study of single nanowire (diameter ~ 45 nm) of half doped manganite. J. Nanomater. 2013: 162315(1-6).

Datta, S., A. Ghatak and B. Ghosh. 2016. Manganite $(La_{1-x}A_xMnO_3; A = Sr, Ca)$ nanowires with adaptable stoichiometry grown by hydrothermal method: Understanding of growth mechanism using spatially resolved techniques. J. Mater. Sci. 51: 9679-9695.

Davies, A.G., A.D. Burnett, W. Fan, E.H. Linfield and J.E. Cunningham. 2008. Terahertz spectroscopy of explosives and drugs. Mater. Today. 11: 18-26.

Deng, J., S. Banerjee, S.K. Mohapatra, Y.R. Smith, M. Misra and J. Fund. 2011. Bismuth iron oxide nanoparticles as photocatalyst for solar hydrogen generation from water. Renew Energy Appl. 1: 1-10.

Deng, Y., J.L. Wang, K.R. Zhu, M.S. Zhang, J.M. Hong, Q.R. Gu, et al. 2005. Synthesis and characterization of single-crystal $PbTiO_3$ nanorods. Mater. Lett. 59: 3272-3275.

Dias, A., F.M. Matinaga and R.L. Moreira. 2007. Raman spectroscopy of $(Ba_{1-x}Sr_x)(Mg_{1/3}Nb_{2/3})O_3$ solid solutions from microwave-hydrothermal powders. Chem. Mater. 19: 2335-2341.

Dias, A., F.M. Matinaga and R.L. Moreira. 2009. Vibrational spectroscopy and electron-phonon interactions in microwave-hydrothermal synthesized $Ba(Mn_{1/3}Nb_{2/3})O_3$ complex perovskites. J. Phys. Chem. B. 113: 9749-9755.

Ding, J., X. Lü, H. Shu, J. Xie and H. Zhang. 2010. Microwave-assisted synthesis of perovskite $ReFeO_3$ (Re: La, Sm, Eu, Gd) photocatalyst. Mater. Sci. Eng. B. 171: 31-34.

Ding, Q.P., Y.P. Yuan, X. Xiong, R.P. Li, H.B. Huang, Z.S. Li, et al. 2008. Enhanced photocatalytic water splitting properties of $KNbO_3$ nanowires synthesized through hydrothermal method. J. Phys. Chem. C. 112: 18846-18848.

Dong, G.H., S.Z. Li, M.T. Yao, Z.Y. Zhou, Y.Q. Zhang, X. Han, et al. 2019. Super-elastic ferroelectric single-crystal membrane with continuous electric dipole rotation. Science 366: 475-479.

Dos Santos-Garcıa, A.J., E. Solana-Madruga, C. Ritter, D. AvilaBrande, O. Fabelo and R. Saez-Puche. 2015. Synthesis, structures and magnetic properties of the dimorphic Mn_2CrSbO_6 oxide. Dalton Trans. 44: 10665-10672.

Du, Z.Z., P. Yang, L, Wang, Y. Lu, J.B. Goodenough, J. Zhang, et al. 2014. Electrocatalytic performances of $LaNi_{1-x}Mg_xO_3$ perovskite oxides as bi-functional catalysts for lithium air batteries. J. Power Sources. 265: 91-96.

Dudric, R., R. Bortnic, G. Souca, R. Ciceo-Lucacel, R. Stiufiuc and R. Tetean. 2019. XPS on $Nd_{0.6-x}Bi_xSr_{0.4}MnO_3$ nanopowders. Appl. Surf. Sci. 487: 17-21.

Dutta, D.P., O.D. Jayakumar, A.K. Tyagi, K.G. Girija, C.G.S. Pillai and G. Sharma. 2010. Effect of doping on the morphology and multiferroic properties of $BiFeO_3$ nanorods. Nanoscale 2: 1149-1154.

Dutta, D.P., B.P. Mandal, R. Naik, G. Lawes and A.K. Tyagi. 2013. Magnetic, ferroelectric, and magnetocapacitive properties of sonochemically synthesized Sc-doped $BiFeO_3$ nanoparticles. J. Phys. Chem. C. 117: 2382-2389.

Dutto, F., C. Raillon, K. Schenk and A. Radenovic. 2011. Nonlinear optical response in single alkaline niobate nanowires. Nano Lett. 11: 2517-2521.

Dyakonov, V., A. Slawska-Waniewska, J. Kazmierczak, K. Piotrowski, O. Iesenchuk, H. Szymczak, et al. 2009. Nanoparticle size effect on the magnetic and transport properties of (La0.7Sr0.3)0.9Mn1.1O3 manganites. Low Temp. Phys. 35: 568-576.

Fan, G.N., L.X. Huangpu and X.G. He. 2005. Synthesis of single-crystal $BaTiO_3$ nanoparticles via a one-step sol-precipitation route. J. Cryst. Growth. 279: 489-493.

Fang, X.Q., J.X. Liu and V. Gupta. 2013. Fundamental formulations and recent achievements in piezoelectric nano-structures: A review. Nanoscale 5: 1716-1726.

Fei, L.F., Y.M. Hu, X. Li, R.B. Song, L. Sun, H.T. Huang, et al. 2015. Electrospun bismuth ferrite nanofibers for potential applications in ferroelectric photovoltaic devices. ACS Appl. Mater. Interfaces. 7: 3665-3670.

Fiebig, M., T. Lottermoser, D. Fro¨hlich, A.V. Goltsev and R.V. Pisarev. 2002. Observation of coupled magnetic and electric domains. Nature 419: 818-820.

Fong, D.D., G.B. Stephenson, S.K. Streiffer, J.A. Eastman, O. Auciello, P.H. Fuoss, et al. 2004. Ferroelectricity in ultrathin perovskite films. Science 304: 1650-1653.

Fox, G.R., J.H. Adair and R.E. Newnham. 1991. Effects of pH and H_2O_2 upon coprecipitated $PbTiO_3$ powders. 2. Properties of calcined powders. J. Mater. Sci. 26: 1187-1191.

Fu, Z.H., X.J. Lin, T. Huang and A.S. Yu. 2012. Nano-sized $La_{0.8}Sr_{0.2}MnO_3$ as oxygen reduction catalyst in non-aqueous Li/O_2 batteries. J. Solid State Electr. 16: 1447-1452.

Gajek, M., M. Bibes, S. Fusil, K. Bouzehouane, J. Fontcuberta, A. Barthelemy, et al. 2007. Tunnel junctions with multiferroic barriers. Nat. Mater. 6: 296-302.

Galasso, F.S. 1969. Structure of perovskite-type compounds. pp. 3-49. In: R. Smoluchowski and N. Kurti [eds.]. Structure, Properties and Preparation of Perovskite-Type Compounds. Pergamon Press, New York, NY, USA.

Ganpule, C.S., A. Stanishevsky, Q. Su, S. Aggarwal, J. Melngailis, E. Williams, et al. 1999. Scaling of ferroelectric and piezoelectric properties in Pt/SrBi$_2$Ta$_2$O$_9$/Pt thin films. Appl. Phys. Lett. 75: 409-411.

Gao, F., Y. Yuan, K.F. Wang, X.Y. Chen, F. Chen, J.-M. Liu, et al. 2006. Preparation and photoabsorption characterization of BiFeO$_3$ nanowires. Appl. Phys. Lett. 89: 102506(1-3).

Gao, T., Z. Chen, Q.L. Huang, F. Niu, X.N. Huang, L.S. Qin, et al. 2015. A review: Preparation of bismuth ferrite nanoparticles and its applications in visible-light induced photocatalyses. Rev. Adv. Mater. Sci. 40: 97-109.

Geneste, G., E. Bousquet, J. Junquera and P. Ghosez. 2006. Finite-size effects in BaTiO$_3$ nanowires. Appl. Phys. Lett. 88: 112906(1-3).

George, G., S.L. Jackson, C.Q. Luo, F. Dong, L. Duan, D.L. Hu, et al. 2018. Effect of doping on the performance of high-crystalline SrMnO$_3$ perovskite nanofibers as a supercapacitor electrode. Ceram. Int. 44: 21982-21992.

Ghosez, P. and K.M. Rabe. 2000. Microscopic model of ferroelectricity in stress-free PbTiO$_3$ thin films. Appl. Phys. Lett. 76: 2767-2769.

Giovannetti, G., S. Kumar, D. Khomskii, S. Picozzi and J. van den Brink. 2009. Multiferroicity in rare-earth nickelates RNiO$_3$. Phys. Rev. Lett. 103: 156401(1-4).

Goldschmidt, V.M. 1926. The laws of crystal chemistry. Die Naturwissenchaften. 14: 477-485.

Gomes, M.A., A.S. Lima, K.I.B. Eguiluz and G.R. Salazar-banda. 2016. Wet chemical synthesis of rare earthdoped barium titanate nanoparticles. J. Mater. Sci. 51: 4709-4727.

Goswami, S., D. Bhattacharya and P. Choudhury. 2011. Particle size dependence of magnetization and noncentrosymmetry in nanoscale BiFeO$_3$. J. Appl. Phys. 9: 07D737(1-3).

Gu, H., Y. Hu, J. You, Z. Hu, Y. Yuan and T. Zhang. 2007. Characterization of single-crystalline PbTiO$_3$ nanowire growth *via* surfactant-free hydrothermal method. J. Appl. Phys. 101: 024319(1-7).

Guo, R., L. Fang, W. Dong, F. Zheng and M. Shen. 2011. Magnetically separable BiFeO$_3$ nanoparticles with a γ-Fe$_2$O$_3$ parasitic phase: Controlled fabrication and enhanced visible-light photocatalytic activity. J. Mater. Chem. 21: 18645-18652.

Guo, R., Y. Guo, H. Duan, H. Li and H. Liu. 2017. Synthesis of orthorhombic perovskite-type ZnSnO$_3$ single-crystal nanoplates and their application in energy harvesting. ACS Appl. Mater. Interfaces 9: 8271-8279.

Haertling, G.H. 1999. Ferroelectric ceramics history and technology. J. Am. Ceram. Soc. 82: 797-818.

Han, H., R. Ji, Y.J. Park, S.K. Lee, G.L. Rhun, M. Alexe, et al. 2009. Wafer-scale arrays of epitaxial ferroelectric nanodiscs and nanorings. Nanotechnology 20: 015301(1-6).

Hao, Y.N., X.H. Wang and L.T. Li. 2014. Highly dispersed SrTiO$_3$ nanocubes from a rapid sol-precipitation method. Nanoscale 6: 7940-7946.

Harnagea, C., M. Alexe, J. Schilling, J. Choi, R.B. Wehrspohn, D. Hesse, et al. 2003. Mesoscopic ferroelectric cell arrays prepared by imprint lithography. Appl. Phys. Lett. 83: 1827-1829.

Hernandez, B.A., K.S. Chang, E.R. Fisher and P.K. Dohout. 2002. Sol-gel template synthesis and characterization of BaTiO$_3$ and PbTiO$_3$ nanotubes. Chem. Mater. 14: 480-482.

Heron, J.T., D.G. Schlom and R. Ramesh. 2014. Electric field control of magnetism using BiFeO$_3$-based heterostructures. Appl. Phys. Rev. 1: 021303(1-18).

Hill, N.A. 2000. Why are there so few magnetic ferroelectrics? J. Phys. Chem. B. 104: 6694-6709.

Hong, S.S., J.H. Yu, D. Lu, A.F. Marshall, Y. Hikita, Y. Cui, et al. 2017. Two-dimensional limit of crystalline order in perovskite membrane films. Sci. Adv. 3: eaao5173(1-5).

Hu, C.G., H. Liu, C.S. Lao, L.Y. Zhang, D. Davidovic and Z.L. Wang. 2006a. Size-manipulable synthesis of single-crystalline BaMnO$_3$ and BaTi$_{1/2}$Mn$_{1/2}$O$_3$ nanorods/nanowires. J. Phys. Chem. B. 110: 14050-14054.

Hu, C.G., Y. Xi, H. Liu and Z.L. Wang. 2009. Composite-hydroxide-mediated approach as a general methodology for synthesizing nanostructures. J. Mater. Chem. 19: 858-868.

Hu, Y., H. Gu, X. Sun, J. You and J. Wang. 2006b. Photoluminescence and Raman scattering studies on PbTiO$_3$ nanowires fabricated by hydrothermal method at low temperature. Appl. Phys. Lett. 88: 193120(1-3).

Huang, F., Z. Wang, X. Lu, J. Zhang, K. Min, W. Lin, et al. 2013a. Peculiar magnetism of BiFeO$_3$ nanoparticles with size approaching the period of the spiral spin structure. Sci. Rep. 3: 2907(1-7).

Huang, L.M., Z.Y. Chen, J.D. Wilson, S. Banerjee, R.D. Robinson, I.P. Herman, et al. 2006. Barium titanate nanocrystals and nanocrystal thin films: Synthesis, ferroelectricity, and dielectric properties. J. Appl. Phys. 100: 034316(1-10).

Huang, L.M., S.Y. Liu, B.J. Van Tassell, X.H. Liu, A. Byro, H.N. Zhang, et al. 2013b. Structure and performance of dielectric films based on self-assembled nanocrystals with a high dielectric constant. Nanotechnology 24: 499601(1-14).

Huczko, A. 2000. Template-based synthesis of nanomaterials. Appl. Phys. A. 70: 365-376.

Hur, N., S. Park, P.A. Sharma, J.S. Ahn, S. Guha and S.W. Cheong. 2004. Electric polarization reversal and memory in multiferroic material induced by magnetic fields. Nature 429: 392-395.

Hwang, U.Y., H.S. Park and K.K. Koo. 2004. Low-temperature synthesis of fully crystallized spherical $BaTiO_3$ particles by the gel-sol method. J. Am. Ceram. Soc. 87: 2168-2174.

Ianculescu, A., D. Berger, M. Viviani, C.E. Ciomaga, L. Mitoseriu, E. Vasile, et al. 2007. Investigation of $Ba_{1-x}Sr_xTiO_3$ ceramics prepared from powders synthesized by the modified Pechini route. J. Eur. Ceram. Soc. 27: 3655-3658.

Ishikawa, K., K. Yoshikawa and N. Okada. 1988. Size effect on the ferroelectric phase transition in $PbTiO_3$ ultrafine particles. Phys. Rev. B. 37: 5852-5855.

Ishiwara, H. 2012. Ferroelectric random access memories. J. Nanosci. Nanotechnol. 12: 7619-7627.

Ishiwata, S., M. Azuma, M. Takano, E. Nishibori, M. Takata, M. Sakata, et al. 2002. High pressure synthesis, crystal structure and physical properties of a new Ni(ii) perovskite $BiNiO_3$. J. Mater. Chem. 12: 3733-3737.

Ishiwata, S., M. Azuma, M. Takano, E. Nishibori, M. Takata and M. Sakata. 2003. Suppression of A site charge disproportionation in $Bi1-xLaxNiO_3$. Phys. B. 329: 813-814.

Ishiwata, S., M. Azuma, M. Hanawa, Y. Moritomo, Y. Ohishi, K. Kato, et al. 2005. Pressure/temperature/ substitution-induced melting of A-site charge disproportionation in $Bi_{1-x}La_xNiO_3$ ($0 \leq x \leq 0.5$). Phys. Rev. B. 72: 045104(1-7).

Ji, D.X., S.H. Cai1, T.R. Paudel, H.Y. Sun, C.C. Zhang, L. Han, et al. 2019a. Freestanding crystalline oxide perovskites down to the monolayer limit. Nature 570: 87-90.

Ji, Q., P.J. Xue, H. Wu, Z.P. Pei and X.H. Zhu. 2019b. Structural characterizations and dielectric properties of sphere- and rod-like $PbTiO_3$ powders synthesized *via* molten salt synthesis. Nanoscale Res. Lett. 14: 62(1-12).

Jia, C.L., S.B. Mi, K. Urban, I. Vrejoiu, M. Alexe and D. Hesse. 2008. Atomic-scale study of electric dipoles near charged and uncharged domain walls in ferroelectric films. Nat. Mater. 7: 57-61.

Jia, C.L., K.W. Urban, M. Alexe, D. Hesse and I. Vrejoiu. 2011. Direct observation of continuous electric dipole rotation in flux-closure domains in ferroelectric Pb(Zr,Ti)O_3. Science 331: 1420-1423.

Jiang, B.B., J. Iocozzia, L. Zhao, H.F. Zhang, Y.W. Harn, Y.H. Chen, et al. 2019. Barium titanate at the nanoscale: Controlled synthesis and dielectric and ferroelectric properties. Chem. Soc. Rev. 48: 1194-1228.

Jiang, J.C., L.L. Henry, K.I. Gnanasekar, C.L. Chen and E.I. Meletis. 2004. Self-assembly of highly epitaxial (La,Sr)MnO_3 nanorods on (001) $LaAlO_3$. Nano Lett. 4: 741-745.

Jiang, J.C. and E.I. Meletis. 2006. Nanofabrication of self-organized, three-dimensional epitaxial oxide nanorods. Thin Solid Films. 515: 39-45.

Jin, C., Z. Yang, X. Cao, F. Lu and R. Yang. 2014. A novel bifunctional catalyst of $Ba_{0.9}Co_{0.5}Fe_{0.4}Nb_{0.1}O_{3-\delta}$ perovskite for lithium-air battery. Int. J. Hydrogen Energy. 39: 2526-2530.

Jinga, C., D. Berger, C. Matei, S. Jinga and E. Andronescu. 2010. Characterization of $BaMg_{1/3}(Ta_{1-x}Nb_x)_{2/3}O_3$ ceramics obtained by a modified Pechini method. J. Alloys Compd. 497: 239-243.

Joshi, U.A. and J.S. Lee. 2005. Template-free hydrothermal synthesis of single-crystalline barium titanate and strontium titanate nanowires. Small 1: 1172-1176.

Joshi, U.A., S. Yoon, S. Baik and J.S. Lee. 2006. Surfactant-free hydrothermal synthesis of highly tetragonal barium titanate nanowires: A structural investigation. J. Phys. Chem. B. 110: 12249-12256.

Joshi, U.A., J.S. Jang, P.H. Borse and J.S. Lee. 2008. Microwave synthesis of single-crystalline perovskite $BiFeO_3$ nanocubes for photoelectrode and photocatalytic applications. Appl. Phys. Lett. 92: 242106(1-3).

Jung, J.H., M. Lee, J.I. Hong, Y. Ding, C.Y. Chen, L.J. Chou, et al. 2011. Lead-free $NaNbO_3$ nanowires for a high output piezoelectric nanogenerator. ACS Nano. 5: 10041-10046.

Junquera, J. and P. Ghosez. 2003. Critical thickness for ferroelectricity in perovskite ultrathin films. Nature 422: 506-509.

Kan, C.Y.D., I. Takeuchi, V. Nagarajan and J. Seidel. 2012. Doping $BiFeO_3$: Approaches and enhanced functionality. Phys. Chem. Chem. Phys. 14: 15953-15962.

Kang, M.G., S.Y. Lee, D. Maurya, C. Winkler, H.C. Song, R.B. Moore, et al. 2017. Wafer-scale single-crystalline ferroelectric perovskite nanorod arrays. Adv. Funct. Mater. 27: 1701542(1-9).

Karmaoui, M., E.V. Ramana, D.M. Tobaldi, L. Lajaunie, M.P. Graca, R. Arenal, et al. 2016. High dielectric constant and capacitance in ultrasmall (2.5 nm) $SrHfO_3$ perovskite nanoparticles produced in a low temperature non-aqueous sol-gel route. RSC Adv. 6: 51493-51502.

Kavian, R. and A. Saidi. 2009. Sol-gel derived $BaTiO_3$ nanopowders. J. Alloy Compd. 468: 528-532.

Kim, J., S.A. Yang, Y.C. Choi, J.K. Han, K.O. Jeong, Y.J. Yun, et al. 2008. Ferroelectricity in highly ordered arrays of ultra-thin-walled $Pb(Zr,Ti)O_3$ nanotubes composed of nanometer-sized perovskite crystallites. Nano Lett. 8: 1813-1818.

Kim, S., J.H. Lee, J. Lee, S.W. Kim, M.H. Kim, S. Park, et al. 2013. Synthesis of monoclinic potassium niobate nanowires that are stable at room temperature. J. Am. Chem. Soc. 135: 6-9.

Kimijima, T., K. Kanie, M. Nakaya and A. Muramatsu. 2014. Solvothermal synthesis of shape-controlled perovskite $MTiO_3$ (M = Ba, Sr, and Ca) particles in H_2O/Polyols mixed solutions. Mater. Trans. 55: 147-153.

Kohlstedt, H. and H. Ishiwara. 2003. Ferroelectric field effect transistors. pp. 389-399. In: R. Waser [ed.]. Nanoelectronics and Information Technology - Advanced Electronic Materials and Novel Devices. WILEY-VCH Verlag, Wienheim, 2003.

Kong, L.B., J. Ma, W. Zhu and O.K. Tan. 2001. Preparation of PMN powders and ceramics via a high-energy ball milling process. J. Mater. Sci. Lett. 20: 1241-1243.

Kong, L.B., T.S. Zhang, J. Ma and F. Boey. 2008. Progress in synthesis of ferroelectric ceramic materials via high-energy mechanochemical technique. Prog. Mater. Sci. 53: 207-322.

Kostopoulou, A., E. Kymakis and E. Stratakis. 2018. Perovskite nanostructures for photovoltaic and energy storage devices. J. Mater. Chem. A. 6: 9765-9798.

Kostopoulou, A., K. Brintakis, N.K. Nasikas and E. Stratakis. 2019. Perovskite nanocrystals for energy conversion and storage. Nanophotonics 8: 1607-1640.

Krogstrup, P., H.I. Jorgensen, M. Heiss, O. Demichel, J.V. Holm, M. Aagesen, et al. 2013. Single-nanowire solar cells beyond the Shockley-Queisser limit. Nat. Photon. 7: 306-310.

Kucharczyk, B., J. Okal, W. Tylus, J. Winiarski and B. Szczygieł. 2019. The effect of the calcination temperature of $LaFeO_3$ precursors on the properties and catalytic activity of perovskite in methane oxidation. Ceram. Int. 45: 2779-2788.

Kudo, A. and Y. Miseki. 2009. Heterogeneous photocatalyst materials for water splitting. Chem. Soc. Rev. 38: 253-278.

Kumar, A. and K.L. Yadav. 2013. Enhanced magnetoelectric sensitivity in $0.3Co_{0.7}Zn_{0.3}Fe_2O_4$-$0.7Bi_{0.9}La_{0.1}FeO_3$ nanocomposites. Mater. Res. Bull. 48: 1312-1315.

Kumar, P. and K.L. Mittal. 1999. Handbook of Microemulsions: Science and Technology. Marcel Dekker, New York.

Kundys, B., M. Viret, D. Colson and D.O. Kundys. 2010. Light-induced size changes in $BiFeO_3$ crystals. Nat. Mater. 9: 803-805.

Lam, S.M., J.C. Sin and A.R. Mohamed. 2017. A newly emerging visible light-responsive $BiFeO_3$ perovskite for photocatalytic applications: A mini review. Mater. Res. Bull. 90: 15-30.

Lan, J., S.H. Xie, L. Tan and J.Y. Li. 2010. Sol-gel based soft lithography and piezoresponse force microscopy of patterned $Pb(Zr_{0.52}Ti_{0.48})O_3$ microstructures. J. Mater. Sci. Technol. 26(5): 439-444.

Lee, C.H., H.S. Shin, D.H. Yeo, G.H. Ha and S. Nahm. 2016. Sintering and microstructure of $BaTiO_3$ nanoparticles synthesized by molten salt method. J. Nanosci. Nanotechnol. 16: 5233-5238.

Lee, C.Y., N.H. Tai, H.S. Sheu, H.T. Chiu and S.H. Hsieh. 2006. The formation of perovskite $PbTiO_3$ powders by sol-gel process. Mater. Chem. Phys. 97: 468-471.

Lee, G.-J., E.K. Park, S.A. Yang, J.J. Park, S.D. Bu and M.K. Lee. 2017. Rapid and direct synthesis of complex perovskite oxides through a highly energetic planetary milling. Sci. Rep. 7: 46241(1-11).

Lee, J.S. 2013. Molten salt synthesis of $YAlO_3$ powders. Mater. Sci-Poland. 31: 240-245.

Lee, J.S. 2017. Preparation of $EuAlO_3$ powders by molten salt method. J. Ceram. Process. Res. 18: 385-388.

Lee, W., H. Han, A. Lotnyk, M.A. Schubert, S. Senz, M. Alexe, et al. 2008. Individually addressable epitaxial ferroelectric nanocapacitor arrays with near Tb inch^{-2} density. Nat. Nanotechnol. 3: 402-407.

Leite, E.R., L.P.S. Santos, N.L.V. Carreno, E. Longo, C.A. Paskocimas, J.A. Varela, et al. 2001. Photoluminescence of nanostructured $PbTiO_3$ processed by high-energy mechanical milling. Appl. Phys. Lett. 78: 2148-2150.

Levin, I., M.G. Tucker, H. Wu, V. Provenzano, C.L. Dennis, S. Karimi, et al. 2011. Displacive phase transitions and magnetic structures in Nd-substituted $BiFeO_3$. Chem. Mater. 23: 2166-2175.

Levy, P., A.G. Leyva, H.E. Troiani and R.D. Sánchez. 2003. Nanotubes of rare-earth manganese oxide. Appl. Phys. Lett. 83: 5247-5249.

Lew, L.C., Y. Voon and M. Willatzen. 2011. Electromechanical phenomena in semiconductor nanostructures. J. Appl. Phys. 109: 031101(1-24).

Li, B.R., W. Shang, Z.L. Hu and N.Q. Zhang. 2014. Template-free fabrication of pure single-crystalline $BaTiO_3$ nano-wires by molten salt synthesis technique. Ceram. Int. 40: 73-80.

Li, B.X. and Y.F. Wang. 2010. Facile synthesis and enhanced photocatalytic performance of flower-like ZnO hierarchical microstructures. J. Phys. Chem. C. 114: 890-896.

Li, F., Y. Liu, R. Liu, Z. Sun, D. Zhao and C. Kou. 2010. Preparation of Ca-doped $LaFeO_3$ nanopowders in a reverse microemulsion and their visible light photocatalytic activity. Mater. Lett. 64: 223-225.

Li, H.P., H. Wu, D.D. Lin and W. Pan. 2009. High Tc in electrospun $BaTiO_3$ nanofibers. J. Am. Ceram. Soc. 92: 2162-2164.

Li, L.L., S.J. Peng, J. Wang, Y.L. Cheah, P.F. Teh, Y. Ko, et al. 2009a. Facile approach to prepare porous $CaSnO_3$ nanotubes *via* a single spinneret electrospinning technique as anodes for lithium ion batteries. ACS Appl. Mater. Interfaces. 4: 6005-6012.

Li, S., J. Zhang, M.G. Kibria, Z. Mi, M. Chaker, D. Ma, et al. 2013a. Remarkably enhanced photocatalytic activity of laser ablated Au nanoparticle decorated $BiFeO_3$ nanowires under visible-light. Chem. Commun. 49: 5856-5858.

Li, T., J. Shen, N. Li and M. Ye. 2013b. Hydrothermal preparation, characterization and enhanced properties of reduced graphene-$BiFeO_3$ nanocomposite. Mater. Lett. 91: 42-44.

Li, W., Z.J. Xu, R.Q. Chu, P. Fu and J.G. Hao. 2009b. Structure and electrical properties of $BaTiO_3$ prepared by sol-gel process. J. Alloy Compd. 482: 137-140.

Li, Z., Y. Shen, C. Yang, Y. Lei, Y. Guan, Y. Lin, et al. 2013c. Significant enhancement in the visible light photocatalytic properties of $BiFeO_3$-graphene nanohybrids. J. Mater. Chem. A. 1: 823-829.

Li, Z.J., W.Y. Zhang, H.Y. Wang and B.C. Yang. 2017. Two-dimensional perovskite $LaNiO_3$ nanosheets with hierarchical porous structure for high-rate capacitive energy storage. Electrochimica Acta 258: 561-570.

Limmer, S.J., S. Seraji, Y. Wu, T.P. Chou, C. Nguyen and G.Z. Cao. 2002. Template-based growth of various oxide nanorods by sol-gel electrophoresis. Adv. Funct. Mater. 12: 59-64.

Liu, B., B. Hu and Z. Du. 2011. Hydrothermal synthesis and magnetic properties of single-crystalline $BiFeO_3$ nanowires. Chem. Commun. 47: 8166-8168.

Liu, L.F., T.Y. Ning, Y. Ren, Z.H. Sun, F.F. Wang, W.Y. Zhou, et al. 2008. Synthesis, characterization, photoluminescence and ferroelectric properties of $PbTiO_3$ nanotube arrays. Mater. Sci. Eng. B. 149: 41-46.

Liu, Y., Q. Qian, Z.G. Yi, L. Zhang, F.F. Min and M.X. Zhang. 2013. Low-temperature synthesis of single-crystalline $BiFeO_3$ using molten KCl-KBr salt. Ceram. Int. 39: 8513-8516.

Lopez-Quitela, M.A. and J. Rivas. 1993. Chemical reactions in microemulsions: A powerful method to obtain ultrafine particles. J. Colloid Interface Sci. 158: 446-451.

Lorena, M., M. Luis, A.A. Pedro, A.R. Luis, M. Csar, M.D.T. Jos, et al. 2014. Enhanced magnetotransport in nanopatterned manganite nanowires. Nano Lett. 14: 423-428.

Lu, H., Z. Du, J. Wang and Y. Liu. 2015. Enhanced photocatalytic performance of Ag-decorated $BiFeO_3$ in visible light region. J. Sol-Gel Sci. Technol. 6: 50-57.

Lu, X., D. Zhang, Q. Zhao, C. Wang, W. Zhang and Y. Wei. 2006. Large-scale synthesis of necklace-like single-crystalline $PbTiO_3$ nanowires. Macromol. Rapid Commun. 27: 76-80.

Luo, J. and P.A. Maggard. 2006. Hydrothermal synthesis and photocatalytic activities of $SrTiO_3$-coated Fe_2O_3 and $BiFeO_3$. Adv. Mater. 18: 514-517.

Luo, Y., I. Szafraniak, N.D. Zakharov, V. Nagarajan, M. Steinhart, R.B. Wehrspohn, et al. 2003. Nanoshell tubes of ferroelectric lead zirconate titanate and barium titanate. Appl. Phys. Lett. 83: 440-442.

Lv, J., T. Kako, Z. Li, Z. Zou and J. Ye. 2010. Synthesis and photocatalytic activities of $NaNbO_3$ rods modified by In_2O_3 nanoparticles. J. Phys. Chem. C. 114: 6157-6162.

Maczka, M., A. Bednarkiewicz, E. Mendoza-Mendoza, A.F. Fuentes and L. Kepinski. 2012. Optical properties of Eu and Er doped $LaAlO_3$ nanopowders prepared by low-temperature method. J. Solid State Chem. 194: 264-269.

Magrez, A., E. Vasco, J.W. Seo, C. Dieker, N. Setter and L. Forró. 2006. Growth of single-crystalline $KNbO_3$ nanostructures. J. Phys. Chem. B. 110: 58-61.

Mao, Y.B., S. Banerjee and S.S. Wong. 2003. Large-scale synthesis of single-crystal line perovskite nanostructures. J. Am. Chem. Soc. 125: 15718-15719.

Mao, Y.B., T.J. Park, F. Zhang, H.J. Zhou and S.S. Wong. 2007. Environmentally friendly methodologies of nanostructure synthesis. Small 3: 1122-1139.

Martin, L.W., S.P. Cane, Y.H. Chu, M.B. Holcomb and M. Gajek et al. 2008. Multiferroics and magneto-electrics: Thin film and nanostructures. J. Phys. Cond. Matt. 20: 434220(1-).

Matei, C., D. Berger, P. Marote, S. Stoleriu and J.P. Deloume. 2007. Lanthanum-based perovskites obtained in molten nitrates or nitrites. Prog. Solid State Chem. 35: 203-209.

Mathews, S., R. Ramesh, T. Venkatesan and J. Benedetto. 1997. Ferroelectric field effect transistor based on epitaxial perovskite heterostructures. Science 276: 238-240.

McDonnell, K.A., N. Wadnerkar, N.J. English, M. Rahman and D. Dowling. 2013. Photoactive and optical properties of bismuth ferrite ($BiFeO_3$): An experimental and theoretical study. Chem. Phys. Lett. 572: 78-84.

Mefford, J.T., W.G. Hardin, S. Dai, K.P. Johnston and K.J. Stevenson. 2014. Anion charge storage through oxygen intercalation in $LaMnO_3$ perovskite pseudocapacitor electrodes. Nat. Mater. 13: 726-732.

Mendoza-Mendoza, E., K.P. Padmasree, S.M. Montemayor and A.F. Fuentes. 2012a. Molten salts synthesis and electrical properties of Sr- and/or Mg-doped perovskite-type $LaAlO_3$ powders. J. Mater. Sci. 47: 6076-6085.

Mendoza-Mendoza, E., S.M. Montemayor, J.I. Escalante-García and A.F. Fuentes. 2012b. A "green chemistry" approach to the synthesis of rare-earth aluminates: Perovskite-type $LaAlO_3$ nanoparticles in molten nitrates. J. Am. Ceram. Soc. 95: 1276-1283.

Miclea, C., C. Tanasoiu, A. Gheorghiu, C.F. Miclea and V. Tanasoiu. 2004. Synthesis and piezoelectric properties of nanocrystalline PZT-based ceramics prepared by high energy ball milling process. J. Mater. Sci. 39: 5431-5434.

Miron, L.V. 2008. Sol-gel process. pp. 119-160. In: G. Ertl, H. Knözinger, F. Schüth, J. Weitkamp [eds.]. Handbook of Heterogeneous Catalysis. Wiley-VCH Verlag GmbH & Co. KGaA.

Miyauchi, M. 2007. Thin films of single-crystalline $SrTiO_3$ nanorod arrays and their surface wettability conversion. J. Phys. Chem. C. 111: 12440-12445.

Moghtada, A. and R. Ashiri. 2015. Nanocrystals of $XTiO_3$ (X = Ba, Sr, Ni, Ba_xTi_{1-x}) materials obtained through a rapid one-step methodology at 50°C. Ultrason. Sonochem. 26: 293-304.

Moghtada, A. and R. Ashiri. 2016. Enhancing the formation of tetragonal phase in perovskite nanocrystals using an ultrasound assisted wet chemical method. Ultrason. Sonochem. 33: 141-149.

Moghtada, A., A. Moghadam and R.A. Shiri. 2018. Tetragonality enhancement in $BaTiO_3$ by mechanical activation of the starting $BaCO_3$ and TiO_2 powders: Characterization of the contribution of the mechanical activation and post-milling calcination phenomena. Int. J. Appl. Ceram. Technol. 15: 1518-1531.

Morelli, A., F. Johann, N. Schammelt, D. McGrouther and I. Vrejoiu. 2013. Mask assisted fabrication of nanoislands of $BiFeO_3$ by ion beam milling. J. Appl. Phys. 113: 154101(1-4).

Muller, O. and R. Roy. 1974. The Major Ternary Structural Families. Springer Verlag, New York, USA.

Nah, S., Y.H. Kuo, F. Chen, J. Park, R. Sinclair and A.M. Lindenberg 2014. Ultrafast polarization response of an optically trapped single ferroelectric nanowire. Nano Lett. 14: 4322-4327.

Naumov, I.I. and H.X. Fu. 2005. Spontaneous polarization in one-dimensional $Pb(ZrTi)O_3$ nanowires. Phys. Rev. Lett. 95: 247602 (1-4).

Nelson, C.T., B. Winchester, Y. Zhang, S.J. Kim, A. Melville, C. Adamo, et al. 2011. Spontaneous vortex nanodomain arrays at ferroelectric heterointerfaces. Nano Lett. 11: 828-834.

Neogi, S., U. Chowdhury, A.K. Chakraborty and J. Ghosh. 2015. Effect of mechanical milling on the structural and dielectric properties of $BaTiO_3$ powders. Micro and Nano Lett. 10: 109-114.

Nguyen, T.V.A., A.N. Hattori, Y. Fujiwara, S. Ueda and H. Tanaka. 2013. Colossal magnetoresistive (La, Pr, Ca) MnO_3 nanobox array structures constructed by the three-dimensional nanotemplate pulsed laser deposition technique. Appl. Phys. Lett. 103: 223105 (1-4).

Niederberger, M., N. Pinna, J. Polleux and M. Antonietti. 2004. General soft-chemistry route to perovskites and related materials: Synthesis of $BaTiO_3$, $BaZrO_3$, and $LiNbO_3$ nanoparticles. Angew. Chem. Int. Ed. 43: 2270-2273.

Niu, F., T. Gao, N. Zhang, Z. Chen, Q.L. Huang and L.S. Qin. 2015. Hydrothermal synthesis of $BiFeO_3$ nanoparticles for visible light photocatalytic applications. J. Nanosci. Nanotechnol. 15: 9693-9698.

Nonomura, H., H. Fujisawa, M. Shimizu, H. Niu and K. Honda. 2003. Self-assembled $PbTiO_3$ nano-islands prepared on $SrTiO_3$ by metalorganic chemical vapor deposition. Jpn. J. Appl. Phys. 42: 5918-5921.

Nourmohammadi, A., M.A. Bahrevar, S. Schulze and M. Hietschold. 2008. Electrodeposition of lead zirconate titanate nanotubes. J. Mater. Sci. 43: 4753-4759.

Nourmohammadi, A., M.A. Bahrevar and M. Hietschold. 2009. Template-based electrophoretic deposition of perovskite PZT nanotubes. J. Alloys Compd. 473: 467-472.

Nuraje, N., K. Su, A. Haboosheh, J. Samson, E.P. Manning, N. Yang, et al. 2006. Room temperature synthesis of ferroelectric barium titanate nanoparticles using peptide nanorings as templates. Adv Mater. 18: 807-811.

Nuraje, N., R. Asmatulu and S.E. Kudaibergenov. 2012. Metal oxide-based functional materials for solar energy conversion: A review. Curr. Inorg. Chem. 2: 124-146.

O'Brien, S., L. Brus and C.B. Murray. 2001. Synthesis of monodisperse nanoparticles of barium titanate: Toward a generalized strategy of oxide nanoparticle synthesis. J. Am. Chem. Soc. 123: 12085-12086.

Oh, Y.S., S. Crane, H. Zheng, Y.H. Chu, R. Ramesh and K.H. Kim. 2010. Quantitative determination of anisotropic magnetoelectric coupling in $BiFeO_3$-$CoFe_2O_4$ nanostructures. Appl. Phys. Lett. 97: 052902 (1-3).

Oka, K., T. Koyama, T. Ozaaki, S. Mori, Y. Shimakawa and M. Azuma. 2012. Polarization rotation in the monoclinic perovskite $BiCo_{1-x}Fe_xO_3$. Angew. Chem. Int. Ed. 51: 7977-7980.

Oka, K., M. Mizumaki, C. Sakaguchi, A. Sinclair, C. Ritter, J.P. Attfield, et al. 2013. Intermetallic charge-transfer transition in $Bi_{1-x}La_xNiO_3$ as the origin of the colossal negative thermal expansion. Phys. Rev. B. 88: 014112 (1-6).

Ortega-San-Martin, L. 2020. Introduction to perovskite: A historical perspective. pp. 1-41. In: N.S. Arul and V.D. Nithya [eds.]. Revolution of Perovskite, Materials Horizons: From Nature to Nanomaterials. Springer Nature, Singapore Pte Ltd., Singapore.

Panomsuwan, G. and H. Manuspiya. 2019. Correlation between size and phase structure of crystalline $BaTiO_3$ particles synthesized by sol-gel method. Mater. Res. Express. 6: 065062(1-8).

Parida, K., K. Reddy, S. Martha, D. Das and N. Biswal. 2010. Fabrication of nanocrystalline $LaFeO_3$: An efficient sol-gel auto-combustion assisted visible light responsive photocatalyst for water decomposition. Int. J. Hydrogen Energy. 35: 12161-12168.

Park, K.I., S. Xu, Y. Liu, G.T. Hwang, S.J. Kang and Z.L. Wang. 2010. Piezoelectric $BaTiO_3$ thin film nanogenerator on plastic substrates. Nano Lett. 10: 4939-4943.

Park, T., J.Y. Mao and S.S. Wong. 2004. Synthesis and characterization of multiferroic $BiFeO_3$ nanotubes. Chem. Commun. 23: 2708-2709.

Park, T.J., G.C. Papaefthymiou, A.J. Viescas, A.R. Moodenbaugh and S.S. Wong. 2007. Size-depend magnetic properties of single-crystalline multiferroic $BiFeO_3$ nanoparticles. Nano Lett. 7: 766-772.

Paul, B.K. and S.P. Moulik. 2001. Uses and applications of microemulsions. Curr. Sci. 80: 990-1001.

Pazik, R., D. Hreniak and W. Strek. 2007. Microwave driven hydrothermal synthesis of $Ba_{1-x}Sr_xTiO_3$ nanoparticles. Mater. Res. Bull. 42: 1188-1194.

Pena, M.A. and J.L.G. Fierro. 2001. Chemical structures and performance of perovskite oxides. Chem. Rev. 101: 1981-2018.

Peng, Q., B. Shan, Y.W. Wen and R. Chen. 2015. Enhanced charge transport of $LaFeO_3$ via transition metal (Mn, Co, Cu) doping for visible light photoelectrochemical water oxidation. Int. J. Hydrogen Energy. 40: 15423-15431.

Perejon, A., N. Murafa, P.E. Sanchez-Jimenez, J.M. Criado, J. Subrt, M.J. Dianez, et al. 2013. Direct mechanosynthesis of pure $BiFeO_3$ perovskite nanoparticles: Reaction mechanism. J. Mater. Chem. 1: 3551-3562.

Phan, T.L., P. Zhang, D.S. Yang, T.D. Thanh, D.A. Tuan and S.C. Yu. 2013. Origin of ferromagnetism in $BaTiO_3$ nanoparticles prepared by mechanical milling. J. Appl. Phys. 113: 17E305(1-3).

Pileni, M.-P. 2003. The role of soft colloidal templates in controlling the size and shape of inorganic nanocrystals. Nat. Mater. 2: 145-150.

Ponzoni, C., R. Rosa, M. Cannio, V. Buscaglia, E. Finocchio, P. Nanni, et al. 2013. Optimization of BFO microwave-hydrothermal synthesis: Influence of process parameters. J. Alloy Compd. 558: 150-159.

Poonam, K. Sharma, A. Arora and S.K. Tripathi. 2019. Review of supercapacitors: Materials and devices. J. Energy Storage. 21: 801-825.

Prado-Gonjal, J., D. Ávila, M.E. Villafuerte-Castrejón, F. González-García, L. Fuentes, R.W. Gómez, et al. 2011. Structural, microstructural and Mossbauer study of $BiFeO_3$ synthesized at low temperature by a microwave-hydrothermal method. Solid State Sci. 13: 2030-2036.

Prashanthi, K., P. Dhandharia, N. Miriyala, R. Gaikwad, D. Barlage and T. Thundat. 2015. Enhanced photo-collection in single BiFeO$_3$ nanowire due to carrier separation from radial surface field. Nano Energy 13: 240-248.

Pugaczowa-Michalska, M. and J. Kaczkowski. 2017. DFT + U studies of triclinic phase of BiNiO$_3$ and La-substituted BiNiO$_3$. Comput. Mater. Sci. 126: 407-417.

Rai, A. and A.K. Thakur. 2017. Effect of Na and Mn substitution in perovskite type LaFeO$_3$ for storage device applications. Ionics 23: 2863-2869.

Rao, S.S., K.N. Anuradha, S. Sarangi and S.V. Bhat. 2005. Weakening of charge order and antiferromagnetic to ferromagnetic switch over in Pr$_{0.5}$Ca$_{0.5}$MnO$_3$ nanowires. Appl. Phys. Lett. 87: 182503(1-3).

Reaney, I.M., E.L. Colla and N. Setter. 1994. Dielectric and structural characteristics of Ba and Sr-based complex perovskites as a function of tolerance factor. Jpn. J. Appl. Phys. 33: 3984-3990.

Retot, H., A. Bessiere, A. Kahn-Harari and B. Viana. 2008. Synthesis and optical characterization of SrHfO$_3$: Ce and SrZrO$_3$: Ce nanoparticles. Opt. Mater. 30: 1109-1114.

Rigoberto, L.J., G. Federico and V.C. Maria-Elena. 2011. Lead-free ferroelectric ceramics with perovskite structure. pp. 318-330. *In*: M. Lallart [ed.]. Ferroelectrics-Material Aspects. InTech Publisher, Rijeka, Croatia.

Rørvik, P.M., K. Tadanaga, M. Tatsumisago, T. Grande and M.A. Einarsrud. 2009. Template-assisted synthesis of PbTiO$_3$ nanotubes. J. Eur. Ceram. Soc. 29: 2575-2579.

Rose, H. 1990. Outline of a spherically corrected semi-aplanatic medium-voltage transmission electron microscope. Optik 85: 19-24.

Rossell, M.D., Q.M. Ramasse, S.D. Findlay, F. Rechberger, R. Erni and M. Niederberger. 2012. Direct imaging of dopant clustering in metal-oxide nanoparticles. ACS Nano. 6: 7077-7083.

Ryu, H., P. Murugavel, J.H. Lee, S.C. Chae, T.W. Noh, Y.S. Oh, et al. 2006. Magnetoelectric effects of nanoparticulate Pb(Zr$_{0.52}$Ti$_{0.48}$)O$_3$-NiFe$_2$O$_4$ composite films. Appl. Phys. Lett. 89: 102907(1-3).

Sagadevan, S., I. Das and J. Podder. 2016. Synthesis of lead titanate nanoparticles *via* sol-gel technique and its characterization. J. Mater. Sci.-Mter. Electron. 27: 13016-13021.

Sahoo, G.K. and R. Mazumder. 2010. Grain size effect on the dielectric properties of molten salt synthesized BaTiO$_3$. Ferroelectrics 402: 193-199.

Saito, K. and A. Kudo. 2010. Niobium-complex-based syntheses of sodium niobate nanowires possessing superior photocatalytic properties. Inorg. Chem. 49: 2017-2019.

Sarkar, A., A.K. Singh, D. Sarkar, G.G. Khan and K. Mandal. 2015. Three-dimensional nanoarchitecture of BiFeO$_3$ anchored TiO$_2$ nanotube arrays for electrochemical energy storage and solar energy conversion. ACS Sustain Chem. Eng. 3: 2254-2263.

Schick, D., M. Herzog, H. Wen, P. Chen, C. Adamo, P. Gaal, et al. 2014. Localized excited charge carriers generate ultrafast inhomogeneous strain in the multiferroic BiFeO$_3$. Phys. Rev. Lett. 112: 097602(1-6).

Schilling, A., R.M. Bowman, G. Catalan, J.F. Scott and J.M. Gregg. 2007. Morphological control of polar orientation in single-crystal ferroelectric nanowires. Nano Lett. 7: 3787-3791.

Schilling, A., D. Byrne, G. Catalan, K.G. Webber, Y.A. Genenko, G.S. Wu, et al. 2009. Domains in ferroelectric nanodots. Nano Lett. 9: 3359-3364.

Schmid, H. 1994. Multiferroic magnetoelectrics. Ferroelectrics 162: 317-338.

Scott, J.F., H.J. Fan, S. Kawasaki, J. Banys, M. Ivanov, A. Krotkus, et al. 2008. THz emission from tubular Pb(Zr,Ti)O$_3$ nanostructures. Nano Lett. 8: 4404-4409.

Selbach, S.M., T. Tybell, M.-A. Einarsrud and T. Grande. 2007. Size-dependent properties of multiferroic BiFeO$_3$ nanoparticles. Chem. Mater. 19: 6478-6484.

Selvaraj, U., A.V. Prasadarao, S. Brooks and S. Kurtz. 1992. Sol-gel processing of PbTiO$_3$ and Pb(Zr$_{0.52}$Ti$_{0.48}$) O$_3$ fibers. J. Mater. Res. 7: 992-996.

Sen, B., M. Stroscio and M. Dutta. 2012. Piezoelectricity in lead zirconate titanate nanowires: A theoretical study. J. Appl. Phys. 201: 024517(1-6).

Seol, K.S., S. Tomita, K. Takeuchi, T. Miyagawa, T. Katagiri, Y. Ohki, et al. 2002. Gas-phase production of monodisperse lead zirconate titanate nanoparticles. Appl. Phys. Lett. 81: 1893-1895.

Setter, N., D. Damjanovic, L. Eng, G. Fox, S. Gevorgian, S. Hong, et al. 2006. Ferroelectric thin films: Review of materials, properties, and applications. J. Appl. Phys. 100: 051606(1-46).

Shandilya, M., R. Rai and J. Singh. 2016. Review: Hydrothermal technology for smart materials. Adv. Appl. Ceram. 115: 354-376.

Shankar, K. and A. Raychaudhuri. 2004. Growth of an ordered array of oriented manganite nanowires in alumina templates. Nanotechnology 15: 1312-1316.

Shannigrahi, S.R. and S.Y. Tan. 2011. Effects of processing parameters on the properties of lanthanum manganite. Mater. Chem. Phys. 129: 15-18.

Shao, T.Y., H.H. You, Z.J. Zhai, T.H. Liu, M. Li and L. Zhang. 2017. Hollow spherical $LaNiO_3$ supercapacitor electrode synthesized by a facile template-free method. Mater. Lett. 201: 122-124.

Shen, H.F., T. Xue, Y.M. Wang, G.Z. Cao, Y.J. Lu, G.L. Fang, et al. 2016. Photocatalytic property of perovskite $LaFeO_3$ synthesized by sol-gel process and vacuum microwave calcination. Mater. Res. Bull. 84: 15-24.

Shen, Z.K., Z.H. Chen, H. Li, X.P. Qu, Y. Chen and R. Liu. 2011. Nanoembossing and piezoelectricity of ferroelectric $Pb(Zr_{0.3}Ti_{0.7})O_3$ nanowire arrays. Appl. Surf. Sci. 257: 8820-8823.

Songwattanasin, P., A. Karaphun, S. Hunpratub, S. Maensiri, E. Swatsitang and V. Amornkitbamrung. 2019. Influence of annealing on microstructure, electrochemical, and magnetic properties of Co-doped $SrTiO_3$ nanocubes. J. Superconductivity Novel Magnetism. 32: 2959-2972.

Sousa, C.T., A.M.L. Lopes, M.P. Proenca, D.C. Leitão, J.G. Correia and J.P. Araújo. 2009. Rapid synthesis of ordered manganite nanotubes by microwave irradiation in alumina templates. J. Nanosci. Nanotechol. 9: 6084-6099.

Spanier, J.E., A.M. Kolpak, J.J. Urban, I. Grinberg, O.Y. Lian, W.S. Yun, et al. 2006. Ferroelectric phase transition in individual single-crystalline $BaTiO_3$ nanowires. Nano Lett. 6: 735-739.

Stojanovic, B.D. 2003. Mechanochemical synthesis of ceramic powders with perovskite structure. J. Mater. Process. Technol. 143-144: 78-81.

Su, K., N. Nuraje and N.-L. Yang. 2007. Open-bench method for the preparation of $BaTiO_3$, $SrTiO_3$ and $Ba_xSr_{1-x}TiO_3$ nanocrystals at 80°C. Langmuir 23: 11369-11372.

Sun, N., H.X. Liu, Z.Y. Yu, Z.N. Zheng and C.Y. Shao. 2014. The $La_{0.6}Sr_{0.4}CoO_3$ perovskite catalyst for Li-O-2 battery. Solid State Ionics. 268: 125-130.

Sun, W.A., C.H. Li, J.Q. Li and W. Liu. 2006. Microwave-hydrothermal synthesis of tetragonal $BaTiO_3$ under various conditions. Mater. Chem. Phys. 97: 481-487.

Sun, Z., T. Liao and L. Kou. 2017. Strategies for designing metal oxide nanostructures. Sci. China Mater. 60: 1-24.

Suryanarayana, C., E. Ivanov and V.V. Boldyrev. 2001. The science and technology of mechanical alloying. Mater. Sci. Eng. A. 304-306: 151-158.

Suryanarayana, S.V. 1994. Magnetoelectric interaction phenomena in materials. Bull. Mater. Sci. 17: 1259-1270.

Swihart, M.T. 2003. Vapor-phase synthesis of nanoparticles. Curr. Opin. Colloid Interface Sci. 8: 127-133.

Szafraniak, I., C. Harnagea, R. Scholz, S. Bhattacharyya, D. Hesse and M. Alexe. 2003. Ferroelectric epitaxial nanocrystals obtained by a self-patterning method. Appl. Phys. Lett. 83: 2211-2213.

Szafraniak, I., M. Połomska, B. Hilczer, A. Pietraszko and L. Kepinski. 2007. Characterization of $BiFeO_3$ nanopowder obtained by mechanochemical synthesis. J. Eur. Ceram. Soc. 27: 4399-4402.

Tagliazucchi, M., R.D. Sanchez, H.E. Troiani and E.J. Calvo. 2006. Synthesis of lanthanum nickelate perovskite nanotubes by using a template-inorganic precursor. Solid State Commun. 137: 212-215.

Tan, K.W., D.T. Moore, M. Saliba, H. Sai, L.A. Estroff, T. Hanrath, et al. 2014. Thermally induced structural evolution and performance of mesoporous block copolymer-directed alumina perovskite solar cells. ACS Nano 8: 4730-4739.

Tang, Y.L., Y.L. Zhu, X.L. Ma, A.Y. Borisevich, A.N. Morozovska, E.A. Eliseev, et al. 2015. Observation of a periodic array of fluxclosure quadrants in strained ferroelectric $PbTiO_3$ films. Science 348: 547-551.

Teowee, G., K. McCarthy, F. McCarthy, T.J. Bukowski, T.P. Alexander and D.R. Uhlmann. 1997. Dielectric and ferroelectric properties of sol-gel derived $BiFeO_3$ films. Integr. Ferroelectr. 18: 329-337.

Thirumal, M., P. Jain and A.K. Ganguli. 2001. Molten salt synthesis of complex perovskite-related dielectric oxides. Mater. Chem. Phys. 70: 7-11.

Thirumalairajan, S., K. Girija and I. Ganesh. 2012. Controlled synthesis of perovskite $LaFeO_3$ microsphere composed of nanoparticles via self-assembly process and their associated photocatalytic activity. Chem. Eng. J. 209: 420-428.

Thomas, J.K., H.P. Kumar, S. Solomon, K.C. Mathai and J. Koshy. 2010. Nanocrystalline $SrHfO_3$ synthesized through a single step auto-igniting combustion technique and its characterization. J. Alloys Compd. 508: 532-535.

Tian, G., F.Y. Zhang, J.X. Yao, H. Fan, P. Li, Z. Li, et al. 2016. Magnetoelectric coupling in well-ordered epitaxial $BiFeO_3/CoFe_2O_4/SrRuO_3$ heterostructured nanodot array. ACS Nano 10: 1025-1032.

Tian, G., D. Chen, J. Yao, Q. Luo, Z. Fan, M. Zeng, et al. 2017. $BiFeO_3$ nanorings synthesized *via* AAO template-assisted pulsed laser deposition and ion beam etching. RSC Adv. 7: 41210-41216.

Tian, Z.Q., H. Wang, W.J. Huang and C.Y. Zhang. 2008. Effect of preparation methods on microstructures and microwave dielectric properties of $Ba(Mg_{1/3}Nb_{2/3})O_3$ ceramics. J. Mater. Sci. - Mater. Electron. 19: 227-232.

Toyoda, M., Y. Hamaji and K. Tomono. 1997. Fabrication of $PbTiO_3$ ceramic fibers by sol-gel processing. J. Sol-Gel Sci. Technol. 9: 71-84.

Tsymbal, E.Y. and H. Kohlstedt. 2006. Tunneling across a ferroelectric. Science 313: 181-183.

Urban, J.J., W.S. Yun, Q. Gu and H. Park. 2002. Synthesis of single-crystalline perovskite nanorods composed of barium titanate and strontium titanate. J. Am. Chem. Soc. 124: 1186-1187.

Urban, K.W. 2008. Studying atomic structures by aberration-corrected transmission electron microscopy. Science 321: 506-510.

Utara, S. and S. Hunpratub. 2018. Ultrasonic assisted synthesis of $BaTiO_3$ nanoparticles at 25°C and atmospheric pressure. Ultrason. Sonochem. 41: 441-448.

Utsunomiya, S. and R.C. Ewing. 2003. Application of high-angle annular dark field scanning transmission electron microscopy, scanning transmission electron microscopy-energy dispersive x-ray spectrometry, and energy-filtered transmission electron microscopy to the characterization of nanoparticles in the environment. Environ Sci. Technol. 37: 786-791.

Varela, M., A.R. Lupini, K.V. Benthem, A.Y. Borisevich, M.F. Chisholm, N. Shibata, et al. 2005. Materials characterization in the aberration-corrected scanning transmission electron microscope. Annu. Rev. Mater. Res. 35: 539-569.

Varma, A., A.S. Mukasyan, A.S. Rogachev and K.V. Manukyan. 2016. Solution combustion synthesis of nanoscale materials. Chem. Rev. 116: 14493-14586.

Vasudevan, R.K., K.A. Bogle, A. Kumar, S. Jesse, R. Magaraggia, R. Stamps, et al. 2011. Ferroelectric and electrical characterization of multiferroic $BiFeO_3$ at the single nanoparticle level. Appl. Phys. Lett. 99: 252905 (1-4).

Wang, J., J.M. Xue, D.M. Wan and W. Ng. 1999. Mechanochemically synthesized lead magnesium niobate. J. Am. Ceram. Soc. 82: 1358-1360.

Wang, J., J.B. Neaton, H. Zheng, V. Nagarajan, S.B. Ogale, B. Liu, et al. 2003. Epitaxial $BiFeO_3$ multiferroic thin film heterostructures. Science 299: 1719-1722.

Wang, J., C.S. Sandu, E. Colla, Y. Wang, W. Ma, R. Gysel, et al. 2007. Ferroelectric domains and piezoelectricity in monocrystalline $Pb(Zr,Ti)O_3$ nanowires. Appl. Phys. Lett. 90: 133107.

Wang, J., A. Durussel, C.S. Sandu, M.G. Sahini, Z.B. He and N. Setter. 2012. Mechanism of hydrothermal growth of ferroelectric PZT nanowires. J. Cryst. Growth 347: 1-6.

Wang, Z.L., J.S. Yin and Y.D. Jiang. 2000. EELS analysis of cation valence states and oxygen vacancies in magnetic oxides. Micron. 31: 571-580.

Wang, Z.Y., A.P. Suryavanshi and M.F. Yu. 2006a. Ferroelectric and piezoelectric behaviors of individual single crystalline $BaTiO_3$ nanowire under direct axial electric biasing. Appl. Phys. Lett. 89: 082903.

Wang, Z.Y., J. Hu and M.F. Yu. 2006b. One-dimensional ferroelectric monodomain formation in single crystalline $BaTiO_3$ nanowire. Appl. Phys. Lett. 89: 263119.

Wen, T., J. Zhang, T.P. Chou, S.J. Limmer and G. Cao. 2005. Template-based growth of oxide nanorod arrays by centrifugation. J. Sol-Gel Sci. Technol. 33: 193-200.

Wu, H., W.R. Xia, P.J. Xue and X.H. Zhu. 2017. Perovskite oxide nanocrystals: Synthesis, characterization, physical properties, and applications. Ferroelectrics 518: 127-136.

Wu, J.A., Z. Fan, D.Q. Xiao, J.G. Zhu and J. Wang. 2016. Multiferroic bismuth ferrite-based materials for multifunctional applications: Ceramic bulks, thin films and nanostructures. Prog. Mater. Sci. 84: 335-402.

Wu, M.K., R.S. Windeler, K.R. Steiner, T. Bors and S.K. Friedlander. 1993. Controlled synthesis of nanosized particles by aerosol processes. Aerosol Sci. Technol. 19: 527-548.

Wu, T., S.B. Ogale, J.E. Garrison, B. Nagaraj, A. Biswas, Z. Chen, et al. 2001. Electroresistance and electronic phase separation in mixed-valent manganites. Phys. Rev. Lett. 86: 5998-6001.

Xia, W.R., Y. Lu and X.H. Zhu. 2020. Microwave-hydrothermal synthesis of perovskite oxide nanomaterials. pp. 1-22. *In*: K.D. Sattler [ed.]. Design Strategies for Synthesis and Fabrication (Vol. 2). Taylor and Francis (CRC Press), London, UK.

Xie, S.H., A. Gannepalli, Q.N. Chen, Y.M. Liu, Y.C. Zhou, R. Proksch, et al. 2012. High resolution quantitative piezoresponse force microscopy of $BiFeO_3$ nanofibers with dramatically enhanced sensitivity. Nanoscale 4: 408-413.

Xu, Q., X.P. Han, F. Ding, L. Zhang, L. Sang, X.J. Liu, et al. 2016. A highly efficient electrocatalyst of perovskite $LaNiO_3$ for nonaqueous $Li-O_2$ batteries with superior cycle stability. J. Alloy Compd. 664: 750-755.

Xu, X., T. Qian, G. Zhang, T. Zhang, G. Li, W. Wang, et al. 2007. Fabrication and magnetic properties of multiferroic $BiFeO_3$ nanotube arrays. Chem. Lett. 36: 112-113.

Xue, J.M., D.M. Wan and J. Wang. 1999a. Mechanochemical synthesis of nanosized lead titanate powders form mixed oxides. Mater. Lett. 39: 364-369.

Xue, J.M., J. Wang, W. Ng and D. Wang. 1999b. Activation-induced pyrochlore-to-perovskite conversion for a lead magnesium niobate precursor. J. Am. Ceram. Soc. 82: 2282-2284.

Xue, J.M., D.M. Wan, S.E. Lee and J. Wang. 1999c. Mechanochemical synthesis of lead zirconate titanate from mixed oxides. J. Am. Ceram. Soc. 82: 1687-1692.

Xue, J.M., J. Wang and T.M. Rao. 2001. Synthesis of $Pb(Mg_{1/3}Nb_{2/3})O_3$ in excess lead oxide by mechanical activation. J. Am. Ceram. Soc. 84: 660-662.

Xue, P.J., Y. Hu, W.R. Xia, H. Wu and X.H. Zhu. 2017. Molten-salt synthesis of $BaTiO_3$ powders and their atomic-scale structural characterization. J. Alloy Compd. 695: 2870-2877.

Xue, P.J., H. Wu, Y. Lu and X.H. Zhu. 2018. Recent progress in molten salt synthesis of low-dimensional perovskite oxide nanostructures, structural characterization, properties, and functional applications: A review. J. Mater. Sci. Technol. 34: 914-930.

Xue, P.J., H. Wu, W.R. Xia, Z.P. Pei, Y. Lu and X.H. Zhu. 2019. Molten salt synthesis of $BaTiO_3$ nanorods: Dielectric, optical properties, and structural characterizations. J. Am. Ceram. Soc. 102: 2325-2336.

Yadav, A.K., C.T. Nelson, S.K. Hsu, Z. Hong, J.D. Clarkson, C.M. Schlepütz, et al. 2016. Observation of polar vortices in oxide superlattices. Nature 530: 198-201.

Yan, L., H. Cao, J.F. Li and D. Viehland. 2009. Triclinic phase in tilted (001) oriented $BiFeO_3$ epitaxial thin films. Appl. Phys. Lett. 94: 132901(1-4).

Yang, G.J. and S.J. Park. 2019. Conventional and microwave hydrothermal synthesis and application of functional materials: A Review. Materials 12: 1177(1-18).

Yang, J., R.S. Li, J.Y. Zhou, X.C. Li, Y.M. Zhang, Y.L. Long, et al. 2010. Synthesis of $LaMO_3$ (M = Fe, Co, Ni) using nitrate or nitrite molten salts. J. Alloy Compd. 508: 301-308.

Yashima, M., T. Hoshina, D. Ishimura, S. Kobayashi, W. Nakamura, T. Tsurumi, et al. 2005. Size effect on the crystal structure of barium titanate nanoparticles. J. Appl. Phys. 98: 014313(1-8).

Yi, X. and J.L. Li. 2010. Synthesis and optical property of $NaTaO_3$ nanofibers prepared by electrospinning. J. Sol-Gel Sci. Technol. 53: 480-484.

Yin, S.G., Y.T. Wu, J.L. Chen, Z.L. Chen, H.Q. Hou, Q.Y. Liu, et al. 2018. Facile hydrothermal synthesis of $BiFeO_3$ nanoplates for enhanced supercapacitor properties. Funct. Mater. Lett. 11:1850013(1-4).

Yoon, K.H., Y.S. Cho and D.H. Kang. 1998. Molten salt synthesis of lead-based relaxors. J. Mater. Sci. 33: 2977-2984.

Yu, J.C., L.Z. Zhang, Q. Li, K.W. Kwong, A.W. Xu and J. Lin. 2003. Sonochemical preparation of nanoporous composites of titanium oxide and size-tunable strontium titanate crystals. Langmuir 19: 7673-7675.

Yu, T., Z.X. Shen, J.M. Xue and J. Wang. 2002. Nanocrystalline $PbTiO_3$ powders form an amorphous Pb-Ti-O precursor by mechanical activation. Mater. Chem. Phys. 75: 216-219.

Yun, K.Y., M. Noda, M. Okuyama, H. Saeki, H. Tabata and K. Saito. 2004. Structural and multiferroic properties of $BiFeO_3$ thin films at room temperature. J. Appl. Phys. 96: 3399-3403.

Yun, W.S., J.J. Urban, Q. Gu and H.K. Park. 2002. Ferroelectric properties of individual barium titanate nanowires investigated by scanned probe microscopy. Nano Lett. 2: 447-450.

Zavaliche, F., T. Zhao, H. Zheng, F. Straub, M.P. Cruz, P.L. Yang, et al. 2007. Electrically assisted magnetic recording in multiferroic nanostructures. Nano Lett. 7: 1586-1590.

Zeng, H., J. Wang, S.E. Lofland, Z. Ma, L. Mohaddes-Ardabill, T. Zhao, et al. 2004. Multiferroic $BaTiO_3$-$CoFe_2O_4$ Nanostructures. Science 303: 661-663.

Zhang, G. and J. Chen. 2005. Synthesis and application of $La_{0.59}Ca_{0.41}CoO_3$ nanotubes. J. Electrochem. Soc. 152: A2069-2073.

Zhang, Q. and F. Saito. 2000. Mechanochemical synthesis of $LaMnO_3$ from La_2O_3 and MnO_3 powders. J. Alloys Compd. 297: 99-103.

Zhang, Q., D. Sando and V. Nagarajan. 2016. Chemical route derived bismuth ferrite thin films and nanomaterials. J. Mater. Chem. C. 4: 4092-4124.

Zhang, T., C.G. Jin, T. Qian, X.L. Lu, J.M. Bai and X.G. Li. 2004a. Hydrothermal synthesis of single-crystalline $La_{0.5}Ca_{0.5}MnO_3$ nanowires at low temperature. J. Mater. Chem. 14: 2787-2789.

Zhang, X.Y., X. Zhao, C.W. Lai, J. Wang, X.G. Tang and J.Y. Dai. 2004b. Synthesis and piezoresponse of highly ordered $Pb(Zr_{0.53}Ti_{0.47})O_3$ nanowire arrays. Appl. Phys. Lett. 85: 4190-4192.

Zhang, X.Y., C.W. Lai, X. Zhao, D.Y. Wang and J.Y. Dai. 2005. Synthesis and ferroelectric properties of multiferroic $BiFeO_3$ nanotube arrays. Appl. Phys. Lett. 87: 143102(1-3).

Zhang, Y., L.Q. Wang and D.F. Xue. 2012. Molten salt route of well dispersive barium titanate nanoparticles. Powder Technol. 217: 629-633.

Zhao, L., M. Steinhart, J. Yu and U. G"osele. 2006. Lead titanate nano- and microtubes. J. Mater. Res. 21: 685-690.

Zheng, H.M., Q. Zhan, F. Zavaliche, M. Sherburne, F. Straub, M.P. Cruz, et al. 2006. Controlling self-assembled perovskite-spinel nanostructures. Nano Lett. 6: 1401-1407.

Zheng, X.H., P.J. Chen, N. Ma, Z.H. Ma and D.P. Tang. 2012. Synthesis and dielectric properties of $BiFeO_3$ derived from molten salt method. J. Mater. Sci. - Mater. Electron. 23: 990-994.

Zhong, C.F., Z.Y. Lu, X.H. Wang and L.T. Li. 2016.Template-based synthesis and piezoelectric properties of $BiScO_3$-$PbTiO_3$ nanotube arrays. J. Alloys Compd. 655: 28-31.

Zhou, D., T. Zhou, Y. Tian, X. Zhu and Y. Tu. 2018. Perovskite-based Solar cells: Materials, methods and future perspectives. J. Nanomaterials 2018: 8148072(1-15).

Zhu, D., H. Zhu and Y.H. Zhang. 2002a. Hydrothermal synthesis of single-crystal $La_{0.5}Sr_{0.5}MnO_3$ nanowire under mild conditions. J. Phys. Condens. Matter. 14: L519-L524.

Zhu, D., H. Zhu and Y. Zhang. 2002b. Hydrothermal synthesis of $La_{0.5}Ba_{0.5}MnO_3$ nanowires. Appl. Phys. Lett. 80: 1634-1636.

Zhu, J.J., H.L. Li, L.Y. Zhong, P. Xiao, X.L. Xu, X.G. Yang, et al. 2014a. Perovskite oxides: Preparation, characterization, and applications in heterogenous catalysis. ACS Catal. 4: 2917-2940.

Zhu, X.H., J.M. Zhu, S.H. Zhou, Z.G. Liu, N.B. Ming and D. Hesse. 2005. $BaTiO_3$ nanocrystals: Hydrothermal synthesis and structural characterization. J. Cryst. Growth. 283: 553-562.

Zhu, X.H., J.Y. Wang, Z.H. Zhang, J.M. Zhu, S.H. Zhou, Z.G. Liu, et al. 2008. Perovskite nanoparticles and nanowires: Microwave-hydrothermal synthesis and structural characterization by high-resolution transmission electron microscopy. J. Am. Ceram. Soc. 91: 2683-2689.

Zhu, X.H., J. Zhou, J.M. Zhu, Z.G. Liu, Y.Y. Li and T.A. Kassab. 2014b. Structural characterization and optical properties of perovskite $ZnZrO_3$ nanoparticles. J. Am. Ceram. Soc. 97: 1987-1992.

Zhu, X.H., J. Zhou, M.C. Jiang, J. Xie, S. Liang, S.Y. Li, et al. 2014c. Molten salt synthesis of bismuth ferrite nano- and microcrystals and their structural characterization. J. Am. Ceram. Soc. 97: 2223-2232.

Zhu, X.H., P.R. Evans, D. Byrne, A. Schilling, C. Douglas, R.J. Pollard, et al. 2006. Perovskite lead zirconium titanate nanorings: Towards nanoscale ferroelectric "Solenoids"? Appl. Phys. Lett. 89: 122913(1-3).

Zhu, X.H., Q.M. Hang, Z.B. Xing, Y. Yang, J.M. Zhu, Z.G. Liu, et al. 2011. Microwave hydrothermal synthesis, structural characterization, and visible-light photocatalytic activities of single-crystalline bismuth ferric nanocrystals. J. Am. Ceram. Soc. 94: 2688-2693.

Zhu, X.H. and Q.M. Hang. 2013. Microscopical and physical characterization of microwave and microwave-hydrothermal synthesis products. Micron. 44: 21-44.

3

Nanostructured Metal Oxides for Hybrid Supercapacitors

Anil Arya[1, 2, *], Anurag Gaur[2], Vijay Kumar[3], Shweta Tanwar[1]
and A.L. Sharma[1]

3.1 INTRODUCTION

Human development and the global economy are mostly governed by energy. Global increasing demand for energy for industries, households, and electric vehicles needs to be addressed owing to the high quality of living standards of humankind. Renewable and non-renewable sources (traditional sources) of energy are key contributors to this. The depletion of conventional energy sources such as fossil fuels has led to various environmental impacts like increased greenhouse gases, global warming, air pollution, etc. Due to this, the environmentally friendly renewable energy sector seems a possible alternative to endorsing zero-emission energy sources. Renewable energy comprises of hydro energy, solar energy, chemical energy, wind energy, etc. The energy generated from them needs to be stored somewhere, and battery and supercapacitor (SC) are two feasible electrochemical devices (store energy based on chemical reactions) for storing energy and may be used as per need. The former one has higher energy density, while the latter one has higher power density (Pal et al. 2019, Arya and Sharma 2020). This strengthens SC candidature for energy storage than a battery and along with this, other features such as fast charge-discharge, zero memory loss, long cyclic stability, and varied configurations make it best for use. Hybrid SC acts as a bridge and fills the void between the traditional capacitors and the battery and provides the ultimate balance of desirable energy and power density (Sharma et al. 2019, Arya and Sharma 2019).

3.1.1 Supercapacitors: An Overview

Parallel to fuel cell and battery, the capacitor emerged as a device for energy storage in 1957 (by H. Becker), and was termed as a low voltage capacitor. Then electrolytic capacitor with an activated carbon electrode was invented in 1970. The difference between the battery and supercapacitor was introduced in 1991 when Evans Conway was working on ruthenium oxide-based electrochemical capacitors. The term 'supercapacitor' was coined for the first time in 1999 by Brian Evans Conway (Conway 1991). Further evolution of supercapacitor occurred up to 2007 and after that development of different materials as an electrode for SC application took place. Later on, strategies to tune the morphology of the electrode material and new SC configuration were introduced. The overall focus of the development for researchers is to lower the cost and enhance the energy density and

[1] Department of Physics, Central University of Punjab, Bathinda-151001, Punjab, India.
[2] Department of Physics, National Institute of Technology, Kurukshetra-136119, Haryana.
[3] Department of Physics, Institute of Integrated and Honors Studies, Kurukshetra-136119, Haryana, India.
[*] Corresponding author: aniljaglan0581@yahoo.com

cyclic stability of the SC device. Figure 3.1 depicts the various phases in the supercapacitor era from evolution to development, followed by current progress and shines a light on the new paths of high-performance supercapacitors (Panda et al. 2020, Ho et al. 2010).

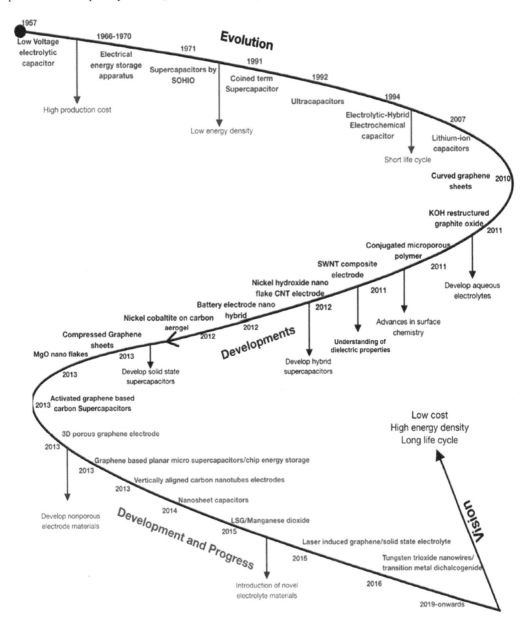

Fig. 3.1 Roadmap of the evolution, progress, and developments in supercapacitors (Reprinted with permission from Panda et al. *Nanoscale Advances, Royal Society of Chemistry, 2020: 70-108*)

The electrodes for SC are mainly classified in the following categories, (i) carbon-based materials, (ii) sulfides/carbides/nitrides, (iii) conducting polymers, (iv) hydroxides, and (v) transition metal oxides.

Supercapacitors are classified based on charge storage mechanism into (i) electric double-layer capacitors (EDLCs), (ii) pseudo capacitors (PC), and (iii) hybrid supercapacitors (HSC). The EDLC deals with carbon based electrodes and charge storage is physically attributed to the double layer formation (adsorption of negative ions on the positive electrode) at the electrode/electrolyte interface. PC deals with transitional metal oxide/sulfide-based electrodes and charge storage occurs via the double-layer formation as well as via redox reactions. The electrolyte ions get inserted in the host active electrode and contribute to charge storage. The PC seems best as compared to EDLC in terms of electrochemical performance. Table 3.1 summarizes the different aspects of various SC.

TABLE 3.1 Comparison of EDLC, pseudocapacitor, and hybrid capacitor

Electrochemical Double-Layer Capacitor (EDLC)	Pseudocapacitor	Hybrid Supercapacitor
Carbon is used as electrode material	MOs and CPs are used as electrode material	A combination of carbon and MOs/CPs is used
Charge storage mechanism is through the electrochemical double layer formation (non-Faradaic process)	The charge is stored through the redox reactions (Faradaic process)	The charge is stored both by Faradaic and non-Faradaic processes
Low E_d, good rate capability, good cyclic stability, low C_s	High C_s, high E_d, high P_d, low rate capability	High E_d, high P_d, good cyclability, polymer/carbon composite has a moderate cost and moderate stability, Li/Carbon capacitors are of high cost

Source: Reprinted with permission from Zhi et al. Nanoscale, Royal Society of Chemistry, 2013 (5): 72.

3.1.2 Hybrid SC

In the HSC, two combinations are possible: (i) one PC electrode and another battery type electrode, and (ii) one carbon-based electrode and another PC electrode. This configuration provides high energy density along with high power density as compared to EDLC and PC. Hybrid SC lies between the EDLC and battery and comprises of high energy density and power density (Fig. 3.2a). Figure 3.2b shows the types of supercapacitor and characteristics for the separator, electrode, electrolyte, and the role of the preparation method.

The existing energy technologies are based on the EDLCs (having activated carbon as an electrode) due to high cyclic stability and power density, and the organic electrolyte with a stability window of 2.0-3.0 V. But, low energy density restricts its application range. So, pseudocapacitors are suitable alternatives to EDLC due to their high specific capacitance, but degradation after a long cycle is a major issue. The hybrid system (having one carbon material as an electrode, and another battery type or pseudo electrode) opens new doors of opportunities to achieve desirable energy density as well as high power density without any degradation issue during term cyclic run (El-Kady et al. 2015).

Two important and significant performance parameters that determine the potential of the electrode for SC is its ability to provide high energy density and high power density. It is optimized by monitoring or controlling the electrode preparation conditions to obtain high electronic and ionic transport within electrodes (Fig. 3.2c) (El-Kady et al. 2012). To achieve enhanced performance, various strategies have been considered: (i) compact thick film formation from metal oxide; (ii) nanostructured metal oxide films; (iii) conductive material's incorporation in nanostructured metal oxide; and (iv) growth of nanostructure on carbon (graphene networks) with high electronic conductivity and surface area. Figure 3.3 compares the energy density and power density for different electrochemical devices and sheds light on the importance of the hybrid supercapacitor. The electrochemical performance of the SC cell and charge storage phenomena (whether EDLC, PC, or HSC) is examined from the nature of the cyclic voltammetry (CV), and galvanostatic charge-discharge (GCD) curve. Figure 3.4(i) shows the CV and GCD curve for the EDLC materials, pseudo capacitor material, and battery like materials (Gogotsi and Penner 2018). The EDLC shows the rectangular nature in CV and voltage varies linearly with discharge time.

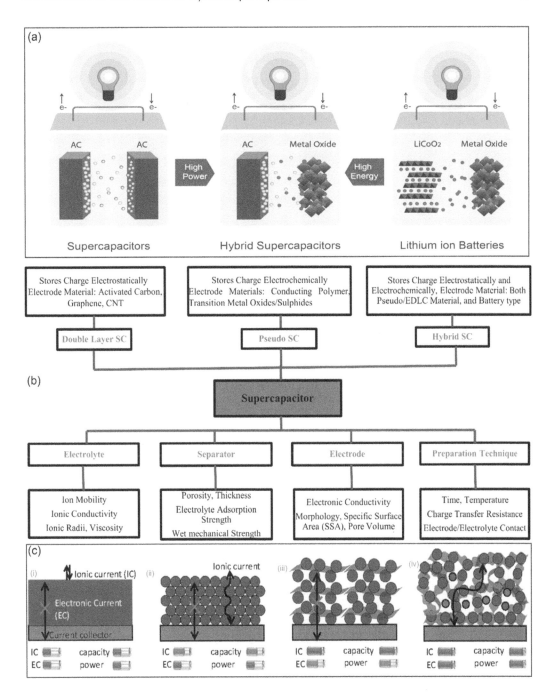

Fig. 3.2 (a) Comparison for EDLC, hybrid and battery device (Source: https://newatlas.com/energy/toomen-powercapacitors-kurt-energy-high-density-supercapacitors/), (b) classification of SC on the basis of charge storage mechanism and important ingredients of SC cell, (c) rational design of high-energy-high-power hybrid supercapacitor electrodes. Improving the ionic current (IC) and electronic current (EC) within the electrode is the key (Reprinted with permission from El-Kady et al. *Proceedings of the National Academy of Sciences, 2015 (112): 4233-4238*)

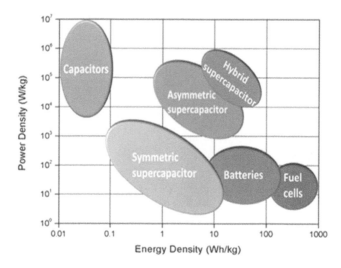

Fig. 3.3 Ragone plot for different energy storage systems (Source: https://gnanomat.com/2019/05/16/innovative-energy-storage-systems/)

The pseudocapacitive material shows peaks corresponding to redox reactions and also comprises of EDLC behaviors. Battery type electrodes CV shows separate redox peaks, and GCD shows the plateau formation and is visible as compared to pseudocapacitive materials (Nan et al. 2019). Due to this, the electrochemical performance for the battery type electrodes is evaluated in terms of the specific capacity instead of specific capacitance that will be discussed in the forthcoming section. Figure 3.4(ii) sheds light on the different approaches to pair the electrodes to fabricate supercapacitor, supercapattery, supercabattery, and battery (Yu and Chen 2016). It also provides a clear difference between the different cell configurations.

3.1.3 SC Device Performance Parameter

The SC cell performance is linked to the electrode material, its thickness, density, active surface area, electrolyte ion, current collector, and the preparation method. The important parameter that determines the SC cell performance is the electronic conductivity of the electrode, equivalent series resistance (ESR), voltage window of the electrolyte, cell capacitance, and the cell weight/volume.

The high specific surface area (SSA) and the porosity allows the accommodation of more electrolyte ions and hence high specific capacitance. The addition of binder and conductive carbon additive during slurry reparation also affects the electrode stability during cell operation for large cycles. The coating of the slurry on the current collector needs to be carefully done to achieve a uniform and dense film without cracks/defects. The high electronic conductivity and low charge transfer resistance (R_{ct}) of the electrode material also support the more charged storage. The energy density of the device is directly proportional to the square of the electrolyte voltage window and it confirms that electrolyte with a broad voltage window is best for achieving high energy density. In brief, the electrode characteristics such as mass loading, coating thickness on the current collector, and morphology (sphere, rod, wire, flower) directly influence the specific capacitance/capacity of the cell (Zhang and Pan 2015a).

As the electrolyte is a key component in an SC cell, it needs to be chosen carefully since it also affects cell performance. The suitable electrolyte should have characteristic features such as broad voltage stability window, high ionic conductivity, low electronic conductivity, high cation transport number, compatibility with electrodes, low viscosity, low cost, and low resistivity. The important electrolytes are aqueous (voltage stability window = 1.2 V), organic liquid (voltage stability window = 2.0-3.0 V), ionic liquid (voltage stability window = 3.0-4.0 V), and polymer electrolyte (voltage stability window = 4-5 V). The energy density of the electrolyte is linked with the voltage window of electrolyte, and power density to ionic conductivity (Zhang and Pan 2015a, Zhong et al. 2015). The proper cell fabrication and packaging in a controlled atmosphere results in superior cyclic stability and the response time of the device determines the cell operation (Kumar et al. 2020). Figure 3.5 provides a clear picture of the interrelation of these factors using different color codes and characteristics influencing each other.

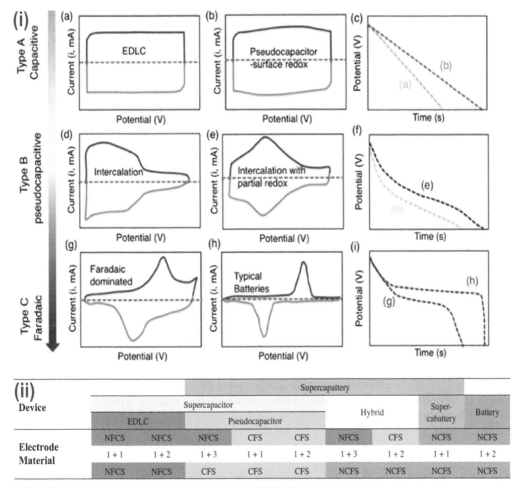

NFCS: Non-Faradaic capacitive storage = EDLC storage
CFS: Capacitive Faradaic storage = pseudocapacitive storage
NCFS: Non-capacitive Faradaic Storage = battery-type storage = Nernstian storage
1 + 1: Symmetrical device of the same electrode material
1 + 2: Asymmetrical device of different materials with the same storage mechanism
1 + 3: Asymmetrical device of different materials with different storage mechanisms

Fig. 3.4 (i) The CV curves of: (a) EDLC materials, (b, d, e) pseudo-capacitors materials, (g) battery-like materials, and (h) battery materials; and (c, f, i) corresponding galvanostatic discharge curves for various types of energy-storage materials (Reprinted with permission from Gogotsi and Penner *ACS Nano, ACS Publications, 2018 (12): 2081-83*), (ii) Ways of pairing the same or different electrode materials into supercapacitor, battery, supercapattery or supercabattery (Reprinted with permission from Yu et al. *J Power Sources, Elsevier Publications, 2016: 604-612, and* Akinwolemiwa et al. *Journal of the Brazilian Chemical Society, Brazilian Chemical Society, 2018, 29: 960-972*)

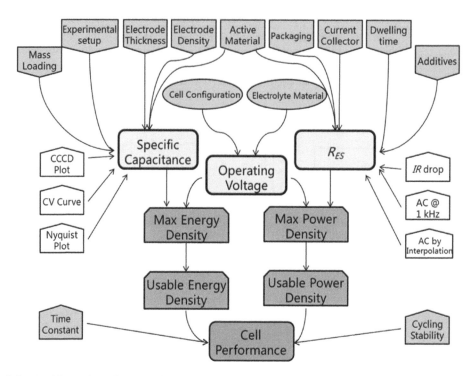

Fig. 3.5 An illustration of key performance metrics, test methods, and major affecting factors for the evaluation of SCs (Reprinted with permission from Zhang and Pan, *Advanced Energy materials, John Wiley and Sons, 2015 (5): 1401401*)

3.2 EXPERIMENTAL SECTION

3.2.1 Preparation Methods

The morphology of the electrode material can be tuned by varying the synthesis method. The variable parameters such as temperature, pressure, reaction time, reaction atmosphere, and heating rate provide control over the nanostructure morphology and growth. Some synthesis methods are described briefly in Table 3.2. Table 3.3 presents the comparison of specific capacitance and capacity retention for different preparation methods.

TABLE 3.2 Synthesis methods for the electrode material

Method	Process
Sol-gel method	In this method, active precursors are mixed to form a gel, and with varying sintering temperature and time varied morphology is obtained. Temperature, solvent, and reaction time affects morphology and area
Electro-polymerization/Electrodeposition	The thickness of the film can be monitored by the rate of polymerization. Uses low chemicals and enables the control of mass loading
In-situ polymerization	Mixing of monomers prepared by sonication followed by the addition of oxidizing agent for polymerization. Nanorod and nanofibers can be obtained
Coating method	The slurry is prepared using active material, conductive carbon and binder, and slurry is coated on the current collector
Chemical vapor deposition (CVD)	To develop porous structures, this method is beneficial
Hydrothermal method	Allows controlled morphology (nanoparticle, nanorod, nanowire, flower) by optimizing reaction time and temperature

TABLE 3.3 Specific capacitance and capacity retention for different preparation methods

Preparation Method	Material	Electrolyte	Potential Window	Sp. Cap. (F g⁻¹)	Capacity Retention	Ref.
SILAR	$MnCO_3$-RGO	1M Na_2SO_4	0.0V-0.8V	157	87% at 1000 cycle	Jana et al. 2017
Self-assembly method	$MnCO_3@MnO_2$	1M Na_2SO_4	−0.2V-0.8V	363	–	Chen et al. 2017
Hydrothermal	$NiFeO_x@MnCO_3$	3M KOH	0.0V-1.0V	283	92.7% at 2000 cycles	Rajendiran et al. 2019
Hydrothermal	$MnCO_3$	0.1M Na_2SO_4	0.0V-1.0V	216	97% at 500 cycle	Devaraj et al. 2014
Precipitation	MnO_2	6.0M KOH	0.0V-0.9V	193	94% at 1300 cycle	Lei et al. 2012
Solvothermal	Mn_3O_4	1M Na_2SO_4	−0.8-0.6V	131	99% at 500 cycle	Subramani et al. 2014
Hydrothermal	MnO_2	0.5M Na_2SO_4	0.0V-1.0V	322	86% at 2000 cycle	Lee et al. 2015

3.2.2 Characterization Techniques

So, before going for supercapacitor fabrication we need to check the following parameters for the electrode material: pore volume, pore size distribution, specific surface area (SSA), active mass loading, electrode thickness, electrolyte composition and volume, separator (if used), and potential window and current for the electrode, while for the device, the crucial parameters that need to be examined are specific capacitance/specific capacity, coulombic efficiency, cyclic stability (preferable via GCD), and capacity retention. Also, the structure, morphology, porosity and the elemental composition of the electrode material need to be analyzed before the fabrication of the complete supercapacitor cell (Wang et al. 2015, Salamne et al. 2016, Arbizzani et al. 2020). This section discusses in detail the above points. So, the important characterization techniques and information obtained from them is given below:

3.2.2.1 Morphological and Surface Area Analysis

(i) *X-ray diffraction (XRD):* It is an important technique to determine the crystalline content, atomic spacing, crystallite size, and form of the carbon, nanocomposite confirmation. The material is in the powder form and needs to be free from any impurity. The Rietveld refinement enables to evaluate the lattice parameters and crystallite size.

(ii) *Raman spectroscopy:* To investigate the structural defects and graphitization level for carbon-based materials, this technique is crucial. The ratio of G-band to D-band (I_G/I_D) gives information about the disorder and graphitization in the carbon materials.

(iii) *Scanning electron microscopy (SEM):* To obtain information regarding nanostructures morphology, i.e. sheet, wire, rod, and flower, SEM is used. It also gives an idea about the porosity in the sample. Along with this, elemental composition and elemental mapping can be checked via energy dispersive spectroscopy (EDS). The material is gold coated before measurement.

(iv) *Transmission electron microscopy (TEM):* This technique is used to extract information about the thickness of the sheets, pore size, and a high view of morphology of the synthesized sample. A thin sample is preferred for better results.

(v) *X-ray photoelectron spectroscopy (XPS):* It is a surface-sensitive spectroscopic technique and is a highly beneficial technique for extracting information about actual elemental composition, chemical/electronic state in the sample.

(vi) *BET-N_2 adsorption-desorption isothermal measurements:* This technique is very beneficial for porous electrodes to examine the pore size, porosity, volume distribution, and the specific surface area (SSA) of the synthesized material.

3.2.2.2 Electrochemical Analysis

The supercapacitor cell fabrication can be done in two ways, (i) symmetric, and (ii) asymmetric. After fabricating the cell, the cell is examined via impedance spectroscopy (IS), cyclic voltammetry (CV) and galvanostatic charge-discharge (GCD) to evaluate the electrochemical performance parameters. First of all, the Open Circuit Voltage (OCV) of the device is checked in the absence of an external load and is referred to as the full voltage of the device. OCV is denoted by V_{oc} and tells about the change in energy of electrode materials. The OCV allows the system to stabilize for a short period before going for measurement.

(i) Complex Impedance Spectroscopy

A plot between the imaginary (Z'') and real part (Z') of the impedance is obtained. The bulk resistance (R_b), charge-transfer resistance (R_{ct}), and equivalent series resistance (ESR) of the SC cell are obtained. The low R_{ct} indicates the high storage capacity for the cell.

(ii) Cyclic Voltammetry (CV)

The nature of the CV curve enables us to validate the charge storage phenomena and is further confirmed by using the power law ($i = av^b$; here, i is the cathodic current (A), v is the scan rate (mV/s), a and b are variables). The value of b differs for different charge storage mechanisms, (i) $b = 1$: capacitive type storage, and (ii) $b = 0.5$: diffusion-controlled charge storage mechanism. Also, the capacitive and diffusion capacitance contribution can be separated using the equation: $i = k_1v + k_2v^{1/2}$, here k_1 and k_2 are values. In this, k_1v is the capacitive contribution and $k_2v^{1/2}$ is the diffusion-limited contribution (Wang et al. 2007, Augustyn et al. 2013).

(iii) Galvanostatic Charge/Discharge (GCD)

This technique is performed to obtain actual storage and indicates the practical storage by the cell. The various electrochemical parameters are obtained from the GCD using the formulas given below (Wang et al. 2014).

➢ **For EDLC and Pseudo Capacitor Electrodes**

Two-Electrode (Symmetric Cell Configuration)

Specific Capacitance

$$C = \frac{2 \times I \times \Delta t}{m \times \Delta V} \tag{1}$$

Here, I is the discharge current, Δt is the discharge time, ΔV is the change in potential during discharge, and m is the active material loading in the single electrode.

Energy Density and Power Density

$$E = \frac{1}{2 \times 3.6} C(\Delta V)^2 \tag{2}$$

$$P = \frac{3600 \times E}{\Delta t} \tag{3}$$

Here, E (Wh/kg) is specific energy, C is specific capacitance, ΔV is potential window, P (W/kg) is specific power and Δt is discharge time.

➢ **For Two-Electrode (Asymmetric Cell Configuration)**

Specific Capacitance

$$C = \frac{I \times \Delta t}{m \times \Delta V} \tag{4}$$

Here I is the discharge current, Δt is the discharge time, ΔV is the potential change during discharge, and m is the active material loading on current collector in both (+ve, −ve) the electrodes.

Energy Density and Power Density

$$E = \frac{1}{2}C(\Delta V)^2 \tag{5}$$

Here, E (Wh/kg) is specific energy, C is specific capacitance, ΔV is potential window, P (W/kg) is specific power and Δt is discharge time.

➢ **For Battery Type Materials**

In hybrid devices, when battery type materials are used as electrodes, CV and GCD nature are not like EDLC and pseudo capacitor. The charge storage mechanism is like battery and in 2008 researchers coined the term supercapattery (= supercapacitor + battery) having behavior identical to SC with high capacity and supercabattery having behavior like a rechargeable battery with high power density (Akinwolemiwa and Chen 2018, Akinwolemiwa et al. 2015, Makino et al. 2012, Chen 2017). So, to calculate the correct electrochemical performance the specific capacity is calculated instead of specific capacitance (Brousse et al. 2015, Oyedotun et al. 2019). In such cases, a single electrode specific capacity is obtained from CV using equation

$$Q_S = \frac{1}{3.6 \times mS_c} \int_{E_1}^{E_2} I \times E dE \tag{7}$$

where, E_1 and E_2 are the peak potentials, I (mA) is the current, $E(V)$ is the potential of electrode, S_c (mVs^{-1}) is scan rate, and m (g) is loaded active material.

The specific capacity, Q_s (mAh g^{-1}) and energy efficiency, η_E (%) of the materials were evaluated by GCD technique using equation 8 and 9 (Oyedotun et al. 2017).

$$Q_S = \frac{1 \times \Delta t}{3.6m} \tag{8}$$

$$\eta_E = \frac{E_d}{E_c} \times 100 \tag{9}$$

Here, I (mA) is the discharge current, Δt (s) is the time taken for one discharge cycle, and m (g) is active material loading. η_E is energy efficiency, $E_c = \left[\frac{I}{3.6m}\int V dt\right]$ represents the charge energy (Wh/kg), while E_d is the discharge energy obtained by integrating the area under the charge-discharge profiles, respectively.

Specific Capacity (C/g) is obtained via GCD using Equation

$$Q = \frac{I \times \Delta t}{m} \tag{10}$$

Here, I(mA) is the discharge current, $\Delta t(s)$ is the discharge time, and m(g) is the mass of the active material.

The optimization of charge storage performance for the positive and negative electrode is crucial to achieving optimum device performance. To balance the charge ($q_- = q_+$) of both positive (q_+) and a negative electrode (q_-), the mass ratio of the positive electrode to the negative electrode in the HSC device is evaluated from CV (Zhao et al. 2017).

$$q = \int \frac{im\,dV}{S_c} \tag{11}$$

Here, where q(C) is the charge, i(Ag^{-1}) is the current density, m(g) is the mass of the active material, V(V) is the voltage, S_C (mV s^{-1}) is the scan rate, and $\int idV$ is the integral area of the CV curve.

and

$$\frac{m_+}{m_-} = \frac{(idV/S_c)_-}{(idV/S_c)_+} \tag{12}$$

3.3   UPDATES AND FABRICATION STRATEGIES ABOUT ELECTRODE MATERIALS FOR HYBRID SC DEVICE

One important characteristic of the electrode material for SC is high electronic conductivity. The TM based electrode possesses low electronic conductivity. So, to eliminate this drawback one effective approach is the incorporation of carbon-based material with the pseudocapacitive material.

Oyedotun et al. (Oyedotun et al. 2019) synthesized the Ti_3C_2-Mn_3O_4 nanocomposite by a two-step synthesis route via a solvothermal process and analyzed the electrochemical performance of SC using this as an electrode and 6M KOH as the electrolyte. XRD diffractogram confirmed the Ti_3C_2-Mn_3O_4 nanocomposite formation. The specific capacity value of the nanocomposite electrode from CV was 98.2 mAh g^{-1} (at 5 mV/s) and from GCD the specific capacity evaluated was 128.0 mAhg^{-1} at 1 A/g. The high performance was attributed to the improved individual material role that results in high electronic conductivity, swift electron dynamics, and enhanced specific surface area to store more electrolyte ions. Also, the TM oxide nanoparticles act as a template to accommodate volume changes (after cation insertion) and secondary current collector to facilitate high conductivity in the composite. The inset of Fig. 3.6(b) indicates the integration of Ti_3C_2 sheets with the addition of Mn_3O_4 and acts as a conductive support base for Mn_3O_4 nanostructures (Tang et al. 2016, Oyedotun et al. 2019). For a single electrode, coulombic efficiency was 83.5% with a capacity retention of 77.7% after 2000 cycles (at 10 A/g). To further evaluate the potential of electrode material, hybrid Ti_3C_2-Mn_3O_4//C-FP supercapattery was fabricated. Figure 3.6(a) shows the charge-discharge profile of the assembled device and non-linear profile confirms the capacity contribution from faradaic/redox reactions. The maximum capacity was 78.9 mAh/g at 1 A/g. The device demonstrated an energy density of 22.2 Wh/kg with a power density of 2285.5 W/kg at 5 A/g. Figure 3.6(b) demonstrates the high energy efficiency of 90.2% with a capacity retention of 92.6% after 10000 cycles (at 3 A/g).

Another interesting carbon-based nanostructure is soft polymer (porous and sheet like) graphitic carbon nitride (g-C_3N_4). The nanocomposite formation using this as carbon source and metal oxide-based another electrode seems to be a fascinating approach to achieve improved performance of SC cell. The silent features of g-C_3N_4 are cost-effective, the presence of highly active nitrogen sites, and exceptional chemical and physical strength (Tahir et al. 2014). Recently, Sharma and Gaur (Sharma and Gaur 2020) synthesized the graphitic-C_3N_4 supported $ZnCo_2O_4$ nanocomposite by a facial hydrothermal method. The morphological analysis shows the presence of clusters with fiber distribution and space between clusters facilitates the electrolyte ion dynamics. Figure 3.7 shows the detailed electrochemical analysis. The CV curve in Fig. 3.7(a) shows the enhanced area as compared to pristine $ZnCo_2O_4$ and peaks correspond to faradic redox reactions. The single electrode specific capacity obtained is highest for the hybrid g-C_3N_4@$ZnCo_2O_4$ electrode and is 157 mAhg^{-1} (4 Ag^{-1})

with a capacity retention of 90% after 2500 cycles (10Ag^{-1}). The hybrid SC cell is fabricated with cell configuration: g-C$_3$N$_4$@ZnCo$_2$O$_4$//gel electrolyte//g-C$_3$N$_4$@ZnCo$_2$O$_4$ and tested in 0-1.5 V voltage range. The specific discharge capacity obtained was 121 mAh/g at 4.6 A/g and capacity retention was 71% after 10000 cycles (at 15 A/g). The maximum energy density of 39 Whkg^{-1} with a power density of 1478 Wkg^{-1} is obtained as shown in the Ragone plot (Fig. 3.7f). Table 3.4 shows the comparison of electrochemical performance parameters for different electrode and electrolyte combinations.

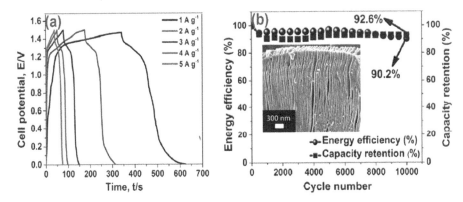

Fig. 3.6 (a) CD profiles of Ti$_3$C$_2$-Mn$_3$O$_4$//C-FP at various specific currents of the asymmetric device, (b) cyclic performance. Inset: Ti$_3$C$_2$-Mn$_3$O$_4$ hybrid at low magnifications (Reprinted with permission from Oyedotun et al. *Electrochimica Acta, Elsevier Publication, 2019 (301): 487-499*)

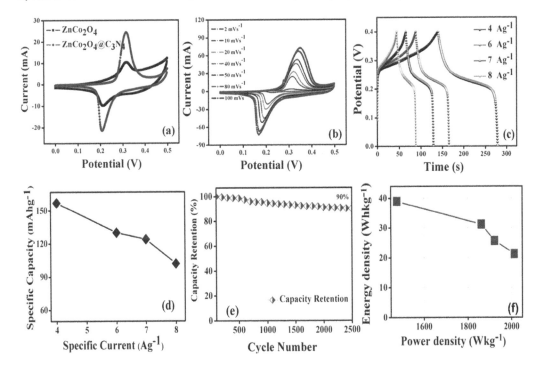

Fig. 3.7 (a) CV curves of ZnCo$_2$O$_4$ and g-C$_3$N$_4$@ZnCo$_2$O$_4$ at a constant scan rate of 20 mVs^{-1}, (b) CV curves for g-C$_3$N$_4$@ZnCo$_2$O$_4$ for different scan rates, (c) GCD curves for g-C$_3$N$_4$@ZnCo$_2$O$_4$ at different specific current, (d) Specific capacity as a function of specific current for g-C$_3$N$_4$@ZnCo$_2$O$_4$ sample, (e) Capacity retention up to 2500 cycles for g-C$_3$N$_4$@ZnCo$_2$O$_4$ sample, (f) Ragone plot for g-C$_3$N$_4$@ZnCo$_2$O$_4$// gel electrolyte//g-C$_3$N$_4$@ZnCo$_2$O$_4$ symmetric device (Reprinted with permission from Sharma et al. *Scientific Reports, Nature, 2020 (10): 1-9*)

TABLE 3.4 Comparison of SC performance parameters based on various electrodes

Electrode	Electrolyte Material	Specific Capacitance	Energy Density	Power Density	Capacitance Retention	Ref.
Treated carbon cloth	PVA–H$_2$SO$_4$	920 mF cm^{-2} at 2 mA cm^{-2}	1.4 mW h cm^{-3}	280 mW cm^{-3}	100% after 10000 cycles	Wang et al. 2017a
a-Fe$_2$O$_3$/rGO	PVA–KOH	32.9 mF cm^{-2} at 1 mA cm^{-2}	1.46 mW h cm^{-3}	2011.8 mW cm^{-3}	79.1% over 1000 cycles (2.5 mA cm^{-2})	Zhang et al. 2017
N/O co-doped graphene quantum dots	PVA–H$_2$SO$_4$	461 mF cm^{-2} at 0.5 mA cm^{-2}	0.032 mW h cm^{-2}	–	87.5% over 2000 cycles (15 mA cm^{-2})	Li et al. 2017c
CNTs/ NiCo$_2$O$_4$	PVA–KOH	337.3 mF cm^{-2} at 0.1 mA cm^{-2}	1.17 mW h cm^{-3}	2430 mW cm^{-3}	95.6% over 2000 cycles (2 A/g)	Zheng et al. 2017
CuO/3D graphene	PVA–LiCl	64 mF cm^{-2} at 0.25 mA cm^{-2}	0.0059 mW h cm^{-2}	110 mW cm^{-2}	86% over 5000 cycles (5 mA cm^{-2})	Li et al. 2017b
MXene/ graphene	PVA–H$_3$PO$_4$	216 F cm^{-3} at 0.1 A cm^{-2}	3.4 mW h cm^{-3}	1600 mW cm^{-3}	85.2% over 2500 cycles (1 A cm^{-3})	Li et al. 2017a
NCCF-rGO	PVA–KOH	200 F/g at 0.1 A/g	20 W h/kg	–	94% over 10000 cycles (3/g)	Fan et al. 2017
MoSe$_2$	PVA–KOH	133 F/g at 2 A/g	36.2 W h/kg	1.4 kW/kg	92% over 2000 cycles (100 mV/s)	Qiu et al. 2017
ZnCo$_2$O$_4$/rGO	PVA–KOH	143 F/g at 1 A/g	11.44 W h/kg	1.382 kW/ kg	93.4% over 5000 cycles (3 A/g)	Moon et al. 2017

Most of the research is focused on the development of metal oxide/sulfide-based electrode materials and shows desirable electrochemical performance. Another important requirement for the device is that it should be flexible and bendable. So, to achieve this (bendable and flexible) device, Subramani et al. (Subramani et al. 2017) prepared the dumb-bell shaped cobalt sulfide (CoS) particles by solvothermal decomposition of cobalt hexacyanoferrate (CoHCF). The FESEM analysis shows the uniformly dispersed dumb-bell shaped particles of length 2.1 to 2.7 μm and each side of the dumb-bell had an average lateral size of 1.3 μm. For CoS (+ve) electrode, highest specific capacitance from GCD was 310 F/g at 5 A/g with 95% capacity retention and coulombic efficiency of 99.5% after 5000 cycles (at 50 A/g). The activated carbon (–ve) electrode shows a specific capacitance of 60 F/g at 5 A/g. Then, a solid-state asymmetric supercapacitor (ASC) was fabricated using CoS and AC electrodes with a PVA/KOH gel polymer electrolyte. The maximum specific capacitance was 47 F/g for voltage window of 1.8 V. Figure 3.8(b) shows the mechanical flexibility of the fabricated device and the CV curve remains almost identical for different bending angles (Fig. 3.8a). After 5000 cycles, capacity retention of the ASC device was 92% and excellent coulombic efficiency was obtained (Fig. 3.8c). The fabricated ASC shows an energy density of about 5.3 W h/kg and power density 1800 W/kg at 2 A/g, and 1.53 W h/kg and power density 9000 W/kg at 10 A/g (Fig. 3.8d).

Recently, Chen et al. (Chen et al. 2019) reported the fabrication of high performance flexible solid-state zinc ion hybrid supercapacitor (ZHC) by derived hollow carbon spheres (acting as cathode) from Co-polymer, polyacrylamide hydrogel (electrolyte), deposited-Zn on carbon cloth (anode). It was concluded from the morphology analysis that the hollow sphere structure will facilitate the ions' adsorption and desorption owing to the enhanced electrode-electrolyte contact area (Fig. 3.9a). The highest capacity of 86.8 mAh/g was obtained at 0.5 A/g, with an energy density of 59.7 Wh /kg, and a power density of 447.8 W/Kg. The capacity retention was 98% after 15,000 cycles with coulombic efficiency of about 100% (at 1.0 A/g). The flexible electrode and electrolyte enable to sustain the performance after deformations (squeezing, twisting, and folding), with minor difference between the specific capacities (Fig. 3.9b).

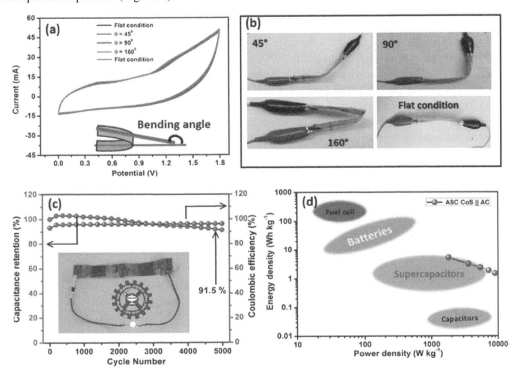

Fig. 3.8 (a) CV profile of the fabricated flexible ASC cell with different bending angles at a scan rate of 100 mV/s, (b) Image of the bent cells, (c) Specific capacitance and coulombic efficiency of the fabricated flexible ASC cell as a function of cycle number, (d) Typical Ragone plot of the fabricated flexible ASC cell and the location, where other energy storage devices would be located on the plot, is highlighted (Reprinted with permission from Subramani et al. *RSC Advances, Royal Society of Chemistry, 2017(7): 6648-6659*)

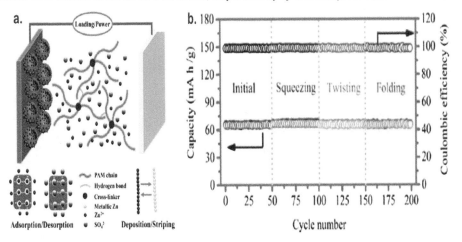

Fig. 3.9 (a) Schematic illustration of the working principle of the solid-state zinc hybrid supercapacitor system. It is fabricated by using nitrogen doped hollow carbon spheres as cathode, polyacrylamide hydrogel as electrolyte and deposited-Zn on carbon cloth as anode, (b) Cycling stability of the flexible solid-state ZHS before and under various deformation conditions (Reprinted with permission from Chen et al. *Journal of materials chemistry, Royal Society of Chemistry, 2019(7): 7784-7790*)

Another report by Zhao et al. (Zhao et al. 2017) presented the fabrication of hybrid supercapacitor based on $Co_xNi_{1-x}(OH)_2$ and reduced graphene oxide (rGO) by the facial process. The specific capacity of the fabricated device from GCD was 743 C/g, and 545 C/g at 1 and 20 A/g, respectively, with long cycling life. The gravimetric energy density of 72 and 44 W h/Kg at a power density of 797 W/Kg and 16.7 kW/Kg are obtained and cell demonstrates excellent cycling stability for 20,000 cycles at 20 A/g.

Surface modification of nanostructures is also an attractive strategy to tune the electrode. Recently, Esfandiar et al. (Esfandiar et al. 2020) reported the preparation of 3-D electrode based on porous Cu_2O-$Cu_{1.8}S$ nanowires via facile fabrication process (wet chemical synthesis followed by post-sulfurization process). The fabricated cell demonstrates the areal and volumetric of about 204.8 $\mu Whcm^{-2}$ and ~ 2.1 $mWhcm^{-3}$ at a power density of 3.1 $mWcm^{-2}$. The capacity retention was 55% at 15.5 $mWcm^{-2}$. The device also demonstrates superior cyclic stability cycling after 15000 cycles, i.e. 91% (at 40 mA/cm^2), and 94% (at 20 mA/cm^2).

Recently, Yang et al. (Yang et al. 2020) reported the preparation of the hybrid supercapacitor with nickel-cobalt oxalate derived bimetal-based nanomaterials, the $CoNi_2S_4$, NiCo-LDH and NiO/Co_3O_4. The FESEM analysis showed the formation of nanoparticle with different nanosheets coating (flower type) outside and will be beneficial for Faradic redox reactions during the cell operation. The CV and GCD analysis evidenced the battery type behavior and $CoNi_2S_4$ shows the specific capacitance of 1836.6 Fg^{-1} (1 A g^{-1}), which increases to 1855.2 Fg^{-1} at 50°C and on decreasing to 0°C the specific capacitance reduces to 1587.6 Fg^{-1}.

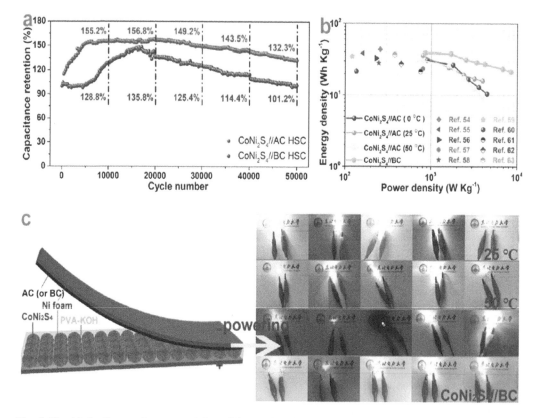

Fig. 3.10 (a) Cycling performance of the solid-state $CoNi_2S_4$//AC HSC and $CoNi_2S_4$//BC HSC at a current density of 5 Ag^{-1}, (b) Ragone plot related to energy and power densities of the solid-state $CoNi_2S_4$//AC HSC, $CoNi_2S_4$//BC HSC and the comparisons with other researches, (c) schematic illustration of the solid-state supercapacitors, and the photos of the LEDs lit up by the as-fabricated solid-state $CoNi_2S_4$//AC HSC (at 25, 50 and 0°C) and $CoNi_2S_4$//BC HSC (Reprinted with permission from Yang et al. *Nanoscale, Royal Society of Chemistry, 2020(12): 1921-1938*)

Then, hybrid SC was fabricated using AC as another electrode and the hybrid SC shows specific capacitance of 100.7 F/g (at 1 A/g), energy density of 35.8 Wh kg^{-1} at a power density of 800.0 W kg^{-1}, and is unchanged with an increase of temperature from 0-50°C. The cyclic stability was 132.3% after 50000 cycles. The HSC with negative electrode prepared from biomass-derived carbon (BC) demonstrates the specific capacitance of 199.6 F/g (at 2 A/g), and the energy density of about 38.9 Wh kg^{-1} with power density of 850.0 W kg^{-1} (capacitance retention = 101.2% after 50000 cycles). Figure 3.10(a) shows the excellent cyclic stability for the $CoNi_2S_4$//AC HSC followed by $CoNi_2S_4$//BC HSC, and also superior energy/power density (Fig. 3.10b). The assembled device was used to demonstrate the glowing of LED, as shown in Fig. 3.10(c) and (d). Two $CoNi_2S_4$//AC HSCs cell in series were sufficient to lighten up the yellow LED (2.0 V) for 25 min (at 25°C and 50°C), and about 30 min (at 0°C). The green LED (3.0 V) glows for more than 10 min by two connected $CoNi_2S_4$//BC cells.

Recently, Alagar et al. (Alagar et al. 2020) prepared the nano/submicrospheres of $Mn_{1-x}Ni_xCO_3$ by co-precipitation method. The $Mn_{0.75}Ni_{0.25}CO_3$ electrode possesses the highest specific capacitance of 364 F/g at 1 A/g and capacity retention was 96% after 7500 cycles (at 5 A/g). Then, a solid-state asymmetric SC was fabricated ($Mn_{0.75}Ni_{0.25}CO_3$//graphene nanosheets) and has specific capacity of 46 Fg^{-1}. The energy density was 25 Wh kg^{-1} and power density was about 499 W kg^{-1}. The capacity retention for the device was 87.7% after 7500 cycles.

Ramesh et al. (Ramesh et al. 2019a) synthesized the SnO_2@NGO composite by thermal reduction process at 550°C in the presence of ammonia and urea as a catalyst. The specific capacitance of about 378 F/g was obtained at 4 A/g and good cyclic stability up to 5000 cycles (89% after 2000 cycles). The high performance was attributed to the contribution from EDLC and pseudocapacitance. Another report by Ramesh et al. (Ramesh et al. 2019b) reports the preparation of the hydroxide nanoparticles (NPs) decorated with nitrogen-doped MWCNT (N-MWCNT) by thermal reduction process. The electrochemical performance was examined and the device showed the highest specific capacitance of 350 Fg^{-1} (at 0.5 A/g), with an energy density of ~43.75 Wkg^{-1} and power density 1500 W kg^{-1}. The device showed 90% capacity retention after 5000 cycles at 0.5 A/g. The same group reported the preparation of the hierarchical 3D flowers like nanostructure based on Co_3O_4 @MnO_2 on nitrogen-doped graphene oxide (NGO) by thermal reduction process at 650°C (in the presence of ammonia and urea). The device demonstrates the specific capacitance e of about 347 F/g at 0.5 A/g, and energy density of 34.83 Wh kg^{-1} with a power density of 820 W/kg. The device shows only 31% capacity loss after 10000 cycles (Ramesh et al. 2018). Nagamuthu et al. (Nagamuthu and Ryu 2019) prepared the metallic silver nickel oxide honeycomb nanoarrays (Ag/NiO NA) by surfactant-assisted hydrothermal route. The maximum specific capacity for a single electrode was 824 C g^{-1} (at 2.5 A g^{-1}), while the asymmetric SC (ASC) cell demonstrates a specific capacity of 204 C g^{-1} (at 2.5 A g^{-1}) with an energy density of 63.75 Wh kg^{-1} and a power density of 2812.5 W kg^{-1}. The ASC cell shows 96% capacity retention after 4000 cycles. Recently, Lee et al. (Lee et al. 2018) prepared the novel electrode using laser scribed graphene (LSG) as the cathode and $AlPO_4$-carbon hybrid coated $H_2Ti_{12}O_{25}$ (LSG/H-HTO) anode. The electrochemical performance of the LSG/H-HTO delivers superior energy and power density of ~70.8 Wh/kg and ~5191.9 W/kg, respectively. The hybrid cell demonstrates 98% cycling stability with a rate capability of 78% (at 3 A/g). Xu and Wang (Xu and Wang 2019) reported the preparation of hierarchical core/shell $MgCo_2O_4$@MnO_2 nanowall arrays (MCMNA) on Ni-foam by facile two-step hydrothermal method. The MCMNA-2 cell demonstrates the specific capacitance of 852.5 F/g at 1 A/g and is attributed to the fast charge transfer during redox reaction and active sites present on nanowall. Then the asymmetric SC was fabricated using AC as another electrode and this device shows an energy density of 67.2 Whkg^{-1} with power density of about 5760.0 Wkg^{-1}.

Das et al. (Das et al. 2017) investigated the influence of (nickel nitrate/citric acid) mole ratio on the end products obtained via sol-gel. Then Ni/NiO@reduced graphene-oxide (rGO) were prepared via probe sonication. The electrode with nickel ion: citric acid (1:1) leads to the highest specific capacity of about 158 C/g as obtained at 1 A/g among all (Ni^{2+}: CA) ratios examined, while the Ni/NiO@reduced graphene-oxide (rGO) based electrode shows an improved specific capacity of 335 C/g (at 1 A/g). The cell demonstrates 95% cell capacity, and 100% Coulombic efficiency after 1000 cycles. The increase in the capacitance was attributed to the increased surface area and enhanced conductivity which results in enhanced electrochemical activity.

The conducting rGO sheets facilitate the fast ion dynamics in the Ni/NiO system, and high surface area allows the storage of more ions. Then, a solid-state hybrid device was fabricated in configuration Ni/NiO@rGO (positive electrode) with rGO (negative electrode). For the device, the energy density of about 14.6 Wh/kg and power density of 4.3 kW/kg was obtained in the voltage window of 0-1.45 V (Fig. 3.11a). The hybrid device shows good stability (60% capacity retention after 1000th cycle) at different current rate and desirable coulombic efficiency (Fig. 3.11b, c). The device compatibility was examined by connecting three hybrid cells in series and was able to glow the LED for 6 minutes after charging once. Table 3.5 summarizes the comparison of hybrid super-capacitor performance for different electrodes.

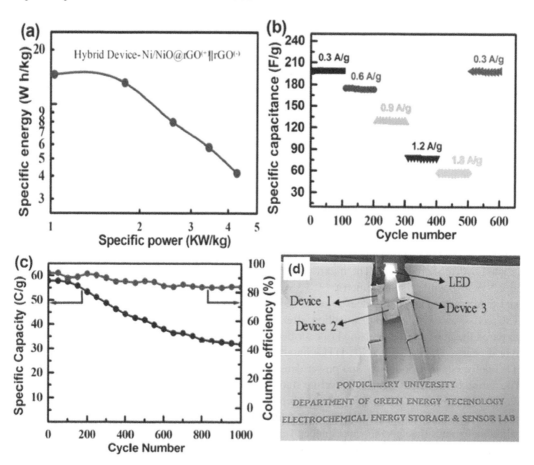

Fig. 3.11 (a) Ragone plot, (b) stability studies at different current densities, (c) cycle-life data and Coulombic efficiency recorded for the hybrid device consisting of the NR2 positive electrode and rGO negative electrode in gel-type electrolyte, (d) photograph of the fabricated three hybrid devices lighting an LED (Reprinted with permission from Chen et al. *Scientific Reports, Nature, 2017(7): 1-14*)

TABLE 3.5 Hybrid supercapacitor performance parameter comparison for different positive and negative electrodes

Electrode (+)	Electrode (−)	Voltage Range	Sp. Capacitance/Capacity	Energy Density	Power Density	Cycling Stability	Ref.
Ti_3C_2-Mn_3O_4	C-FP	–	78.9 mAh/g at 1 A/g	22.2 Wh/kg	2285.5 W/kg	92.6% after 10000 cy.	Oyedotu et al. 2019
g-C_3N_4@$ZnCo_2O_4$	g-C_3N_4@$ZnCo_2O_4$	0-1.5 V	121 mAh/g at 4.6 A/g	39 Whk/g	1478 W/kg	71% after 10000 cy.	Sharma and Gaur 2020
CoS	AC	1.8 V	47 F/g	5.3 W h/kg (at 2 A/g)	1800 W/kg (at 2 A/g)	92% after 5000 cy.	Subramani et al. 2017
Hollow carbon spheres	deposited-Zn on carbon cloth	–	86.8 mAh/g was obtained at 0.5 A/g	59.7 Wh/kg	447.8 W/kg	98% after 15,000 cy.	Chen et al. 2019
$Co_xNi_{1-x}(OH)_2$	reduced graphene oxide (rGO)		743 C/g at 1 A/g	44 Wh/kg	16.7 kW/kg	Stable for 20,000 cy.	Zhao et al. 2017
$CoNi_2S_4$	AC		100.7 F/g (at 1 A/g)	35.8 Wh/kg	800.0 W/kg	–	Yang et al. 2020
$CoNi_2S_4$	BC		199.6 F/g (at 2 A/g)	38.9 Wh/kg	850.0 W/kg	101.2% after 50000 cy.	Lu et al. 2017
$Ni(OH)_2$/rGO/Ni	rGO aerogel/Ni	KOH/PVA, 1.6 V	69 F/g at 2 A/g	24.5 W h/kg	10.3 kW/kg	83% after 6000 cy.	Yang et al. 2017
ZnO@C@CoNi-LDH	Fe_2O_3@C	KOH/PVA, 1.6 V	30.9 mF cm^{-2} at 1 mA cm^{-2}	1.078 mW h cm^{-3}	0.4 Wcm^{-3}	95% after 10000 cy.	Pan et al. 2018
MnO_2@CNTs@3D graphene foams	Ppy@CNTs@3D graphene foams	Na_2SO_4/PVA 1.8 V	8.56 F cm^{-3} at 1 mA cm^{-2}	3.85 mW h cm^{-3}	630 mW cm^{-3}	84.6% after 20000 cy.	Cheng et al. 2017
$NiCo_2S_4$@$NiCo_2O_4$	AC	KOH/PVA 1.6 V	0.41 F cm^{-2} at 2 mA cm^{-2}	44.6 W h/kg	6.4 kW/kg	92.5% after 6000 cy.	Wang et al. 2017b
$CoMoO_4$@$NiMoO_4$_xH_2O	Fe_2O_3	KOH/PVA 1.6 V	153.6 F/g at 1 A/g	41.8 W h/kg	12 kW/kg	89.3% after 5000 cy.	Liu et al. 2017
Ni-Co@Ni-Co LDH	Carbon fibers	KOH/PVA 1.5 V	319 F/g at 2 A/g	100 W h/kg	15 kW/kg	98.6% after 3000 cy.	Pang et al. 2017
Co_3O_4-nanocube/$Co(OH)_2$-nanosheet	AC	KOH/PVA 1.4 V	210 mF cm^{-2} at 0.3 mA cm^{-2}	9.4 mW h cm^{-3}	354 mWcm^{-3}	97.6% after 5000 cy.	Feng et al. 2015
Mn_3O_4/NGP	$Ni(OH)_2$/NGP	NaOH/PVA 1.3 V	1.96 F cm^{-3} at 50 mV/s	0.35 mW h cm^{-3}	32.5 mWcm^{-3}	83% after 12000 cy.	Kong et al. 2015
Co_3O_4@C@Ni_3S_2	AC	KOH/PVA 1.8 V		1.52 mW h cm^{-3}	60000 mWcm^{-3}	91.43% after 10000 cy.	Zhang et al. 2015b
$Co_{11}(HPO_3)_8(OH)_6$-Co_3O_4	Graphene	KOH/PVA 1.38 V	1.84 F cm^{-3} at 0.5 mA cm^{-3}	0.48 mW h cm^{-3}	105 mWcm^{-3}	98.7% after 2000 cy.	Zhang et al. 2015c
CF@RGO@MnO_2	CF@TRGO	KCl/PAAK 1.6 V		1.23 mW h cm^{-3}	270 mWcm^{-3}	91% after 10000 cy.	Shinde et al. 2016
CuO	Fe_2O_3	Na_2SO_4/CMC 2.0	79 F/g at 2 mA cm^{-2}	23 W h/kg	19 kW/kg	90% after 1000 cy.	Wang et al. 2017c
$NiCo_2Se_4$	$NiCo_2O_4$@PPy	KOH/PVA 1.7 V	14.2 F cm^{-3} at 63.7 mA cm^{-3}	5.18 mW h cm^{-3}	260 Wcm^{-3}	94% after 5000 cy.	Xie et al. 2015
Cobalt carbonate hydroxide/ N-doped graphene	N-Doped graphene	KOH/PVA 1.9 V	153.5 mF cm^{-2} at 1.0 mA cm^{-2}	0.077 mW h cm^{-3}	2.5 mWcm^{-2}	93.6% after 2000 cy.	

3.4　SUMMARY

The hybrid supercapacitor (HSC) has emerged as an attractive energy storage candidate having the potential to replace the batteries in different applications. The HSC strengthens its candidature due to high energy density, high power density, fast charge-discharge, and long term cyclic stability. Further research is required to further boost the energy density and ion storage capacity so that it can be used in transportation as well as in the portable electronics sector. The main focus is to enhance the energy density without any loss of power density and may be achieved by tuning the morphology of the electrode material and by forming composites using nanostructures. The optimum combination of the positive and negative electrode and electrolyte will lead to fulfilling the goal. Still, challenges are there that need to be tackled. First, the insights into the ion storage mechanics in battery type electrode materials need to be explored. Second, the operating voltage window of the hybrid device needs to be improved by the selection of suitable electrolyte. Also, one important research path is towards the development of flexible HSC for portable electronics and can be achieved by adopting polymer electrolytes for device fabrication. So, in summary, new dimensions or paths need to be explored for further enhancement of hybrid supercapacitor performance by tailoring the device architecture and components (electrode, electrolyte).

3.5　ACKNOWLEDGMENT

One of the authors Shweta Tanwar is thankful to CSIR India for JRF fellowship.

3.6　REFERENCES

Akinwolemiwa, B., C. Peng and G.Z. Chen. 2015. Redox electrolytes in supercapacitors. J. Electrochem. Soc. 162: A5054-A5059.

Akinwolemiwa, B. and G.Z. Chen. 2018. Fundamental consideration for electrochemicalengineering of supercapattery. J. Braz. Chem. Soc. 29(5): 960-972.

Alagar, S., R. Madhuvilakku, R. Mariappan, C. Karuppiah, C.C. Yang and S. Piraman. 2020. Ultra-stable $Mn_{1-x}Ni_xCO_3$ nano/sub-microspheres positive electrodes for high-performance solid-state asymmetric supercapacitors. Sci. Rep. 10: 1-13.

Arbizzani, C., Y. Yu, J. Li, J. Xiao, Y.Y. Xia, Y. Yang, et al. 2020. Good practice guide for papers on supercapacitors and related hybrid capacitors for the Journal of Power Sources 450: 227636.

Arya, A. and A.L. Sharma. 2019. Electrolyte for energy storage/conversion (Li+, Na+, Mg2+) devices based on PVC and their associated polymer: A comprehensive review. J. Solid State Electrochem. 23: 997-1059.

Arya, A. and A.L. Sharma. 2020. A glimpse on all-solid-state Li-ion battery (ASSLIB) performance based on novel solid polymer electrolytes: A topical review. J. Mater. Sci. 55: 6242-6304.

Augustyn, V., J. Come, M.A. Lowe, J.W. Kim, P.L. Taberna, S.H. Tolbert, et al. 2013. High-rate electrochemical energy storage through Li+ intercalation pseudocapacitance. Nat. Mater. 12: 518.

Brousse, T., D. Bélanger and J.W. Long. 2015. To be or not to be pseudocapacitive? J. Electrochem. Soc. 162: A5185-A5189.

Chen, G.Z. 2017. Supercapacitor and supercapattery as emerging electrochemical energy stores. Int. Mater. Rev. 62: 173-202.

Chen, H., Z. Yan, X.Y. Liu, X.L. Guo, Y.X. Zhang and Z.H. Liu. 2017. Rational design of microsphere and microcube $MnCO_3@MnO_2$ heterostructures for supercapacitor electrodes. J. Power Sources. 353: 202-209.

Chen, S., L. Ma, K. Zhang, M. Kamruzzaman, C. Zhi and J.A. Zapien. 2019. A flexible solid-state zinc ion hybrid supercapacitor based on co-polymer derived hollow carbon spheres. J. Mater. Chem. A. 7: 7784-7790.

Cheng, S., T. Shi, Y. Huang, X. Tao, J. Li, C. Cheng, et al. 2017. Rational design of nickel cobalt sulfide/oxide core-shell nanocolumn arrays for high-performance flexible all-solid-state asymmetric supercapacitors. Ceram. Int. 43: 2155-2164.

Conway, B.E. 1991. Transition from "supercapacitor" to "battery" behavior in electrochemical energy storage. J. Electrochem. Soc. 138: 1539.

Das, H.T., K. Mahendraprabhu, T. Maiyalagan and P. Elumalai. 2017. Performance of solid-state hybrid energy-storage device using reduced graphene-oxide anchored sol-gel derived Ni/NiO nanocomposite. Sci. Rep. 7: 1-14.

Devaraj, S., H.Y. Liu and P. Balaya. 2014. $MnCO_3$: A novel electrode material for supercapacitors. J. Mater. Chem. A. 2: 4276.

El-Kady, M.F., V. Strong, S. Dubin and R.B. Kaner 2012. Laser scribing of high-performance and flexible graphene-based electrochemical capacitors. Science 335(6074): 1326-1330.

El-Kady, M.F., M. Ihns, M. Li, J.Y. Hwang, M.F. Mousavi, L. Chaney, et al. 2015. Engineering three-dimensional hybrid supercapacitors and microsupercapacitors for high-performance integrated energy storage. Proceedings of the National Academy of Sciences 112: 4233-4238.

Esfandiar, A., M. Qorbani, I. Shown and B.O. Dogahe. 2020. A stable and high-energy hybrid supercapacitor using porous Cu_2O-$Cu_{1.8}S$ nanowire arrays. J. Mater. Chem. A. 8: 1920-1928.

Fan, Y.M., W.L. Song, X. Li and L.Z. Fan. 2017. Assembly of graphene aerogels into the 3D biomass-derived carbon frameworks on conductive substrates for flexible supercapacitors. Carbon 111: 658-666.

Feng, J.X., S.H. Ye, X.F. Lu, Y.X. Tong and G.R. Li. 2015. Asymmetric paper supercapacitor based on amorphous porous Mn_3O_4 negative electrode and $Ni(OH)_2$ positive electrode: A novel and high-performance flexible electrochemical energy storage device. ACS Appl. Mater. Interfaces. 7: 11444-11451.

Gogotsi, Y. and R.M. Penner. 2018. Energy storage in nanomaterials – capacitive, pseudocapacitive, or battery-like? ACS Nano. 12: 2081-2083.

Ho, J., T.R. Jow and S. Boggs. 2010. Historical introduction to capacitor technology. IEEE Electr. Insul. Mag. 26: 20-25.

Jana, M., P. Samanta, N. Chandra Murmu and T. Kuila. 2017. Morphology controlled synthesis of $MnCO_3$ – RGO materials and their supercapacitor applications. J. Mater. Chem. A. 5: 12863-12872.

Kong, D., C. Cheng, Y. Wang, J.I. Wong, Y. Yang and H.Y. Yang. 2015. Three-dimensional Co_3O_4@C@Ni_3S_2 sandwich-structured nanoneedle arrays: Towards high-performance flexible all-solid-state asymmetric supercapacitors. J. Mater. Chem. A. 3: 16150-16161.

Kumar, A., M. Madaan, A. Arya, S. Tanwar and A.L. Sharma. 2020. Ion transport, dielectric, and electrochemical properties of sodium ionconducting polymer nanocomposite: Application in EDLC. J. Mater. Sci.: Mater. Electron. 31: 10873-10888.

Lee, S.H., Y. Kwon, S. Park, M. Cho and Y. Lee. 2015. Facile synthesis of $MnCO_3$ nanoparticles by supercritical CO_2 and their conversion to manganese oxide for supercapacitor electrode materials. J. Mater. Sci. 50: 5952-5959.

Lee, S.H., J.H. Kim and J.R. Yoon. 2018. Laser scribed graphene cathode for next generation of high performance hybrid supercapacitors. Sci. Rep. 8: 1-9.

Lei, Z., F. Shi and L. Lu. 2012. Incorporation of MnO_2-coated carbon nanotubes between graphene heets as supercapacitor electrode. Appl. Mater. Interfaces. 4: 1058-1064.

Li, H., Y. Hou, F. Wang, M.R. Lohe, X. Zhuang, L. Niu, et al. 2017a. Flexible all-solid-state supercapacitors with high volumetric capacitances boosted by solution processable MXene and electrochemically exfoliated graphene. Adv. Energy Mater. 7: 1601847.

Li, Y., X. Wang, Q. Yang, M.S. Javed, Q. Liu, W. Xu, et al. 2017b. Ultra-fine CuO nanoparticles embedded in three-dimensional graphene network nano-structure for high-performance flexible supercapacitors. Electrochim. Acta. 234: 63-70.

Li, Z., Y. Li, L. Wang, L. Cao, X. Liu, Z. Chen. et al. 2017c. Assembling nitrogen and oxygen co-doped graphene quantum dots onto hierarchical carbon networks for all-solid-state flexible supercapacitors. Electrochim. Acta 235: 561-569.

Liu, Y., N. Fu, G. Zhang, M. Xu, W. Lu, L. Zhou, et al. 2017. Design of hierarchical Ni-Co@Ni-Co layered double hydroxide core-shell structured nanotube array for high-performance flexible all-solid-state battery-type supercapacitors. Adv. Funct. Mater. 27: 1605307.

Lu, K., J. Zhang, Y. Wang, J. Ma, B. Song and H. Ma. 2017. Interfacial deposition of three-dimensional nickel hydroxide nanosheet-graphene aerogel on Ni wire for flexible fiber asymmetric supercapacitors. ACS Sustain. Chem. Eng. 5: 821-827.

Makino, S., Y. Shinohara, T. Ban, W. Shimizu, K. Takahashi, N. Imanishi, et al. 2012. 4 V class aqueous hybrid electrochemical capacitor with battery-like capacity. RSC Advances 2(32): 12144-12147.

Moon, I.K., S. Yoon and J. Oh. 2017. Three-dimensional hierarchically mesoporous $ZnCo_2O_4$ nanowires grown on graphene/sponge foam for high-performance, flexible, all-solid-state supercapacitors. Chem. Eur. J. 23: 597-604.

Nagamuthu, S. and K.S. Ryu. 2019. Synthesis of Ag/NiO honeycomb structured nanoarrays as the electrode material for high performance asymmetric supercapacitor devices. Sci. Rep. 9: 1-11.

Nan, H.S., X.Y. Hu and H.W. Tian. 2019. Recent advances in perovskite oxides for anion-intercalation supercapacitor: A review. Mat. Sci. Semicon. Proc. 94: 35-50.

Oyedotun, K.O., D.Y. Momodu, M. Naguib, A.A. Mirghni, T.M. Masikhwa, A.A. Khaleed, et al. 2019. Electrochemical performance of two-dimensional Ti_3C_2-Mn_3O_4 nanocomposites and carbonized iron cations for hybrid supercapacitor electrodes. Electrochim. Acta. 301: 487-499.

Oyedotun, K.O., M.J. Madito, A. Bello, D.Y. Momodu, A.A. Mirghni and N. Manyala. 2017. Investigation of graphene oxide nanogel and carbon nanorods as electrode for electrochemical supercapacitor. Electrochim. Acta. 245: 268-278.

Pal, B., S. Yang, S. Ramesh, V. Thangadurai and R. Jose. 2019. Electrolyte selection for supercapacitive devices: A critical review. Nanoscale Adv. 1: 3807-3835.

Pan, Z., H. Zhi, Y. Qiu, J. Yang, L. Xing, Q. Zhang, et al. 2018. Achieving commercial-level mass loading in ternary-doped holey graphene hydrogel electrodes for ultrahigh energy density supercapacitors. Nano Energy 46: 266-276.

Panda, P.K., A. Grigoriev, Y.K. Mishra and R. Ahuja. 2020. Progress in supercapacitors: Roles of two dimensional nanotubular materials. Nanoscale Adv. 2: 70-108.

Pang, H., X. Li, Q. Zhao, H. Xue, W.Y. Lai, Z. Hu, et al. 2017. One-pot synthesis of heterogeneous Co_3O_4-nanocube/$Co(OH)_2$-nanosheet hybrids for high-performance flexible asymmetric all-solid-state supercapacitors. Nano Energy 35: 138-145.

Qiu, Y., X. Li, M. Bai, H. Wang, D. Xue, W. Wang, et al. 2017. Flexible full-solid-state supercapacitors based on self-assembly of mesoporous $MoSe_2$ nanomaterials. Inorg. Chem. Front. 4: 675-682.

Rajendiran, R., D. Chinnadurai, A.R. Selvaraj, R.K. Gunasekaran, H.J. Kim, S. Karupannan, et al. 2019. Nickel self-doped iron oxide/manganese carbonate hierarchical 2D/3D structures for electrochemical energy storage. Electrochim. Acta. 297: 77-86.

Ramesh, S., K. Karuppasamy, H. Seok Kim, H. Soo Kim and J.H. Kim. 2018. Hierarchical flowerlike 3D nanostructure of Co_3O_4@MnO_2/N-doped graphene oxide (NGO) hybrid composite for a high-performance supercapacitor. Sci. Rep. 8: 1-11.

Ramesh, S., H.M. Yadav, Y.J. Lee, G.W. Hong, A. Kathalingam, A. Sivasamy, et al. 2019a. Porous materials of nitrogen doped graphene oxide@SnO_2 electrode for capable supercapacitor application. Sci. Rep. 9: 1-10a.

Ramesh, S., K. Karuppasamy, H.M. Yadav, J.J. Lee, H. Seok Kim, H. Soo Kim, et al. 2019b. $Ni(OH)_2$-decorated nitrogen doped MWCNT nanosheets as an efficient electrode for high performance supercapacitors. Sci. Rep. 9: 1-10b.

Salanne, M., B. Rotenberg, K. Naoi, K. Kaneko, P.L. Taberna, C.P. Grey, et al. 2016. Efficient storage mechanisms for building better supercapacitors. Nat. Energy 1: 16070.

Sharma, K., A. Arora and S.K. Tripathi. 2019. Review of supercapacitors: Materials and devices. J. Energy Storage 21: 801-825.

Sharma, M. and A. Gaur. 2020. Designing of carbon nitride supported $ZnCO_2O_4$ hybrid electrode for high-performance energy storage applications. Sci. Rep. 10: 1-9.

Shinde, A.V., N.R. Chodankar, V.C. Lokhande, A.C. Lokhande, T. Ji, J.H. Kim, et al. 2016. Highly energetic flexible all-solid-state asymmetric supercapacitor with Fe_2O_3 and CuO thin films. RSC Advances 6: 58839-58843.

Subramani, K., D. Jeyakumar and M. Sathish. 2014. Manganese hexacyanoferrate derived Mn_3O_4 nanocubes–reduced graphene oxide nanocomposites and their charge storage characteristics in supercapacitors. Phys. Chem. Chem. Phys. 16: 4952.

Subramani, K., N. Sudhan, R. Divya and M. Sathish. 2017. All-solid-state asymmetric supercapacitors based on cobalt hexacyanoferrate-derived CoS and activated carbon. RSC Advances 7: 6648-6659.

Tahir, M., C. Cao, N. Mahmood, F.K. Butt, A. Mahmood, F. Idrees, et al. 2014. Multifunctional g-C_3N_4 nanofibers: A template-free fabrication and enhanced optical, electrochemical, and photocatalyst properties. ACS Appl. Mater. Interfaces. 6: 1258-1265.

Tang, Y., J. Zhu, C. Yang and F. Wang. 2016. Enhanced capacitive performance based on diverse layered structure of two-dimensional Ti_3C_2 MXene with long etching time. J. Electrochem. Soc. 163: A1975.

Wang, H., H. Yi, X. Chen and X. Wang. 2014. Asymmetric supercapacitors based on nano-architectured nickel oxide/graphene foam and hierarchical porous nitrogen-doped carbon nanotubes with ultrahigh-rate performance. J. Mater. Chem. A. 2: 3223.

Wang, H., J. Deng, C. Xu, Y. Chen, F. Xu, J. Wang, et al. 2017a. Ultra-microporous carbon cloth for flexible energy storage with high areal capacitance. Energy Storage Mater. 7: 216-221.

Wang, J., J. Polleux, J. Lim and B. Dunn. 2007. Pseudocapacitive contributions to electrochemical energy storage in TiO_2 (anatase) nanoparticles. J. Phys. Chem. C. 111: 14925.

Wang, J., L. Zhang, X. Liu, X. Zhang, Y. Tian, X. Liu, et al. 2017b. Assembly of flexible $CoMoO_4@NiMoO_4 \cdot xH_2O$ and Fe_2O_3 electrodes for solid-state asymmetric supercapacitors. Sci. Rep. 7: 41088.

Wang, J.G., F. Kang and B. Wei. 2015. Engineering of MnO_2-based nanocomposites for high-performance supercapacitors. Prog. Mater. Sci. 74: 51.

Wang, Q., Y. Ma, Y. Wu, D. Zhang and M. Miao. 2017c. Flexible asymmetric threadlike supercapacitors based on $NiCo_2Se_4$ nanosheet and $NiCo_2O_4$/polypyrrole electrodes. Chem. Sus. Chem. 10: 1427-1435.

Xie, H., S. Tang, J. Zhu, S. Vongehr and X. Meng. 2015. A high energy density asymmetric all-solid-state supercapacitor based on cobalt carbonate hydroxide nanowire covered N-doped graphene and porous graphene electrodes. J. Mater. Chem. A. 3: 18505-18513.

Xu, J. and L. Wang. 2019. Fabrication of hierarchical core/shell $MgCo_2O_4@MnO_2$ nanowall arrays on Ni-foam as high-rate electrodes for asymmetric supercapacitors. Sci. Rep. 9: 1-11.

Yang, L., X. Lu, S. Wang, J. Wang, X. Guan, X. Guan, et al. 2020. Designed synthesis of nickel-cobalt-based electrode materials for high-performance solid-state hybrid supercapacitors. Nanoscale 12: 1921-1938.

Yang, Q., Z. Li, R. Zhang, L. Zhou, M. Shao and M. Wei. 2017. Carbon modified transition metal oxides/hydroxides nanoarrays toward high-performance flexible all-solid-state supercapacitors. Nano Energy 41: 408-416.

Yu, L. and G.Z. Chen. 2016. Redox electrode materials for supercapatteries. J. Power Sources. 326: 604-612.

Zhang, C., W. Zhang, S. Yu, D. Wang, W. Zhang, W. Zheng, et al. 2017. Unlocking the electrocatalytic activity of chemically inert amorphous carbon-nitrogen for oxygen reduction: Discerning and refactoring chaotic bonds. ChemElectroChem. 4: 1-6.

Zhang, S. and N. Pan. 2015a. Supercapacitors performance evaluation. Adv. Energy Mater. 5: 1401401.

Zhang, Y., M. Zheng, M. Qu, M. Sun and H. Pang. 2015b. Core-shell $Co_{11}(HPO_3)_8(OH)_6$–Co_3O_4 hybrids for high-performance flexible all-solid-state asymmetric supercapacitors. J. Alloys Compd. 651: 214-221.

Zhang, Z., F. Xiao, J. Xiao and S. Wang. 2015c. Functionalized carbonaceous fibers for high performance flexible all-solid-state asymmetric supercapacitors. J. Mater. Chem. A. 3: 11817-11823.

Zhao, B., D. Chen, X. Xiong, B. Song, R. Hu, Q. Zhang, et al. 2017. A high-energy, long cycle-life hybrid supercapacitor based on graphene composite electrodes. Energy Storage Mater. 7: 32-39.

Zheng, Y., Z. Lin, W. Chen, B. Liang, H. Du, R. Yang, et al. 2017. Flexible, sandwich-like CNTs/$NiCo_2O_4$ hybrid paper electrodes for all-solid state supercapacitors. J. Mater. Chem. A 5: 5886-5894.

Zhi, M., C. Xiang, J. Li, M. Li and N. Wu. 2013. Nanostructured carbon-metal oxide composite electrodes for supercapacitors: A review. Nanoscale 5: 72-88.

Zhong, C., Y. Deng, W. Hu, J. Qiao, L. Zhang and J. Zhang. 2015. A review of electrolyte materials and compositions for electrochemical supercapacitors. Chem. Soc. Rev. 44: 7484-7539.

4

Nanocontainers to Increase the Absorption of Energy and Heat Conversion

George Kordas

4.1 INTRODUCTION

The energy dependence on oil, combined with its limited reserves and the burden of the environment caused by combustion, require the development of new technologies for absorption of solar energy and the creation of methods of storage. We know that energy is not created or destroyed, but it is preserved, while it varies from one form to another. This is the principle that energy storage technology exploits from a period of low demand in a high demand period. This also applies to the absorption of solar energy from selective absorption and heat conversion surfaces. Energy storage is essential in processes where energy supply and demand vary. The power supply is manifested when there is a supply and its demand is required by the needs of the network. Renewable energy sources can make up for high-cost fossil fuels, as storage capacity makes them viable solutions. The reduced energy limits takes place in various ways, as with the compensation of previously discarded energy that can be stored for future handling. The production of energy from intermittent renewable sources (e.g. wind, solar) without storage capacity will not be technically applicable unless at the same time ways of storing the energy produced are found. In other words, thermal storage is essential for the utilization of renewable energy sources. This saves energy and increases the performance of systems in different ways. Excessive heating and cooling leads to lost energy that can be saved by the thermal storage that is achieved with Phase Change Materials (PCM). The application of this technology is particularly important in areas where electricity costs are high, the electrical supply is impractical, etc. The application of PCMs brings financial revenues. In recent years, PCMs are applied to buildings that can store a large amount of heating and cooling, making use of latent heat gained/released when changing the phase. Thermal energy storage is achieved by changing the internal energy of a material as a heat transfer, latent heat and thermochemical or a combination of these. Phase Change Slurries (PCS) are liquids with PCMs which have developed over the last few years for closed heat transfer circuits. PCSs provide liquidity and ability to accumulate heat in a heating and cooling circuit. To date, few PCS applications have been implemented and exploited commercially. A PCM with a melting temperature of 60°C has showed a smaller natural coefficient of thermal permeability compared to water. PCS can contain up to 40% nanocapsules.

Peter the Great St. Petersburg Polytechnic University, St. Petersburg, Russian Federation.
E-mail: gckordas@gmail.com

The inclusion of storage systems in the market imposes accurate knowledge of how PCSs operate within commercial-scale applications. Microcapsules can also be used as sun light microtraps to increase the absorption of solar energy from a solar-absorbing surface. This technology can be used in water heaters. Devises are now in the market in operating conditions. Some reviews present such applications (Kenisarin and Mahkamov 2007, Nader et al. 2015).

Here, we review phase change memory change (PCM) materials and how nano/micro-containers can be used to encapsulate these PCMs to enhance performance and stability of PCMs. Furthermore, the use of light microtraps is reported as a means to enhance the absorption efficiency of the spinel coatings.

4.2   THERMAL ENERGY STORAGE

The storage of thermal energy is achieved by changing the internal energy of a material as heat transmission, latent heat and thermochemical or blend therefrom. Figure 4.1 briefly describes the methods of thermal energy storage that we describe in the following text.

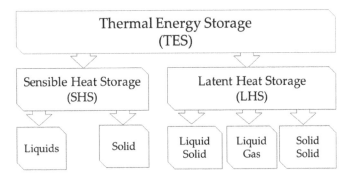

Fig. 4.1   Categories of materials used for energy storage

4.2.1   Sensible Heat Storage (SHS)

The method is based on storing energy by increasing the temperature of a solid or liquid. Storage is done with the heat capacity and the change in the temperature of the material through the charging and discharging process.

$$Q = m C_p (T_f - T_i) \tag{1}$$

where the parameters are: Q is the heat, T_i the initial temperature, T_f the final temperature, m the mass of the material, and C_p the specific heat of the material.

The water has very good storage capacity but up to 100°C. For temperatures higher than 100°C, salts and liquid metals are employed. Table 4.1 gathers some examples of this type of material.

TABLE 4.1   Examples of Sensible Heat Storage (SHS)

Material	Temperature°C	Density (kg/m³)	Specific Heat Capacity (J/kg K)
Stone	20	2560	879
Water	0-100	1000	4190
Engine oil	Till 160	888	1880
Propanol	Till 97	800	2500
Octane	Till 126	704	2400

4.2.2   Latent Heat Storage (LHS)

Latent heat storage is founded on heat capture or release via PCM phase change. The capacity of the system with moderate PCM is given by the formula:

$$Q = m \left[C_{sp}(T_m - T_i) + a_m \Delta h_m + C_{lp}(T_f - T_m) \right] \tag{2}$$

where the parameters of the equation are: m is the mass of PCM melted, Δh_m, C_{sp} and C_{lp} the specific heat of solid and liquid PCM at temperatures $T_m - T_l$ and $T_h - T_m$, respectively.

The equation shows the benefits of LHS systems with the utmost significant being the maximum energy density per unit of mass and volume.

4.2.2.1   Implications

If we take a solar collector and heat a liquid, the heat can be transferred to a system like a water boiler, an energy storage house, a cooking stove, etc. and we'll take a temperature increase equal to T_{f_1} in the temperature device (Fig. 4.2A) determined by the equation (1). In the case where the liquid is enriched with PCM, the condition changes and in the energy release, system (Fig. 4.2B) will acquire T_{f_2} temperature imposed by the equation (2). With such liquids, which we call nanofluids, we can create many applications that can be an air-conditioned house, or new building materials for the construction of housing that will climate the building in the desired comfort temperature avoiding the energy charge from energy sources such as electricity, oil, etc.

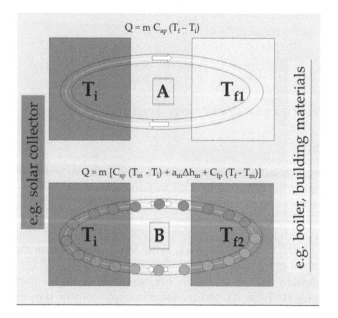
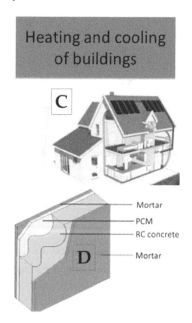

Fig. 4.2   (A) Transport of heat from a solar collector to a device, (B) Transport of heat from a solar collector to a device using PCM, (C) Heating and cooling a house via PCM, (D) Building block using PCM

4.2.2.2   Paraffins

Paraffin consists of a chain CH_3-(CH_2)-CH_3 that the section (CH_3) crystallizes releasing a large latent heat. The melting point and fusion latent heat depends on the length of the chain as shown in Fig. 4.3.

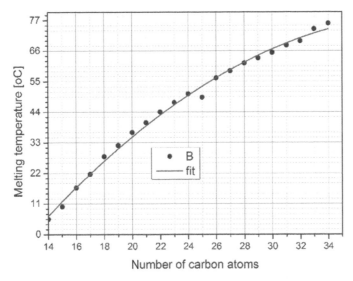

Fig. 4.3 Paraffin melting temperature depending on carbon atoms' number

4.2.2.3 Non-paraffins

Non-paraffin is a PCM that becomes energy storage. They can be esters, fatty acids, alcohols and glycols. Some of the characteristics of these organic materials are: (i) high heat of melting, (ii) some are flammable, (iii) have a low thermal conductivity, (iv) have low ignition points, (v) have a different level of toxicity, and (vi) the instability at high temperatures. Table 4.2 gives some data of these materials.

TABLE 4.2 Examples of non-paraffin a PCM that becomes energy storage

	Compound	T Melting [°C]	Heat of Fusion [kJ/kg]	Thermal Conductivity [W/m K]	Density [kg/m³]
Inorganic	$MgCl_2 6H_2O$	117	168.6	0.570 (liquid, 120°C), 0.694 (solid, 90°C)	1450 (liquid, 120°C), 1569 (solid, 20°C)
	$Mg(NO_3)_2 6H_2O$	89	162.8	0.490 (liquid, 95°C), 0.611 (solid, 37°C)	1550 (liquid, 94°C), 1636 (solid, 25°C)
	$Ba(OH)_2 8HO$	48	265.7	0.653 (liquid, 85.7°C), 1.225 (solid, 23°C)	1937 (liquid, 84°C), 2070 (solid, 24°C)
	$CaCl_2 6H_2O$	29	190.8	0.540 (liquid, 38.7°C), 1.088 (solid, 23°C)	1562 (liquid, 32°C), 1802 (solid, 24°C)
Organics	Paraffin wax	64	173.6	0.167 (liquid, 63.5°C), 0.346 (solid, 33.6°C)	790 (liquid, 65°C), 916 (solid, 24°C)
	Polyglycol E600	22	127.2	0.189 (liquid, 38.6°C)	1126 (liquid, 25°C), 1232 (solid, 4°C)
Fatty acids	Palmitic acid	64	185.4	0.162 (liquid, 68.4°C)	850 (liquid, 25°C), 989 (solid, 4°C)
	Capric acid	32	152.7	0.153 (liquid, 38.5°C)	878 (liquid, 25°C), 1004 (solid, 4°C)
	Caprylic acid	16	148.5	0.149 (liquid, 38.6°C)	901 (liquid, 25°C), 981 (solid, 4°C)
Aromatics	Naphthalene	80	147.7	0.132 (liquid, 83.8°C), 0.341 (solid, 49.9°C)	976 (liquid, 25°C), 1145 (solid, 4°C)

4.2.2.4 *Inorganic Phase Change Materials*

Inorganic materials fall further into hydrated salt and minerals. These phase-change materials are not significantly thawed and the melting heat is not altered by available transition. Salt hydrates contain a crystalline solid of the generic type $AB-nH_2O$. The mechanism is based on the dehydration of salt water that resembles thermodynamic melting or cooling.

4.2.2.5 *Phase Change Materials (PCM)*

Heat storage in materials based on "phase" change is called "latent" heat storage materials and occurs when a material is transformed from solid to liquid or liquid into the solid. The temperature in these materials increases as they absorb heat. PCM absorbs and emits heat 5-14 times more per unit volume than water, masonry, or rock (Fig. 4.4A). Figure 4.4A shows the energy stored in a PCM material, in water and in rock. While in water and in rock the energy stored varies linearly with the heat (Fig. 4.4A), in a PCM at the melting point a phase change ratio jump is presented demonstrating the material's ability to store the energy corresponding to the latent heat. For the use of these materials, some desirable thermodynamic, kinetic and chemical properties must be present. In addition, there are economic criteria for their use. Financial estimates and easy availability of these materials should be taken into account. The appropriate phase transition temperature is determined by the respective application where there must be high latent heat of transition and good heat transmission (Fig. 4.4B). Apart from thermodynamic parameters, that must be met, some physical properties of which are favorable phase equilibrium, high density and small volume changes. Another criterion is that paraffins do not leak from the porosity of nanocontainers encapsulated during the stay in the wet phase. The nanocontainers must be sealed to keep the paraffin inside. Also, the paraffins create a thermal insulating wall on the surface of the nanocontainers during cooling, thus this wall prevents the heat flow from the liquid phase towards the outside of the nanocontainers. These materials should not be supercooling and have an adequate crystallization rate. They must not be toxic, have long-term chemical stability and not be ignited. The main factor for their application is their abundance and the low cost of production.

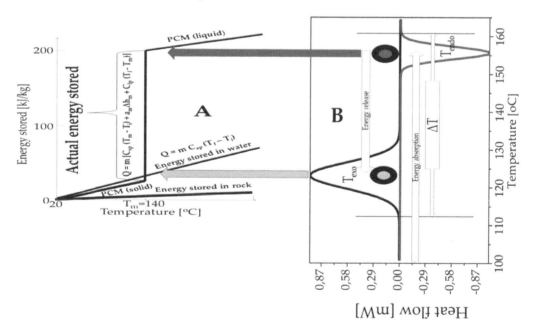

Fig. 4.4 Energy stored in a PCM material

4.2.2.6 Sorting PCM Materials

PCM is classified into organic, inorganic and eutectic. Figure 4.5 illustrates categories of materials.

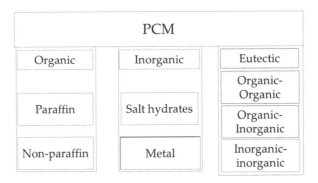

Fig. 4.5 Categories of PCM materials

4.2.2.7 Latent Heat Determination

The techniques used to measure latent heat and melting temperature of PCMs are the Differential Thermal Analysis (DTA), and the Differential Scan Calorimetry (DSC). The sample and a material are heated at a constant rate and the temperature difference between the sample and reference material is measured and the DSC curve is recorded. The reference material must be alumina. Latent heat of fusion is calculated using the area below the top of the melting curve and calculated from the tangent at the base of the curve (Fig. 4.4B).

4.3 SOLAR WATER HEATING SYSTEMS

PCMs can be applied to solar water heaters as they are quite cheap materials without prohibitively increasing the cost of installation. The PCM system only increases the efficiency of water heater and insulation in relation to polyurethane. When the water is heated during the hours of sunshine, the excess heat is stored in the PCM when the temperature reaches the melting point of the material. Excess energy stored in PCMs is done in the form of latent heat and melts. During the long sunshine, the hot water subsides and is replaced by cold water, which gains energy from PCM. The energy released by PCM leads to a change of phases from liquid to solid. PCM can be integrated into the thermal insulation system. Hot water in larger systems can be transferred to a heat storage system which will return to hours without sun where needed (Fig. 4.6).

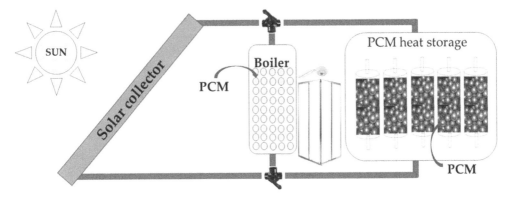

Fig. 4.6 Solar water PCM heating system

4.4   BUILDING

PCMs will be developed for thermal storage in buildings. PCM will be integrated into walls, plasterboard, shutters, insulating materials, and floor heating systems as portion of the structure for heating and cooling applications. The purpose is to shift the peak load and utilize the solar energy. The claim of PCMs in the building can have two dissimilar purposes. First, the use of natural heat which is solar energy for heating or cold at night for cooling. Second, taking advantage of man-made heat or cold sources. In any case, storing heat or cold is necessary to match availability and demand in terms of time. PCMs will be used to build a wall, other elements of the building outside the walls, in storage units, on plasterboard, and in colors. Figure 4.7A shows a relative application with the corresponding energy gain (Fig. 4.7B).

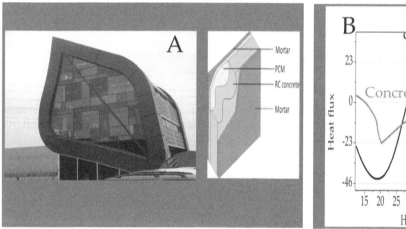
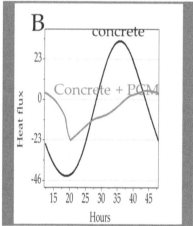

Fig. 4.7   PCMs in buildings

4.5   SOLAR GREEN LODGE

Phase change materials will be produced to be used in closed-type greenhouses and fish farms for the storage of solar energy for the treatment and drying process and plant production, spirulina production and shrimp. PCM type $CaC_{12}\text{-}6H_2O$ in aerosol cans will be used for energy storage, both inside and outside the installation. Buildings will use these facilities for air conditioning.

4.6   SOLAR COOKERS

Solar cookers are one of the most important applications of PCM that can be used for cooking and at times when there is no sun. The kitchen will use PCM with latent heat materials with storage materials in a solar kitchen type box to cook food late at night. PCM can be hexahydrate magnesium nitrate $(Mg\,(NO_3)^2\text{-}6H_2O)$ as a PCM for heat storage. A sketch of a solar cooking device is shown in Fig. 4.8.

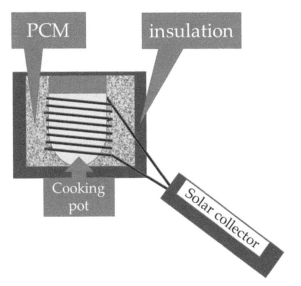

Fig. 4.8   Solar cooking system

4.7   THEORETICAL STUDIES

The various structures will be accompanied with similar theoretical calculations for the optimum use of constructions. For a phase-change process involving either melting or cooling, energy conservation will be expressed with enthalpy and temperature. For stable thermophysical properties of refrigerants, it will be expressed as follows:

$$\frac{\partial H}{\partial \tau} = \Delta(k(\Delta T)) \tag{3}$$

where the parameters are: k is the thermal conductivity of the material in the gradient of the heat ΔT.

Commercial fluid engineering programs will be used to simulate the quality of systems and optimize them.

4.8   PCM PRODUCTION METHODS

The implementation of the material phase change (PCMs) in the storage of thermal energy gained great attention due to the increase of energy consumption and the rescue of the environment from pollution. PCMs absorb, store and release large amounts of latent heat at specified temperature range while phase changes improve device energy efficiency. Depending on the application, the size of PCMs is selected. Typically, EPCMs are classified in NanoPCMs, MicroPCMs and MacroPCMs depending on the diameter. The size of the MicroPCMs usually varies from 1 mm to 1 mm, while capsules less than 100 nm are classified as NanoPCMs and capsules greater than 1 mm as MacroPCMs. Encapsulated Phase Change Materials (EPCMs) consist of PCMs with polymer core and inorganic shell. Microcapsules and nanocapsules containing N-Octadecane in the melamine-formaldehyde shell are manufactured from spot polymerization. The effects of stirring, the content of the emulsifier and contents of the cyclohexane in diameters, morphology, phase change properties and thermal stability of PCMs are studied using FT-IR, SEM, DSC and TG. For mass production, techniques such as spay drying will be used. The capsules can also be produced with sol-gel techniques (Belessiotis et al. 2018). Sizes of capsules can be obtained around 300 nm or 600 nm depending on the synthesis conditions. Paraffin can be used to fill the containers demonstrating PCM behavior. Long-term use requires a great stability of their properties after a long cycle of heat changes. PCM must have geometric and thermal stability as PCMs must be able to undergo repeated cycles of heating and cooling. This test requires special devices that automatically change the temperature for many cooling – heating cycles. Many applications require the use of PCM with different melting temperatures. The heat-transfer coefficient of a single-phase liquid is increased by the addition of phase-change nanoparticles (nano-PCMs) that absorb thermal energy during solid-liquid phase changes. Silica encapsulated indium nanoparticles and polymer encapsulated paraffin particles in the range between 100 and 300 mm increase the nano-liquid heat transfer factor by 1,6 and 1,75 times compared to those of the corresponding single-phase liquids. Encapsulation seals the nanoparticles to prevent paraffin from being eliminated and the process is repeated for many cycles (Li et al. 2020). This improvement in the heat transfer coefficient of nano fluids has an impact on the size of the absorbent surface, water heating time in a water heater, etc. An innovative method was described to encapsulate high temperature PCM (salts and eutectics, $NaNO_3$, KNO_3, $NaNO_3$-KNO_3, $NaNO_3$-KNO_3-$LiNO_3$) that melt in the 120-350°C temperature range (Roget 2013). The study was started to manufacture encapsulated PCMs that can endure the highly corrosive environment of molten alkali metal nitrate-based salts and their eutectic. The established technique does not need a sacrificial layer to lodge the volumetric expansion of the PCMs on melting and reduces the chance of metal corrosion inside the capsule. The encapsulation consists of coating a non-reactive polymer over the PCM pellet followed by deposition of a metal layer by a novel non-vacuum (more practical and economically feasible) metal deposition technique (for large-scale fabrication of capsules utilizing commercially available electroless and electroplating chemistry). The fabricated capsules have survived more than 2200 thermal cycles (5133 h, equivalent to about seven years of power plant service). Thermal cycling test showed no significant degradation in thermophysical properties of the capsules and PCM on cycling at any stage of testing.

4.9   SOLAR ABSORBING MATERIALS

The solar selective surfaces on metals are widely used to absorb solar energy and convert it into heat in solar water heaters. In these applications, the selective surfaces must be high absorption, α, in the spectral range from 0.4 to 2.5 μm, low induction, ε, in the infrared range from 2.5 and 50 μm, and high selectivity, α/ε, to 100°C. Such a system should ideally have reflection in the area of the visible range $R = 0$ and in infrared $R = 1$ equivalent to the theoretical spectrum of black body radiation. The solar radiation extends from the wavelength of 250-2500 nm, while most of the thermal radiation is located in the infrared with wavelengths above 2500 nm. Figure 4.9 shows such an ideal case.

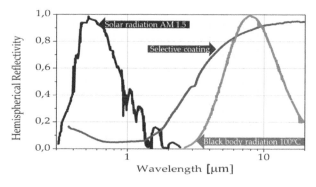

Fig. 4.9   Ideal case of a solar absorbing system

In 1955, Tabor produced an absorber coating by electro galvanization. Since that patent, the "black chrome" has become the standard product on the market (Tabor 1956, Wäckelgård 1998). Today, absorbing coatings are produced by vacuum-deposited coatings (Tabor 1956). Commercial products (sunselect) appeared in the market exhibiting α_{sol} and $\varepsilon_{100°C}$ equal to 0.95 ± 0.02 and 0.05

± 0.02, respectively (Shanker and Holloway 1985). The PVD production processes avoid wastes (no galvanic solutions necessary) and are reproducible produced of high quality. Coatings based on e.g. cobalt containing aluminium oxide have been prepared (https://alanod.com/. 2020. ALANOD GmbH and Co. KG, Germany). Other coatings were developed using nanostructured silver layers (Niklasson and Granqvist 1984), titanium oxynitride (Nishimura and Ishiguro 2004), chromium oxynitride (Lazarov et al. 1995), or metal containing amorphous hydrogenated carbon (Graf et al. 1997). In the last years, different routes have been chosen for the production of such coatings including sputtering (Frei and Brunold 2000), e-beam and evaporation techniques (Yin et al. 2009), and electrodeposition method (Barshilia et al. 2008). Absorber coatings have also been prepared via the sol–gel process, based on e.g. CoCuMnO$_x$ spinels (Kaluža et al. 2001). Spectral selective coatings of Ni-Al$_2$O$_3$/SiO$_2$ on aluminium have been prepared by the sol-gel method for use as solar absorbers in low temperature photo-thermal conversion (Lampert 1987). Other examples are coatings based on cobalt oxide and tin oxide (Vince et al. 2003) and on titanium-doped and non-doped CuCoMnO$_x$ spinel materials (Boström et al. 2003). Multiple absorber coatings produced by sol-gel method makes use of two surfaces: (i) the front surface of the coating with high absorptance in the solar wavelength range (0.3 to 3 µm) and low emittance (due to high transmittance) in the infrared wavelength range (3 to 50 µm), and (ii) the surface beneath with low emittance (due to high reflectance) in the same infrared wavelength range (Barrera et al. 2005). The tandem structure was optimized theoretically using the materials' parameters. Silica AR thin films as well as selectively absorbing films with varying nickel-to-alumina ratios were prepared by sol-gel processes. Selective coatings consist of nickel nanoparticles embedded in a dielectric alumina matrix on aluminum substrate.

A recent publication produced CuCoMnO$_x$ spinels over aluminum substrates via sol-gel processing (Boström et al. 2011). Before deposition, the metal was anodized to produce an uneven surface that would present diffuse reflection in the incident sunlight. In CuCoMnO$_x$ spinels coating, CuO@SiO$_2$ microspheres were incorporated with a total diameter of 980 ± 30 nm. The coatings were evaluated as α/ε ratio in relation to the content of the CuO@SiO$_2$ microspheres at concentrations between 0 and 1wt.%. Figure 4.10A shows the dependence of α/ε ratio on the concentration of light nanotraps showing a linear dependence. The value of α/ε ratio is double in a concentration of 1wt.% from that to zero concentration. The increase in the α/ε was explained by trapping light inside the spheres by converting its energy into thermal energy according to Fig. 4.10B (Kordas 2020). A similar case was certified in hollow ZnO microspheres that demonstrated multiple internal light reflections, thus increasing the UV/Vis diffuse reflectance spectra that was proportional to the number of internal microspheres (Dong et al. 2012).

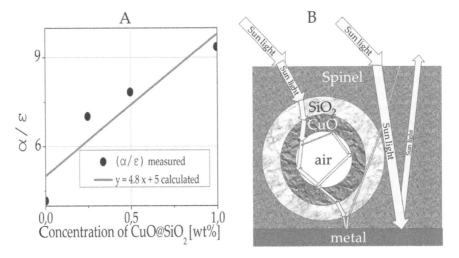

Fig. 4.10 α/ε of light microtraps

4.10 SUMMARY

The use of nano/microcontainers for PCM encapsulation and absorption of solar energy gives a new dimension to nanotechnology with unimaginable capabilities. PCM production in form of microcontainers entrapping paraffin can be efficient to capture much of the solar radiation on the walls, building paints, etc. These materials with their thermodynamic physical properties promise many applications in the construction space, heat transfer and storage, household appliances, etc. Light microtraps may soon give spinel coatings an increase in the efficiency of solar devices by potentially minimizing absorbent surfaces.

4.11 REFERENCES

Barrera, E., L. Huerta, S. Muhl and A. Avila. 2005. Optical conduct of nanostructure Co_3O_4 rich highly Doping Co_3O_4 Zn alloys. Sol. Energy Mater. Sol. Cells. 88: 179.

Barshilia, H., N. Selvakumar and K. Rajam. 2008. Structure and optical properties of pulsed sputter deposited $Cr_xO_y/Cr/Cr_2O_3Cr_xO_y/Cr/Cr_2O_3$ solar selective coatings. J. Appl. Phys. 103: 23.

Belessiotis, G.V., K.G. Papadokostaki, E.P. Favvas, E.K. Efthimiadou and S. Karellas. 2018. Preparation and investigation of distinct and shape stable paraffin/SiO_2 composite PCM nanospheres. Energ. Convers. Manage. 168: 382.

Boström, T., S. Valizadeh, J. Lu, J. Jensen, G. Westin and E. Wäckelgård. 2011. Structure and morphology of nickel-alumina/silica solar thermal selective absorbers. J. Non-Cryst. Solids. 357: 1370.

Boström, Y., E. Wäckelgård and W. Westin. 2003. Solution-chemical derived nickel-alumina coatings for thermal solar absorbers. Sol. Energy. 74: 497.

Dong, Z., X. Lai, J.E. Halpert, N. Yang, L. Yi, J. Zhai, et al. 2012. Accurate control of multishelled ZnO hollow microspheres for dye-sensitized solar cells with high efficiency. Adv. Mater. 24: 1046.

Frei, U. and S. Brunold. 2000. Materials in high performance solar collectors. pp. 241-252. In: A.A.M. Sayigh [ed.]. World Renewable Energy Congress VI. Renewables: The Energy for the 21st Century World Renewable Energy Congress VI Brighton, UK.

Graf, W., F. Brucker, M. Köhl, T. Troscher, V. Wittwer and L. Herlitze. 1997. Development of large area sputtered solar absorber coatings. J. Non-Cryst. Solids. 218: 380. https://alanod.com/.2020. ALANOD GmbH and Co. KG, Germany.

Kaluža, L., B. Orel, and G. DražiandKöhl. 2001. Sol-gel derived $CuCoMnO_x$ spinel coatings for solar absorbers: Structural and optical properties. Sol. Energy Mater. Sol. Cells. 70: 187.

Kenisarin, M. and K. Mahkamov. 2007. Solar energy storage using phase change materials. Renew. Sust. Energ. Rev. 11: 1913.

Kordas, G. 2020. Incorporation of spherical shaped $CuO@SiO_2$ light microtraps into $CuCoMnO_x$ spinels to enhance solar absorbance. J. Amer. Cer. Soc. 103: 1536.

Lampert, C.M. 1987. Advanced optical materials for energy efficiency and solar conversion. pp. 277-346. In: H.P. Garg, M. Dhayal, G. Furlan and A.A.M. Sayigh [eds.]. Physics and Technology of Solar Energy volume 2 Photovoltaics and Solar Energy Materials. D. Reidel Publishing Company, Dordrecht, Holland.

Lazarov, M., R. Raths, H. Metzger and W. Spirkl. 1995. Optical constants and film density of TiN_xO_y solar selective absorbers. J. Appl. Phys. 77: 2133.

Li, R., Y. Zhou and X. Duan. 2020. Nanoparticle enhanced paraffin and tailing ceramic composite phase change material for thermal energy storage. Sustain. Energy Fuels. 4: 4547.

Nader, N.A., B. Bulshlaibi, M. Jamil, M. Suwaiyah and Uzair M. 2015. Application of phase-change materials in buildings. Am. J. Energy Eng. 3: 46.

Niklasson, G.A. and C.G. Granqvist. 1984. Optical properties and solar selectivity of co-evaporated Co-Al_2O_3 composite films. J. Appl. Phys. 55: 3382.

Nishimura, M. and T. Ishiguro. 2004. Solar selective absorber coating composed of aluminum and nitrogen with high performance induced by surface roughness. Jpn. J. Appl. Phys. 43: 757.

Roget, F. 2013. Study of the KNO_3–$LiNO_3$ and KNO_3–$NaNO_3$–$LiNO_3$ eutectics as phase change materials for thermal storage in a low-temperature solar power plant. Sol. Energy. 95: 155.

Shanker, K., and P.H. Holloway. 1985. Electrodeposition of black chrome selective solar absorber coatings with improved thermal stability. Thin Solid Films 127: 181.

Tabor, H.Z. 1956. Receiver for solar energy collectors, United States Patent, 2: 917.

Vince, J., A. Surca-Vuk, U. Opara-Krasovec, B. Orel, M. Kohl and M. Heck. 2003. Nanostructured CuO thin films prepared through sputtering for solar selective absorbers. Sol. Energy Mater. 79: 330.

Wäckelgård, E. 1998. Characterization of black nickel solar absorber coatings electroplated in a nickel chlorine aqueous solution. Sol. Energy Mater. Sol. Cells. 56: 35.

Yin, Y., Y. Pana, L.H. Hang, D.R. McKenzien and M.M.M Bilek. 2009. Direct current reactive sputtering $Cr-Cr_2O_3$ cermet solar selective surfaces for solar hot water applications. Thin Solid Films 517: 1601.

5

Nanostructured Oxide Based Ceramic Materials for Light and Mechanical Energy Harvesting Applications

Priyanka Bamola[1], Shilpa Rana[2], Bharti Singh[2],
Charu Dwivedi[3] and Himani Sharma[1,*]

5.1 INTRODUCTION

The world energy crisis has sparked more focus towards energy conversion and energy saving systems with higher efficiency (Wang et al. 2020). The world's energy production is mainly based on fossil fuels, which affect both the climate and environment (Ibrahim et al. 2019). It is expected that by 2050, half of all energy in industrial countries needs to be carbon free (Lund 2013). This requires an energy revolution: massive investments both in clean energy and efficient energy use (Lund 2013). Research relating to alternative forms of electricity production is a great deal of interest towards the sustainable energy future. In this regard, the renewable energy sources available in our environment, such as solar, thermal, wind and mechanical can be converted into electricity by energy harvesters (Ibrahim et al. 2019). Based on the nature of the input energy source, specific materials are required for the fabrication of energy harvesters, which can scavenge the ambient energy efficiently.

It has been reported in literature that advanced ceramic materials can play a major role towards harvesting renewable energy for conversion into electricity (Guo et al. 2019, Chevalier et al. 2005). In this regard, nanostructured ceramic materials which include, oxides, carbides and glasses have received major attention of the scientific community in the field of energy applications (Serrapede et al. 2019). These energy harvesting devices include lithium ion batteries, supercapacitors, solar and fuel cells, sensors, catalysis, optoelectronic devices, hydrogen generation and storage. Piezo electric nanogenerators have been found to be strongly dependent on the active surface area and interfaces for their operation (Hu et al. 2010, Ibrahim et al. 2019). The study shows that ceramic based nanostructured materials having high porosity and large specific surface area play an important role in energy harvesting devices by enhancing charge transfer and improving efficiency, in comparison to their bulk counterpart (Wang et al. 2020).

Nanostructure-based energy harvesting devices may help bringing down the cost of new energy technologies and enable their mass production because they typically have a few or no moving

---

[1] Department of Physics, School of Physical Sciences, Doon University Dehradun, Uttarakhand 248001, India.
[2] Department of Applied Physics, Delhi Technological University, Rohini, Delhi, 110042, India.
[3] Department of Chemistry, School of Physical Sciences, Doon University Dehradun, Uttarakhand 248001, India.
* Corresponding author: himanitiet427@gmail.com

parts (Hu et al. 2010). Design complexity is not an issue for nanoenergy based devices (e.g. a solar or fuel cell resembles a transistor more than an engine) (Hu et al. 2010, Lund 2013). Therefore, nano scaling provides essentially a better chance for rapid cost reduction. Bringing nanomaterials to large-scale in energy harvesting field will necessitate the better linking of the scale differences in energy systems (~1:10^6) and nanoenergy and integrating these materials into workable energy applications (Fig. 5.1) (Lund 2013).

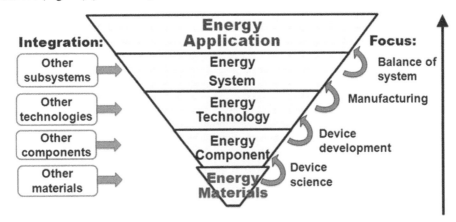

Fig. 5.1 Hierarchical system of energy which interacts with different technologies and components to create a full application. (Reprinted with permission from Lund, copyright (2013) Elsevier)

Modern scientific developments over the past two centuries have led to a more fundamental understanding of the composition and property relations of nanostructured ceramics (Guo et al. 2019). It permits the rational design and efficient preparation of novel nanostructured ceramic materials, including high temperature resistant ceramics (e.g. ZrC, HfC, WC and Ta_4HfC_5), structural ceramics (e.g. Al_2O_3, mullite, ZrO_2, SiC, and Si_3N_4, $BaTiO_3$ for ferroelectricity, ZnO for piezoelectricity and GaN for semiconducting device applications), functional ceramics (e.g. S TiO_2 for photocatalysis) and many others. Compared with these modern nanostructured ceramics, nanostructured TiO_2 ceramics demonstrate not only largely improved mechanical properties and high-temperature stability, but also sensitive responses to external stimuli, such as pressure, temperature, light, or electricity, which enables numerous energy applications (Rani et al. 2018).

5.2 NANOSTRUCTURED TIO₂ CERAMIC MATERIAL

With the rapid development of nanotechnology, nanostructured TiO_2 ceramics (TNSs) in various ways are finding wider applications because of their size-related properties (Chevalier et al. 2005). The energy band structure is discrete for TiO_2 nanometer scale, and due to the quantum size effect its photophysical, photochemical, and surface properties are somewhat different from those of the bulk structures (Wang et al. 2020). Therefore, various studies have focused on the synthesis of nanocrystalline TiO_2 with the large specific surface area (Cai et al. 2018). Synthesis of zero dimensional (0 D), one dimensional (1 D) and two dimensional (2 D) nanostructures of TiO_2 has been widely reported owing to their enhanced charge separation and transport properties (Tiwari et al. 2012). However, TNSs respond primarily to UV light which only accounts for less than 5% of total solar radiation. Under visible light, low electron transfer and high electron/pair recombination rates limit the energy harvesting applications (Nasir et al. 2020).

To overcome these limitations, modification in the chemical and optical properties of the shape adapted titanium dioxide and its heterostructures (HSs) is quite essential (Ali et al. 2018). As a result, the efficiency of the energy harvesting applications has been enhanced by tuning the electronic band structure and light harvesting range of TiO_2 to the visible and NIR region through various strategies

(Bagheri et al. 2014). These include deposition of metal nanoparticles (MNPs), coupling with a narrow band gap semiconductor (type-IIHSs), co-doping with multiple foreign elements, doping with metal and non-metal ions, surface fluorination and surface sensitization with organic dyes etc. (Bagheri et al. 2014, Ali et al. 2018).

In the earlier reports, secondary materials (MNPs, non-metals, and coupling with narrow band gap semiconductors) at the TNSs' interface were explored as a method for tailoring the structure and energy harvesting applications (Bagheri et al. 2014). Structural modifications of TNS-based hybrids introduce interface defects, which are primarily vacancies, interstitial vacancies and covalent intershell bonds, which may affect bonding between the Ti-O atoms (Bamola et al. 2020). The defects introduced at the interface of secondary materials and TNSs affect bonding between Ti-O atoms and induce stresses in Ti-O bonds and lattices, resulting in change in their structure and, therefore, efficiency of energy harvesting applications. One of the most important defects in TiO_2 is the oxygen vacancy (Nowotny et al. 2007). These vacancies in TiO_2 not only act as adsorption and active sites for energy harvesting applications but also affect the reactivity of TiO_2 (Polarz et al. 2006). Apart from generating oxygen vacancies, the secondary materials also induce strains in TiO_2 and have a significant effect on the electronic interaction at the interface of the hybrids (Bamola et al. 2020). The effect of strains on TNSs is important since the nanomaterials react to strain differently from their counterparts in bulk (Yu et al. 2016). Strain induces shifts in the intrinsic inter atomic distance and may alter the structure of material (Tavares et al. 2008). The modified interface changes the electronic properties of the hybrids at the interface, which could be related to light energy harvesting (Yu et al. 2016).

In above context, Section 3 will discuss about how the interfacial interaction in the nanostructured TiO_2 ceramics based hybrids affect the energy harvesting applications such as lithium ion batteries, supercapacitors, solar cells and fuel cells, sensors, catalytic reaction for fuel productions, optoelectronics devices and hydrogen generation.

5.3   ROLE OF INTERFACE IN APPLICATIONS OF NANOSTRUCTURED TIO$_2$ CERAMICS BASED HYBRIDS

5.3.1   Lithium Ion Batteries

The battery is a vital component of energy storage which can store electrical energy through electrochemical reactions in the form of chemical energy and then discharge this energy by reverse reactions (Deng et al. 2009). Batteries have achieved incredible success in research and commercialization over the last few decades. In present times, common problems and changes in technology raise higher demands on batteries. Therefore, advanced batteries require better performance with high power density, good safety, high energy density and long cycle life (Deng et al. 2009). In the meantime, comfort and convenience are another evolutionary way for batteries requiring flexibility, wearability and lightweightedness. Key components of a battery consist of positive electrodes and negative electrodes, a separator and an electrolyte. To ensure the above requirements, lithium ion batteries (LIBs) have definitely proven to be the first alternative for electrical energy storage as primary power sources whether for mobile or stationary applications (Madien et al. 2018).

The battery's energy storage follows the ion oxidation/reduction reactions between the active materials (anode/cathode) that apparently vary in the presence of an electrolyte. As illustrated in Fig. 5.2, a LIB consists of three main components: two electrodes with different electrical potential, and a separator that electrically isolates the electrodes and allows the Li ions to migrate between the electrodes (Madian et al. 2018). The electrolyte is primarily a liquid negotiating medium between the electrodes that allows Li ions to migrate, while the separator is fully saturated with the liquid to also avoid being a Li ion barrier composite. The electrolyte is usually a solution that contains a lithium salt (i.e. $LiPF_6$) and is mixed with liquid alkyl carbonates (Xu et al. 2016). The electrochemical reactions of the electrodes undergo a method in which Li ions are transferred between the anode and the cathode (Fig. 5.2) (Madian et al. 2018). Li ions de-intercalate during discharge from the anode material and the ions are transported through the electrolyte to intercalate into the cathode, which also serves as electron acceptors (Assat and Tarascon 2018). The electrons are therefore shuttled through the external circuit which provides the current flow. As for the charging phase, both electrodes respond reversibly. The functions of the electrode material, especially the electronic and ionic conductivity, therefore evaluate the electrode reactions (Deng et al. 2009). Electrode materials with efficient ion migration channels, abundant reactive sites and high electronic conductivity are promising part to achieve high efficiency. In addition, the insertion/extraction of ions can deform the structure of electrode materials, resulting in volume changes, or even the channels will collapse, resulting in a decay of capacity (Madian et al. 2018).

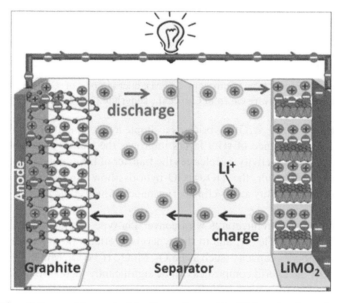

Fig. 5.2 Schematic of the working principle for rechargeable lithium ion batteries. (Reprinted with permission from Madian et al. copyright (2018) Multidisciplinary Digital Publishing Institute)

Previous studies show that TNSs based hybrids play an increasingly essential role as electrode materials because of their superior coulombic efficiency and cycling stability (Zhao et al. 2019, Nam et al. 2010). Additionally, they are abundant, obtained by cost-effective production processes and environmentally friendly (Zhao et al. 2019). Several studies found that TNSs based hybrids as anodes demonstrate the crystallite size of TiO_2 as the key factor, which defines reversible capacity and rate performance (Wagemaker et al. 2007). As size of particle decreases to nanometer levels, the surface area increases as a consequence and the diffusion path of the lithium ion is shortened, resulting in an enhanced Li ion penetration capability. In addition, a large number of experimental results have also shown that secondary material at the interface of TNSs can effectively improve the performance of LIBs. Tao et al. used an *in-situ* technique to fabricate the TiO_2-graphene composites (Tao et al. 2012). For Li ion intercalation, the obtained TiO_2-graphene composites were tested as anodes showing a reversible capacity of 60 mAh.g^{-1} at a high current rate of 5 A.g^{-1}. The unique graphene sheet conductivity at the TNSs interface has led to superior electrochemical efficiency in Li ion storage.

Carbon nanotubes (CNTs) were also used as an excellent candidate for the synthesis of highly valuable TiO_2 composite anodes, due to their high conductivity, stability and ease of manufacturing (Cao et al. 2010). Core-shell CNTs/TiO_2 (Fig. 5.3a) were synthesized by hydrolysis of tetrabutyl titanate and showed a maximum capacity of ~240 mAh.g^{-1} at a current density of 5 A.g^{-1}, exhibiting a 3 fold increase in capacity compared to bare TiO_2 (Fig. 5.3b). The increase in capacity was attributed to better electron supply because of the CNTs that allow the TiO_2 shells to store higher concentrations of lithium ions.

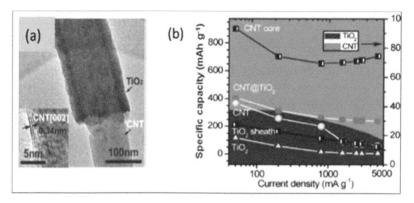

Fig. 5.3 (a) Transmission electron microscopy (TEM) micrographs of CNTs/TiO_2, (b) their cycling performance. Reprinted with permission from Cao et al., copyright (2010) American Chemical Society)

Zhao et al. synthesized TiO_2@rGO hybrids by a simple low temperature hydrothermal approach (Zhao et al. 2019). The presence of rGO layers improves the charge separation at the interface, enhances the electronic conductivity, accelerates the transfer of lithium ions, and provides structure stability protection. As a result, the TiO_2@rGO hybrids show high surface area, excellent rate property, good structural stability and cycling performance, which could be the ideal anode material for LIBs.

The heterogeneous nano architectures with conversion-type anodes based on TiO_2 were more reported and are listed in Table 5.1; due to the synergistic effects of stable substrate of TiO_2 and high conversion-type anodes capacity (such as Fe_3O_4, MoO_2, α-Fe_2O_3, MnO_2, Co_3O_4, $NiFe_2O_4$, FeS, MoS_2), the capacity of the hybrid composites can be significantly increased up to 820 mAh/g (after 1000 cycles at 500 mA/g) (Wang et al. 2019).

TABLE 5.1 Comparison of the electrochemical performance of Ti based g = hybrid nanocomposites with conversion type anodes

Electrode Material	Electrode Composition	Loading Density (mg/cm²)	Capacity after Cycles (Current Density)
TiO_2@MoO_2/C heterostructures	80:10:10	~1.0	400, 450 cycles (500 mA/g)
TiO_2-C/MnO_2 core-double-shell nanowire arrays	100:0:0	1.5	218, 150 cycles (3350 mA/g)
MoO_2-modified TiO_2 nanofibers	80:10:10	N/A	514.5, 50 cycles (0.2 C)
Core/shell TiO_2-MnO_2/MnO_2 heterostructures	80:10:10	N/A	185, 500 cycles (2000 mA/g)
Fe_3O_4-TiO_2-carbon hierarchical nanofibrous	80:10:10	1-2	525, 100 cycles (100 mA/g)
TiO_2@α-Fe_2O_3 core-shell nanostructures	100:0:0	~0.6	820, 1000 cycles (500 mA/g)
TiO_2/Fe_3O_4-PG ternary heterostructures	70:20:10	N/A	524, 200 cycles (1000 mA/g)
Co_3O_4/TiO_2 hierarchical heterostructures	70:20:10	N/A	602.8, 480 cycles (200 mA/g)
Core-shell NiF_2O_4@TiO_2 nanorods	70:20:10	N/A	321, 100 cycles (100 mA/g)
TiO_2-B/MoS_2 nanowire arrays	100:0:0	1:2	350, 100 cycles (20 mA/g)
FeS@TiO_2 nanostructures	70:20:10	N/A	430, 500 cycles (400 mA/g)
$Li_4Ti_5O_{12}$-TiO_2@MoO_2/C 3D heterostructures	80:10:10	~1.0	413, 500 cycles (1000 mA/g)

Source: Reprinted with permission from Wang et al., copyright (2019) Springer Nature.

5.3.2 Supercapacitors

Supercapacitors are devices proficient in managing high power rates compared to batteries (Chu and Braatz 2002). Supercapacitors require high energy density, high charging-discharge levels and a long cycle life (Peng et al. 2016). They act as a link for energy power gap between conventional capacitors and secondary batteries (Peng et al. 2016). Supercapacitors store energy using either ion adsorption or redox reactions, which transfer most of its charge near the surface of the electrode material (Chen et al. 2018). Supercapacitors are suitable for applications requiring at least 10 kW.kg^{-1} high power density, one order of magnitude greater than lithium-ion batteries (Chen 2017). With the secondary material, the electrochemical efficiency of symmetrical supercapacitors based on TNSs can be considerably improved (Li et al. 2019). The secondary material is stated to alter the charge separation at the TNS interface and thus the efficiency of supercapacitors.

Kolathodi et al. synthesized MnO_2 coated TiO_2 nanofibers by electrospinning and post-hydrothermal process (Kolathodi et al. 2019). The MnO_2 shell with an average thickness of about 10 nm contributed to the high electrochemical efficiency of charge storage via redox reaction and intercalation mechanisms, while the TiO_2 anatase phase core provided a simple route for electronic transport with additional electrochemical stability over thousands of charge-discharge cycles. They further stated that an asymmetric supercapacitor designed from the MnO_2-TiO_2 nanofiber electrode showed a high operating voltage (2.2 V) window with a maximum gravimetric capacitance of 111.5.

The improved and superior electrochemical performance of MnO_2-encapsulated TiO_2 fibers could be due to the highly electrochemical crystalline phases of TiO_2 and MnO_2, as well as the nanofibers' high surface area and aspect ratio. A thin layer of MnO_2 increases ionic conductivity and charge storage sites, while the inner TiO_2 core promotes electronic conductivity and provides chemical stability for a large number of charge-discharge cycles.

Kim et al. fabricated a high-energy, power-density hybrid supercapacitor (Kim et al. 2013a). A new concept, i.e. a hybrid supercapacitor was proposed to combine the advantages of LIBs and supercapacitors. The hybrid supercapacitor uses a Faradaic lithium-intercalation cathode (such as $LiMn_2O_4$) or anode (such as $Li_4Ti_5O_{12}$), commonly used as LIB electrodes, and combines it with a non-Faradaic capacitive anode or cathode (usually a carbonaceous material) used in a supercapacitor. The hybrid supercapacitor stores charges asymmetrically and simultaneously on one electrode by surface ion adsorption/desorption and on the other electrode by lithium de/intercalation. The combination of the Faradaic intercalation with the non-Faradaic surface reaction provides an opportunity to increase energy and power densities effectively.

A recent hybrid supercapacitor comprising an anatase TiO_2 (reduced graphene oxide) anode and an activated carbon (AC) cathode exhibited promising energy-storage capabilities that could bridge the gap between traditional LIBs and supercapacitors. The new system's energy density (42 Wh.kg^{-1}) was one of the highest among hybrid supercapacitor systems based on intercalation compounds.

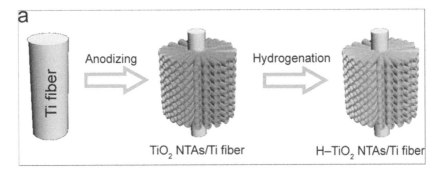

Fig. 5.4 A schematic diagram showing the fabrication of H−TiO_2 NTAs. (Reprinted with permission from Lu et al., copyright (2012) American Chemical Society).

Lu et al. presented a new and general strategy to improve the capacitive properties of TiO_2 materials for supercapacitors, involving the synthesis of hydrogenated TiO_2 nanotube arrays (TiO_2 NTAs), shown in Fig. 5.4 (Lu et al. 2012). H-TiO_2 NTAs are good supporting scaffold for MnO_2 nanoparticles. The capacitor electrodes produced on metal nanoparticles deposited H-TiO_2 NTAs through electrochemical deposition achieve a remarkable specific capacitance of 912 F.g^{-1} at a scan rate of 10 mV.s^{-1}. The ability of TiO_2 electrode materials to improve their capacitive properties should open up new opportunities for high-performance supercapacitors.

5.3.3 Solar Cells and Fuel Cells

5.3.3.1 Solar Cells

The key advantages of solar cells are the internal photocurrent efficiency, the portion of absorbed photons transformed into electrical current, the exterior quantum efficiency or the portion of incident photons converted into electrical current and the efficiency of energy conversion (Chen 2017). Although several advances in photovoltaic (PV) devices have been made, efforts still need to be made to significantly improve the conversion performance of solar cells (Peng et al. 2016). Silicon is the most commonly used absorber so far and currently dominates the market for PV devices (Nayak et al. 2019). State-of-the-art silicon-PV devices based on p-n junctions, also referred to as solar cells of the first generation, have an efficiency of up to 25 percent. Based on thin-film technologies, the production of second-generation PVs was driven by the need to increase performance. However, the efficiency of second-generation PVs is less than that of silicon (Husain et al. 2018). PVs of the third generation rely on the exploitation of emerging organic PV cells, dye-sensitized solar cells and solar cells of the quantum dot (Mingsukang et al. 2017).

TNSs have emerged in recent years as promising materials for low cost, versatile, and highly effective solar cells (Hu et al. 2020). Doping is easily performed with bulk semiconductors and has also been limited to nanoscale materials recently (Feng et al. 2009). Feng et al. prepared tantalum-doped TiO_2 nanowire arrays by a low-temperature hydrothermal method. They found that doping is easily expanded to allow homogeneous doping of other metal cations to Ti^{4+} sites in nanowires, which will be of incredible use when applying in dye-sensitized solar cells (DSSCs).

Su et al. used a novel and water-soluble precursor to synthesize Nb-doped TiO_2 nanostructure and were successfully used as photoanode materials in DSSCs (Su et al. 2015). We observed that Nb doping significantly improves the nanostructure's conductivity and the energy conversion efficiency obtained for a DSSC was 18.2%. A systematic investigation on the effect of Nb doping provides a valuable insight for the designing of high-performance DSSCs.

In Cu(In, Ga)Se$_2$ solar cells, TiO_2-SiO_2 core-shell nanostructured anti-reflective coatings were also introduced and could increase the optical path length by a scattering effect or by a refractive index gradient formation (Chen et al. 2019). The suppression of reflectance resulted in an improved power conversion efficiency from 6.32% to 7.00% after applying the TiO_2-SiO_2 core-shell nanostructures.

Zhao et al. constructed a highly efficient quantum dot sensitized solar cell (QDSC) based on TiO_2/ZnO nanorod arrays (NAs) decorated with Ag nanoparticles (NPs) (Zhao et al. 2016). The QDSCs using TiO_2/ZnO NAs modified with Ag NPs have achieved considerably higher power conversion efficiency (PCE) than the QDSCs without Ag NPs. The deposition of Ag NPs into TiO_2/ZnO NAs photoelectrode can offer several advantages: (1) enhancing light absorption, (2) increasing electron injection rate, (3) suppressing recombination of charges, and (4) extending the lifetime of electrons.

5.3.3.2 Fuel Cells

Fuel cells convert chemical energy from a fuel into electricity through an oxygen reaction or other oxidizing agents (Chen 2017). The application of fuel cells in electronics faces many challenges: (i) identifying suitable electrodes for versatile electronics; (ii) removing expensive noble metals such as platinum, ruthenium, gold and their electrocatalyst alloys; and (iii) preventing metal electrode poisoning. To overcome these challenges, low-cost, high-efficiency and long-lasting TiO_2 ceramic materials need to be produced before fuel cells can be used as a serious energy conversion technology in electronic devices (Peng et al. 2016). Fuel cells are currently receiving considerable attention as an alternative source of energy due to their greater efficiency without pollutant emissions (Staffell et al. 2019). Due to its excellent power density and low operating temperature, direct methanol fuel cells (DMFCs) and proton exchange membrane fuel cells (PEMFCs) are commonly regarded as the most promising for use in portable devices such as mobile phones, laptops and personal digital assistants (Peng et al. 2016). Nanomaterials consisting of platinum and its alloys have traditionally been used in fuel cells as catalysts, due to their high catalytic activity (Qiao and Li 2011). However, one of the biggest barriers to the commercialization of DMFCs is the high cost of the catalyst. It is therefore one of the major efforts of scientists to increase platinum's catalytic efficiency while at the same time reducing the amount of catalyst required. One approach is to use a supporting material to boost catalyst efficiency. When used in fuel cells, catalytic supporting materials must be stable and uniformly distributed. Researchers have found in recent years that novel nanostructured TNSs can be used as superior catalyst supports in fuel cells because of their good electrical conductivity and low cost.

Tian et al. investigated plasmon-induced photoelectrochemistry in the visible region at gold nanoparticle nanoporous TiO_2 composites (Au-TiO_2) prepared by photocatalytic deposition of Au in a porous TiO_2 film (Tian and Tatsuma 2005). They found that photoaction spectra were in strong agreement with the absorption spectrum of the gold nanoparticles in the TiO_2 film for both the open-circuit potential and the short-circuit current. Gold nanoparticles are photoexcited due to plasmon resonance, and charge separation at the interface is achieved by the transfer of photoexcited electrons from the gold particle to the TiO_2 conduction band and simultaneously transfer compensatory electrons from a donor in the gold particle solution. A photovoltaic cell with the optimized electron mediator ($Fe^{2+/3+}$) exhibits an optimal incident photon to current conversion efficiency (IPCE) of 26%. In addition, the Au-TiO_2 can photocatalytically oxidize ethanol and methanol under visible light, at the expense of reducing oxygen; it is theoretically applicable to a new generation of photovoltaic fuel cells.

Huang et al. synthesized a novel Pt electrocatalyst supported by TiO_2 and explored it as a potential cathode catalyst for fuel cells (Huang et al. 2011). The Pt/TiO_2 electrocatalyst's excellent fuel cell performance was attributed to the low mass transport limit in the cathode catalyst layer. Based on the experimental results, Pt/TiO_2 electrocatalyst can be considered as an alternative cathode electrocatalyst to increase fuel cell efficiency and durability.

5.3.4 Sensors

As public awareness of environmental issues increases and governments commit to reducing pollution globally, sensor arrays (electronic noses) will play a very important role in protecting our environment (Bourgeois et al. 2003). Sensors capable of detecting and even quantifying both simple and complex gas mixtures present a far more facile analytical method than analysis and processing of samples using traditional equipment. The electronic noses are especially attractive for on-site monitoring of priority pollutants and for addressing other environmental needs. Despite its huge potential for environmental monitoring, wide-ranging applications of electronic nose technology for air and water quality control are still in demand (Capelli et al. 2014). In particular, recent developments in new nanostructured materials that serve as sensing units will likely expand the sensors towards a broad range of organic and inorganic pollutants (Hu et al. 2010).

TNSs have been widely explored as sensor units for various gases (Wang et al. 2017a). Many researchers claim that doping TiO_2 with transition metals is an appropriate way of improving its efficiency as it affects the interface charge separation (Eadi et al. 2017). Xiang et al. developed a TiO_2 nanotube film doped by Pd nanoparticles for preparation of hydrogen sensor with improved performance (Xiang et al. 2014). The detection of hydrogen using anodic TiO_2 nanotubes is greatly enhanced by employing Pd nanoparticles in the system leading to Schottky barrier formation at the interface. The aforementioned device is found to be able to provide rapid response and selective detection with good reproducibility and fast recovery time at room temperature. The detection was found to be non-inferential due to the presence of other gases such as CO_2, CH_4 and NH_3, which could be very useful in real-life applications. The results obtained also showed that the Pt-doped TiO_2 film reduces sensing response time, Ag nanoparticles modifies TiO_2 structures and improves gas sensing efficiency, and Cr-doped TiO_2 thin films improve gas sensitivity.

Buvailo et al. demonstrated an easy and straightforward process to deposit nanostructured thin films on glass and $LiNbO_3$ sensor substrates based on LiCl-doped TiO_2 (Buvailo et al. 2011). In addition to gas sensing, the nanostructured material and sensor response times were also analyzed using conductometric and surface acoustic wave (SAW) sensor techniques, showing reversible signals with strong reproducibility and quick response times of around 0.75 s, as shown in Fig. 5.5. The applicability of this nanostructured film has been demonstrated for the construction of rapid humidity sensors.

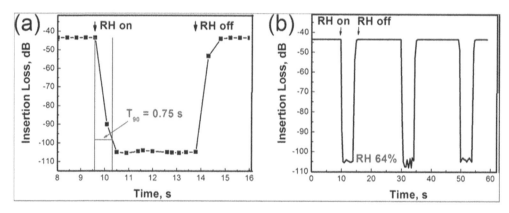

Fig. 5.5 (a) LiCl doped TiO_2 nanomaterials film SAW response dynamics to 64% RH, (b) repeatability of the sensor response. (Reprinted with permission from Buvailo et al., copyright (2011) American Chemical Society)

Farzaneh et al. explored the semiconducting ceramic based humidity sensors that are of great interest for many industrial automation applications (Farzaneh et al. 2019). Regarding these demands, Copper (Cu) doped TiO_2 films are potential candidates for humidity sensors that are fabricated using sol-gel method. From their work, Cu-doping in TiO_2 significantly changes the optical and structural properties of the films. The obtained samples suggested that the Cu doping decreases the optical band gap and semiconductor activity. The Cu doped TiO_2 has also been found to enhance the water molecule's energy adsorption and charging transfer values. Meanwhile, the band gap value of Cu doped TiO_2 showed a comparatively greater sensitivity towards the water molecule compared with undoped TiO_2.

5.3.5 Catalytic Reaction for Fuel Productions

5.3.5.1 Catalytic Reaction for CO₂ Reduction

In recent years, the catalytic conversion of carbon dioxide (CO_2) into renewable energy sources is seen as a viable way of addressing both environmental problems and energy losses (Gao et al. 2020). Several attempts have been made to increase the CO_2 conversion efficiency by developing the photocatalysts. TiO_2 is the most reported photocatalyst for CO_2 reduction due to its high chemical stability, non-toxicity and low cost (Nguyen et al. 2020). TiO_2 is a distinctive d^0 metal oxide with appropriate conduction band to impel the reduction of CO_2 to hydrocarbons (Peng et al. 2017). As a result, various efforts have been made to enhance the quantum efficiency of TiO_2 and its visible-light absorption activity (Sharma et al. 2020, Ola and Maroto-Valer 2015). Initially, decoration of non-metals such as sulfur (S), carbon (C), fluorine (F) and nitrogen (N) over TNs could lead to a narrow band gap and consequently to a shift in the absorption band of TiO_2 towards the visible light (Emeline et al. 2008). Li et al. reported high absorption in the visible light and increased catalytic activity of CO_2 to CH_4 under visible light illumination for N doped TiO_2, while bare TiO_2 showed no photocatalytic activity (Li et al. 2012).

Ong et al. reported that the N doped TiO_2 NPs with (001) facets are photocatalytically active in visible light region with 11 times higher activity than bare TiO_2 (Ong et al. 2014). Zhang et al. synthesized iodine (I)-doped TiO_2 with mixed phase of anatase and brookite showing enhanced reduction activity of CO_2 under visible light irradiation (Zhang et al. 2011). Apart from non-metals, doping with transient metals (such as Cr, V, Fe, Mn etc.) is another successful technique for expanding the TiO_2 light adsorption spectrum (Stepanov 2012). Co-doping of metal and non-metal with TiO_2 in visible light irradiation was also found to be effective in photoreduction of CO_2 (Fan et al. 2011). In CO_2 photocatalytic reduction, the doping of N and Ni^{2+} could improve TiO_2's illumination response and enhanced photocatalytic activity of TiO_2 by reducing charge carrier recombination.

Plasmonic NPs of Ag and Au decorated over TNSs reveal positive effects by (i) increasing the charge carrier separation, (ii) elongating the light absorption in visible light region and increasing the excitation of electrons by their surface plasmon resonance (SPR) effects (Tatsuma et al. 2017). A number of studies have recently been reported using plasmonic nanoparticles decorated TiO_2 nanostructures for CO_2 reduction under visible light and UV light illumination (Tatsuma et al. 2017, Matsubara and Tatsuma 2007). Cronin et al. synthesized the TiO_2-Au material for photocatalytic reduction of CO_2 under UV light and visible light (Hou et al. 2011). They found that under visible light of illumination, the photoreduction of CO_2 produced only methane with 24 fold enhancement due to strong electric fields produced by the SPR effect of Au plasmonic nanoparticles (Fig. 5.6).

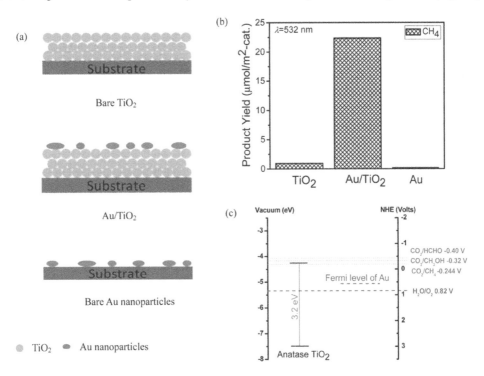

Fig. 5.6 (a) Schematic diagram of three types of photocatalysts, (b) photocatalytic product yields (after 15 h of visible irradiation) over three different catalysts, (c) energy band alignment of anatase TiO_2 and Au, and the relevant redox potentials of CO_2 and H_2O under visible illumination. (Reprinted with permission from Hou et al., copyright (2011) American Chemical Society)

Conversely, under UV light of irradiation, three different products such as C_2H_6, HCHO and CH_3OH were found. Furthermore, Liu et al. reported that under the irradiation of UV and visible light, composite of Ag and TiO_2 exhibited 9.4 times higher methanol development rate than bare TiO_2 (Liu et al. 2015).

5.3.5.2 Hydrogen Generation

In many applications, hydrogen (H_2) is commonly considered as the potential renewable energy carrier, such as in domestic heating, immobile power generation and environmentally friendly cars (Ahmed and Xinxin 2016). Despite the ongoing global energy crisis, solar light photocatalytic H_2 development as a clean and sustainable source of energy is rising as one of the most challenging areas. For three reasons, photocatalytic H_2 water production is one of the most exciting ways to recognize a hydrogen economy. Photocatalytic H_2 production is based on energy of photon, photocatalytic conversion of solar energy into a storable energy (i.e. hydrogen) and environmentally safe (Liao et al. 2012). Conversely, it has been complicated to get a perfect photocatalyst with all the necessities such as corrosion resistance, chemical stability, and visible light harvesting that would make the improvement of photocatalytic H_2 a feasible alternative (Zhu and Zach 2009). However, nanotechnology has facilitated modification of vacant photocatalysts. Many papers studied the impact of various nanostructures and nanomaterials on photocatalyst results, since their energy conversion efficiency is mainly determined by nanoscale properties.

Recently, TNSs have been widely explored for hydrogen generation (Naldoni et al. 2019). The structural and electronic properties of TNSs in the conversion of photocatalytic energy are significant because of their absorption in the UV-Vis region. However, the utilization of solar light absorption is limited only to UV range. Further, the ascendancy of defects in TNSs for progress of H_2 is convoluted.

It is observed that the defects may initiate new states into the band gap, resulting in a recombination of charge carriers and a dwindling of the reduction and oxidation capacities of carrier (Kang et al. 2019). Further, delicately introducing particular defects assist the charge carrier's separation. Recently, Wu et al. synthesized yellow TiO_2 nanoparticles with ultra-small size of ~3 nm (Wu et al. 2016). They found that H_2 production for yellow TiO_2 was ~3.7 fold higher than the pure TiO_2. They proposed that the incorporation of the defect states by the elemental doping will be more easily compared to the complicated steps needed to precisely get control of the defects. In order to improve light absorption to visible light for applications in the solar spectrum, researchers have changed photocatalysts using doping in the lattice.

Recent research shows that the doping of Mg could remove the built-in defect states and deteriorate the defect state of TiO_2 material (Wang et al. 2017b). The outcomes were confirmed by the transient spectroscopic analysis of the energy absorption-excitation in infrared. The photocatalytic overall measurements of water splitting showed that the evolution rates of H_2 and O_2 were 850 and 425 $\mu mol.h^{-1} \cdot g^{-1}$ under air mass illumination.

Naik et al. reported hydrogen-generating photocatalytic activity of Pt nanoparticles decorated N-doped TiO_2 nanostructures (Naik et al. 2015). The interstitial N doping increases the absorption of light in the visible field. From their work, it was found that the porosity of N-doped TiO_2 nanostructures can be modified using different temperatures for reinforcement. They also found that the co-doping of Pt displays a 30% higher catalytic activity than undoped catalysts. The increased catalytic activity is accredited to the synergistic effects of N doping, porosity and the transport of charges amid the TiO_2 nanostructures and the co-catalyst of Pt.

Su et al. measured the kinetics of the photogenerated electrons on metal semi-conductor photocatalyst. They found that the Au-Pd nanoparticles' structure modifies the electronic properties of the co-catalyst by fine tuning (Su et al. 2014). In particular, Pd shell-Au immobilized core nanoparticles on TiO_2 revealed enormously high quantum efficiencies for H_2 development.

High-performance photocatalytic hydrogen evolution can be accomplished by producing hetero-or homo-interfaces (Wang et al. 2017b). Compared to the crystalline and amorphous TiO_2 film, high mobility and concentration of electrons can simultaneously be obtained in a bilayer TiO_2 thin film at the homo-interface amid amorphous and crystalline layers. Extraordinary properties could therefore be explored in well-designed interfaces with homogeneous chemical composition. Moreover, extensive research has been done on the hetero-interfaces between TiO_2 materials and other semiconducting nanostructures such as plasmonic nanoparticles, carbon, NiO, Si, CdS, MoS_2, ZnS, MoC_2 and double-layered hydroxides. Wu et al. reported that selective spatial congregation and successive hydrolysis could attain anisotropic TiO_2 overgrowth on Au nanorods (Wu et al. 2016). Plasmon enhanced H_2 evolution has been demonstrated under visible light illumination. The interface of Au nanorod and TiO_2 with the Au nanorod side unveiled as a Schottky junction can filter out the Au nanorods hot resonance surface electrons, which is necessary to improve the efficiency of the H_2 evolution.

5.3.6 Optoelectronic Devices

TNSs were used in many optoelectronic systems as transparent electrodes; for example, organic emitting diodes systems (OLEDs), dye synthesized solar cells (DSSC) and liquid crystal display (LCDs) (Rasheed and Barille 2017). Zhu et al. reported that Nb doped 20 nm thin TiO_2 films on two different substrates of glass and PET by DC sputtering technique. Size of particles differs on both the substrates. The optical constants of films have a typical transmission value of approximately 85-91% for films decorated on glass substrates and 81-85% for PET substrates in the visible region. Such transmission qualities make the films extremely appropriate for optoelectronic applications.

5.3.7 Piezoelectric Nanogenerator (PENG)

Mechanical energy is ubiquitously available in our surroundings and is helpful for continuous operation of self-powered system, where other sources of energy are not available. There are number of ambient sources of mechanical energy (kinetic and potential energies), such as vibration associated with operating machinery or induced by flow motion, acoustic noise, deformation and pressure fluctuations, civil infrastructure, transportation, movements associated with human body such as walking, breathing, talking, typing and many more, that can be used to harvest energy to power micro-/nano devices (Sun et al. 2010, Mitcheson 2010, Orrego et al. 2017, Priya 2007). The device which is capable of harnessing this ambient mechanical energy into the useful form of electricity, based on the intrinsic properties of the specific materials, is known as the Piezoelectric Nanogenertaor (PENG). This PENG works on the principle of the piezoelectric effect, which states that application of force/pressure on certain (piezoelectric) materials deforms the crystal structure, which results in the generation of piezoeletric polarization charges at the two ends of the material, which is known as Direct Piezoelectric effect; greater the application of force, higher will be the polarization charges. The electrical potential created by the polarization charges is balanced by the flow of electrons from the external circuit. Piezoelectric effect is a reversible process, therefore contrary of piezoelectric effect also exhibit, i.e. application of the electric field/voltage results in the deformation of the shape of the material, known as Converse/Indirect Piezoelectric effect (Bhalla et al. 2016, Jaffe and Cook 1971). For fabricating the energy harvesters, specialized kind of materials, which have the ability of converting mechanical energy into electric energy, are required and are known as Piezoelectric materials. In this category, several oxides and ceramics materials have been investigated, such as cadmium sulfide (CdS), zinc oxide (ZnO), zinc stannate ($ZnSnO_3$), lithium niobate ($LiNbO_3$), potassium niobate ($KNbO_3$), barium titnate ($BaTiO_3$), lead zirconate titnate (PZT), polyvinylidene fluoride (PVDF), polyvinylidene fluoride-trifluoro ethylene (PVDF-TrFE), polyamides (PA), polyactic acids (PLA), cellulose etc. (Priya and Inman 2009, Lin et al. 2008,

Alam and Mandal 2016). The Piezoelectric Nanogenerator have ample advantages due to its ease of fabrication at macro and micro-scale, high output current, high power density, flexibility, and wide bandwidth, making them suitable for low frequency application, wireless sensor network (WSN), wearable electronics, e-skin, internet of things (IoT), structural health monitoring, militiary, biomedical applications etc. (Zhang et al. 2015, Ciofani and Menciassi 2012, Sun et al. 2015, Wang and Shi 2012b).

Several bulk materials have been explored for designing PENG; however, one dimensional (1D) nanomaterials (nanowires, nanorods) are more frequently used for energy harvesting due to (i) the presence of strain gradient in ferroelectric nanomaterials which enhances the piezoelectric effect upto 400-500%, (ii) lattice perfection in 1D materials which provide superior critical strain, high flexibility, extended lifetime of device, (iii) 1D nanomaterials are more sensitive to the small force (Wang et al. 2012b).

5.3.7.1 Material Selection for Energy Harvesting

There are various types of the piezoelectrical materials such as ceramics, single crystals, polymers, their composite etc. and the selection of the materials for a specific application such as sensors, actuators, and energy harvesters depends upon the characteristic of material and their functionality in the application sector (Li and Deng 2014). Ceramics with randomly oriented grain will exhibit the piezoelectricity only if they are ferroelectric in nature because internal dipoles in the ferroelectric material can be reoriented by the application of the external field and exhibit a remanent polarization even when we remove the electric field but in inorganic crystalline materials, such as aluminium nitride (AlN) and polycrystalline ZnO which does not exhibit piezoelectric effect because of arrangements of ions in crystal structure, the piezoelectric coeffecient in these materials can not be changed after deposition (Jaffe and Cook 1971, Setter et al. 2006, Marauska et al. 2012).

ZnO is considered as one of the promising materials which exhibits both piezoelectric and semiconducting properties and has the numerous configurations of nanostructure, such as nanobelts, nanowire, nanocomb, nanohelixs, nanospring, nanobows, nanocages, nanosphere, nanowhiskers, and flower like nanopropellers, which can be synthesized by various methods under some specific growth condition, and therefore belongs to the richest family in terms of their properties as well as structure (Wang and Song 2006, Wang 2004). Because of the excellent properties of ZnO, such as semiconducting, piezoelectric, pyroelectric, catalytic, strong luminescence, high thermal conductivity, high radiation hardness, and flexible morphology, it has novel application in transistors, sensors, transducers, solar cell, and light emitting diode (Janotti and Van 2009, Williander et al. 2010, Ellmer et al. 2007, Uthirakumar et al. 2009). The first piezoelectric nanogenerator based on the ZnO nanowire array was fabricated by Wang group (Wang et al. 2006) in 2006 in which NWs are deflected by conductive atomic force microscope (AFM) tip in contact mode creating a strain field; as a result, outer stretched surface has positive potential and inner compressed surface has negative potential. Therefore, when schottky barrier between tip and ZnO NW becomes forward bias, it allows electron to flow from semiconductor ZnO NW to the metal tip. This piezoelectric nanogenerator has low efficiency; therefore, to improve power generation capibilities, the AFM tip is replaced by ultrasonic wave to actuate all NWs simultaneously and continuosly (Wang et al. 2007).

Kim et al. showed that formation of the good Schottky contact with Au electrode on the ZnO nanosheet results in high voltage and current response corresponds to the applied force in comparison to the weak schottky contact and (ITO or graphene) the ohmic contacts (with Al), and fabricates the direct current piezoelectric power generator with 2D ZnO nanosheets and an anionic nanoclay layer heterojunction, which is promising for the minituarization of the device (Kim et al. 2013b).

Further, the mechanical stabililiy of two distinct nanostructures of ZnO (nanosheet and nanorods) is evaluated by using strain density concept under different loads, which shows that ZnO nanosheets are more mechanically stable due to their networked structure. The piezoelectric properties in the

ZnO are attributed to their crystallographic nature, which arise from the non-centrosymmetric lattice structure in the hexagonal Wurtzite structure, combined with large electromechanical coupling and have application in actuators, sensors, and nanogenerators. The piezoelectric response of a material is highly dependent on its dimension and shows an increment, while going from macro to nanoscale dimension (Agrawal and Espinosa 2011, Goel and Kumar 2020). The value of d_{33} for bulk ZnO in its pure form is ~9.9 pm/V and is found to be frequency independent, while ZnO nanobelts are frequency dependent (with increase in frequency, the value of piezoelectric coefficient d_{33} decreases) and varies from 14.3 to 26.7 pm/V (Zhao et al. 2004). Ghosh and Rao synthesized the ZnO nanorods by low temperature sol gel method, which showed a piezoelectric coefficient d_{33} 44.33 pm/V (Ghosh and Rao 2013). Christian et al. (Christian et al. 2016) reported the piezoelectric coefficient d_{33} = 15 pm/V for ZnO nanowire fabricated by wet chemical growth method. The interface between the two materials plays an important role for their design and the performance of the integrated material. Seongcheol et al. reported that coating of ZnO nanolayer by solution based hydrothermal process on the cellulose film exhibits 3.5 times higher piezoelectric charge constant without any variation in its transparency and flexibility (Mun et al. 2016). ZnO NWs can be utilized for strengthening of the interface, as mechanical strength is an important parameter for continuous power generation of the energy harvesting device (Zhu et al. 2010). Mohammad et al. showed that with the growth of the vertically aligned ZnO, NW array on aramid fabrics will create a strong mechanical interlocking interface between fiber and the material which improves the mechanical property and result in the enhancement of the elastic modulus and the tensile strength of composite by 34.3% and 18.4% (Malakooti et al. 2016). Usage of AlN interlayer in ZnO vertically integrated NG (VINGs) will enhance the output voltage significantly up to 200 times as compared to the ZnO based VING without AlN interlayer as the insulating AlN interlayer acts as electron blocking layer in device because of its high dielectric constant and large Young's modulus (Lee et al. 2015). Baek et al. reported that construcing the pn hetrojunction between the copper oxide layer (CuO) and ZnO diminishes the electron screening effect under external strain and enhances the performance of the ZnO based nano generator (Baek et al. 2016). In addition to the simple binary metal oxide based ceramics material, barium titanate (BTO) which belongs to the perovskite family and is known for its unique dielectric, ferroelectric, piezoelectric and electro optical properties having wide application in multilayer ceramic capacitors (MLCs), positive temperature coefficient (PTC) thermistors, temperature sensors, stabilizer etc. (Ertug 2013, Vijatovic et al. 2008, Heywang 1971), has been explored for energy harvesting devices. The piezoelectric coefficients for the pure barium titanate is found to be d_{33} = 191 pC/N, d_{31} = −79 pC/N and d_{15} = 270 pC/N when synthesized by conventional solid state reaction method (Bechmann 1956). The grain size has considerable influence on the properties of the material; it has been shown by Huan et al. that decrease in the grain size will result in the increase in the piezoelectric coefficient (d_{33} = 519 pC/N), electromechanical coupling factor (k_p = 39.5%), and relative permittivity (ε_r = 6079) with maximum value, when grain size ~1 μm and further decrease with decrease in the grain size (Huan et al. 2014).

Cho et al. reported that incorporation of $BaTiO_3$ hollow nanosphere enhances the β phase content in the PVDF. To prevent the agglomeration and increase the surface area for interaction, these barium titanante hollow spheres are treated with silane coupling agent, which increases the nucleation of β phase by suppressing α phase formation (Cho et al. 2015). Koka et al. showed that the vertically aligned $BaTiO_3$ nanowire based piezoelectric sensor have high sensitivity as compared to the ZnO nanowire based sensor having application in accelerometer system (Koka and Sodano 2013). Mahadeva et al. fabricated the piezoelectric hybrid paper with $BaTiO_3$ nanostructure and fiber functionalization of wood cellulose using layer by layer assembly as shown in Fig. 5.7, which is promising for low cost substrate and useful in developing affordable, eco-friendly, and versatile device for various applications (Mahadeva et al. 2014).

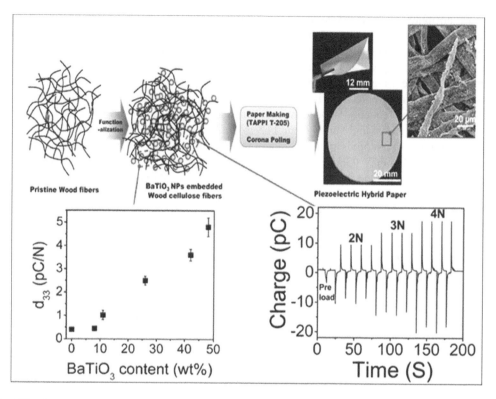

Fig. 5.7 Schematic representation of the piezoelectric hybrid paper fabrication process. (Reprinted with permission from Mahadeva et al., copyright (2014) American Chemical Society)

5.4 SUMMARY

The nanostructured oxides ceramic based hybrids and the piezoelectric nanogenerator based on nanostructured zinc oxide and barium titanate for various energy harvesting applications have been discussed in this chapter. It is inferred that interfacial interaction plays an important role in the field of energy related applications by enhancing the charge transfer and efficiency of energy devices. Also, stretchable and flexible piezoelectric energy harvesters are briefly discussed, which facilitate a significant role not only in self-powered system but also in sensor system. The chapter further concisely discusses the selection of nanostructured oxides based hybrids ceramic materials and their applications in the field of various energy fields.

5.5 REFERENCES

Agrawal, R. and H. Espinosa. 2011. Giant piezoelectric size effects in zinc oxide and gallium nitride nanowires: A first principles investigation. Nano Lett. 11: 786-790.

Ahmed, M. and G. Xinxin. 2016. A review of metal oxynitrides for photocatalysis. Inorg. Chem. Front. 3: 578-590.

Alam, M.M. and D. Mandal. 2016. Native cellulose microfiber-based hybrid piezoelectric generator for mechanical energy harvesting utility. ACS Appl. Mater. Interfaces. 8: 1555-1558.

Ali, I., M. Suhail, Z.A. Alothman and A. Alwarthan. 2018. Recent advances in syntheses, properties and applications of TiO$_2$ nanostructures. RSC Adv. 8: 30125-30147.

Assat, G. and J. Tarascon. 2018. Fundamental understanding and practical challenges of anionic redox activity in Li-ion batteries. Nat. Energy. 3: 373-386.

Baek, S.K., S.S. Kwak, J.S. Kim, S.W. Kim and H. Cho. 2016. Binary oxide *pn* heterojunction piezoelectric nanogenerators with an electrochemically deposited high *p*-type Cu_2O layer. ACS Appl. Mater. Interfaces. 8: 22135-22141.

Bagheri, S., N.M. Julkapli, S. Bee and A. Hamid. 2014. Titanium dioxide as a catalyst support in heterogeneous catalysis. Sci. World J. 2014: 14.

Bamola, P., C. Dwivedi, A. Gautam, M. Sharma, S. Tripathy, A. Mishra, et al. 2020. Strain-induced bimetallic nanoparticles-TiO_2 nanohybrids for harvesting light energy. Appl. Surf. Sci. 511: 45416.

Bechmann, R. 1956. Elastic, piezoelectric, and dielectric constants of polarized barium titanate ceramics and some applications of the piezoelectric equations. J. Acoust. Soc. Am. 28: 347-350.

Bhalla, S., S. Moharana, V. Talakokula and N. Kaur. 2016. Piezoelectric Materials: Applications in SHM. Energy Harvesting and Biomechanics: John Wiley and Sons. 4010007.

Bourgeois, W., A.C. Romain, J. Nicolas and R.M. Stuetz. 2003. The use of sensor arrays for environmental monitoring: Interests and limitations. J. Environ. Monit. 5: 852-860.

Buvailo, A.I., Y. Xing, J. Hines, N. Dollahon and E. Borguet. 2011. TiO_2/LiCl-based nanostructured thin film for humidity sensor applications. ACS Appl. Mater. Interfaces. 3: 528-533.

Cai, J., J. Shen, X. Zhang, Y.H. Ng, J. Huang, W. Guo, et al. 2018. Light-driven sustainable hydrogen production utilizing TiO_2 nanostructures: A review. Small Methods 1800184: 1-24.

Cao, F.F., Y.G. Guo, S.F. Zheng, X.L. Wu, L.Y. Jiang, R.R. Bi, et al. 2010. Symbiotic coaxial nanocables: Facile synthesis and an efficient and elegant morphological solution to the lithium storage problem. Chem. Mater. 22: 1908-1914.

Capelli, L., S. Sironi and R. Del Rosso. 2014. Electronic noses for environmental monitoring applications. Sensors 14: 19979-20007.

Chen, C.W., H.W. Tsai, Y.C. Wang, T.Y. Su, C.H. Yang, W.S. Lin, et al. 2019. Design of novel TiO_2-SiO_2 core-shell helical nanostructured anti-reflective coatings on Cu(In,Ga)Se$_2$ solar cells with enhanced power conversion efficiency. J. Mater. Chem. A. 7: 11450-11459.

Chen, G.Z. 2017. Supercapacitor and supercapattery as emerging electrochemical energy stores. Int. Mater. Rev. 62: 173-202.

Chen, W., G. Li, A. Pei, Y. Li, L. Liao, H. Wang, et al. 2018. A manganese–hydrogen battery with potential for grid-scale energy storage. Nat. Energy. 3: 428-435.

Chevalier, J., S. Deville and G. Fantozzi. 2005. Nanostructured ceramic oxides with a slow crack growth resistance close to covalent materials. Nano Lett. 5: 1297-1301.

Cho, S., J.S. Lee and J. Jang. 2015. Enhanced crystallinity, dielectric, and energy harvesting performances of surface-treated barium titanate hollow nanospheres/PVDF nanocomposites. Adv. Mater. Inetrfaces. 2: 1500098.

Christian, B., J. Volk, I.E. Lukács, E. Sautieff, C. Sturm, A. Graillot, et al. 2016. Piezo-force and vibration analysis of ZnO nanowire arrays for sensor application. Procedia Eng. 168: 1192-1195.

Chu, A. and P. Braatz. 2002. Comparison of commercial supercapacitors and high-power lithium-ion batteries for power-assist applications in hybrid electric vehicles I. Initial characterization. J. Power Sources. 112: 236-246.

Ciofani, G. and A. Menciassi. 2012. Piezoelectric Nanomaterials for Biomedical Applications. Springer.

Deng, D., G. Kim, Y. Lee and J. Cho. 2009. Green energy storage materials: Nanostructured TiO_2 and Sn-based anodes for lithium-ion batteries. Energy Environ. Sci. 2: 818-837.

Eadi, S.B., S. Kim, S.W. Jeong and H.W. Jeon. 2017. Novel preparation of Fe doped TiO_2 nanoparticles and their application for gas sensor and photocatalytic degradation. Adv. Mater. Sci. Eng. 2017: 6.

Ellmer, K., A. Klein and B. Rech. 2007. Transparent Conductive Zinc Oxide: Basics and Applications in Thin Film Solar Cells. Springer Science and Business Media.

Emeline, A.V., V.N. Kuznetsov, V.K. Rybchuk and N. Serpone. 2008. Visible-light-active titania photocatalysts: The case of N-doped TiO_2–properties and some fundamental issues. Int. J. Photoenergy 2008: 1-19.

Ertuğ, B. 2013. The overview of the electrical properties of barium titanate. American J. Eng. Res. 2: 1-7.

Fan, J., E.Z. Liu, L. Tian, X.Y. Hu, Q. He and T. Sun. 2011. Synergistic effect of N and Ni^{2+} on nanotitania in photocatalytic reduction of CO_2. J. Environ. Eng. 137: 171-176.

Farzaneh, A., A. Mohammadzadeh, M.D. Esrafili and O. Mermer. 2019. Experimental and theoretical study of TiO_2-based nanostructured semiconducting humidity sensor. Ceram. Int. 45: 8362-8369.

Feng, X., K. Shankar, M. Paulose and C.A. Grimes. 2009. Tantalum-doped titanium dioxide nanowire arrays for dye-sensitized solar cells with high open-circuit voltage. Angew. Chemie – Int. Ed. 48: 8095-8098.

Gao, Y., K. Qian, B. Xu, Z. Li, J. Zheng, S. Zhao, et al. 2020. Recent advances in visible-light-driven conversion of CO_2 by photocatalysts into fuels or value-added chemicals. Carbon Resour. Convers 3: 46-59.

Ghosh, M. and M. Rao. 2013. Growth mechanism of ZnO nanostructures for ultra-high piezoelectric d33 coefficient. Mater. Express. 3: 319-327.

Goel, S. and B. Kumar. 2020. A review on piezo-/ferro-electric properties of morphologically diverse ZnO nanostructures. J. Alloys and Compd. 816: 152491.

Guo, Q., B. Zhang, X. Feng, X. Yan, Z. Su, S.Z.D. Cheng, et al. 2019. Controlling the periodically ordered nanostructures in ceramics: A macromolecule-guided strategy. Macromol. Rapid Commun. 1900534: 1-15.

Heywang, W. 1971. Semiconducting barium titanate. J. Mater. Sci. 6: 1214-1224.

Hou, W., W.H. Hung, P. Pavaskar, A. Goeppert, M. Aykol and S.B. Cronin. 2011. Photocatalytic conversion of CO_2 to hydrocarbon fuels via plasmon-enhanced absorption and metallic interband transitions. ACS Catal. 1: 929-936.

Hu, X., G. Li and J.C. Yu. 2010. Design, fabrication, and modification of nanostructured semiconductor materials for environmental and energy applications. Langmuir 26: 3031-3039.

Hu, Z., J.M. García-Martín, Y. Li, L. Billot, B. Sun, F. Fresno, et al. 2020. TiO_2 nanocolumn arrays for more efficient and stable perovskite solar cells. ACS Appl. Mater. Interfaces. 12: 5979-5989.

Huan, Y., X. Wang, J. Fang and L. Li. 2014. Grain size effect on piezoelectric and ferroelectric properties of $BaTiO_3$ ceramics. J. Eur. Ceram. Soc. 34: 1445-1448.

Huang, S.Y., P. Ganesan and B.N. Popov. 2011. Mechanisms and applications of plasmon-induced charge separation at TiO_2 films loaded with gold nanoparticles. Appl. Catal. B. Environ. 102: 171-177.

Husain, A.A.F., W.Z.W. Hasan, S. Shafie, M.N. Hamidon and S.S. Pandey. 2018. A review of transparent solar photovoltaic technologies. Renew. Sustain. Energy Rev. 94: 779-791.

Ibrahim, I.D., E.R. Sadiku, T. Jamiru, Y. Hamam, Y. Alayli and A.A. Eze. 2019. Prospects of nanostructured composite materials for energy harvesting and storage. J. King Saud Univ. Sci. 32: 758-764.

Jaffe, B. and J. Cook. 1971. The Piezoelectric Effect in Ceramics. Academic Press, 7-21.

Janotti, A. and de Van. 2009. Fundamentals of Zinc Oxide as a Semiconductor. Rep. Prog. Phys. 72: 126501.

Kang, X., S. Liu, Z. Dai, Y. He, X. Song and Z. Tan. 2019. Titanium dioxide: From engineering to applications. Catalysts 9: 191.

Kim, H., M. Cho, M. Kim, K. Park, H. Gwon, Y. Lee, et al. 2013a. A novel high-energy hybrid supercapacitor with an anatase TiO_2-reduced graphene oxide anode and an activated carbon cathode. Adv. Energy Mater. 3: 1500-1506.

Kim, K.H., B. Kumar, K.Y. Lee, H.K. Park, J.H. Lee, H.H. Lee, et al. 2013b. Piezoelectric two-dimensional nanosheets/anionic layer heterojunction for efficient direct current power generation. Sci. Rep. 3: 2017.

Koka, A. and H.A. Sodano. 2013. High-sensitivity accelerometer composed of ultra-long vertically aligned barium titanate nanowire arrays. Nat. Commun. 4: 2682.

Kolathodi, M.S., M. Palei, T.S. Natarajan and G. Singh. 2019. MnO_2 encapsulated electrospun TiO_2 nanofibers as electrodes for asymmetric supercapacitors. Nanotehnology 31: 12.

Lee, E., J. Park, M. Yim, Y. Kim and G. Yoon. 2015. Characteristics of piezoelectric ZnO/AlN−stacked flexible nanogenerators for energy harvesting applications. Appl. Phys. Lett. 106: 023901.

Li, H., C. Tian and Z. Deng. 2014. Energy harvesting from low frequency applications using piezoelectric materials. Appl. Phys. Rev. 14: 041301.

Li, Q., M. Horn, Y. Wang, J. MacLeod, N. Motta and J. Liu. 2019. A review of supercapacitors based on graphene and redox-active organic materials. Materials 12: 703.

Li, X., Z. Zhuang, W. Li and H. Pan. 2012. Photocatalytic reduction of CO_2 over noble metal-loaded and nitrogen-doped mesoporous TiO_2. Appl. Catal. A. Gen. 429-430: 31-38.

Liao, C.H., C.W. Huang and J.C.S. Wu. 2012. Hydrogen production from semiconductor-based photocatalysis via water splitting. Catalysts 2: 490-516.

Lin, Y.F., J. Song, Y. Ding, S.Y. Lu and Z. Wang. 2008. Piezoelectric nanogenerator using CdS nanowires. App. Phys. Lett. 92: 022105.

Liu, E., L. Qi, J. Bian, Y. Chen, X. Hu, J. Fan, et al. 2015. A facile strategy to fabricate plasmonic Cu modified TiO_2 nano-flower films for photocatalytic reduction of CO_2 to methanol. Mater. Res. Bull. 68: 203-209.

Lu, X., G. Wang, T. Zhai, M. Yu, J. Gan, Y. Tong, et al. 2012. Hydrogenated TiO_2 nanotube arrays for supercapacitors. Nano Lett. 12: 1690-1696.

Lund, P.D. 2013. Nanostructured materials for energy applications. Microelectron. Eng. 108: 84-85.

Madian, M., A. Eychmüller and L. Giebeler. 2018. Current advances in TiO_2-based nanostructure electrodes for high performance lithium ion batteries. Batteries 4: 7.

Mahadeva, S.K., K. Walus and B. Stoeber. 2014. Piezoelectric paper fabricated via nanostructured barium titanate functionalization of wood cellulose fibers. ACS Appl. Mater. Interfaces. 6: 7547-7553.

Malakooti, M.H., B.A. Patterson, H.S. Hwang and H. Sodano. 2016. ZnO nanowire interfaces for high strength multifunctional composites with embedded energy harvesting. Energy Environ. Sci. 9: 634-643.

Marauska, S., V. Hrkac, T. Dankwort, R. Jahns, H. Quenzer, R. Knöchel, et al. 2012. Sputtered thin film piezoelectric aluminum nitride as a functional MEMS material. Procedia. Eng. 18: 787-795.

Matsubara, K. and T. Tatsuma. 2007. Morphological changes and multicolor photochromism of Ag nanoparticles deposited on single-crystalline TiO_2 surfaces. Adv. Mater. 19: 2802-2806.

Mingsukang, M.A., M.H. Buraidah and A.K. Arof. 2017. Third-generation-sensitized solar cells. pp. 7-31. *In*: N. Das [ed.]. Nanostructured Solar Cells. IntechOpen.

Mitcheson, P.D. 2010. Energy harvesting for human wearable and implantable bio-sensors. IEEE. 3432-3436.

Mun, S., H.U. Ko, L. Zhai, S.K. Min, H.C. Kim and J. Kim. 2016. Enhanced electromechanical behavior of cellulose film by zinc oxide nanocoating and its vibration energy harvesting. Acta Materialia. 114: 1-6.

Naik, B., S.Y. Moon, S.H. Kim and J.Y. Park. 2015. Enhanced photocatalytic generation of hydrogen by Pt-deposited nitrogen-doped TiO_2 hierarchical nanostructures. Appl. Surf. Sci. 354: 347-352.

Naldoni, A., M. Altomare, G. Zoppellaro, N. Liu, Š. Kment, R. Zbořil, et al. 2019. Photocatalysis with reduced TiO_2: From black TiO_2 to Co catalyst-free hydrogen production. ACS Catal. 9: 345-364.

Nam, S.H., H. Shim, Y. Kim, M.A. Dar, J.G. Kim and W.B. Kim. 2010. Ag or Au nanoparticle-embedded one-dimensional composite TiO_2 nanofibers prepared via electrospinning for use in lithium-ion batteries. ACS Appl. Mater. Interfaces. 2: 2046-2052.

Nasir, M.S., G. Yang, I. Ayub, S. Wang and W. Yan. 2020. Tin diselinide a stable co-catalyst coupled with branched TiO_2 fiber and g-C_3N_4 quantum dots for photocatalytic hydrogen evolution. Appl. Catal. B. Environ. 270: 118900.

Nayak, P.K., S. Mahesh, H.J. Snaith and D. Cahen. 2019. Photovoltaic solar cell technologies: Analysing the state of the art. Nat. Rev. Mater. 4: 269-285.

Nguyen, T.P., D.L.T. Nguyen, V.H. Nguyen, T.H. Le, D.V.N. Vo, Q.T. Trinh, et al. 2020. Recent advances in TiO_2-based photocatalysts for reduction of CO_2 to fuels. Nanomaterials 10: 1-24.

Nowotny, J., T. Bak, M.K. Nowotny and L.R. Sheppard. 2007. Titanium dioxide for solar-hydrogen II. Defect Chemistry Int. J. Hydrogen Energy. 32: 2630-2643.

Ola, O. and M.M. Maroto-Valer. 2015. Review of material design and reactor engineering on TiO_2 photocatalysis for CO_2 reduction. J. Photochem. Photobiol. C. Photochem. Rev. 24: 16-42.

Ong, W.J., L.L. Tan, S.P. Chai, S.T. Yong and A.R. Mohamed. 2014. Self-assembly of nitrogen-doped TiO_2 with exposed {001} facets on a graphene scaffold as photo-active hybrid nanostructures for reduction of carbon dioxide to methane. Nano Res. 7: 1528-1547.

Orrego, S., K. Shoele, A. Ruas, K. Doran, B. Caggiano, R. Mittal, et al. 2017. Harvesting ambient wind energy with an inverted piezoelectric flag. Appl. Energy. 194: 212-222.

Peng, C., G. Reid, H. Wang and P. Hu. 2017. Perspective: Photocatalytic reduction of CO_2 to solar fuels over semiconductors. J. Chem. Phys. 147: 030901-030914.

Peng, S., G. Jin, L. Li, K. Li, M. Srinivasan, S. Ramakrishna, et al. 2016. Multi-functional electrospun nanofibres for advances in tissue regeneration, energy conversion and storage, and water treatment. Chem. Soc. Rev. 45: 1225-1241.

Polarz, S., J. Strunk, V. Ischenko, M.W.E.V.D. Berg, O. Hinrichsen, M. Muhler, et al. 2006. On the role of oxygen defects in the catalytic performance of zinc oxide. Angew. Chemie — Int. Ed. 45: 2965-2969.

Priya, S. 2007. Advances in energy harvesting using low profile piezoelectric transducers. J. Electroceram. 19: 167-184.

Priya, S. and D.J. Inman. 2009. Energy Harvesting Technologies. Springer.

Qiao, Y. and C.M. Li. 2011. Nanostructured catalysts in fuel cells. J. Mater. Chem. 21: 4027-4036.

Rani, A., R. Reddy, U. Sharma, P. Mukherjee, P. Mishra, A. Kuila, et al. 2018. A review on the progress of nanostructure materials for energy harnessing and environmental remediation. J. Nanostructure Chem. 8: 255-291.

Rasheed, M. and R. Barillé. 2017. Optical constants of DC sputtering derived ITO, TiO_2 and TiO_2:Nb thin films characterized by spectrophotometry and spectroscopic ellipsometry for optoelectronic devices. J. Non. Cryst. Solids. 476: 1-14.

Serrapede, M., U. Savino, M. Castellino, J. Amici, S. Bodoardo, E. Tresso, et al. 2019. Nanostructured ceramic oxides with a slow crack growth resistance close to covalent materials. Materials 13: 21.

Setter, N., D. Damjanovic, L. Eng, G. Fox, S. Gevorgian, S. Hong, et al. 2006. Ferroelectric thin films: Review of materials, properties, and applications. J. Appl. Phys. 100: 051606.

Sharma, E., V. Thakur, S. Sangar and K. Singh. 2020. Recent progress on heterostructures of photocatalysts for environmental remediation. Mater. Today Proc. 7: 1700006.

Staffell, I., D. Scamman, A. Velazquez Abad, P. Balcombe, P.E. Dodds, P. Ekins, et al. 2019. The role of hydrogen and fuel cells in the global energy system. Energy Environ. Sci. 12: 463-491.

Stepanov, A.L. 2012. Applications of ion implantation for modification of TiO$_2$: A review. Rev. Adv. Mater. Sci. 30: 150-165.

Su, H., Y.T. Huang, Y.H. Chang, P. Zhai, N.Y. Hau, P.C.H. Cheung, et al. 2015. The synthesis of Nb-doped TiO$_2$ nanoparticles for improved-performance dye sensitized solar cells. Electrochim. Acta. 182: 230-237.

Su, R., R. Tiruvalam, A.J. Logsdail, Q. He, C.A. Downing, M.T. Jensen, et al. 2014. Designer titania-supported Au-Pd nanoparticles for efficient photocatalytic hydrogen production. ACS Nano 8: 3490-3497.

Sun, C., J. Shi and X. Wang. 2010. Fundamental study of mechanical energy harvesting using piezoelectric nanostructures. J. Appl. Phys. 108(3): 034309.

Sun, Q., W. Seung, B.J. Kim, S. Seo, S.W. Kim and J. Cho. 2015. Active matrix electronic skin strain sensor based on piezopotential-powered graphene transistors. Adv. Mater. 27: 3411-3417.

Tao, H.C, L.Z. Fan, X. Yan and X. Qu. 2012. *In situ* synthesis of TiO$_2$-graphene nanosheets composites as anode materials for high-power lithium ion batteries. Electrochim. Acta. 69: 328-333.

Tatsuma, T., H. Nishi and T. Ishida. 2017. Plasmon-induced charge separation: Chemistry and wide applications. Chem. Sci. 8: 3325-3337.

Tavares, C.J., S.M. Marques, S. Lanceros-Méndez, V. Sencadas, V. Teixeira, J.O. Carneiro, et al. 2008. Strain analysis of photocatalytic TiO$_2$ thin films on polymer substrates. Thin Solid Films 516: 1434-1438.

Tian, Y. and T. Tatsuma. 2005. Mechanisms and applications of plasmon-induced charge separation at TiO$_2$ films loaded with gold nanoparticles. J. Am. Chem. Soc. 127: 7632-7637.

Tiwari, J.N., R.N. Tiwari and S.K. Kwang. 2012. Zero-dimensional, one-dimensional, two-dimensional and three-dimensional nanostructured materials for advanced electrochemical energy devices. Prog. Mater. Sci. 57: 724-803.

Uthirakumar, P., H.G. Kim and C. Hong. 2009. Zinc oxide nanostructures derived from a simple solution method for solar cells and LEDs. Chem. Eng. J. 155: 910-915.

Vijatović, M., J. Bobić and B. Stojanović. 2008. History and challenges of barium titanate: Part II. Sci. Sinter. 40: 235-244.

Wagemaker, M., W.J.H. Borghols and F.M. Mulder. 2007. Large impact of particle size on insertion reactions. A Case for Anatase LixTiO$_2$. J. Am. Chem. Soc. 129: 4323-4327.

Wang, H., X. Liang, J. Wang, S. Jiao and D. Xue. 2020. Multifunctional inorganic nanomaterials for energy applications. Nanoscale 12: 14-42.

Wang, S., Y. Yang, Y. Dong, Z. Zhang and Z. Tang. 2019. Recent progress in Ti-based nanocomposite anodes for lithium ion batteries. J. Adv. Ceram. 8: 1-18.

Wang, X., J. Song, J. Liu and Z. Wang. 2007. Direct-current nanogenerator driven by ultrasonic waves. Science 316: 102-105.

Wang, X. 2012a. Piezoelectric nanogenerators—harvesting ambient mechanical energy at the nanometer scale. Nano Energy 1: 13-24.

Wang, X. and J. Shi. 2012b. Piezoelectric Nanogenerators for Self-powered Nanodevices. Piezoelectric Nanomaterials for Biomedical Applications. Springer 135-172.

Wang, X., Y. Zhao, K. Mølhave and H. Sun. 2017b. Engineering the surface/interface structures of titanium dioxide micro and nano architectures towards environmental and electrochemical applications. Nanomaterials 7: 382.

Wang, Y., T. Wu, Y. Zhou, C. Meng, W. Zhu and L. Liu. 2017a. TiO$_2$-based nanoheterostructures for promoting gas sensitivity performance: Designs, developments, and prospects. Sensors 17: 1971.

Wang, Z. 2004. Nanostructures of zinc oxide. Mater. Today. 7: 26-33.

Wang, Z.L. and J. Song. 2006. Piezoelectric nanogenerators based on zinc oxide nanowire arrays. Science 312: 242-246.

Willander, M., O. Nur, J.R. Sadaf, M.I. Qadir, S. Zaman, A. Zainelabdin, et al. 2010. Luminescence from zinc oxide nanostructures and polymers and their hybrid devices. Materials 3: 2643-2667.

Wu, B., D. Liu, S. Mubeen, T.T. Chuong, M. Moskovits and G.D. Stucky. 2016. Ultra-small yellow defective TiO_2 nanoparticles for co-catalyst free photocatalytic hydrogen production. J. Am. Chem. Soc. 138: 1114-1117.

Xiang, C., Z. She, Y. Zou, J. Cheng, H. Chu, S. Qiu, et al. 2014. A room-temperature hydrogen sensor based on Pd nanoparticles doped TiO_2 nanotubes. Ceram. Int. 40: 16343-16348.

Xu, Z., L. Gao, Y. Liu and L. Li. 2016. Review—recent developments in the doped $LiFePO_4$ cathode materials for power lithium ion batteries. J. Electrochem. Soc. 163: 2600-2610.

Yu, X., H. Dong, L. Wang and Y. Li. 2016. Strain effects on the electronic and transport properties of TiO_2 nanotubes. RSC Adv. 6: 80431-80437.

Zhang, M., T. Gao, J. Wang, J. Liao, Y. Qiu, Q. Yang, et al. 2015. A hybrid fibers based wearable fabric piezoelectric nanogenerator for energy harvesting application. Nano Energy 13: 298-305.

Zhang, Q., Y. Li, E.A. Ackerman, M. Gajdardziska-Josifovska and H. Li. 2011. Visible light responsive iodine-doped TiO_2 for photocatalytic reduction of CO_2 to fuels. Appl. Catal. A. Gen. 400: 195-202.

Zhao, H., F. Huang, J. Hou, Z. Liu, Q. Wu, H. Cao, et al. 2016. Efficiency enhancement of quantum dot sensitized TiO_2/ZnO nanorod arrays solar cells by plasmonic Ag nanoparticles. ACS Appl. Mater. Interfaces. 8: 26675-26682.

Zhao, M.H., Z.L. Wang and S. Mao. 2004. Piezoelectric characterization of individual zinc oxide nanobelt probed by piezoresponse force microscope. Nano Lett. 4: 587-590.

Zhao, X., H. Liu, M. Ding and Y. Feng. 2019. *In-situ* constructing of hollow TiO_2@rGO hybrid spheres as high-rate and long-life anode materials for lithium-ion batteries. Ceram. Int. 45: 12476-12483.

Zhu, G., R. Yang, S. Wang and Z. Wang. 2010. Flexible high-output nanogenerator based on lateral ZnO nanowire array. Nano Lett. 10: 3151-3155.

Zhu, J. and M. Zach. 2009. Nanostructured materials for photocatalytic hydrogen production. Curr. Opin. Colloid Interface Sci. 14: 260-269.

6

Titanium Oxide-Based Noble Metal-Free Core-Shell Photocatalysts for Hydrogen Production

Sara El Hakim, Tony Chave and Sergey I. Nikitenko*

6.1 INTRODUCTION

The present global energy consumption is greater than 17 TW (IEA 2019) and is estimated to be raised to 26.6 TW by 2050 (Arnell et al. 2002). On the other hand, the theoretically attained solar power has a potential of about 21,840 TW (Hermann 2006), dwarfing by many orders of magnitude the growing energy demand. However, a large-scale use of solar energy is still challenging and one of the most serious issues is solar energy storage. Photocatalytic and photoelectrocatalytic production of hydrogen from water is considered as a promising sustainable solution of this problem. The feasibility of both technologies depends on the performance of catalysts, which should meet several strict requirements. They should be active under the solar light, chemically and photochemically stable, nontoxic, and cost-effective. In this view, preparation of effective photocatalysts from earth-abundant elements is of great importance for sustainable energetics.

Among large variety of photocatalysts studied in the past decades, the catalysts based on titanium oxide, TiO_2, have attracted special attention. The discovery of the photoelectrocatalytic water splitting on a TiO_2 electrode (Fujishima and Honda 1972) has initiated a huge amount of studies focused on photocatalytic hydrogen generation. Being cheap, stable, nontoxic, and environmentally benign, TiO_2 could be the best choice for clean hydrogen production on a large scale using solar energy. However, TiO_2 can absorb only around 6% of the sunlight, owing to a quite large bandgap of 3.2 eV for anatase phase (Ghosh 2018). Another serious bottleneck of TiO_2 is the rapid electron-hole recombination leading to the decrease in photocatalytic activity. About 90% of the electron-hole pairs photogenerated in colloidal TiO_2 particles recombine within 10 ns (Serpone et al. 1995). Therefore, tremendous efforts have been undertaken to narrow bandgap of TiO_2 and to improve photogenerated charge separation.

The recent advances in photocatalysis with TiO_2-based materials (Wang and Domen 2020, Ghosh 2018, Li et al. 2018, Ma et al. 2014) pointed out several strategies of TiO_2 bandgap engineering: doping of TiO_2 with cations or anions, co-doping with cations and anions, self-doping of TiO_2 with Ti^{3+}, and surface sensitization with organic dyes or transition metal complexes. On the other hand, fabrication of TiO_2 heterojunctions with other semiconductors, anatase-rutile phase junctions, and TiO_2 loading with co-catalysts, often noble metal nanoparticles, allows to minimize electron-hole recombination during the photocatalytic process. Design of catalyst morphology is another important strategy to reach maximal photocatalytic activity. A large number of studies have been dedicated to the synthesis of TiO_2-based photocatalysts having 1D (Lee et al. 2014, Wang et al.

ICSM, Univ Montpellier, UMR 5257, CEA-CNRS-UM-ENSCM, Marcoule, France.
* Corresponding author: serguei.nikitenko@cea.fr

2014a, b), 2D (Faraji et al. 2019, Heard et al. 2018, Yang et al. 2019), and 3D (Li et al. 2018, Gawande et al. 2015, Zhang et al. 2012) morphologies. 3D core-shell nanoparticles have attracted a great deal of attention as promising photocatalysts for hydrogen production. The core-shell nanoparticles can be broadly defined as composite material comprising an inner core and outer layer shell. Various classes of core-shell nanoparticles have been described in the literature: single core-single shell, multiple core-single shell, single core-multiple shell including concentric shell (nanomatryushka) particles, and hollow shell particles (Chaudhuri and Paria 2012). TiO_2-based photocatalysts loaded with plasmonic and non-plasmonic noble metals are already largely reviewed in the literature. This chapter is primarily focused on titania-based core-shell photocatalysts fabricated from earth-abundant elements for efficient and cost-effective hydrogen production. Special attention will also be paid to the photothermal effect, which has recently been proven to be promising not only for steam generation but also for more efficient photocatalytic solar light conversion (Zhu et al. 2018, Song et al. 2017, Nikitenko et al. 2015, 2018).

6.2 STRUCTURE AND ELECTRON PROPERTIES OF TIO₂

6.2.1 Structure of TiO₂

The anatase and rutile polymorphs of TiO_2 are widely studied, and even used in photocatalysis at industrial scale. Therefore, the understanding of the stability of the different TiO_2 polymorphs is of prime importance in order to control the particle size, structure properties, predominance of specific crystal facets and phase composition of the materials which can have a direct impact on the electronic properties of the various catalysts and their photocatalytic performances. Titanium dioxide is a naturally occurring mineral observed under three main polymorphs: anatase, brookite and rutile. All these compounds are based on the spatial arrangement of distorted TiO_6 octahedrons linked together by edges, or corners, in three different manners, leading to different crystal structures as depicted in Fig. 6.1 with structural data given in Table 6.1 (Cromer and Herrington 1955, Baur 1961). Anatase and rutile exhibit both a tetragonal crystal structure but they belong to different space groups which are, respectively, *I41/amd* and *P42/mmm*. On the other hand, brookite shows an orthorhombic structure with the space group *Pbca*.

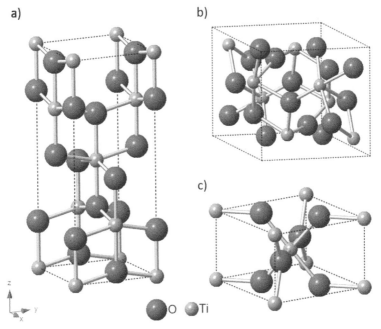

Fig. 6.1 Schematic representation of the crystal cells of the three main TiO_2 polymorphs (a) anatase, (b) brookite, (c) rutile

TABLE 6.1 Structural data and properties of the three main TiO₂ polymorphs

Polymorph		Anatase	Brookite	Rutile
Crystal system		Tetragonal	Orthorhombic	Tetragonal
Space group		*I41/amd*	*Pbca*	*P42/mmm*
Number of molecule/cell		4	8	2
Unit cell data (nm)	a	0.379	0.917	0.459
	b		0.546	
	c	0.951	0.514	0.296
Density (g.cm⁻³)		3.79	3.99	4.13

These differences in crystal structure will be at the origin of some material macroscopic properties, like the density, but also of the surface energy of the various crystal facets (Batzill 2011). For TiO₂ polymorphs, the variation of surface energy for the different crystal facets is related to the percentage of under-coordinated Ti atoms. For instance, for anatase phase, the facet with the highest energy will be the {110} exhibiting 4-fold coordinated Ti atoms (Ti_{4C}) and the most stable will be the {101} consisting of half Ti_{6C} and half Ti_{5C} atoms (Lazzeri et al. 2001, 2002). This surface energy aspect is important in order to understand the morphology of the material but also to tailor the reactivity of the various polymorphs used as catalysts (Liu et al. 2014). The evolution of the low index facet surface energies for the anatase is reported to be {110} > {001} > {101}, whereas for rutile the most stable facet is {110}, followed by {100}, and finally {001} having the highest surface energy (Lazzeri et al. 2001, 2002, Perron et al. 2007). In the same way, some data were also obtained for brookite showing the following order of surface energies: (010) < (110) < (100) (Beltrán et al. 2006).

Other structures have also been reported within the literature such as TiO₂ (B), which can be found in nature, along with other synthetic high pressure phases like TiO₂ (II) or TiO₂ (H) for a sum of at least 11 different polymorphs (Zhu and Gao 2014, Zhang and Banfield 2014a). Among the main three polymorphs, at a macroscopic scale, rutile is known to be the thermodynamically stable phase, whereas anatase and brookite are metastable but can readily be obtained at low temperature in the case of anatase or under more specific conditions in the case of brookite (Paola et al. 2013, Hanaor and Sorrell 2011). As a matter of fact, anatase is considered as a kinetic product and the transformation from anatase, or brookite, into rutile can be easily achieved by heating leading to an irreversible conversion.

Nevertheless, when considering nanomaterials this statement on the stability of the different polymorphs should be tempered since recent studies on the surface enthalpies of anatase, brookite and rutile pointed out that the thermodynamic stability was actually size-dependent (Zhang and Banfield 2000). Thus, for particles within the range of 11 nm to 35 nm, brookite seems to be the stable phase, whereas anatase was preferentially formed for particles below 11 nm and rutile for particles above 35 nm (Zhang and Banfield 2000). Note that this observation is in agreement with the thermal stability of the different polymorphs and the general phase transformation from anatase to rutile during temperature increase above 600°C. Actually, this phase transformation implies for nanoparticles within the same size range the structural change from anatase to brookite, which can be done at lower temperature since the activation energy is quite low (around 12 kJ/mol), whereas the brookite transformation into rutile requires much higher activation energy above 160 kJ/mol (Zhang and Banfield 2000). Thus, the initial anatase to brookite transformation, which starts from the interfaces of the agglomerated particles, doesn't lead to a drastic modification of the material structure, with a moderate particle size increase, whereas the second transformation step induces a reconstructive phase transition with higher coarsening due to the temperature treatment (Hanaor and Sorrell 2011, Zhang and Banfield 2014a).

Other experimental conditions have to be taken into account when speaking about the structure of the various TiO₂ polymorphs. Since these compounds are synthetized and used mainly in aqueous solutions, the phase stability and the modification of surface properties in water were the focus of several studies (Barnard and Curtiss 2005a, b, Kumar et al. 2014, Zhang and Banfield 2014a). Thus, it appears that under very acidic conditions the rutile phase is the most stable phase whereas anatase is the one preferentially formed and stable under basic conditions. Brookite, which is the least studied of the three main naturally occurring TiO_2 polymorph, was for a long time a byproduct observed during synthesis of anatase and mainly obtained under neutral to acidic conditions (Zhang and Banfield 2014a, Paola et al. 2013).

In the same way, thanks to a thermodynamic approach based on surface free energies and surface tensions, Barnard et al. could predict the morphology of rutile and anatase from very acidic to very basic conditions as depicted in Fig. 6.2 (Barnard and Curtiss 2005a). Insofar as the various crystal facets of the TiO_2 polymorphs exhibit different surface energies, this observation implies that the reactivity of the material will be modified and can actually be tailored by adopting a surface engineering approach (Liu et al. 2014, Batzill 2011, Fujishima et al. 2008).

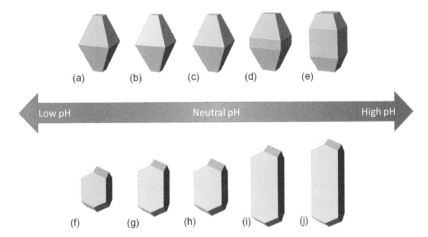

Fig. 6.2 Predicted morphology evolution at various pH for anatase (top - red) with (a) hydrogenated surfaces, (b) hydrogen-rich surface adsorbates, (c) hydrated surfaces, (d) hydrogen-poor adsorbates, (e) oxygenated surfaces and for rutile (bottom – grey) with (f) hydrogenated surfaces, (g) hydrogen-rich surface adsorbates, (h) hydrated surfaces, (i) hydrogen-poor adsorbates, (j) oxygenated surface. (Reprinted with permission from Barnard, A. S. and L. A. Curtiss. 2005a. Prediction of TiO_2 nanoparticle phase and shape transitions controlled by surface chemistry. Nano Letters 5: 1261-1266)

6.2.2 Electronic Properties

Titanium dioxide is one of the most widely studied materials because of its major role as photocatalyst. Generally speaking, TiO_2 polymorphs are considered as n-type semiconductors, which is an electron donor, since TiO_2 at room temperature exhibits oxygen vacancies. The various TiO_2 polymorphs can readily absorb solar light within the UV range leading to the formation of charge carriers, namely an electron (e^-)-hole (h^+) pair, in the bulk or at the surface of the material. The various TiO_2 polymorphs do not exhibit the same ability to absorb light and they are actually characterized by different band gap energies with corresponding wavelength ranging from 387 nm to 410 nm. Anatase, which is nowadays the most industrially used and studied photocatalyst, has an indirect band gap with a value determined at 3.23 eV. On the other hand, rutile and brookite show a direct band gap with values respectively around 3.03 and 3.14 eV (Grätzel and Rotzinger 1985).

Despite the larger band gap, anatase is generally considered to have the highest photocatalytic activity compared to the other TiO_2 polymorphs (Zhang et al. 2014b). One usually admitted explanation is related to the electron-hole lifetime, which is longer for anatase than rutile due to its indirect band gap recombination since the electron hole recombination cannot be done directly. The notion of direct and indirect band gap is schematized in Fig. 6.3. Contrary to direct band gap transition, the electron transfer from the valence bond (VB) to the conduction band (CB), or the reverse, is observed along with a crystal momentum transfer (phonon) leading to a longer electron-hole lifetime (Xu et al. 2011b, Jung and Kim 2009). Thus, in comparison with rutile or brookite, the e^--h^+ pairs formed within the anatase bulk material would have a longer time to diffuse towards the surface where various reactions could take place (Zhang et al. 2014b).

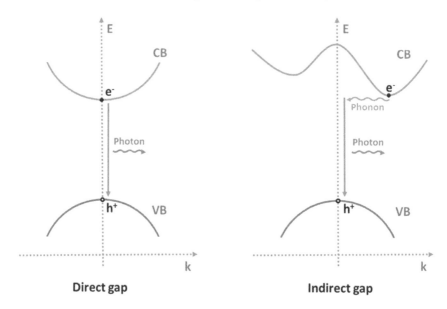

Fig. 6.3 Schematic representation of direct and indirect band gap electron hole recombination

The overall reactions that take place with the photogenerated species on the surface of the TiO_2 crystals or particles depend strongly on the surface properties of the material itself. Actually, it was demonstrated that anatase shows a better adsorption affinity of reactants compared to rutile, leading to a better ability to trap charge carriers at the material surface (Sclafani and Herrmann 1996). Nevertheless, even if rutile is considered as a less active photocatalyst than anatase, it was demonstrated that the combination of both phases leads to a higher global photocatalytic activity compared to the pristine phases. This so called "synergistic effect" is usually attributed to charge separation at the phase junction and to an extended photoactivity within the visible range due to the presence of rutile (Hurum et al. 2003). Noteworthy, there is yet no consensus upon the band alignment between rutile and anatase even if a global tendency is in agreement with an electron flow from rutile to anatase and with a hole migration in the reverse direction in the case of anatase-rutile heterojunction as depicted in Fig. 6.4 (Shen et al. 2014, Scanlon et al. 2013).

Finally, it is important to note that the electronic properties and global photocatalytic performances of titanium dioxide polymorphs are actually not only dependent on intrinsic TiO_2 properties but also on morphological and structural aspects like crystal facets, grain size, phase junction or the presence of dopants, as it will be discussed later in this chapter. All these factors can actually have a direct influence on the generation and lifetime of photoexcited charge carriers, band-gap energy values and surface reactant adsorption ability, which are key parameters when designing photocatalysts for hydrogen production from water splitting or sacrificial reagents.

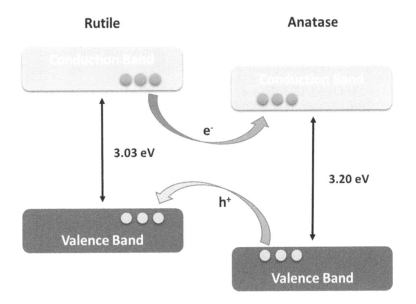

Fig. 6.4 Schematic representation of one possible band alignment between anatase and rutile with the electron-hole transfer between both phases

6.3 MECHANISM OF PHOTOCATALYTIC WATER SPLITTING WITH TIO$_2$

Water splitting occurs naturally in photosynthesis when the solar light is absorbed and converted into the chemical energy through a complex biological pathway. However, this process is incompatible with the requirements of large-scale hydrogen production. In fact, water splitting is a strongly endothermic, non-spontaneous uphill reaction with a $\Delta G° = 238$ kJ·mol^{-1}. Various energy driven water splitting routes are possible throughout the use of thermal, electric, and photonic energy or sometimes the combination of these (Tee et al. 2017). Among the proposed routes, water splitting by means of photonic energy, in other words, via utilization of solar light is known to be an outstanding sustainable approach towards clean energy production that requires only the presence of light harvesting photocatalysts to drive the reaction. Theoretically, the minimum photonic energy for water splitting should be 1.23 eV, which corresponds to 1100 nm, i.e. NIR light. However, water itself does not absorb appreciable radiation in visible and NIR ranges of the optical spectrum. Early in the 1970s, TiO$_2$ semiconductor electrode was studied as photocatalyst for water splitting with the assistance of external bias (Fujishima and Honda 1972). Titanium oxide meets the basic requirements of a semiconductor to be engaged in the photocatalytic water splitting. It possesses a wide band gap and a sustainable band energy potential, which means that the bottom level of the conduction band is more negative than the reduction potential of H$^+$/H$_2$ (0 V vs. NHE) and the top level of the valence band is more positive than the oxidation potential of O$_2$/H$_2$O (1.23 V vs. NHE) (Joy et al. 2018).

Four-electron process of water splitting over TiO$_2$ leading to H$_2$ and O$_2$ emission is initiated by the generation of the electron-hole pair via the absorption of photons with an energy greater than the band gap, E_g, of TiO$_2$ equal to 3 eV for rutile and 3.2 eV for anatase, respectively, as it shown in Fig. 6.5 (Murphy et al. 2006). After excitation, the photogenerated charge carriers migrate to the surface of the photocatalyst, and this step strongly depends on the structure, crystallinity and size of TiO$_2$ photocatalyst. A high degree of crystallinity is often required for activating water splitting and inhibiting any possible electron-hole recombination. The final step involves the surface chemical reactions which are greatly affected by the presence and number of active sites ready to carry out photoreduction and photo-oxidation liberating H_2 and O_2 (Kudo and Miseki 2009). It is noteworthy that water molecule can also be dissociated via one- or two-electron processes shown by the equations (1) and (2), respectively, but the four-electron reaction pathway (reaction 3) is thermodynamically more favorable (Siahrostami et al. 2017).

$$H_2O \rightarrow OH^\bullet + (H^+ + e^-) \qquad E^\circ = 2.73 \text{ V} \tag{1}$$

$$2H_2O \rightarrow H_2O_2 + 2(H^+ + e^-) \qquad E^\circ = 1.76 \text{ V} \tag{2}$$

$$2H_2O \rightarrow O_2 + 4(H^+ + e^-) \qquad E^\circ = 1.23 \text{ V} \tag{3}$$

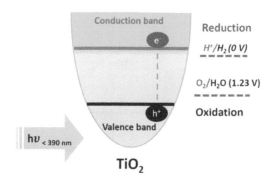

Fig. 6.5 Photocatalytic mechanism of water splitting over TiO_2 photocatalyst

Water adsorption onto a TiO_2 surface is one of the most important parameters determining the photocatalytic activity. Molecular modelling pointed out the importance of the five-fold coordinated Ti(IV) sites (TiN5) and the bridging oxygen sites in the mechanism of photocatalytic water splitting (Lindan et al. 1998). Water molecule is adsorbed first at TiN5 sites and then moves to the bridging oxygen via hydrogen bonding. The H-bonding becomes increasingly stronger and eventually results in complete heterolytic dissociation of water molecule into OH^- and H^+ ions, respectively. This process leads to a steady-state between the H^+ on the bridging oxygen and the OH^- coordinated by TiN5. It has also been reported that the photogenerated hole lifetime is important for the efficiency of TiO_2 photocatalyst (Tang et al. 2008). The transient absorption spectra provided direct observation of photogenerated charge carriers. It was shown that in the presence of electron scavengers, photohole exhibits an absorption band at 460 nm. By contrast, in the presence of hole scavengers photoelectron shows an absorption band at 800 nm. The photoholes and photoelectrons have the identical dynamics attributed to the electron-hole recombination in a microsecond time scale. Methanol can trap photoholes very rapidly (< 10 ns), which is much higher than for water molecule (ca. 100 μs).

In 1980, Kawai and Sakata succeeded in being the first researchers to introduce the use of sacrificial reagents in photocatalytic processes (Kawai and Sakata 1980). Before going forward, it is crucial to remember that once sacrificial reagents are incorporated in the photocatalytic H_2 or O_2 production, the process is no longer called water splitting. Nevertheless, this field of study is of great importance as the true water splitting process in the absence of sacrificial reagents and stable photocatalysts seems to be inefficient (Pan 2016).

Sacrificial reagents are classified either as reducing reagents or as oxidizing ones. Normally, reducing reagents are also known as sacrificial electron donors (SED); they are capable of scavenging the photogenerated holes, concentrating the electrons in a photocatalyst to further enhance H_2 evolution reaction. There exists a wide variety of SED reagents either of organic or inorganic nature; however, only the ones derived from biomass and abundant in nature and industries are interesting to be engaged in hydrogen evolution reactions (Kawai and Sakata 1980). On the other hand, oxidizing sacrificial reagents are also referred as electron acceptors; they are targeted to consume the photogenerated electrons, enhancing the yield of O_2 evolution (Kudo and Miseki 2009). In response to the growing demands for clean energy resources, photocatalytic hydrogen production reactions were investigated vigorously in the past years. Cargnello et al. investigated photo-reforming of bio-derived sacrificial reagents over natural metal oxides (Cargnello et al. 2011); then, a similar work was reported by Puga using synthetic photocatalysts (Puga 2016). More recently, Yasuda et al. reported a review that summarized previous studies of photocatalytic hydrogen production with TiO_2-based photocatalysts in the presence of the various types of sacrificial reagents, such as polyols, carboxylic acids and saccharides (Yasuda et al. 2018). Herein, we will highlight the processes of photocatalytic H_2 production in the presence of organic sacrificial reagents, including alcohols, carboxylic acids, and sugars.

Methanol is a typical alcoholic sacrificial reagent used in photocatalytic H_2 production systems, often employed to study the photocatalytic behavior of photocatalysts. Even a small amount of methanol is sufficient to increase significantly the rate of hydrogen production upon efficient scavenging the photogenerated holes, enhancing the overall electron-hole separation and thereby increasing the quantum efficiency toward better H_2 production. For instance, small amount of methanol elevated the hydrogen production rate up to 6644 $\mu mol.h^{-1}.g^{-1}$ upon the photocatalytic treatment of aqueous methanol solution in the presence of Cu_2O mesoporous beads of TiO_2 (Cheng et al. 2014a). The Gibbs free energy of methanol decomposition ($\Delta G° = 16.1$ kJ mol^{-1}) is much lower than that of water splitting ($\Delta G° = 237.1$ kJ mol^{-1}), which can explain high efficiency of methanol in photocatalytic H_2 production (Kawai and Sakata 1980). Two other works reported that the decomposition of methanol over platinum modified TiO_2 photocatalyst that derives either from the direct contact of methanol with the photogenerated free or trapped holes at the TiO_2 surface (Chen et al. 1999), or indirectly upon reacting with the hydroxyl radical ($\cdot OH$) produced primarily once hydroxy-groups are oxidized by the photogenerated holes (Turchi and Ollis 1990). The decomposition of methanol throughout direct reaction with photogenerated hole yields strong electron donor $\cdot CH_3OH$ radical that further enhances the yield of H_2 production. This effect, called "double effect", was also reported for formate radical $\cdot CO_2^-$ (Turchi and Ollis 1990). The reaction mechanism involving a combination of both possible decomposition routes was reported for the H_2 photocatalytic generation in D_2O or H_2O mixtures with MeOH in the gaseous phase (Chiarello et al. 2011). It was shown that the variation of methanol concentration determines its decomposition pathway. At low concentration, the oxidation of methanol occurs indirectly by $\cdot OH$ radical, whereas at higher concentration direct decomposition of methanol was carried out by the hole transfer from the photoexcited TiO_2. However, the nature and mechanism of formation of the final products obtained, besides H_2, cannot be generalized because they are greatly dependent on the nature of modified TiO_2 utilized in the photoreaction. For example, it was shown that in the presence of Pd-loaded TiO_2, the CO_2 formed during methanol decomposition originates from the oxidation of the initially produced CO by the O^- anion-radical formed upon the photoexcitation of TiO_2 (Dickinson et al. 1999).

In 1981, Sakata and Kawai reported for the first time H_2 production throughout ethanol decomposition over metal-modified TiO_2 (Sakata and Kawai 1981). Methane and acetaldehyde were identified as principal products of this photocatalytic reaction. In a more recent review, Yasuda et al. reported photocatalytic hydrogen production with other alcohols as sacrificial reagents, including 1-propanol, 2-propanol, 2-butanol and 1,2 ethanediol (Yasuda et al. 2018). Velàzquez et al. reported the kinetics of H_2 production with Pt-loaded TiO_2 photocatalysts in the presence of MeOH, EtOH, and i-PrOH sacrificial reagents in aqueous solutions (Velázquez et al. 2017). The obtained H_2 formation rates summarized in Table 6.2 indicate the highest values for MeOH in agreement with other works. It was shown that H_2 production rate is strongly dependent on the reaction temperature, increasing continuously in the temperature range from 5 to 60°C.

TABLE 6.2 H_2 production rate achieved for TiO_2 photocatalysts at 25°C in different water-alcohol solutions under irradiation with UV light (Velázquez et al. 2017). 2 wt%Pt-TiO_2-1S and 6 wt%Pt-TiO_2-2S catalysts were prepared by Pt(IV) photoreduction without and with drying, respectively. 2 wt%Pt-TiO_2 catalyst has been prepared by incipient wetness impregnation technique

Photocatalyst	Alcohol	mmol.g^{-1}.h^{-1}
2 wt%Pt-TiO_2	MeOH	16.4
	EtOH	11.5
	i-PrOH	9.4
	H_2O	1.1
2 wt%Pt-TiO_2-1S	MeOH	7.6
	EtOH	6.3
	i-PrOH	2.9
6 wt%Pt-TiO_2-2S	MeOH	4.0
	EtOH	2.6
	i-PrOH	2.0
P25 TiO_2	MeOH	0.6

Reddy et al. have investigated the effect of different sacrificial reagents on H_2 production rate using Cu-Ag quantum dots embedded into TiO_2 nanotubes (Reddy et al. 2017). They concluded that tri hydroxyl alcohols like glycerol are more efficient for H_2 production than mono and di-hydroxyl alcohols. The highest activity was reported for glycerol, followed by ethylene glycol, ethanol and methanol. Photocatalytic reforming of ethylene glycol over Pt/TiO_2 catalyst leads to the formation of not only H_2 and CO_2, but also CO, CH_4, C_2H_6, and C_2H_4 (Li et al. 2017). The different polarity of the sacrificial reagents strongly influences the surface properties of the catalyst and thereby the overall electron donation property of the catalyst, resulting in different efficiency of sacrificial reagents in hydrogen evolution (Yasuda et al. 2018).

6.4 CORE-SHELL NANOPARTICLES

6.4.1 TiO$_2$-Core Photocatalysts

The pristine TiO_2 is known to be a photocatalyst of moderate activity. Coating of TiO_2 particles with an appropriate shell is believed to alleviate this problem. Self-doped $TiO_2@TiO_{2-x}$ core-shell nanoparticles have been prepared by TiO_2 partial reduction with hydrogen at 400-600°C (Hu et al. 2020). It was found that the original band gap of TiO_2 after hydrogenation remained unchanged about 3 eV. However, a new absorption band was observed at ca. 1425 nm, which was assigned to oxygen vacancies at the surface shell of TiO_2 with a thickness of ca. 2 nm. The authors reported high photocatalytic activity of prepared material in the reaction of methylene blue photocatalytic degradation under the irradiation with Xe lamp. By contrast, the performance of this photocatalyst for hydrogen production is unknown. Core-shell nanoparticles composed of nanocrysralline rutile doped with Ti(III) core and amorphous Ti(IV) shell have been obtained by combination of TiO_2 hydrogenation and hydrothermal treatment of partially reduced $TiO_2{}^{Ti(III)}$ (Yang et al. 2016). These particles exhibit photocatalytic activity in hydrogen generation from aqueous methanol solution. It should be noted, however, that maximal photocatalytic activity can be reached after 1% of Pt loading only. This is a general problem of self-doped TiO_2, since, despite good vis/NIR light absorption, oxygen vacancies in such materials create electron-hole recombination centers and noble metal loading is required to provide efficient charge separation (Ma et al. 2014). Recently, TiO_2@C nanoparticles, where C is a graphitic carbon shell, have been obtained by polymeric nano-encapsulation process using Ti@PAN, where PAN is a polyacrylonitrile, followed by PAN pyrolysis (Vasei et al. 2014). Spherical particles are composed from ca. 17 nm anatase core coated with uniform ca. 2 nm carbon shell. Electrochemical characterization revealed that the impedance of TiO_2@C nanoparticles is much lower than that of pristine TiO_2. Furthermore, irradiation of TiO_2@C sample with UV light resulted in a further decrease of impedance and in a significant enhancement of the photocurrent indicating that it could be a promising material for the photoelectrochemical water splitting. Nanographene (NGO) coated TiO_2 (TiO_2@NGO) has been synthesized by a two-step oxidation process (Kim et al. 2011). Photocatalytic and photoelectrochemical measurements revealed higher activity of TiO_2@NGO compared to TiO_2 nanoparticles deposited on micrometric GO sheets in the process of hydrogen generation. The authors suggested that the electrons photogenerated in TiO_2 can be more easily transferred to NGO than micrometric GO because of a very close contact of the TiO_2 surface with NGO in core-shell nanoparticles. However, as in the case of self-doped TiO_2, platinum loading is required to provide high photocatalytic activity indicating the problems of charge separation in TiO_2@NGO nanoparticles. The core-shell nanocomposite TiO_2@$ZnIn_2S_4$ has been prepared by a hydrothermal method from low-cost precursors (Yuan et al. 2013). Ternary semiconductor chalcogenide $ZnIn_2S_4$ has a band gap in a visible range (2.34-2.48 eV) and exhibits high photocatalytic activity (Gao et al. 2013). It was found that core-shell TiO_2@$ZnIn_2S_4$ nanoparticles have better photocatalytic performance than TiO_2 and $ZnIn_2S_4$ under visible light irradiation, which has been attributed to the electron transfer between the conduction bands of both components.

The hollow 3D TiO_2 particles have attracted a lot of attention last decade as promising photocatalytic systems due to their high surface area and strong light harvesting. Stable TiO_2 nanostructured hollow microspheres have been prepared by template-free hydrothermal method using the mixtures of $(NH_4)_2TiF_6$ and glucose (Yu and Wang 2008) or $TiOSO_4$ and HBF_4 (Wang et al. 2014a). The size of hollow TiO_2 spheres composed of plate-like pristine anatase or anatase/rutile nanoparticles varies in the range of 0.6-5.0 μm. The study of nitrogen adsorption/desorption isotherms revealed the presence of mesopores (2-50 nm) in TiO_2 shell of the hollow spheres (Yu and Wang 2008). Photocatalytic study of hydrogen production over Pt-loaded (1 wt.%) TiO_2 samples in methanol aqueous solutions pointed out better performance of the hollow microspheres compared to non-spherical TiO_2 nanoparticles (Wang et al. 2014a). This effect has been assigned to the multireflections of light inside the interior cavities leading to the enhanced light harvesting. Alternatively, it can also be attributed to microsphere mesoporosity, which provides higher percentage of exposed reactive TiO_2 facets. Loading of nanostructured TiO_2 hollow spheres with Cu, Co, Cr, Ag, and Ni using the incipient wetness impregnation technique and thermal treatment in the presence of hydrogen allowed the assessment of hollow core-shell particles coated with corresponding metal nanoparticles (Seadira et al. 2018). It was found that the photocatalytic activity in the process of glycerol reforming for prepared catalysts follows the order of Cu>Ag>Ni>Co>Cr in correlation with electron/hole separation efficiency. The efficient in photocatalysis type II heterojunction has been obtained in the noble metal-free core-shell particles comprised TiO_2 hollow spheres and $ZnIn_2S_4$ flower-like nanocrystals (Xia et al. 2017). The SEM/TEM images of these particles with a quite unique morphology are shown in Fig. 6.6. It was shown that TiO_2@$ZnIn_2S_4$ exhibits high photocatalytic activity. However, $ZnIn_2S_4$ suffers a serious drawback of photocorrosion (Chen et al. 2015). More stable TiO_2@g-C_3N_4 hollow microspheres have been obtained by hydrothermal treatment as described previously (Yu and Wang 2008) followed by the ultrasonically assisted g-C_3N_4 deposition (Ma et al. 2018). After coating, g-C_3N_4 appeared at the surface of TiO_2 spheres as stacked lamellar structures. It was reported that TiO_2@g-C_3N_4 particles are stable under Xe lamp irradiation and exhibit higher photocatalytic activity than non-coated TiO_2 hollow microspheres. The photocatalytic performance of TiO_2@g-C_3N_4 has been attributed to three phenomena: higher specific area of core-shell nanoparticles, lower bandgap of g-C_3N_4 (2.76 eV) compared to pristine TiO_2 (3.2 eV), and three-dimensional heterojunction between the TiO_2 core and g-C_3N_4 shell.

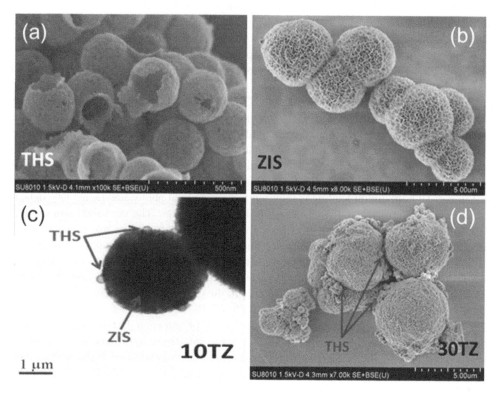

Fig. 6.6 (a) SEM and TEM images of the TiO_2 hollow spheres (THSs), (b) $ZnIn_2S_4$ (ZIS) nanoflowers, (c) TEM image of $TiO_2@ZnIn_2S_4$ composite (xTZ) where x corresponds to TiO_2 wt%, (d) SEM images of $TiO_2@ZnIn_2S_4$ obtained at different experimental conditions. (Reprinted with permission from Xia, Y., Q. Li, K. Lv and M. Li. 2017. Heterojunction construction between TiO_2 hollow sphere and $ZnIn_2S_4$ flower for photocatalysis application. Appl. Surf. Sci. 398: 81-88)

Doping of TiO_2 with Bi atoms causes a red shift in photoadsorption spectrum of TiO_2, which in return enhances its photoresponse under visible light (Reddy et al. 2011). Since Bi atom (106 pm) is bigger than Ti atom (61 pm), most of the studied Bi-TiO_2 coupled photocatalysts consist of the Bi_2O_3 as a separate phase which might limit the catalyst's durability due to the possible leaching of Bi_2O_3 species. However, Bian et al. reported a promising method of incorporating Bi dopants into anatase TiO_2 lattice throughout alcoholysis under solvothermal conditions (Bian et al. 2009), generating more durable and visible light active $Bi_xTi_{1-x}O_2$ hollow spheres with mesoporous outer surface. The photocatalytic activity of this nanocomposite was tested for the degradation of 4-chlorophenol (4-CP) with a 500 W Xe lamp and a UV light filter. The increased photocatalytic activity was assigned to the presence of intermediate energy levels formed after doped making $Bi_xTi_{1-x}O_2$ easier to be activated. The photocatalytic activity of bismuth (Bi) doped hollow titania using carbon nanosphere as a template (Bi-H-TiO_2) was studied for the first time by Xu et al. in the process of photocatalytic degradation of 4-chlorophenol (4-CP) and methylene blue (MB) (Xu et al. 2011a). Bi-H-TiO_2 micro-hollow spheres with improved photocatalytic activity have been synthesized by relatively simple three-stage process (Yu et al. 2007). At the first stage, porous carbon spheres were prepared using glucose precursor via a hydrothermal treatment. Next, Bi-doped titania was obtained from titanium (IV) butoxide precursors dissolved in i-PrOH and added dropwise to bismuth nitrate and kept under reflux conditions for 24 h at 75°C. The final photoactive Bi-H-TiO$_2$ was obtained by simple stirring of the two previous precursors. It was suggested that the doping with Bi atoms inhibits the grain growth and enhances the absorption of visible light, confirmed by the red shift for Bi-H-TiO$_2$ (ca. 380 nm) when compared to pure titania (ca. 320 nm). The visible light absorption was enhanced with the increase in bismuth concentration up to 4%, increasing the overall photocatalytic degradation of 4-CP and MB. However, data for photocatalytic hydrogen production was not reported.

In principle, formation of *p-n* heterojunction between *n*-type TiO$_2$ and suitable *p*-type earth-abundant metal oxide semiconductor could be very beneficial for electron/hole separation, leading to the increase in photocatalytic activity of TiO$_2$. Recently, TiO$_2$@Co$_3$O$_4$ core-shell nanoparticles have been prepared by plasma-enhanced atomic layer deposition (ALD) technique (Zhao et al. 2017). In the experiments, volatile cyclopentadienyl cobalt (CoCp(CO)$_2$) was used as a cobalt precursor and oxygen plasma as an oxygen source. High-resolution TEM analysis revealed formation of 25-26 nm TiO$_2$ nanocrystals coated by ultra-small (ca. 2 nm) Co$_3$O$_4$ nanoparticles. Mott-Schottky plots pointed out the *p-n* heterojunction formation in TiO$_2$@Co$_3$O$_4$ nanocomposite. Upon UV-light irradiation, electron-hole pairs can be formed in both TiO$_2$ and Co$_3$O$_4$. However, photogenerated electrons would move preferably to the conduction band of TiO$_2$ due to much lower bandgap of Co$_3$O$_4$ (ca. 2.4 eV) compared to that of TiO$_2$ (ca. 3.2 eV) providing effective charge carrier separation. Consequently, TiO$_2$@Co$_3$O$_4$ nanoparticles exhibit noticeable photocatalytic activity in the process of methyl orange degradation under visible light ($\lambda \geq$ 420 nm) irradiation. However, this catalyst was not tested for photocatalytic hydrogen generation. The combination of hydrothermal and chemical-bath deposition has been applied for the synthesis of TiO$_2$@NiO core-shell nanorod array (Cai et al. 2014). It was shown that NiO formed nanoflake film on TiO$_2$ nanorod array. Deposition of NiO film causes the decrease of a bandgap until ca. 3.0 eV. The photocatalytic study has not been performed. However, the enhanced electrochemical activity of TiO$_2$@NiO structure was confirmed by cyclic voltammetry technique. In addition, obtained materials showed promising electrochromic properties, which was attributed to the synergetic contribution from the core-shell heterostructure.

The activity of TiO$_2$@Cu$_2$O core-shell composite has been mostly studied for the photocatalytic degradation of organic pollutants (Liu et al. 2016, Ding et al. 2018). Cu$_2$O is a semiconductor with a small band gap around 2 eV and it can be coupled with other semiconductors of large band gaps to promote electron-hole separation and ultimately improve the photocatalytic efficiency of the material. However, the inadequate structural design of anatase TiO$_2$ in TiO$_2$@Cu$_2$O composites limited its photocatalytic activity (Barreca et al. 2009, 2011). TiO$_2$@Cu$_2$O composite with improved H$_2$ production activity was reported by Cheng et al. (Cheng et al. 2014b). In this work, sub-micron-sized mesoporous TiO$_2$ beads (MTBs) were obtained via a simple metal-salt-based hydrothermal process, and further decorated with Cu$_2$O nanocrystals through a fast and low-cost chemical deposition process. An ultra-high H$_2$ evolution rate of 223 mmolh^{-1}g^{-1} is achieved at 20°C upon irradiation of water-methanol-solution containing 0.01 g of TiO$_2$@Cu$_2$O composite with Cu$_2$O precursor concentration of 0.2 M with a high-pressure mercury lamp of 400 W. This ultra-high H$_2$ evolution rate is one order of magnitude higher than those achieved by commercial P25 TiO$_2$. The improved photocatalytic activity TiO$_2$@Cu$_2$O composites is related to the improved structural properties of MTBs including their high specific area of 189 m^{2}g^{-1}, large pore volume of 0.43 cm^{3}g^{-1}, and a suitable pore size of 8.9 nm in combination with the excellent charge separation maintained by the presence of Cu$_2$O semi-conductor.

Use of magnetic nanocatalysts allows to avoid filtration or centrifugation operations after completion of the reaction. Both hydrothermal and solvothermal processes were employed to prepare the 3D-TiO$_2$ particles fully covered with magnetic cobalt ferrites (CoFe$_2$O$_4$) and coated with hierarchical porous graphene aerogels (HPGA) (Santhosh et al. 2018). Images obtained from the field emission scanning electron microscopy (FE-SEM) confirmed the flower-like structure of the magnetic 3D-TiO$_2$@HPGA nanocomposite and X-ray diffraction patterns confirmed the existence of TiO_2 in its rutile phase. These particles showed good results toward Bisphenol A (BPA) degradation during the irradiation by UV light at 254 nm and 365 nm due to the action of the photogenerated $OH \cdot$ and h^+ species.

In general, TiO_2-core photocatalysts can be successfully obtained without further incorporation of noble metals. However, their efficiency towards photocatalytic hydrogen production has not yet been reported and in most of the cases was not feasible without the presence of noble metals. Table 6.3 provides information concerning the photocatalytic H_2 generation efficiency of doped catalyst over TiO_2-core photocatalyst. Yet, it remains difficult to compare the rates of H_2 production between two systems presented by different authors as most of their works lack precision with respect to the nature and power intensity of their light source. Without such crucial information, it is unfair to conclude which catalyst has the highest photocatalytic activity.

TABLE 6.3 Photocatalytic hydrogen production rate presented in $\mu mol.g^{-1}.h^{-1}$ upon light irradiation of aqueous solutions over $TiO_2@X$ core-shell photocatalyst

X	Light Source	H_2 $\mu mol.g^{-1}.h^{-1}$	Ref.
Amorphous Ti(IV)*	300 W Xe lamp	5630	Yang et al. 2016
Cr**		149	
Ni**		368	
Co**	Solar light	465	Seadira et al. 2018
Ag**		1593	
Cu**		4194	

* Core TiO_2 (Rutile) was doped with Ti(III). Amorphous Ti(IV) shell was loaded with 1% of Pt. **Core of the particles was TiO_2 hollow sphere.

6.4.2 TiO₂-Shell Photocatalysts

In aqueous solutions, it is extremely difficult to coat nanoparticles with TiO_2 because of high sensitivity of titanium precursors, usually Ti(IV) alkoxides, to hydrolysis. The core-shell Ag@TiO_2 nanoparticles have been prepared for the first time by reflux of $AgNO_3$ and Ti(IV) butoxide in a mixture of dimethylformamide (DMFA) and ethanol in the presence of small amounts of water and acetylacetone as a cupping reagent (Pastoriza-Santos et al. 2000). DMFA plays a role of reducing reagent for Ag(I) and also provides sufficiently slow Ti(IV) butoxide hydrolysis to form a shell around the Ag core. The authors observed two different populations of nanoparticles: (I) Ag particles with an average size of ca. 20 nm coated by amorphous TiO_2 shell with a thickness of 2 nm and (II) smaller (ca. 4 nm) non-coated Ag particles. The prepared Ag@TiO_2 nanoparticles have been used for the preparation of thin films by layer-by-layer deposition with a polyelectrolyte. UV-photoexcitation of Ag@TiO_2 nanoparticles with an amorphous TiO_2 shell results in accumulation of the electrons in the Ag core leading to the shift in the surface plasmon band from 460 to 420 nm (Hirakawa and Kamat 2005). It was shown that photoexcited Ag@TiO_2 particles enable the reduction of C_{60} molecules in ethanol solutions. The core-shell Ag@TiO_2 particles with 50-100 nm Ag core and nanocrystalline (3-10 nm) anatase shell have been prepared by polyol synthesis in ethylene glycol solutions using titanium (IV) glycolate as titanium precursor (Yang et al. 2013). The Ag@TiO_2 particles show high photocatalytic performance in the process of organic dyes degradation in aqueous solutions under UV-light irradiation. An alternative approach to access M@TiO_2 nanoparticles, where M is Pt, Au, and Ni, reported recently (Liu and Sen 2012) is based on the oxidation of Ti(III) and followed hydrolysis of Ti(IV) species on nanostructured metal surfaces, which catalyze TiO_2 growth. The deposited amorphous TiO_2 can be further annealed at 400-500°C to form anatase nanocrystalline shell. It was shown that the core-shell Ni@TiO$_2$ nanoparticles can be utilized as magnetic recyclable photocatalyst active under UV-light irradiation. Recently, Ti@TiO$_2$ core-shell nanoparticles have been prepared by sonohydrothermal (SHT) treatment of metallic titanium, Ti^0, nanoparticles in pure water (Nikitenko et al. 2015, 2018). The SHT treatment is an innovative technique allowing synthesis of materials with advanced properties by simultaneous hydrothermal heating and action of power ultrasound. The TEM images of Ti^0 show quasi-spherical metallic particles with a diameter in the range of 20-80 nm (Fig. 6.7a). The SHT treatment at 200°C (P = 13 bar, τ = 6 h) using 20 kHz ultrasound (P$_{ac}$ = 10 W) leads to the formation of 5-15 nm anatase crystals coating the metallic surface (Fig. 6.7b). A solid-state reflectance spectrum of Ti^0 NPs exhibits a broad continuum in the UV/vis/NIR spectral range resulting from interband transitions of non-plasmonic titanium particles. The optical spectrum of Ti@TiO$_2$ NPs shows an absorption band centered at 308 nm in addition to broad continuum. This band is typical for the bandgap transition of the pristine anatase bandgap. It was found that Ti@TiO$_2$ NPs demonstrate strong photothermal effect in hydrogen production from the aqueous solutions of sacrificial reagents. The photothermal effect will be discussed below.

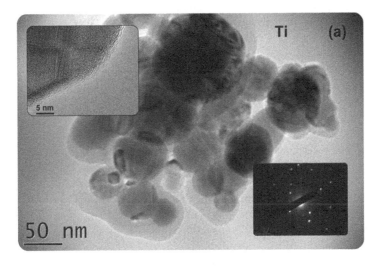

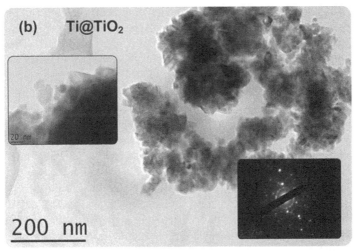

Fig. 6.7 Typical TEM images of titanium nanoparticles before and after sonohydrothermal treatment. (a) Ti^0 NPs, (b) Ti@TiO$_2$ NPs. (Reprinted with permission from Nikitenko, S.I., T. Chave, C. Cau, H.-P. Brau and V. Flaud. 2015. Photothermal hydrogen production using noble-metal-free Ti@TiO$_2$ core-shell nanoparticles under visible-NIR light irradiation. ACS Catal. 5: 4790-4795)

The hydrothermal carbonization (HTC) process of biomass is employed for the synthesis of novel carbon-based materials. When other reagents are present within the HTC process of carbohydrates or other organic monomers, novel carbon-encapsulated core-shell composites can be synthesized. Owing to the ease of removing these carbonaceous materials, hollow spheres with tunable void space can be easily fabricated, e.g. Fe_2O_3, Ni_2O_3, Co_3O_4, CeO_2, MgO, and CuO were successfully synthesized using this approach (Hu et al. 2010). Based on this, TiO_2 hollow spheres with a tunable diameter and shell thickness were obtained upon adding the hydrothermally prepared carbon spheres and $TiCl_4$ precursors into mixture of $(NH_4)_2SO_4$ and HCl with a molar ratio of 1:5 (Zhang et al. 2011). By the combination of acid catalyzed hydrolysis process and hydrothermal treatment, hollow TiO_2 spheres were obtained. Unfortunately, the photocatalytic activity of these hollow spheres was evaluated according to their ability to decompose phenol under UV light and no records were found for the hydrogen production process. The author suggested that the improved activity was related to the bimodal meso-macroporous structure and the multiple reflection of the light in the interior of the hollow spheres.

Heterogeneous coupling is one of the most common strategies adopted for lowering the rate of the recombination processes of the photogenerated charge carriers. The junctions formed between magnetic materials and TiO_2 in a core-shell structure and their effect on the recombination rate or in general on the overall photocatalytic process was of great interest for several researchers concerned in this field. Recently, Fu et al. managed to develop an accessible coating method to elaborate porous 1D core-shell structures, with an alpha iron oxide (α-Fe_2O_3) core coated with TiO_2 shell, α-Fe_2O_3@TiO_2 (Fu et al. 2018). Initially, spindle-shaped α-Fe_2O_3 cores with a diameter of ~100 nm and ~300 nm length has been prepared via a hydrothermal method and then uniform TiO_2 shell has been obtained by a controlled condensation process of tetrabutyltitanate (TBT) in acetone solution. The heterojunction of α-Fe_2O_3-TiO_2 interface is shown in the HRTEM images (Fig. 6.8). By this method, the thickness of the shell can be controlled and no calcination step is required for the crystallization into anatase phase. According to literature, pure α-Fe_2O_3 does not possess surface plasmon resonance in the range of 200 to 800 nm (Hou et al. 2012). However, compared to pure TiO_2 nanoparticles, a red shift in the bandgap of TiO_2 was observed for α-Fe_2O_3@TiO_2 nanoparticles. Unfortunately, the photocatalytic hydrogen production is not yet tested for these particles, only the photocatalytic degradation of methyl orange (MO) was investigated. Thanks to the narrow band gap semiconductor α-Fe_2O_3 solar light was absorbed and the photogenerated electrons migrate to the conduction band of TiO_2 under the action of internal electric field. Then, the electrons interact with the dissolved oxygen species to produce superoxide ion O_2^- and hydrogen peroxide H_2O_2, while the photogenerated holes accumulate in the conduction band of α-Fe_2O_3 to react with the surface existing OH^- to produce reactive hydroxyl radicals $OH^{\bullet}$. The ultrasonically assisted sol gel synthesis was used for the preparation of Fe_2O_3@TiO_2 flower-like core-shell nanoparticles (Abdel-Wahab et al. 2017). Their photocatalytic activity was demonstrated by the degradation of paracetamol once the aqueous solutions are illuminated with a 450 W medium pressure mercury vapor lamp. The proposed mechanism of paracetamol degradation was attributed to the activity of the photogenerated $OH^{\bullet}$ radicals.

Liu et al. studied the photocatalytic activity of bismuth ferrite ($BiFeO_3$) nanoparticles coated with TiO_2 that were obtained by the initial citrate-nitrate combustion method to form the $BiFeO_3$ followed by the hydrolysis of $Ti(OBu)_4$ to elaborate the $BiFeO_3$@TiO_2 core-shell photocatalyst (Liu et al. 2017). Unlike $BiFeO_3$ and TiO_2 separately, the nanocomposite was found to be more active under visible light towards the degradation of different organic dyes using a 400 W Xe lamp. The higher photocatalytic efficiency was related to the red shift in the photoadsorption spectrum of core-shell nanoparticles and to the presence of the electron accumulation centers formed by the electronic interactions of the two semiconductors. The presence of the p-n junctions induces the formation of internal electric fields (Gao et al. 2014), inducing the flow of electron and holes in two opposing direction, thus limiting the rate of the recombination process and improving the photocatalytic activity.

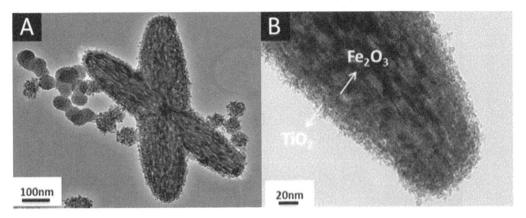

Fig. 6.8 (A) TEM image of α-Fe$_2$O$_3$ nanorods coated with TiO$_2$ nanoparticles, (B) TEM image of α-Fe$_2$O$_3$@TiO$_2$ core-shell nanostructure in high magnification. (Reprinted with permission from Fu, H., S. Sun, X. Yang, W. Li, X. An, H. Zhang et al. 2018. A facile coating method to construct uniform porous α-Fe$_2$O$_3$@TiO$_2$ core-shell nanostructures with enhanced solar light photocatalytic activity. Powder Technol. 328: 389-396)

The design and synthesis of multi-shell burr-shaped hierarchical plasmonic enhanced photocatalyst Ag@Fe$_3$O$_4$@SiO$_2$@TiO$_2$ has been reported recently (Su et al. 2014). The synthesis has been performed by the combination of solvothermal and modified Stöber methods. The particles show good photocatalytic activity in the processes of organic dye degradation and Cr(VI) reduction. The enhanced photocatalytic performance has been attributed to the hierarchical micro/nano structures with high surface area and profuse interfacial active sites, the enhanced absorption of visible light due to the surface plasmon effect of Ag core, and positively synergetic effect between the plasmonic and multilayer interference.

6.5 PHOTOTHERMAL EFFECT

During the last decade, heating of colloidal nanoparticles in water with solar light has been intensively studied for steam generation rather than for photochemical transformations (Neumann et al. 2013, Zhu et al. 2018). Light-absorbing particles being irradiated with concentrated sunlight can reach temperature well above the boiling point of water, creating non-equilibrium conditions between the particle surface and surrounding fluid (Govorov and Richardson 2007). More recently, it was reported that light-heated plasmonic nanoparticles enable the acceleration of heterogeneous catalytic processes (Adleman et al. 2009). Thermal mediation of photocatalytic oxidative degradation of a variety of organics is known since 1980s (Tang et al. 2017). Nevertheless, thermal effects are often ignored in the processes of photocatalytic hydrogen production. Recently, several studies have clearly demonstrated a beneficial effect of bulk temperature on H$_2$ photocatalytic formation over Pt/TiO$_2$ (Kim et al. 2016, Song et al. 2017, Velázquez et al. 2017) and Ti@TiO$_2$ (Nikitenko et al. 2015, 2018) nanocatalysts in aqueous solutions of sacrificial reagents. The latter system is an interesting example of the noble metal-free core-shell nanocatalyst showing strong photothermal effect. Figure 6.9 shows H$_2$ emission profile from the aqueous 40 vol% glycerol solution in the presence of Ti@TiO$_2$ nanoparticles under the irradiation with Xe lamp. Increase of bulk temperature leads to the progressive enhancement of H$_2$ production. However, under dark conditions H$_2$ emission is not observed even at quite high temperature indicating a photonic activation of the process.

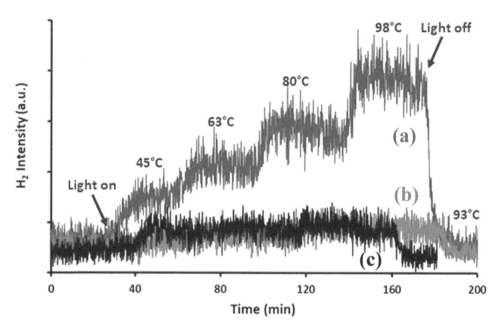

Fig. 6.9 (a) Profile of H_2 emission measured by mass spectrometry in Ar-flow photocatalytic reactor filled with 40 vol% glycerol solution and irradiated with a white light of Xe lamp in the presence of Ti@TiO$_2$, (b) TiO$_2$ anatase, (c) air-passivated Ti^0. (Reprinted with permission from Nikitenko, S.I., T. Chave and X. Le Goff. 2018. Insights into the photothermal hydrogen production from glycerol aqueous solutions over noble metal-free Ti@TiO$_2$ core-shell nanoparticles. Part. Part. Syst. Charact. 35: 1800265)

The origin of photothermal effect in photocatalysis is still not fully understood. The first step of the photochemical process involves formation of the electron-hole pairs with an energy largely overcoming activation barriers of thermal chemical processes and the activation energy would need to be zero. However, temperature-dependent surface adsorption-desorption and/or migration of intermediates could influence the reaction kinetics in heterogeneous systems. The apparent activation energy of the reaction with Ti@TiO$_2$ photocatalyst in glycerol solutions (E_{act} = 27±2 kJ·mol^{-1}) is much lower than typical activation energy of the chemical bonds, which is in line with the reaction mechanism involving dynamics of reaction intermediates at the catalyst surface (Nikitenko et al. 2018). At the same time, the activation energy obtained for Ti@TiO$_2$ core-shell nanoparticles is close to the E_{act} = 28 kJ·mol^{-1} reported for the photoelectrocatalytic oxidation of pure water in the presence of Au/TiO$_2$ nanocatalyst (Nishijima et al. 2012). In principal, such similarity could indicate the influence of water dynamics on the reaction kinetics in both systems, which is in agreement with the temperature dependence of photocatalytic oxidation of organic molecules in water over TiO$_2$, interpreted as the effect of the adsorbed water on the activation energy of a photocatalytic reaction (Parrino et al. 2017). On the other hand, photoelectrochemical analysis of Pt/TiO$_2$ catalyst in aqueous EDTA solutions revealed a positive relationship between reaction temperature and photocurrent generation (Kim et al. 2016). In addition, the study of electrochemical impedance spectroscopy performed by the same authors has shown a decrease in charge-transfer resistance with temperature increase. These results suggest enhanced charge carrier mobility and advanced interfacial charge transfer at higher temperature in photocatalytic systems. This is a quite surprising conclusion since usually the charge carrier mobility in semiconductors decreases with temperature increase due to the strong phonon scattering as it follows from Bose-Einstein distribution (Bianconi and Barabási 2001). There is no doubt that the mechanism of photothermal effect in the systems with TiO$_2$-based catalysts requires further investigation.

6.6 SUMMARY

In general, the preparation of core-shell nanoparticles is a rapidly growing research area at the frontiers of advanced materials chemistry. More specifically, core-shell TiO_2-based photocatalysts exhibit promising light-harvesting properties due to the smart bandgap engineering and more efficient photogenerated charge separation even without loading with highly expensive noble metals. Combination of TiO_2 with magnetic component, like magnetite core or shell, allows to obtain easily recyclable photocatalysts using external magnetic field.

Several traditional synthetic strategies to access TiO_2 core-shell nanocatalysts fabricated from earth-abundant elements and exhibiting advanced photocatalytic properties have been overviewed in this chapter, such as sol-gel process, hydro/solvothermal synthesis, and plasma-enhanced atomic layer deposition. In addition, innovative approach based on simultaneous hydrothermal and ultrasonic (sonohydrothermal) treatment has been highlighted. This technique allows to accelerate the synthesis and to improve the catalytic properties of nanocrystalline materials. Synthesis of some TiO_2-based core-shell photocatalysts can be performed by sonohydrothermal treatment in pure water.

It should be noted that despite the impressive progress in the synthesis of functional photocatalysts for hydrogen generation achieved in the last decade, an economically feasible genuine photocatalytic water splitting still remains a challenge. Practically all studied photocatalytic reactions of hydrogen production involve various kinds of sacrificial reagents. Realistically speaking, it is difficult to expect solar photocatalytic hydrogen production at large scale without the use of sacrificial reagents. This is often considered as a weak point of the solar fuel processes. However, such apparent drawback could bring, in fact, some advantages. As it was shown in this chapter, some photocatalytic processes in the solutions of organic sacrificial reagents lead to effective H_2 emission without CO_2 or CO formation indicating photocatalytic formation of some secondary products, which could be valuable for several chemical and biochemical applications. On the other hand, aqueous pollutants can be chosen as sacrificial reagents and, thus, simultaneous hydrogen generation and pollutant removal can be achieved. This strategy fits together the concepts of "positive" and "negative" photocatalysis (Lanterna and Scaiano 2017), where "positive" and "negative" photocatalysis means production of valuable chemicals and water remediation, respectively (Corma and Garcia 2004).

At near room temperature, the energy efficiency of photocatalytic hydrogen generation under the sunlight remains quite low in the range of 1.6-2.5% even in the presence of sacrificial reagents (Salgado et al. 2016). In most of the cases, it is hard to compare the reported efficiency of H_2 production since the results are often presented as H_2 formation rate for various light sources without precise indication of input power. In principal, boosting of the hydrogen yield is possible due to the photothermal effect discussed in this chapter. Using the concentrated solar light would allow heating of the photocatalytic cell without external input of thermal energy. Photothermal catalysis provides the optimal utilization of the full solar spectrum: UV/vis light generates chemically active electron-hole pairs and the action of NIR light is somewhat similar to that in volumetric systems for photothermal evaporation. It is worth noting that, in general, photoexcitation by NIR light only without the support of UV/vis light does not occur (Neelgund and Oki 2018). In these terms, the future efficient photothermal systems for hydrogen production should merge efficient photocatalysts and advanced materials developed for photothermal steam generation.

6.7 REFERENCES

Abdel-Wahab, A.M., A.S. Al-Shirbini, O. Mohamed and O. Nasr. 2017. Photocatalytic degradation of paracetamol over magnetic flower-like TiO_2/Fe_2O_3 core-shell nanostructures. J. Photochem. Photobiol. A. Chem. 347: 186-198.

Adleman, J.R., D.A. Boyd, D.G. Goodwin and D. Psaltis. 2009. Heterogeneous catalysis mediated by plasmon heating. Nano Lett. 9: 4417-4423.

Arnell, N., M. Cannel, M. Hulme, J. Kovats, R. Mitchell, M. Nicholls, et al. 2002. The consequences of CO_2 stabilization for the impacts of climatic change. Climatic Change 53: 413-446.

Barnard, A.S. and L.A. Curtiss. 2005a. Prediction of TiO_2 nanoparticle phase and shape transitions controlled by surface chemistry. Nano Letters 5: 1261-1266.

Barnard, A.S., P. Zapol and L.A. Curtiss. 2005b. Modeling the morphology and phase stability of TiO_2 nanocrystals in water. J. Chem. Theory Comput. 1: 107-116.

Barreca, D., P. Fornasiero, A. Gasparotto, V. Gombac, C. Maccato, T. Montini, et al. 2009. The potential of supported Cu_2O and CuO nanosystems in photocatalytic H_2 production. Chem. Sus. Chem. 2: 230-233.

Barreca, D., G. Carraro, V. Gombac, A. Gasparotto, C. MacCato, P. Fornasiero, et al. 2011. Supported metal oxide nanosystems for hydrogen photogeneration: Quo vadis? Adv. Funct. Mater. 21: 2611-2623.

Batzill, M. 2011. Fundamental aspects of surface engineering of transition metal oxide photocatalysts. Energy Environ. Sci. 4: 3275-3286.

Baur, W. 1961. Atomabstande und bindungswinkel im brookit, TiO_2. Acta Cryst. 14: 214-216.

Beltrán, A., L. Gracia and J. Andrés. 2006. Density functional theory study of the brookite surfaces and phase transitions between natural titania polymorphs. J. Phys. Chem. B. 110: 23417-23423.

Bian, Z., J. Ren, J. Zhu, S. Wang, Y. Lu and H. Li. 2009. Self-assembly of $Bi_xTi_{1-x}O_2$ visible photocatalyst with core-shell structure and enhanced activity. Appl. Catal. B. Environ. 89: 577-582.

Bianconi, G. and A.L. Barabási. 2001. Bose-Einstein condensation in complex networks. Phys. Rev. Lett. 86: 5632.

Cai, G., J. Tu, D. Zhou, L. Li, J. Zhang, X. Wang, et al. 2014. Constructed TiO_2/NiO core/shell nanorod array for efficient electrochromic application. J. Phys. Chem. C. 118: 6690-6696.

Cargnello, M., A. Gasparotto, V. Gombac, T. Montini, D. Barreca and P. Fornasiero. 2011. Photocatalytic H_2 and added-value by-products-the role of metal oxide systems in their synthesis from oxygenates. Eur. J. Inorg. Chem. 2011: 4297-4302.

Chaudhuri, R.G. and S. Paria. 2012. Core/shell nanoparticles: Classes, properties, synthesis mechnaismes, characterization, and applications. Chem. Rev. 112: 2373-2433.

Chen, J., D.F. Ollis, W.H. Rulkens and H. Bruning. 1999. Photocatalyzed oxidation of alcohols and organochlorides in the presence of native TiO_2 and metallized TiO_2 suspensions. Part (II): Photocatalytic mechanisms. Water Res. 33: 669-676.

Chen, J., H. Zhang, P. Liu, Y. Li, X. Liu, G. Li, et al. 2015. Cross-linked $ZnIn_2S_4$/rGO composite photocatalyst for sunlight-driven photocatalytic degradation of 4-nitrophenol. Appl. Cat. B: Environ. 168-169: 266-273.

Cheng, W.Y., T.H. Yu, K.J. Chao and S.Y. Lu. 2014a. Cu_2O-decorated mesoporous TiO_2 beads as a highly efficient photocatalyst for hydrogen production. Chem. Cat. Chem. 6: 293-300.

Cheng, W.Y., T.H. Yu, K.J. Chao and S.Y. Lu. 2014b. Cu_2O-decorated mesoporous TiO_2 beads as a highly efficient photocatalyst for hydrogen production. Chem. Cat. Chem. 6: 293-300.

Chiarello, G.L., D. Ferri and E. Selli. 2011. Effect of the CH_3OH/H_2O ratio on the mechanism of the gas-phase photocatalytic reforming of methanol on noble metal-modified TiO_2. J. Catal. 280: 168-177.

Corma, A. and H. Garcia. 2004. Zeolite-based photocatalysts. Chem. Commun. 13: 1443-1459.

Cromer, D.T. and K. Herrington. 1955. The structures of anatase and rutile. J. Am. Chem. Soc. 77: 4708-4709.

Dickinson, A., D. James, N. Perkins, T. Cassidy and M. Bowker. 1999. The photocatalytic reforming of methanol. J. Mol. Cat. A: Chemical. 146: 211-221.

Ding, R.C., Y.Z. Fan and G.S. Wang. 2018. High efficient Cu_2O/TiO_2 nanocomposite photocatalyst to degrade organic pollutant under visible light irradiation. ChemistrySelect 3: 1682-1687.

Faraji, M., M. Yousefi, S. Yousefzadeh, M. Zirak, N. Naseri, T. Hwa Jeon, et al. 2019. Two-dimensional materials in semiconductor photoelectrocatalytic systems for water splitting. Energy Environ. Sci. 12: 59-95.

Fu, H., S. Sun, X. Yang, W. Li, X. An, H. Zhang, et al. 2018. A facile coating method to construct uniform porous α-Fe_2O_3@TiO_2 core-shell nanostructures with enhanced solar light photocatalytic activity. Powder Technol. 328: 389-396.

Fujishima, A. and K. Honda. 1972. Electrochemical photolysis of water at a semiconductor electrode. Nature 238: 37-38.

Fujishima, A., X. Zhang and D.A. Tryk. 2008. TiO_2 photocatalysis and related surface phenomena. Surf. Sci. Rep. 63: 515-582.

Gao, B., L. Liu, J. Liu and F. Yang. 2013. Photocatalytic degradation of 2,4,6,-tribromophenol over Fe-doped $ZnIn_2S_4$: Stable activity and enhanced debromination. Appl. Cat. B: Environ. 129: 89-97.

Gao, T., Z. Chen, Y. Zhu, F. Niu, Q. Huang, L. Qin, et al. 2014. Synthesis of $BiFeO_3$ nanoparticles for the visible-light induced photocatalytic property. Mater. Res. Bull. 59: 6-12.

Gawande, M.B., A. Goswami, T. Asefa, H. Guo, A.V. Biradar, D.-L. Peng, et al. 2015. Core-shell nanoparticles: Synthesis and applications in catalysis and electrocatalysis. Chem. Soc. Rev. 44: 7540-7590.

Ghosh, S. [ed.] 2018. Visible-Light-Active Photocatalysis: Nanostructured Catalyst Design, Mechanisms, and Applications. Wiley, Weinheim, Germany.

Govorov, A.O. and H.H. Richardson. 2007. Generating heat with metal nanoparticles. Nano Today 2: 30-38.

Grätzel, M. and F.P. Rotzinger. 1985. The influence of the crystal lattice structure on the conduction band energy of oxides of titanium (IV). Chem. Phys. Lett. 118: 474-477.

Hanaor, D.A.H. and C.C. Sorrell. 2011. Review of the anatase to rutile phase transformation. J. Mat. Sci. 46: 855-874.

Heard, C.J., J. Čejka, M. Opanasenko, P. Nachtigall, G. Centi and S. Perathoner. 2018. 2D oxide nanomaterials to address the energy transition and catalysis. Adv. Mater. 31: 1801712.

Hermann, W.A. 2006. Quantifying global energy resources. Energy 31: 1349-1366.

Hirakawa, T. and P.V. Kamat. 2005. Charge separation and catalytic activity of $Ag@TiO_2$ core-shell composite clusters under UV-irradiation. J. Am. Chem. Soc. 127: 3928-3934.

Hou, Y., F. Zuo, A. Dagg and P. Feng. 2012. Visible light-driven α-Fe_2O_3 nanorod/graphene/$BiV_{1-x}Mo_xO_4$ core/shell heterojunction array for efficient photoelectrochemical water splitting. Nano Lett. 12: 6464-6473.

Hu, B., K. Wang, L. Wu, S.H. Yu, M. Antonietti and M.M. Titirici. 2010. Engineering carbon materials from the hydrothermal carbonization process of biomass. Adv. Mater. 22: 813-828.

Hu, H., Y. Lin and Y.H. Hu. 2020. Core-shell structured TiO_2 as highly efficient visible light photocatalyst for dye degradation. Catal. Today 341: 90-95.

Hurum, D.C., A.G. Agrios, K.A. Gray, T. Rajh and M.C. Thurnauer. 2003. Explaining the enhanced photocatalytic activity of Degussa P25 mixed-phase TiO_2 using EPR. J. Phys. Chem. B. 107: 4545-4549.

IEA. World Energy Outlook 2019. International Energy Agency: 2019.

Joy, J., J. Mathew and S.C. George. 2018. Nanomaterials for photoelectrochemical water splitting – review. Int. J. Hydrogen Energ. 43: 4804-4817.

Jung, H.S. and H. Kim. 2009. Origin of low photocatalytic activity of rutile TiO_2. Electron. Mater. Lett. 5: 73-76.

Kawai, T. and T. Sakata. 1980. Conversion of carbohydrate into hydrogen fuel by a photocatalytic process. Nature 286: 474-476.

Kim, G., H.J. Choi, H. Kim, J. Kim, D. Monllor-Satosa and H. Park. 2016. Temperature-boosted photocatalytic H_2 production and charge transfer kinetics on TiO_2 under UV and visible light. Photochem. Photobiol. Sci. 15: 1247-1253.

Kim, H., G. Moon, D. Monllor-Satoca, Y. Park and W. Choi. 2011. Solar photoconversion using graphene/TiO_2 composites: Nanographene shell on TiO_2 core versus TiO_2 nanoparticles on graphene sheet. J. Phys. Chem. C. 116: 1535-1543.

Kudo, A. and Y. Miseki. 2009. Heterogeneous photocatalyst materials for water splitting. Chem. Soc. Rev. 38: 253-278.

Kumar, S.G. and K.S.R.K. Rao. 2014. Polymorphic phase transition among the titania crystal structures using a solution-based approach: From precursor chemistry to nucleation process. Nanoscale 6: 11574-11632.

Lanterna, A.E. and J.C. Scaiano. 2017. Photoinduced hydrogen fuel production and water decontamination technologies. Orthogonal strategies with a parallel future? ACS Energy Lett. 2: 1909-1910.

Lazzeri, M., A. Vittadini and A. Selloni. 2001. Structure and energetics of stoichiometric TiO_2 anatase surfaces. Phys. Rev. B. 63: 155409.

Lazzeri, M., A. Vittadini and A. Selloni. 2002. Erratum: Structure and energetics of stoichiometric TiO_2 anatase surfaces [Phys. Rev. B. 63, 155409 (2001)]. Phys. Rev. B. 65: 119901.

Lee, K., A. Mazare and P. Schmuki. 2014. One-dimentional titanium dioxide nanomaterials: Nanotubes. Chem. Rev. 114: 9385-9454.

Li, F., Q. Gu, Y. Niu, R. Wang, Y. Tong, S. Zhu, et al. 2017. Hydrogen evolution from aqueous-phase photocatalytic reforming of ethylene glycol over Pt/TiO_2 catalysts: Role of Pt and product distribution. Appl. Surf. Sci. 391: 251-258.

Li, W., A. Elzatahry, D. Aldhayan and D. Zhao. 2018. Core-shell structured titanium dioxide nanomaterials for solar energy utilization. Chem. Soc. Rev. 47: 8203-8237.

Lindan, P.J.D., N.M. Harrison and M.J. Gillan. 1998. Mixed dissociative and molecular adsorption of water on the rutile (110) surface. Phys. Rev. Lett. 80: 762-765.

Liu, G., H.G. Yang, J. Pan, Y.Q. Yang, G.Q. Lu and H.-M. Cheng. 2014. Titanium dioxide crystals with tailored facets. Chem. Rev. 114: 9559-9612.

Liu, R. and A. Sen. 2012. Controlled synthesis of heterogeneous metal-titania nanostructures and their applications. J. Am. Chem. Soc. 134: 17505-17512.

Liu, X., L. Cao, W. Sun, Z. Zhou and J. Yang. 2016. A P/N type compounded Cu_2O/TiO_2 photo-catalytic membrane for organic pollutant degradation. Res. Chem. Intermed. 42: 6289-6300.

Liu, Y., S. Ding, J. Xu, H. Zhang, S. Yang, X. Duan, et al. 2017. Preparation of a *p-n* heterojunction $BiFeO_3$@ TiO_2 photocatalyst with a core-shell structure for visible-light photocatalytic degradation. Cuihua Xuebao/ Chinese J. Catal. 38: 1052-1062.

Ma, L., G. Wang, C. Jiang, H. Bao and Q. Xu. 2018. Synthesis of core-shell TiO_2@g-C_3N_4 hollow microspheres for efficient photocatalytic degradation of rhodamine B under visible light. Appl. Surf. Sci. 430: 263-272.

Ma, Y., X. Wang, Y. Jia, X. Chen, H. Han and C. Li. 2014. Titanium dioxide-based nanomaterials for photocatalytic fuel generations. Chem. Rev. 114: 9987-10043.

Murphy, A.B., P.R.F. Barnes, L.K. Randeniya, I.C. Plumb, I.E. Grey, M.D. Horne, et al. 2006. Efficiency of solar water splitting using semiconductor electrodes. Int. J. Hydrogen Energ. 31: 1999-2017.

Neelgund, G.M. and A. Oki. 2018. Photothermal effect: An important aspect for the enhancement of photocatalytic activity under illumination by NIR radiation. Mater. Chem. Front. 2: 64-75.

Neumann, O., A.S. Urban, J. Day, S. Lai, P. Nordlander and N.J. Halas. 2013. Solar vapor generation enabled by nanoparticles. ACS Nano 7: 42-49.

Nikitenko, S.I., T. Chave, C. Cau, H.-P. Brau and V. Flaud. 2015. Photothermal hydrogen production using noble-metal-free Ti@TiO_2 core-shell nanoparticles under visible-NIR light irradiation. ACS Catal. 5: 4790-4795.

Nikitenko, S.I., T. Chave and X. Le Goff. 2018. Insights into the photothermal hydrogen production from glycerol aqueous solutions over noble metal-free Ti@TiO_2 core-shell nanoparticles. Part. Part. Syst. Charact. 35: 1800265.

Nishijima, Y., K. Ueno, Y. Kotake, K. Murakoshi, H. Inoue and H. Misawa. 2012. Near-infrared plasmon-assisted water oxidation. J. Phys. Chem. Lett. 3: 1248-1252.

Pan, H. 2016. Principles on design and fabrication of nanomaterials as photocatalysts for water-splitting. Renew. Sustain. Energy Rev. 58: 584-601.

Paola, A.D., M. Bellardita and L. Palmisano. 2013. Brookite, the least known TiO_2 photocatalyst. Catalysts 3: 36-73.

Parrino, F., P. Conte, C. De Pasquale, V.A. Laudicina, V. Loddo and L. Palmisano. 2017. Influence of adsorbed water on the activation energy of model photocatalytic reactions. J. Phys. Chem. 121: 2258-2267.

Pastoriza-Santos I., D.S. Koktysh, A.A. Mamedov, M. Giersig, N.A. Kotov and L.M. Liz-Marzán. 2000. One-pot synthesis of Ag@TiO_2 core-shell nanoparticles and their layer-by-layer assembly. Langmuir 16: 2731-2735.

Perron, H., C. Domain, J. Roques, R. Drot, E. Simoni and H. Catalette. 2007. Optimisation of accurate rutile TiO_2 (110), (100), (101) and (001) surface models from periodic DFT calculations. Theor. Chem. Acc. 117: 565-574.

Puga, A.V. 2016. Photocatalytic production of hydrogen from biomass-derived feedstocks. Coord. Chem. Rev. 315: 1-66.

Reddy, N.L., S. Kumar, V. Krishnan, M. Sathish and M.V. Shankar. 2017. Multifunctional Cu/Ag quantum dots on TiO_2 nanotubes as highly efficient photocatalysts for enhanced solar hydrogen evolution. J. Catal. 350: 226-239.

Reddy, P.A.K., B. Srinivas, P. Kala, V.D. Kumari and M. Subrahmanyam. 2011. Preparation and characterization of Bi-doped TiO_2 and its solar photocatalytic activity for the degradation of isoproturon herbicide. Mater. Res. Bull. 46: 1766-1771.

Sakata, T. and T. Kawai. 1981. Heterogeneous photocatalytic production of hydrogen and methane from ethanol and water. Chem. Phys. Lett. 80: 341-344.

Salgado, S.Y.A., R.M.R. Zamora, R. Zanella, J. Peral, S. Malato and M.I. Maldonado. 2016. Photocatalytic hydrogen production in a solar pilot plant using a Au/TiO_2 photocatalyst. Int. J. Hydrogen Energy 41: 11933-11940.

Santhosh, C., A. Malathi, E. Daneshvar, P. Kollu and A. Bhatnagar. 2018. Photocatalytic degradation of toxic aquatic pollutants by novel magnetic 3D-TiO_2@HPGA nanocomposite. Sci. Rep. 8: 15531.

Scanlon, D.O., C.W. Dunnill, J. Buckeridge, S.A. Shevlin, A.J. Logsdail, S.M. Woodley, et al. 2013. Band alignment of rutile and anatase TiO_2. Nat. Mater. 12: 798-801.

Sclafani, A. and J.M. Herrmann. 1996. Comparison of the photoelectronic and photocatalytic activities of various anatase and rutile forms of titania in pure liquid organic phases and in aqueous solutions. J. Phys. Chem. 100: 13655-13661.

Seadira, T.W.P., G. Sadanandam, T. Ntho and C.M. Masuku. 2018. Preparation and characterization of metals supported on nanostructured TiO_2 hollow spheres for production of hydrogen via photocatalytic reforming of glycerol. Appl. Cat. B: Environ. 222: 133-145.

Serpone, N., D. Lawless, R. Khairutdinov and E. Pelizzetti. 1995. Subnanosecond relaxation dynamics in TiO_2 colloidal sols (particle sizes R_p = 1.0-13.4 nm). Relevance to heterogeneous photocatalysis. J. Phys. Chem. 99: 16655-16661.

Shen, S., X. Wang, T. Chen, Z. Feng and C. Li. 2014. Transfer of photoinduced electrons in anatase-rutile TiO_2 determined by time-resolved mid-infrared spectroscopy. J. Phys. Chem. C. 118: 12661-12668.

Siahrostami, S., G.L. Li, V. Viswanathan and J.K. Nørskov. 2017. One- or two-electron water oxidation, hydroxyl radical, or H_2O_2 evolution. J. Phys. Chem. Lett. 8: 1157-1160.

Song, R., B. Luo, M. Liu, J. Geng, D. Jing and H. Liu. 2017. Synergetic coupling of photo and thermal energy for efficient hydrogen production by formic acid reforming. AIChE J. 63: 2916-2925.

Su, J., Y. Zhang, S. Xu, S. Wang, H. Ding, S. Pan, et al. 2014. Highly efficient and recyclable triple-shelled Ag@Fe_3O_4@SiO_2@TiO_2 photocatalysts for degradation of organic pollutants and reduction of hexavalent chromium ions. Nanoscale 6: 5181-5192.

Tang, J., J.R. Durrant and D.R. Klug. 2008. Mechanism of photocatalytic water splitting in TiO_2. Reaction of water with photoholes, importance of charge carrier dynamics, and evidence for four-hole chemistry. J. Am. Chem. Soc. 130: 13885-13891.

Tang, S., J. Sun, H. Hong and Q. Liu. 2017. Solar fuel from photo-thermal catalytic reactions with spectrum-selectivity: A review. Front. Energy. 11: 437-451.

Tee, S.Y., K.Y. Win, W.S. Teo, L.D. Koh, S. Liu, C.P. Teng, et al. 2017. Recent progress in energy-driven water splitting. Adv. Sci. 4: 1600337.

Turchi, C.S. and D.F. Ollis. 1990. Photocatalytic degradation of organic water contaminants: Mechanisms involving hydroxyl radical attack. J. Catal. 122: 178-192.

Vasei, M., P. Das, H. Cherfouth, B. Marsan and J.P. Claverie. 2014. TiO_2@C core-shell nanoparticles formed by polymeric nano-encapsulation. Front. Chem. 2: 47.

Velázquez, J.J., R. Fernández-González, L. Díaz, E. Pulido Melián, V.D. Rodríguez and P. Núñez. 2017. Effect of reaction temperature and sacrificial agent on the photocatalytic H_2-production of Pt-TiO_2. J. Alloys Compd. 721: 405-410.

Wang, B., X.-Y. Lu, L.K. Yu, J. Xuan, M.K.H. Leung and H. Guo. 2014a. Facile synthesis of TiO_2 hollow spheres composed of high percentage of reactive facets for enhanced photocatalytic activity. Cryst. Eng. Comm. 16: 10046-10055.

Wang, Q. and K. Domen. 2020. Particulate photocatalysis for light-driven water splitting: Mechanisms, challenges, and design strategies. Chem. Rev. 120: 919-985.

Wang, X., Z. Li, J. Shi and Y. Yu. 2014b. One-dimensional titanium dioxide nanomaterials: Nanowires, nanorods, and nanobelts. Chem. Rev. 114: 9346-9384.

Xia, Y., Q. Li, K. Lv and M. Li. 2017. Heterojunction construction between TiO_2 hollow sphere and $ZnIn_2S_4$ flower for photocatalysis application. Appl. Surf. Sci. 398: 81-88.

Xu, J., M. Chen and D. Fu. 2011a. Study on highly visible light active Bi-doped TiO_2 composite hollow sphere. Appl. Surf. Sci. 257: 7381-7386.

Xu, M., Y. Gao, E.M. Moreno, M. Kunst, M. Muhler, Y. Wang, et al. 2011b. Photocatalytic activity of bulk TiO_2 anatase and rutile single crystals using infrared absorption spectroscopy. Phys. Rev. Lett. 106(13): 138302.

Yang, X.H., H.T. Fu, K. Wong, X.C. Jiang and A.B. Yu. 2013. Hybrid Ag@TiO_2 core-shell nanostructures with highly enhanced photocatalytic performance. Nanotechnology 24: 415601.

Yang, Y., G. Liu, J.T.S. Irvine and H.-M. Cheng. 2016. Enhanced photocatalytic H_2 production in core-shell engineered rutile TiO_2. Adv. Mater. 28: 5850-5856.

Yang, Y., X. Li, C. Lu and W. Huang. 2019. G-C_3N_4 nanoshhets coupled with TiO_2 nanosheets as 2D/2D heterojunction photocatalysts toward high photocatalytic activity for hydrogen production. Catal. Lett. 149: 2930-2939.

Yasuda, M., T. Matsumoto and T. Yamashita. 2018. Sacrificial hydrogen production over TiO_2-based photocatalysts: Polyols, carboxylic acids, and saccharides. Renew. Sustain. Energy Rev. 81: 1627-1635.

Yu, J., S. Liu and H. Yu. 2007. Microstructures and photoactivity of mesoporous anatase hollow microspheres fabricated by fluoride-mediated self-transformation. J. Catal. 249: 59-66.

Yu, J. and G. Wang. 2008. Hydrothermal synthesis and photocatalytic activity of mesoporous titania hollow microspheres. J. Phys. Chem. Solids. 69: 1147-1151.

Yuan, W.-H., Z.-L. Xia and L. Li. 2013. Synthesis and photocatalytic properties of core-shell TiO_2@$ZnIn_2S_4$ photocatalyst. Chin. Chem. Lett. 24: 984-986.

Zhang, H. and J.F. Banfield. 2000. Understanding polymorphic phase transformation behavior during growth of nanocrystalline aggregates: Insights from TiO_2. J. Phys. Chem. B. 104: 3481-3487.

Zhang, H. and J.F. Banfield. 2014a. Structural characteristics and mechanical and thermodynamic properties of nanocrystalline TiO_2. Chem. Rev. 114: 9613-9644.

Zhang, J., P. Zhou, J. Liu and J. Yu. 2014b. New understanding of the difference of photocatalytic activity among anatase, rutile and brookite TiO_2. Phys. Chem. Chem. Phys. 16: 20382-20386.

Zhang, N., S. Liu and Y.-J. Xu. 2012. Recent progress on metal@semiconductor shell nanocomposites as a promising type of photocatalyst. Nanoscale 4: 2227-2238.

Zhang, Q., W. Li and S. Liu. 2011. Controlled fabrication of nanosized TiO_2 hollow sphere particles via acid catalytic hydrolysis/hydrothermal treatment. Powder Technol. 212: 145-150.

Zhao, X.-R., Y.-Q. Cao, J. Chen, L. Zhu, X. Qian, A.-D. Li, et al. 2017. Photocatalytic properties of Co_3O_4-coated TiO_2 powders prepared by plasma-enhanced atomic layer deposition. Nanoscale Res. Lett. 12: 497.

Zhu, L., M. Gao, C.K. Nuo Peh and G. Wei Ho. 2018. Solar-driven photothermal nanostructured materials designs and prerequisites for evaporation and catalysis applications. Mater. Horiz. 5: 323-343.

Zhu, T. and S.-P. Gao. 2014. The stability, electronic structure, and optical property of TiO_2 polymorphs. J. Phys. Chem. C. 118: 11385-11396.

7

Graphene Based Nanocomposites for Energy Applications

Deepakjot Singh[1], Sanjay Kumar[2] and Anup Thakur[3,*]

7.1 INTRODUCTION

Graphene revolution started in October 2004, when Novoselov et al. successfully prepared graphene sheets (first 2-D layered material) of carbon atoms and investigated the electric field effect on these sheets (Novoselov et al. 2004). Novoselov and Geim frequently held 'Friday night experiments'. On one of the Fridays, while exfoliating graphite with sticky tape, they observed that some of the flakes were thinner with respect to the others. By repeating this and reducing the number of flakes, it was reduced to just one atomic layer thick and subsequent material was named as "graphene". Graphene has a hexagonal (honeycomb) structure of carbon atoms held together by the π-cloud overlapping sp² hybrid carbon atoms (Duplock et al. 2004). Graphene is also called mother of all graphitic forms as all other dimensionalities are shown in Figure 7.1. Over the years, graphene has gained interest in energy applications due to its high electrical/thermal conductivities, high chemical/electrochemical stability, high flexibility and elasticity, and 97.7% transparency to white light (Sheehy and Schmalian 2009).

The thickness or layers of graphene holds vital consideration while commencing study on any particular properties for targeted application. It is usual practice to state the layers/thickness of graphene sheets used in any study. In general, thickness of mono-layered, bi-layered, tri-layered, four-layered, and seven-layered graphene is 0.335 nm, 0.81 nm, 1.285 nm, 1.76 nm and 3.185 nm, respectively, with interlayer distance of 0.14 nm for multi-layered graphene (Lu 1997, Ye et al. 2014, Zhang and Pan 2012). The number of layers has a significant impact on properties and performance of graphene-based material; for instance, hardness and elastic modulus show a linearly decreasing trend on increasing the number of graphene layers. This trend proved itself as a template to determine the number of layers with the help of nanoindentation (Zhang and Pan 2012). Apart from mechanical properties, increasing layers also hinder optical transmittance of graphene-based material, whereas bi-layered graphene finds its promising applicability in tunnel field effect transistor, by offering non-zero band gap alike mono-layered graphene (Kang et al. 2015, Sheehy and Schmalian 2009, Zhou et al. 2007). Hence, layer selectivity is subject to application, rather than describing any particular thickness/number as ideal model.

[1] Department of Mechanical Engineering, Punjabi University, Patiala-147 002, Punjab, India.
[2] Department of Chemistry, Multani Mal Modi College, Patiala-147 001, Punjab, India.
[3] Advanced Materials Research Lab, Department of Basic and Applied Sciences, Punjabi University, Patiala-147 002, Punjab, India.
* Corresponding author: dranupthakur@gmail.com

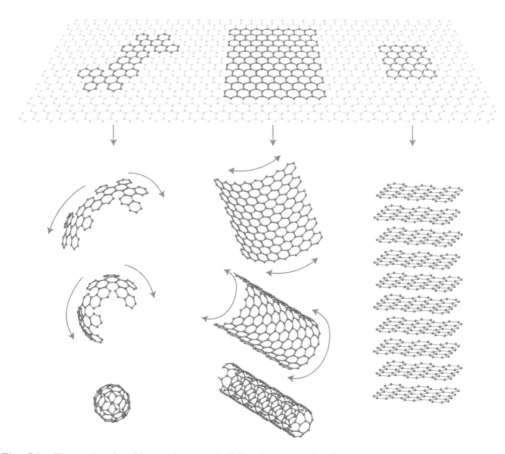

Fig. 7.1 Figure showing 2D graphene as building block for 0D fullerene, 1D carbon nanotubes and stacking up for 3D graphene. (Reprinted with permission from Novoselov, K. S. and A. Geim. 2007. The rise of graphene. Nat. Mater., 6(3): 183-191.) Copyright (2007) Nature Publishing Group

Though graphene has unique properties and immense potential, two major drawbacks of cumbersome mass production and limited application in its pristine form are associated with it. Therefore, graphene derivatives, namely, graphene oxide (GO), reduced graphene oxide (rGO) and their nanocomposites with SnO_2, ZnO, TiO_2, are being explored. Graphene can be produced by exfoliation of graphite, whereas GO is produced by exfoliation of graphite oxide and rGO is synthesized by the use of reducing agent over GO. There is a difference between graphene and rGO as even after reduction, there is still oxygen left in rGO. Hence, rGO is treated as lower quality graphene in some instances in some perspectives. The synthetic methods based on the approach can be usually sorted into top-down or bottom-up approach. The top-down approach involves exfoliating graphite and reducing number of layers up to desired level and obtaining graphene, whereas bottom-up approach involves growing graphene sheets from organic precursors. Preparation of graphene can be carried out by various techniques, such as mechanical exfoliations, liquid phase exfoliations, electrochemical exfoliations, chemical vapour deposition (CVD), and reduction of graphene oxide (Papageorgiou et al. 2017).

Materials used play a crucial role in the development of energy conversion and storage systems as their performance is directly related to the properties of those materials. Out of different types of materials, carbon materials draw interest due to their natural abundance and stability. They maintain high chemical stability in both acidic and basic media throughout a broad temperature range, promoting their use as electrodes in energy devices. In energy applications, graphene-based nanocomposites have become a subject of intensive research. This chapter makes the reader familiar with modern advances in the field of graphene derivative and its nanocomposites emphasizing energy applications. In batteries, graphene has been extensively used as both anode and cathode additives in lithium ion batteries owing to its high charge carrier mobility ($20 \, m^2 \, V^{-1} \, s^{-1}$) (Bolotin et al. 2008, Morozov et al. 2008), broad electrochemical window (Wei et al. 2012) and theoretical surface area of $2630 \, m^2 \, g^{-1}$ (Zhu et al. 2010). Optoelectronics is a field where transparency and flexibility along with electronic properties of graphene are exploited. Utilization of graphene in flexible electrodes and some devices such as field-effect transistors and light emitting diodes (LEDs) has been discussed. Solar cells in apartment windows is an idea which can be made possible with graphene being transparent and having high charge carrier mobility. Large surface area and discharge rates are desired specification for supercapacitors and sensors. Various types of sensors such as gas sensors, biochemical sensors and strain sensors benefit from the high electron mobility and sustain high levels of load due to high modulus of elasticity (1 TPa) (Lee et al. 2008) of graphene.

7.2 BATTERIES

Batteries are comprised of three parts: an anode (–), a cathode (+) and an electrolyte. Electrons get accumulated at the anode due to the chemical reactions in a battery and are prevented from directly reaching the cathode by the electrolyte shown in Figure 7.2. When the battery is in use, these accumulated electrons are transferred from anode to cathode through a conductive path (wire) in a closed circuit. Lithium ion battery is one of the most employed battery technologies in the world, having its utility in electronic devices as well as in electric vehicle industry. The basic factors of consideration for these batteries are cyclability, safety, thermal stability, energy density, power density and pocket friendliness. Graphene has proven itself as one of the most suitable candidates as conductive additive due to its high surface/mass ratio for better contact with active material and exceptional electron conductivity due to hexagonally arranged carbon atoms in 2D lattice (Cooper et al. 2012).

Cathode

In designing the batteries, cathode is the heaviest and ranks second in cost among various components of lithium ion batteries, due to which research on cathode material holds a big stake. When mixed with metal oxides, graphene develops a 3D electron conducting network which significantly improves the transport of electron in material. This increases electron conductivity and hence rate capability and cyclability of the materials shown in Figure 7.3. Various composites such as $LiFePO_4$/graphene ($LiFePO_4$/G), $Li_3V_2(PO_4)_3$/G, $LiMn_2O_4$ and sulphur can be used for such applications. $LiFePO_4$/G have better rate capability due to improvement in electron conductivity, reduced charge transfer resistance, and very high cyclability due to improved inter-particle electronic contacts, although better performance is observed at low content of graphene due to stacking of graphene nanosheets at high concentration (Su et al. 2010).

An optimum carbon coating helps in forming effective conductive network in $LiFePO_4$/G. Instead of carbon coating, optimized addition of amorphous carbon/glucose derived amorphous carbon can be done to form $LiFePO_4$/G (Zhou et al. 2011). This possesses one of the best rate capabilities due to formation of small grain size providing short diffusion path to Li^+ ions, increased surface area, increased electron conductivity than that from carbon coating and formation of 3D conducting network encapsulating $LiFePO_4$ units. Other $LiFePO_4$/G type is $Li_3V_2(PO_4)_3$/G which is vastly studied for cathode material in lithium ion batteries. Similar results of carbon coating were found to be enhancing the performance of $Li_3V_2(PO_4)$/G due to formation of barrier, preventing contact of active material and electrolyte, eventually restraining dissolution of vanadium into electrolyte (Wang et al. 2010).

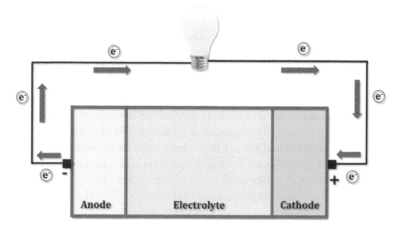

Fig. 7.2   Schematics and working of a battery showing movement of electrons from anode to cathode

Similarly, $LiMn_2O_4$ is another cathode material studied extensively due to having low cost of production, safe and environmental friendly. The dissolution of Mn^{2+} in electrolyte results in sacrifice of cyclability of material, which was significantly improved after composite formation with graphene (Zhang et al. 2012). However, when connected with well dispersed reduced graphene oxide sheets, $LiMn_2O_4$ is reported to be showing outstanding electrochemical properties due to formation of 3D network caused by reduced graphene nano sheets. $LiMn_2O_4/G$ composite have better rate capability than commercially available $LiMn_2O_4$ nanoparticles (Bak et al. 2011).

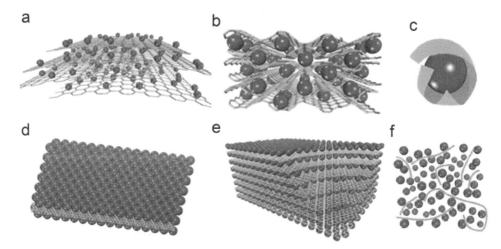

Fig. 7.3   Schematic of structural models of graphene/metal oxide composites: (a) Anchored model: nanosized oxide particles are anchored on the surface of graphene, (b) Wrapped model: metal oxide particles are wrapped by graphene, (c) Encapsulated model: oxide particles are encapsulated by graphene, (d) Sandwich-like model: graphene serves as a template for the creation of a metal oxide/graphene/metal oxide sandwich-like structure, (e) Layered model: a structure composed of alternating layers of metal oxide nanoparticles and graphene, (f) Mixed model: graphene and metal oxide particles are mechanically mixed and graphene forms a conductive network among the metal oxide particles. Dark circles: metal oxide particles; Light: graphene sheets. (Reprinted with permission from Wu, Z.-S., G. Zhou, L.-C. Yin, W. Ren, F. Li, and H.-M. Cheng. 2012. Graphene/metal oxide composite electrode materials for energy storage. Nano Energy 1(1): 107-131.) Copyright 2011, Elsevier)

Sulphur has extremely high theoretical capacity (1672 mAh/g) and being cost effective at the same time makes it a potential cathode material for the lithium-sulphur (Li-S) batteries (Manthiram et al. 2014). The sulphur enriched polymers (lithium-polysulphur (Li-polyS)) were found to be effective in restricting the sulphide deposition on the electrodes and subsequently, when used as active material batteries, had better performance in comparison to commercial Li-S batteries (Simmonds et al. 2014). However, according to Chang and Manthiram (Chang and Manthiram 2017), the moderately low sulphur doping (<2 mg cm^{-2}) in the present Li-polyS batteries somewhat concedes the areal capacity (capacity per unit area), which suggests a new cathode active material 'polysulphur-graphene nanocomposite (polySGN)' for higher doping Li-polyS batteries, doping up to 10.5 mg cm^{-2} and areal capacity of almost 12 mAh cm^{-2}.

Anode

Among the wide array of anode materials, SnO$_2$/graphene nanocomposites have gained interest owing to its high theoretical specific capacities combined with low charge potential. The electrochemical performances of SnO$_2$/graphene nanocomposites can be improved by altering chemical compositions, crystal structure and surface features. According to Deng et al. (Deng et al. 2016), the graphene sheets could prevent the stacking of SnO$_2$ particles (and restacking of graphene) and restrict the volume change during charge/discharge processes and better the electronic conductivity of the electrode.

Natural abundance of some elements from alkali metals group (same as Li) such as Na (6th most abundant) and K (8th most abundant) has paved the way for K-ion batteries (KIBs) and Na-ion batteries (NIBs) as a supplement of lithium batteries. SnS$_2$/graphene thin films have been used by Bin et al. (Bin et al. 2019) for high-performance KIBs anode (with Al as base metal) through combined efforts of targeting both the active material and the resulting electrode film. The electrochemically-active species of SnS$_2$ was converted into nanoparticles with size below 5 nm to provide the required high number of reactive sites for K$^+$ storage. SnO$_2$/graphene anodes hold great potential for reversible batteries in both Li-ion batteries and NIBs due to synergistic effect of a couple of reactions (Chen et al. 2017). The reactions to produce the desired synergistic effect were production of rGO and SnO$_2$ nanoparticles, and the resulting NIBs performed almost as good as their Li counterpart.

7.3 OPTOELECTRONIC APPLICATIONS

Optoelectronics is a branch of photonics which deals with electronic instruments that source, detect and control light. The optoelectronic devices work on the quantum mechanical effects of light in materials used in the electrical industries (usually semiconductors). Optoelectronic devices can be broadly classified in two groups:

1. Emitters
2. Detectors

Emitters are devices which take voltage or current as input and radiate electromagnetic waves as output, whereas detectors take electromagnetic radiation as input and produce a voltage or current in return. Optoelectronic devices include lasers, optical fibres, photo diode, LEDs, field-effect transistors, and remote sensing systems.

7.3.1 Flexible Transparent Electrodes

Flexible electronics demand compact, light and handy material having appreciable mechanical durability upon bending, for which flexible and stretchable organic materials have proven themselves superior over brittle inorganic materials. Serving for decades in the field of transparent conducting oxides for organic light emitting diodes (OLEDs), organic solar cells, and indium tin oxide (ITO) have serious drawbacks like poor mechanical stress tolerance, expensive and chemically instable. To overcome these drawbacks, graphene has emerged as a promising substitute for ITO in flexible electronics. Single layer graphene exhibits $\approx$ 97.7% transparency which is very advantageous for its use in optoelectronic devices (Sheehy and Schmalian 2009). Nair et al. (Nair et al. 2008) elucidated the trade-off between number of graphene layers and transmittance through fine structure coefficient ($\alpha = e^2/\hbar c$). This parameter describes light-electron coupling in relativistic frame in quantum electrodynamics. Being single atom thick, graphene here presents considerable analogy from material science. Thus, single layered graphene absorbs 2.3% ($= \pi\alpha$) of light. For Dirac fermions of graphene, high frequency conductivity is considered as universal constant, yielding out the relation between transmittance (T) and α as

$$(1 - T) \approx \pi\alpha \tag{7.1}$$

$$T \approx (100 - 2.3)\% = 97.7\% \tag{7.2}$$

Hence, generalizing above equation for N layers, the expression took the form as

$$T \approx (100 - 2.3 \times N) \tag{7.3}$$

where N is the number of layers. This equation tells that per layer increase in graphene thickness would reduce the transparency by 2.3%.

Apart from excellent transparency, single layered graphene offers breaking strength ($\sim$ 42 N/m) corresponding to intrinsic strength of $\sim$ 125 GPa and Young's modulus of $\sim$ 1100 GPa, making it obvious choice for transparent and flexible electrode (Lee et al. 2008).

Besides excellent transparency and sheet resistance offered by ITO, its brittleness, high cost, and chemical sensitivity imposed limitation on its application and hence paved way for exploration of its replacement. Due to good transparency and appreciable conductivity, graphene poses itself as a promising candidate for optoelectronic devices. Graphene electrodes could also be developed at low prices using inkjet printing technology bestowing the ease of dispersion of GO (Wan et al. 2012).

7.3.2 Field-Effect Transistors

A transistor is an electronic component that can be used as an amplifier or a switch as per requirement. A transistor which uses an electric field to control the flow of current in a circuit is called field-effect transistor (FET). FETs have three terminals: source, gate, and drain. The current flows from the source to the drain on application of voltage to gate. Most commonly used field-effect transistor is the metal-oxide-semiconductor field-effect transistor (MOSFET) (refer Figure 7.4). Transistor's ability depends upon their charge carrier mobility (electrons or holes). It is desirable to have thin gates and high mobility across the channel in order to achieve FETs with high response rates. However, FETs with thin gates are known to have constraints, such as decrease of threshold voltage (voltage across gate and source required to turn it on) and drain current saturation (additional drain-to-source voltage across depletion layer of the drain end of the gate) (Taur and Ning 2013). Combined, these are known as short-channel effects. Threshold voltages are also increased by source, drain and channel resistances being in series. The effect becomes more prominent as the length of the gate decreases in MOSFETs (Thompson et al. 2005).

Graphene having superior electron mobility to silicon semiconductors and at the same time being flexible and transparent has wide range of possibilities in field-effect transistors (FETs). But zero band gap of pristine graphene limits its usage in this field (Zhou et al. 2007). The problem with zero band gap transistor would be no control over switching it on or off, which is possible in case of semiconductors. A band gap for graphene can be created in three ways:

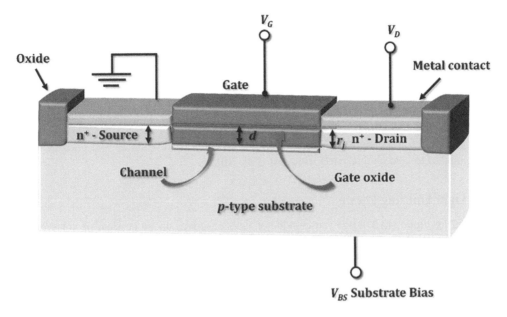

Fig. 7.4   Schematic diagram of a p-type MOSFET

1. Decreasing the area of graphene in one dimension by creating nanoribbons.
2. Creating asymmetrical bilayer graphene.
3. Applying strain to graphene.

Apparently, all these methods have their own limitations when considering applications to real world scenarios. The mobility of graphene and its nanocomposites decreases with defects, increase of layers, wrinkling and doping (Choi et al. 2012). Prime advantage of graphene to be used in FETs is to place as thin channels (gate-controlled region) that are just one atomic layer thick, which can enable healthier transistor (Schwierz 2010).

Graphene devices based on the conventional FET principle suffer from some fundamental problems. Alternative concepts such as tunnel field-effect transistor (TFET), first discussed by Chang and Esaki (Chang and Esaki 1977), can be considered. In a TFET, quantum tunnelling is modulated across the source-channel junction by varying the gate-source voltage. Iannaccone et al. (Iannaccone et al. 2009) used simulations to show potential of bilayer graphene TFET by highlighting problems faced while manufacturing narrow nanoribbons and suggest generation of band gap by applying vertical field and lowering sub-threshold swing below 60 mV per decade at room temperature. Sub-threshold swing value (S_{sth}) is the reciprocal value of the subthreshold slope (obtained from graph of drain current versus gate voltage while keeping other voltages constant). Sub-threshold swing numerically is denoted by (Sze and Ng 2006):

$$S_{\text{sth}} = (\ln 10)\frac{dV_G}{d(\ln I_D)} = (\ln 10)\left(\frac{kT}{q}\right)\left(\frac{C_{ox} + C_D}{C_{ox}}\right) \tag{7.4}$$

where, V_G is the gate voltage, I_D is the diffusion current, $\dfrac{kT}{q}$ is thermal voltage at temperature T, C_D is capacitance of depletion layer and C_{ox} is capacitance of gate-oxide.

A vertical TFET made by stacking double bilayer graphene (BLG) and hexagonal boron nitride heterostructure was operated at room temperature by Kang et al. (Kang et al. 2015). The TFET shows a couple of tunnelling resonances having negative differential resistance due to relative alignment of the lower or upper bands of the double BLG. The low quantum capacitance of BLG-TFETs allow large on-off ratios at low supply voltage and obtaining an extremely small sub-threshold swing (20 mV per decade at room temperature) at room temperature (Fiori and Iannaccone 2009). Small sub-threshold swing values allow us to eliminate short-channel effects.

Also, concomitantly, expensive metals are extensively in use as the source/drain electrodes for the fabrication of FETs on SiO_2/Si substrates. Solution-processed rGO based transistor have been made and are a viable option for transparent and flexible electronics (He et al. 2011). The devices with graphene and graphene oxide electrodes have less contact resistances and higher mobilities when compared to gold counterparts (Becerril et al. 2010).

7.3.3 Light Emitting Diode

Light-emitting diode (LED) is a semiconductor device which, under passage of current, allows the electrons and holes to recombine in order to emit light. Semiconductor devices are usually comprised of 3rd and 5th groups of the periodic table along with some other elements, common examples include InGaN, AlGaInP, AlGaAs and GaP. The desirable property for an optically conductive electrode is to have low sheet resistance and high work function. The sheet resistance characterizes the conductive nature of device, i.e. it is the measure of current collected at applied bias and work function is energy barrier between anode/transparent conducting electrode (TCE) and overlaying layer which is the measure of hole injection in OLEDs. Basically, an OLED consists of an anode (+ve, transparent), cathode (–ve, metal or metal composite) and an active region which is sandwiched by electrodes as shown in Figure 7.5. A transparent anode like graphene or ITO is required to emit light generated by the recombination of electrons and holes. Being a good prospect for TCE, graphene found its use in lighting devices. A conductive polymer of gradient work function modified four layered graphene surface which resulted in high current and luminous efficiency in lighting devices, thereby replacing conventional ITO (Rana et al. 2014). In terms of layers, using monolayer graphene for preparing flexible OLED permitted significant hole injection from graphene electrode to active layer, where usage of multilayer graphene underwent self-absorption of 2-3% per layer (Rana et al. 2014). Pristine graphene has high sheet resistance and lower work function in comparison to ITO and these drawbacks are widely addressed by doping of pristine graphene by various doping and surface modification methods, discussed by Han et al. (Han et al. 2017). Graphene TCE being flexible, cost-effective and chemically stable is basis for production of next generation flexible OLED displays (Rana et al. 2014). Multilayer graphene electrode has lower efficiency in functional device than ITO electrode. This was due to inefficient charge injection from electrode into organic layer (Rana et al. 2014).

The device fabricated using solution-based graphene anode showed higher operating voltage and lower luminous efficacy, which was improved significantly by alternative deposition of negatively and positively charged rGO at the cost of transparency and performance (Lee et al. 2011, Wu et al. 2010). This problem was addressed by producing solution processed graphene composite electrodes via new way, i.e. oxidizing graphene to GO followed by stabilization using small molecule surfactant with subsequent reduction using hydrazine (Chang et al. 2010). The general problem encountered with ITO was etching of ITO during solution processing by acidic polymer dispersion, which resulted in diffusion of In/Sn into OLEDs, forming trapping sites and exciton quenching sites which degrade charge injection, charge transport and luminous efficiency of OLEDs (Lee et al. 1999, Sharma et al. 2011). Graphene based conducting electrodes have proven themselves as appreciable solution above problem.

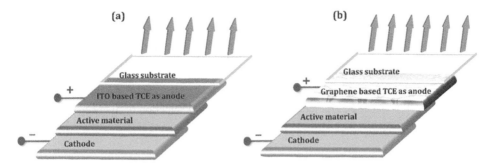

Fig. 7.5 An exploded-view of an OLED using a glass substrate. (a) OLED using ITO as an anode, (b) OLED using a graphene based anode. Graphene based anode will be able to transmit more light due to superior transparency

7.4   SOLAR CELLS

Solar cell or photovoltaic cell is an electrical device that deals with the conversion of energy from incident light into electrical energy via the photovoltaic effect (upon exposure to light, when a material is induced with voltage and electric current, it is known as photovoltaic effect). Silicon solar cells are created by stacking of silicon-based n-type and p-type layers on top of one other. The electricity is produced due to photovoltaic effect in the form of direct current (DC) depicted in Figure 7.6.

Poly (3,4-ethylenedioxythiophene) polystyrene sulfonate (PEDOT:PSS) is a transparent and conductive polymer with high ductility having additional properties such as low contact resistance. These properties enable PEDOT:PSS to be used as a material in flexible solar cells. In a dye-sensitized solar cell (DSSC), Lee et al. (Lee et al. 2017) suggested that after bending the paper electrodes considerable number of times, the performance of the cell with the GD/PEDOT:PSS paper electrode was still not degraded, whereas platinum based device lost its cell performance suffering the bending cycles.

Polymer solar cells (PSCs) is an alternative technology with advantages such as being flexible and cost effective with solution processability. The key factors in determining power conversion efficiency (PCE) are not the donor and acceptor materials in the active layer alone, but equal importance is held by the anode/cathode interlayer (AIL/CIL). Interlayer can be inserted between active layer and ITO to enhance the performance of polymer solar cells. Graphene quantum dot (GQD) are used as interlayer material for PSCs. The primary advantage of cathode interlayer (CIL) is lower work function. The work function of ITO electrode can be modulated by edge-carboxylated GQD (ECGQD) as shown by Znag et al. (Zhang et al. 2017), with the low work function ECGQD derivatives being used as CILs for PSCs as their performances were almost as good as CILs made of ZnO.

Fill Factor

The term used for determining or comparing quality and overall behaviour of a solar cell is the fill factor (FF). This is the ratio of available power at the maximum power point (P_m) divided by the open circuit voltage (V_{OC}, potential difference across terminals of a device while it's not connected to any circuit) and the short circuit current (I_{SC}, current through the solar cell at zero voltage):

$$FF = \frac{P_m}{V_{OC} \times I_{SC}} = \frac{h \times A_c \times G}{V_{OC} \times I_{SC}} \qquad (7.5)$$

Typical fill factors fall in the range 50% to 82%. The fill factor for a normal silicon photovoltaic cell is 80%. The photovoltaics prepared with chemical vapour deposition (CVD) graphene showed efficiency comparable to ITO accompanied by remarkable response under bending conditions (Gomez De Arco et al. 2010). This was done by comparing fill factor (FF) of both ITO and graphene at different bend angles. Poor performances were noted in even acute angles of bend (2Θ) in case of ITO whereas graphene showed satisfactory results at obtuse angles as well. FF being directly proportional to cell conversion efficiency plays a vital role in selection of material.

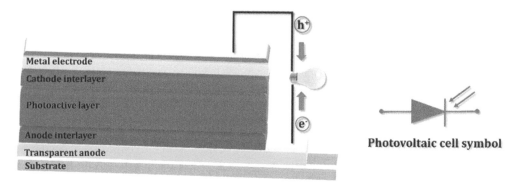

Fig. 7.6   Schematic of an organic solar cell

Balance Limit of Efficiency of *p-n* Junction Solar Cells

The maximum theoretical efficiency of a solar cell is often set by the Shockley-Queisser (SC) limit, which considers single junction solar cell yielding only radiative recombination and no other radiative loss (Shockley and Queisser 1961). The limit is formulated as follows:

$$n = t_s u(x_g) v(f, x_g, x_c) m(v x_g / x_c) \qquad (7.6)$$

where $x_g = v_g/v_s$, $x_c = v_c/v_s$, m is impedance matching factor, u is ultimate efficiency factor and v is ratio of open circuit voltage to band-gap voltage.

7.4.1   Perovskite Solar Cell

Perovskite solar cells are named after the ABX_3 crystal structure of the absorber materials, which is also known as the perovskite structure. For example, methylammonium lead trihalide ($CH_3NH_3PbX_3$, where X is a halogen) is a commonly studied perovskite structure for absorber material. Perovskite solar cells have gained significant attention as they pose power conversion efficiencies comparable to traditional solar cells and include the benefits of being lightweight, flexible and having relatively cheaper production cost.

In perovskite solar cells, varying halide content can help in tuning and modification of band gap to make it compatible with solar spectrum. A perovskite solar cell having band gap 1.55 eV has SC radiative efficiency limit of about 31%, corresponding to Am1.5G solar spectrum at 100 mW/cm² (Rühle 2016). However, this limit is small as compared to that for gallium arsenide, which shows 33% efficiency with 1.42 eV band gap (GaAs, having internal luminescence efficiency of 99%). Ren et al. (Ren et al. 2017), through their Drift diffusion model, suggest two prerequisite procedures to predict and proceed for SC efficiency limit, stated as follows:

1. The rectification of intrinsic radiative recombination is vital after assumption of optical design, which will leave significant impact on open circuit voltage at SC limit.
2. The careful engineering of contact characteristics of electrodes is necessary to avoid surface recombination and charge accumulation at the electrodes.

7.5 CAPACITORS

Capacitors are widely used devices that accumulate energy by creating an electrostatic field across electrodes. Supercapacitors (also known as ultracapacitors or electrochemical capacitors) are capacitors with advantages such as high charge-discharge rate, longer life and high power density. Production of sustainable supercapacitors is an enormous challenge faced by the industry. Supercapacitors in daily life undergo fair amount of stress cycles and might not be able to maintain proper functionality throughout. Wang et al. (Wang et al. 2017) suggest rGO based springs as electrodes to sustain stretching and even heal on breaking while maintaining appreciable capacitances. They did so by coating the rGO springs with stretchable carboxylated polyurethane. The arrangement did manage to reconnect the hydrogen bonds initially broken and restore the functionality of the supercapacitor. Le et al. (Le et al. 2017) made efforts to incorporate sodium in electrochemical capacitors by producing anatase TiO_2 assembled onto graphene which would act as sodium storage material. This nanocomposite managed to restore 90% of the initial capacity at high cycling rates (about 10,000); the possible charge storage type and structure are suggested by Le et al. (Le et al. 2017).

7.6 SENSORS

A sensor is a transducer that senses or detects a physical quantity and converts it into useful signals. Sensors could be made for temperature, light, heat, motion, moisture, blood pressure, humidity, speed, pressure, or any one of a great number of other environmental phenomena.

7.6.1 Gas Sensors

Gas sensors work on the principle of varying resistance under different concentrations of a certain gas. Certain parameters are used to test the capability of a gas sensor. The response of the gas sensor is defined as the ratio of the resistance of the sensor in the reference gas (R_x) to that in air (R_a). For

$$\text{oxidizing reference gas, Response} = R_x/R_a$$
$$\text{reducing reference gas, Response} = R_a/R_x$$

The response time is defined as the time taken by the sensor to gain 90% of the total resistance change in the case of adsorption.

The recovery time is defined as the time taken by the sensor to lose 10% of the total resistance change in the case of desorption (refer Figure 7.7). The figure shows cycles of varying resistance with and without the influence of the test gas. The SnO_2/rGO composites show high response and reproducibility towards detection of NO_2 with the response of the synthesized sensor to 1 ppm NO_2 at 75°C which was higher than that of SnO_2 sensors alone, and the detection limit was increased to 50 ppb bestowed to the new potential barrier at the SnO_2/rGO assemblage (Xiao et al. 2016). ZnO nanorods can be used to synthesize ultrathin ZnO nanorods/rGO which show amazing response and sensitivity to NO_2 while being inert to other gases (Xia et al. 2016). The sensitivity of ZnO/graphene nanocomposite gas sensors can be enhanced further by microwave irradiation as after irradiation defects form finer nanoparticles and increases the surface area (Kim et al. 2017). Nanorods of other materials such as WO_3 with graphene can also prove to be useful in gas sensors because graphene improves electron transfer rate resulting in faster response as well as lowering detection limit (An et al. 2012). Sensing a flammable as such as H_2 in time can sometimes be enough to avoid a disaster. A Micro-Electro-Mechanical Systems (MEMS) based graphene-Pd-Ag-gate FET (GPA-FET) was produced by Sharma and Kim (Sharma and Kim 2018). MEMS are miniaturized structures, sensors, actuators, and microelectronics, which are transducers and are used in the form of microchips and sometimes do outperform their macroscale counter-parts. This MEMS comprises of 2 FETs, Pt-gate FET as a platform to sensor and GPA-FET to sense H_2.

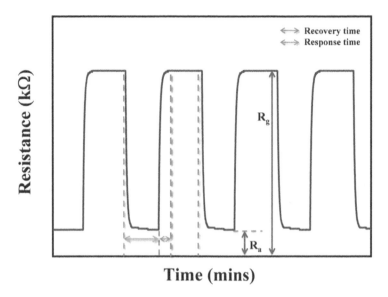

Fig. 7.7   Resistance v/s time graph, showing general trend observed while studying gas sensor characteristics

7.6.2   Electrochemical Biosensors

Biosensors are used to detect an analyte by using a physiochemical transducer. The biosensors which convert biochemical information into a useful electrical signal are known as electrochemical biosensors. Nanocomposites of GO/Au nanocluster modified electrode and TiO_2/graphene modified glassy carbon electrodes were used to detect *l*-cysteine in human urine (Ge et al. 2012) and dopamine in an analyte (Fan et al. 2011), respectively.

7.6.3   Strain Sensors

A strain sensor is a device used to measure strain (ratio of change in dimension to original dimension) by converting force from physical parameters to variation in resistance. Most commonly used types of strain sensors include strain gauge and piezoelectric sensors. In case of strain gauge, when subjected to tension the resistance increases, whereas in case of compression there is a decrease in resistance; this happens as there is inverse relation between resistivity and area (refer Figure 7.8). Piezoelectric sensors comprise of two piezoelectric crystals which sandwich an electrode foil. When force is applied to a piezoelectric crystal, there is generation of voltage across it; this effect is known as piezoelectric effect and subsequent current produced is known as piezoelectric current. This current can be calibrated to measure the applied force as it is directly proportional to it. Strain sensors primarily find their requirement while detecting deformations or structural changes in structures and precision tools where tolerances are low. Gauge factor, also known as strain factor of a strain gauge, can be defined as the ratio of corresponding change in electrical resistance under mechanical strain. The gauge factor may be expressed as:

$$\text{Gauge factor} = \frac{\Delta R/R}{\Delta L/L_0} \tag{7.7}$$

ΔR is change in resistance due to strain, R is resistance without any strain, ΔL is change in length and L_0 denotes the original length.

Fig. 7.8 Working concept behind the strain gauge demonstrated on a bar (bending is exaggerated for visualization). Reproduced with edits from Wikipedia under CC0 license (Commons 2015)

In piezoelectric materials, resistivity changes under strain resulting in a change in gauge factor by a factor $\dfrac{\Delta\rho/\rho}{\Delta L/L_0}$, new gauge factor can be obtained by adding this factor to equation 7.7 (Rolnick 1930). Here, ρ stands for resistivity of material and $\Delta\rho$ is for change in resistivity. After addition of this factor to equation 7.7:

$$\text{Gauge factor} = \frac{\Delta R/R}{\Delta L/L_0} + \frac{\Delta\rho/\rho}{\Delta L/L_0} \tag{7.8}$$

Further, we can use the relation,

$$\frac{\Delta R}{R} = (1 + 2v)\frac{\Delta L}{L_0}$$

where v is the Poisson's ratio for the material. Hence,

$$\text{Gauge factor} = 1 + 2v + \frac{\Delta\rho/\rho}{\Delta L/L_0} \tag{7.9}$$

Equation 7.9 is the general and most commonly used equation to find out the gauge factor, and can be reduced back to equation 7.7 in cases where $\Delta\rho = 0$. Higher gauge factor indicates more sensitivity while high strain percentage means larger range.

Graphene-based strain sensors have the ability to replace conventional strain sensors by offering more cycles before failure due to its high Young's modulus value ~ 1 TPa (Lee et al. 2008). In addition to having extraordinary mechanical properties, graphene also offers versatility as its transparency, flexibility and ability to have high electronic conductivity make it a suitable material for nanocomposites in sensors.

Piezoresistive sensors, being robust, find their applications where sufficient power supply is available and sensors might be subjected to high force. CVD can be used to have graphene foams which maintain high electrical conductivity by virtue of conductive networks and making composites like graphene foams/poly(dimethyl siloxane) which can sustain high strains without breaking (Chen et al. 2011). A more cost-efficient possibility is the use of graphene along with a stretchable nanopaper where high strain rates are to be measured as it can sustain strain up to 100% (Yan et al. 2014).

Conductive polymer composites (CPCs) have gained interest in the field of strain sensors as the material shows rapid variations in electrical resistance when under tensile or compressive strain (Bilotti et al. 2010, Costa et al. 2013, Gao et al. 2019, Lin et al. 2013, Liu et al. 2016, Pham et al. 2008, Ponnamma et al. 2013, Wang et al. 2018). CPCs have high recoverability and reproducibility, also withstanding good sensing stability after initial cycles as the strain patterns stabilize. CPCs composed of the thermoplastic polyurethane (TPU) matrix and graphene with extremely low percolation had high strain amplitudes indicating good sensitivity (Liu et al. 2016). rGO/TPU strain sensors were produced by Wang et al. (Wang et al. 2018) which had appreciable gauge factor of 79 at 100% strain and sustained a very high response even after a large number of stretch/release cycles (6000). High precision biological applications of strain sensors have been made in order to obtain movements of elbow, neck, fingers and also obtain stress at various joints and muscles of the body by using stretchability of polyurethane (PU) (Gao et al. 2019, Wang et al. 2018). Superhydrophobic and conductive nanofiber composite using PU as base and SiO_2/graphene as an outer layer was prepared by Gao et al. (Gao et al. 2019) with high cyclability and very high corrosion resistance. This nanocomposite had high performance and a rapid response to slightest variations (angular and longitudinal strain in sensor due to human movements) and managed high Young's modulus due to SiO_2/graphene and without compromising the ability of PU as a stretchable nanofiber (Gao et al. 2019).

Performance of the strain sensors depends upon the method of preparation used. A comparison of Raman 2D band shift rates between exfoliated graphene and CVD based graphene showed results in favour of exfoliated graphene by at least 35% (Raju et al. 2014). Post production, the CVD based sensors had pre-compression stresses which do even out after initial cycles.

7.7 FUEL CELLS

Fuel cells consist of an anode (–), a cathode (+) and an electrolyte, similar to the battery (refer Figure 7.9), although there is a key difference in electrolyte, i.e. it allows ions to move through it, which is exactly the opposite of electrolyte for batteries. An anode catalyst oxidises the fuel (first redox reaction), which generates ions and electrons. The ions generated reach the cathode passing through the electrolyte, whereas electrons flow from an external circuit enabling generation of DC (direct current) which is the desired output. Further, a cathode catalyst permits the reduction of oxygen (the second of the two redox reactions), forming water and by-products like heat (if any). So, fuel cell may be defined as an electrochemical cell, where with the help of a pair of redox reactions, chemical energy is converted into electrical energy without the release of harmful pollutants.

Fuel cells are being thought of as an alternative to the burning of fossil fuels. They use a fuel (such as hydrogen) to produce electricity whilst producing no carbon footprints, which are a major cause for global warming. Graphene is a promising base material for fuel cells as its high surface area offers high catalytic activity. Specific surface area (SSA) is defined as the surface area of a material per unit of mass (in m^2/g usually). The theoretical SSA value for ideally separated/isolated single layer graphene was predicted to be 2630 m^2/g (Peigney et al. 2001), although there have been results with higher SSA values, obtained by perforating graphene sheets (Baburin et al. 2015, Klechikov et al. 2015). As the production of quality graphene is an expensive process, GO is used by majority of articles in this field due to low-cost and scalability. Cathode materials play a crucial role towards the cost of fuel cell as they are usually made up of expensive metals like platinum (Pt), gold (Au) and their alloys, which are advantageous for oxygen reduction reaction (ORR).

Lack of availability of these materials have ignited the thoughts of having an alternative approach. A couple of possible scenarios can be minimising the use of these expensive metals by increasing the surface area (hence increasing the catalytic activity), or else even better, find efficient catalysts that can replace the existing ones (Guo et al. 2010, Liang et al. 2011, Qu et al. 2010). Although material like GO suffers from poor electrical conductivity and surface area with respect to graphene, having oxygen-containing groups allows it to catalyze the ORR. This is done by making more oxygen atoms being available at the electrode/electrolyte interface but compromising the electrical conductivity, hence density of these groups needs to be evaluated for honing the use of GO in fuel cells (Choi et al. 2012).

As discussed earlier, decrease in Pt usage will result in a cheaper fuel cell, making it commercially feasible. Honma et al. (Yoo et al. 2009) performed tests by using lower concentrations of Pt (20%) braced on graphene nanosheets (GNS) and found out that it had superior catalytic activity in comparison to Pt-Ru/carbon black methanol oxidation reaction electrocatalyst. The preparation of

Pt/GNS nanoparticle involves using $Pt(NO_2)_2 \cdot (NH_3)_2$ as precursor along with GNS powder, followed by heat-treatment in Ar/H_2 (4:1 v/v) steam at 400°C for 2 h in a furnace. Pt/GNS had 4 higher current density (0.12 mA/cm²) in comparison with Pt/carbon black (0.03 mA/cm²), which was interpreted from I-T curve result at 0.6 V versus reference hydrogen electrode (RHE) for 30 min. In addition to these excellent properties, Pt/GNS had outstanding resistance to poisoning by carbon monoxide (CO). Poisoning reduces the number of active sites (example: CO by binding to Pt tightly occupies active sites which were to be occupied by hydrogen atoms), hence reducing the catalytic activity. Li et al. (Li et al. 2009) prepared Pt/rGO nanocomposites by reducing graphite oxide (formed by Hummer's method and $NaBH_4$ as reducing agent) to obtain rGO and using it along H_2PtCl_6. The peak current densities for Pt/rGO was almost twice the commercial Pt/C (Pt/Vulcan, using Vulcan XC-72) catalysts indicating better performance. Key reason for better catalytic activity of graphene braced Pt can be chalked up to high interaction between GNS and Pt particles, which made these nanocomposites smaller and increased electrochemical active surface area (ECSA) (Choi et al. 2012).

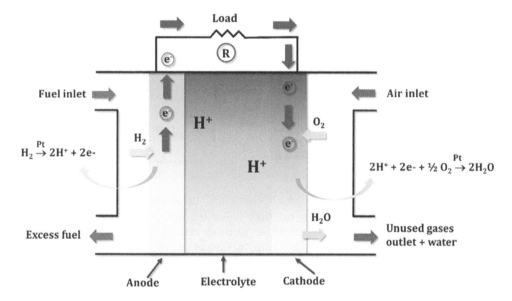

Fig. 7.9 Schematic diagram of a fuel cell using hydrogen as fuel and Pt as catalyst

For long, Pt nanoparticles have been deemed the best option as a catalyst for the ORR in fuel cells; yet, as seen till now, Pt-based electrodes, apart from being expensive, also suffer from shortcomings like CO poisoning and degradation of catalyst (Yu and Ye 2007). Nitrogen doped graphene (N-graphene) can be used as an alternative to obtain metal free catalysts for carbon materials, as it offers 3 times higher steady-state catalytic current density in comparison to traditional Pt/C electrodes in alkaline solutions (Ref. Figure 7.10(a)) (Qu et al. 2010). The highly stable amperometric response from the ORR on the N-graphene electrode was unaltered upon the addition of hydrogen gas, glucose and methanol which can be seen in Figure 7.10(b). This N-graphene electrode, on being exposed to 10% CO in air (v/v), was found to be inert to CO; contrary to this, there were drastic effects on commercial Pt/C electrode (Ref. Figure 7.10(c)). There was negligible drop in current density even after 2,00,000 cycles between −1.0 and 0 V (Ref. Figure 7.10(d)). Hence, these N-graphene electrodes are not only cheaper to manufacture due to processes like CVD, but also have endurance and tolerance to poisoning effect and fuel crossover. The phenomenon where fuel manages to permeate through electrolyte and reach the other electrode is known as fuel crossover. The commercially available Pt-based catalysts show poor tolerances to fuel crossover and poisoning effects while being expensive at the same time, which means they ought to be replaced. Hence, the graphene based nanocomposites have a future in fuel cells as they had better performance in comparison to commercial Pt/C, like having higher current density, stability and ECSA.

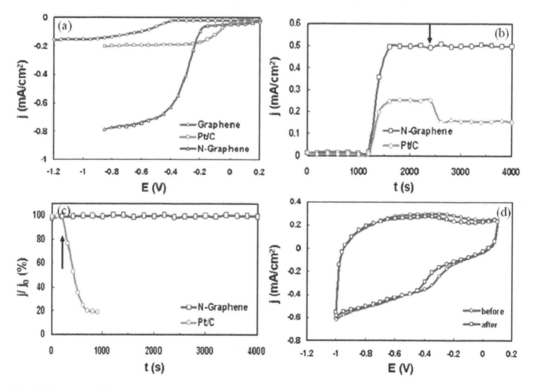

Fig. 7.10 (a) RRDE voltammograms for the ORR in air-saturated 0.1 M KOH at the C-graphene electrode, Pt/C electrode, and N-graphene electrode. Electrode rotating rate: 1000 rpm. Scan rate: 0.01 V/s. $Mass_{(graphene)} = Mass_{(PtC)} = Mass_{(N-graphene)}$ 7.5 μg, (b) Current density (j)-time (t) chronoamperometric responses obtained at the Pt/C and N-graphene electrodes at –0.4 V in air saturated 0.1 M KOH. The arrow indicates the addition of 2% (w/w) methanol into the air-saturated electrochemical cell, (c) Current (j)-time (t) chronoamperometric response of Pt/C and N-grapheneelectrodes to CO. The arrow indicates the addition of 10% (v/v) CO into air saturated 0.1 M KOH at –0.4 V; j_0 defines the initial current, (d) Cyclic voltammograms of N-graphene electrode in air saturated 0.1 M KOH before and after a continuous potentiodynamic swept for 2,00,000 cycles at room temperature (25°C). Scan rate: 0.1 V/s. (Reprinted with permission from Qu, L., Y. Liu, J.-B. Baek, and L. Dai. 2010. Nitrogen-doped graphene as efficient metal-free electrocatalyst for oxygen reduction in fuel cells. ACS Nano, 4(3): 1321-1326. Copyright (2010) American Chemical Society)

7.8   SUMMARY

The energy-based applications of graphene and graphene-based nanocomposites have been discussed in this chapter. It has been shown that the possibilities and opportunities of engineering these applications are practically endless. When used as flexible electrode, graphene found its use in transparent optoelectronic devices, solar panels and sensors. Graphene offers exceptional utility in lithium ion batteries, solar cells, LEDs and sensors due to its applaudable role as conductive additive in cathode/anode material, and low work function in composite enables better charge transfer to luminescent counter-parts and fast response, respectively. Graphene based nanocomposites pose themselves as an alternative to traditional Pt based fuel cells by providing high current density and electrochemical active surface area, making it commercially viable. Keeping in view its splendid properties, graphene will be attracting interest of research community as a potential candidate for energy applications in future as well.

7.9 ACKNOWLEDGMENTS

Thanks to Ms. Manju from Advanced Materials Research Lab for invaluable discussion regarding content of the chapter. Her assistance in language editing and chapter formatting is highly acknowledged.

7.10 REFERENCES

An, X., C.Y. Jimmy, Y. Wang, Y. Hu, X. Yu and G. Zhang. 2012. WO_3 nanorods/graphene nanocomposites for high-efficiency visible-light-driven photocatalysis and NO_2 gas sensing. J. Mater. Chem. 22(17): 8525-8531.

Baburin, I.A., A. Klechikov, G. Mercier, A. Talyzin and G. Seifert. 2015. Hydrogen adsorption by perforated graphene. Int. J. Hydrog. Energy 40(20): 6594-6599.

Bak, S.-M., K.-W. Nam, C.-W. Lee, K.-H. Kim, H.-C. Jung, X.-Q. Yang, et al. 2011. Spinel $LiMn_2O_4$/reduced graphene oxide hybrid for high rate lithium ion batteries. J. Mater. Chem. 21(43): 17309-17315.

Becerril, H.A., R.M. Stoltenberg, M.L. Tang, M.E. Roberts, Z. Liu, Y. Chen, et al. 2010. Fabrication and evaluation of solution-processed reduced graphene oxide electrodes for *p*-and *n*-channel bottom-contact organic thin-film transistors. ACS Nano 4(11): 6343-6352.

Bilotti, E., R. Zhang, H. Deng, M. Baxendale and T. Peijs. 2010. Fabrication and property prediction of conductive and strain sensing TPU/CNT nanocomposite fibres. J. Mater. Chem. 20(42): 9449-9455.

Bin, D.-S., S.-Y. Duan, X.-J. Lin, L. Liu, Y. Liu, Y.-S. Xu, et al. 2019. Structural engineering of SnS_2/Graphene nanocomposite for high-performance K-ion battery anode. Nano Energy 60: 912-918.

Bolotin, K.I., K.J. Sikes, Z. Jiang, M. Klima, G. Fudenberg, J. Hone, et al. 2008. Ultrahigh electron mobility in suspended graphene. Solid State Commun. 146(9-10): 351-355.

Chang, C.-H. and A. Manthiram. 2017. Covalently grafted polysulfur–graphene nanocomposites for ultrahigh sulfur-loading lithium–polysulfur batteries. ACS Energy Lett. 3(1): 72-77.

Chang, H., G. Wang, A. Yang, X. Tao, X. Liu, Y. Shen, et al. 2010. A transparent, flexible, low temperature, and solution-processible graphene composite electrode. Adv. Funct. Mater. 20(17): 2893-2902.

Chang, L. and L. Esaki. 1977. Tunnel triode—a tunneling base transistor. Appl. Phys. Lett. 31(10): 687-689.

Chen, W., K. Song, L. Mi, X. Feng, J. Zhang, S. Cui, et al. 2017. Synergistic effect induced ultrafine SnO_2/graphene nanocomposite as an advanced lithium/sodium-ion batteries anode. J. Mater. Chem. A 5(20): 10027-10038.

Chen, Z., W. Ren, L. Gao, B. Liu, S. Pei and H.-M. Cheng. 2011. Three-dimensional flexible and conductive interconnected graphene networks grown by chemical vapour deposition. Nat. Mater. 10(6): 424.

Choi, H.-J., S.-M. Jung, J.-M. Seo, D.W. Chang, L. Dai and J.-B. Baek. 2012. Graphene for energy conversion and storage in fuel cells and supercapacitors. Nano Energy 1(4): 534-551.

Commons, W. 2015. File:straingaugevisualization.svg — wikimedia commons, the free media repository.

Cooper, D.R., B.D'Anjou, N. Ghattamaneni, B. Harack, M. Hilke, A. Horth, et al. 2012. Experimental review of graphene. ISRN Condens. Matter Phys. 2012: 1-56.

Costa, P., A. Ferreira, V. Sencadas, J.C. Viana and S. Lanceros-Mendez. 2013. Electromechanical properties of triblock copolymer styrene-butadiene-styrene/carbon nanotube composites for large deformation sensor applications. Sens. Actuator A. Phys. 201: 458-467.

Deng, Y., C. Fang and G. Chen. 2016. The developments of SnO_2/graphene nanocomposites as anode materials for high performance lithium ion batteries: A review. J. Power Sources 304: 81-101.

Duplock, E.J., M. Scheffler and P.J. Lindan. 2004. Hallmark of perfect graphene. Phys. Rev. Lett. 92(22): 225502.

Fan, Y., H.-T. Lu, J.-H. Liu, C.-P. Yang, Q.-S. Jing, Y.-X. Zhang, et al. 2011. Hydrothermal preparation and electrochemical sensing properties of TiO_2-graphene nanocomposite. Colloids Surf. B: Biointerfaces 83(1): 78-82.

Fiori, G. and G. Iannaccone. 2009. Ultralow-voltage bilayer graphene tunnel FET. IEEE Electron Device Lett. 30(10): 1096-1098.

Gao, J., B. Li, X. Huang, L. Wang, L. Lin, H. Wang, et al. 2019. Electrically conductive and fluorine free superhydrophobic strain sensors based on SiO_2/graphene-decorated electrospun nanofibers for human motion monitoring. Chem. Eng. J. 373: 298-306.

Ge, S., M. Yan, J. Lu, M. Zhang, F. Yu, J. Yu, et al. 2012. Electrochemical biosensor based on graphene oxide-Au nanoclusters composites for *l*-cysteine analysis. Biosens. Bioelectron. 31(1): 49-54.

Gomez De Arco, L., Y. Zhang, C.W. Schlenker, K. Ryu, M.E. Thompson and C. Zhou. 2010. Continuous, highly flexible, and transparent graphene films by chemical vapor deposition for organic photovoltaics. ACS Nano 4(5): 2865-2873.

Guo, S., S. Dong and E. Wang. 2010. Three-dimensional Pt-on-Pd bimetallic nanodendrites supported on graphene nanosheet: Facile synthesis and used as an advanced nanoelectrocatalyst for methanol oxidation. ACS Nano 4(1): 547-555.

Han, T.-H., H. Kim, S.-J. Kwon and T.-W. Lee. 2017. Graphene-based flexible electronic devices. Mater. Sci. Eng. R. Rep. 118: 1-43.

He, Q., S. Wu, S. Gao, X. Cao, Z. Yin, H. Li, et al. 2011. Transparent, flexible, all-reduced graphene oxide thin film transistors. ACS Nano 5(6): 5038-5044.

Iannaccone, G., G. Fiori, M. Macucci, P. Michetti, A. Cheli, A. Betti, et al. 2009. Perspectives of graphene nanoelectronics: probing technological options with modeling. pp. 1-4. *In*: M.D. Baltimore [ed.]. IEEE International Electron Devices Meeting (IEDM), IEEE, USA.

Kang, S., B. Fallahazad, K. Lee, H. Movva, K. Kim, C.M. Corbet, et al. 2015. Bilayer graphene-hexagonal boron nitride heterostructure negative differential resistance interlayer tunnel FET. IEEE Electron Device Lett. 36(4): 405-407.

Kim, H.W., Y.J. Kwon, A. Mirzaei, S.Y. Kang, M.S. Choi, J.H. Bang, et al. 2017. Synthesis of zinc oxide semiconductors-graphene nanocomposites by microwave irradiation for application to gas sensors. Sens. Actuator B. Chem. 249: 590-601.

Klechikov, A., G. Mercier, T. Sharifi, I.A. Baburin, G. Seifert and A.V. Talyzin. 2015. Hydrogen storage in high surface area graphene scaffolds. Chem. Comm. 51(83): 15280-15283.

Le, Z., F. Liu, P. Nie, X. Li, X. Liu, Z. Bian, et al. 2017. Pseudocapacitive sodium storage in mesoporous single-crystal-like TiO_2-graphene nanocomposite enables high-performance sodiumion capacitors. ACS Nano 11(3): 2952-2960.

Lee, C., X. Wei, J.W. Kysar and J. Hone. 2008. Measurement of the elastic properties and intrinsic strength of monolayer graphene. Science 321(5887): 385-388.

Lee, C.-P., K.-Y. Lai, C.-A. Lin, C.-T. Li, K.-C. Ho, C.-I. Wu, et al. 2017. A paper-based electrode using a graphene dot/PEDOT: PSS composite for flexible solar cells. Nano Energy 36: 260-267.

Lee, D.W., T.-K. Hong, D. Kang, J. Lee, M. Heo, J.Y. Kim, et al. 2011. Highly controllable transparent and conducting thin films using layer-by-layer assembly of oppositely charged reduced graphene oxides. J. Mater. Chem. 21(10): 3438-3442.

Lee, S., Z. Gao and L. Hung. 1999. Metal diffusion from electrodes in organic light-emitting diodes. Appl. Phys. Lett. 75(10): 1404-1406.

Li, Y., L. Tang and J. Li. 2009. Preparation and electrochemical performance for methanol oxidation of Pt/graphene nanocomposites. Electrochem. Commun. 11(4): 846-849.

Liang, Y., Y. Li, H. Wang, J. Zhou, J. Wang, T. Regier, et al. 2011. Co_3O_4 nanocrystals on graphene as a synergistic catalyst for oxygen reduction reaction. Nat. Mater. 10(10): 780-786.

Lin, L., S. Liu, Q. Zhang, X. Li, M. Ji, H. Deng, et al. 2013. Towards tunable sensitivity of electrical property to strain for conductive polymer composites based on thermoplastic elastomer. ACS Appl. Mater. Interfaces. 5(12): 5815-5824.

Liu, H., Y. Li, K. Dai, G. Zheng, C. Liu, C. Shen, et al. 2016. Electrically conductive thermoplastic elastomer nanocomposites at ultralow graphene loading levels for strain sensor applications. J. Mater. Chem. C. 4(1): 157-166.

Lu, J.P. 1997. Elastic properties of carbon nanotubes and nanoropes. Phys. Rev. Lett. 79(7): 1297.

Manthiram, A., Y. Fu, S.-H. Chung, C. Zu and Y.-S. Su. 2014. Rechargeable lithium-sulfur batteries. Chem. Rev. 114(23): 11751-11787.

Morozov, S., K. Novoselov, M. Katsnelson, F. Schedin, D. Elias, J.A. Jaszczak, et al. 2008. Giant intrinsic carrier mobilities in graphene and its bilayer. Phys. Rev. Lett. 100(1): 016602.

Nair, R.R., P. Blake, A.N. Grigorenko, K.S. Novoselov, T.J. Booth, T. Stauber, et al. 2008. Fine structure constant defines visual transparency of graphene. Science 320(5881): 1308.

Novoselov, K.S., A.K. Geim, S.V. Morozov, D. Jiang, Y. Zhang, S.V. Dubonos, et al. 2004. Electric field effect in atomically thin carbon films. Science 306(5696): 666-669.

Papageorgiou, D.G., I.A. Kinloch and R.J. Young. 2017. Mechanical properties of graphene and graphene-based nanocomposites. Prog. Mater. Sci. 90: 75-127.

Peigney, A., C. Laurent, E. Flahaut, R. Bacsa and A. Rousset. 2001. Specific surface area of carbon nanotubes and bundles of carbon nanotubes. Carbon 39(4): 507-514.

Pham, G.T., Y.-B. Park, Z. Liang, C. Zhang and B. Wang. 2008. Processing and modeling of conductive thermoplastic/carbon nanotube films for strain sensing. Compos. Part B: Eng. 39(1): 209-216.

Ponnamma, D., K.K. Sadasivuni, M. Strankowski, Q. Guo and S. Thomas. 2013. Synergistic effect of multi walled carbon nanotubes and reduced graphene oxides in natural rubber for sensing application. Soft. Matter. 9(43): 10343-10353.

Qu, L., Y. Liu, J.-B. Baek and L. Dai. 2010. Nitrogen-doped graphene as efficient metal-free electrocatalyst for oxygen reduction in fuel cells. ACS Nano 4(3): 1321-1326.

Raju, A.P.A., A. Lewis, B. Derby, R.J. Young, I.A. Kinloch, R. Zan, et al. 2014. Wide-area strain sensors based upon graphene-polymer composite coatings probed by raman spectroscopy. Adv. Funct. Mater. 24(19): 2865-2874.

Rana, K., J. Singh and J.-H. Ahn. 2014. A graphene-based transparent electrode for use in flexible optoelectronic devices. J. Mater. Chem. C. 2(15): 2646-2656.

Ren, X., Z. Wang, W.E. Sha and W.C. Choy. 2017. Exploring the way to approach the efficiency limit of perovskite solar cells by drift-diffusion model. ACS Photonics 4(4): 934-942.

Rolnick, H. 1930. Tension coefficient of resistance of metals. Phys. Rev. 36(3): 506.

Ru‥hle, S. 2016. Tabulated values of the shockley-queisser limit for single junction solar cells. Sol. Energy 130: 139-147.

Schwierz, F. 2010. Graphene transistors. Nat. Nanotechnol. 5(7): 487.

Sharma, A., G. Andersson and D.A. Lewis. 2011. Role of humidity on indium and tin migration in organic photovoltaic devices. Phys. Chem. Chem. Phys. 13(10): 4381-4387.

Sharma, B. and J.-S. Kim. 2018. MEMS based highly sensitive dual FET gas sensor using graphene decorated Pd-Ag alloy nanoparticles for H_2 detection. Sci. Rep. 8(1): 1-9.

Sheehy, D.E. and J. Schmalian. 2009. Optical transparency of graphene as determined by the fine-structure constant. Phys. Rev. B. 80(19): 193411.

Shockley, W. and H.J. Queisser. 1961. Detailed balance limit of efficiency of *p-n* junction solar cells. J. Appl. Phys. 32(3): 510-519.

Simmonds, A.G., J.J. Griebel, J. Park, K.R. Kim, W.J. Chung, V.P. Oleshko, et al. 2014. Inverse vulcanization of elemental sulfur to prepare polymeric electrode materials for Li-S batteries. ACS Macro Lett. 3(3): 229-232.

Su, F.-Y., C. You, Y.-B. He, W. Lv, W. Cui, F. Jin, et al. 2010. Flexible and planar graphene conductive additives for lithium-ion batteries. J. Mater. Chem. 20(43): 9644-9650.

Sze, S.M. and K.K. Ng. 2006. Physics of Semiconductor Devices. John Wiley and Sons.

Taur, Y. and T.H. Ning. 2013. Fundamentals of Modern VLSI Devices. Cambridge University Press.

Thompson, S.E., R.S. Chau, T. Ghani, K. Mistry, S. Tyagi and M.T. Bohr. 2005. In search of "Forever," continued transistor scaling one new material at a time. IEEE Trans. Semicond. Manuf. 18(1): 26-36.

Wan, X., Y. Huang and Y. Chen. 2012. Focusing on energy and optoelectronic applications: A journey for graphene and graphene oxide at large scale. Acc. Chem. Res. 45(4): 598-607.

Wang, L., X. Zhou and Y. Guo. 2010. Synthesis and performance of carbon-coated $Li_3V_2(PO_4)_3$ cathode materials by a low temperature solid-state reaction. J. Power Sources 195(9): 2844-2850.

Wang, S., N. Liu, J. Su, L. Li, F. Long, Z. Zou, et al. 2017. Highly stretchable and self-healable supercapacitor with reduced graphene oxide based fiber springs. ACS Nano 11(2): 2066-2074.

Wang, Y., J. Hao, Z. Huang, G. Zheng, K. Dai, C. Liu, et al. 2018. Flexible electrically resistive-type strain sensors based on reduced graphene oxide-decorated electrospun polymer fibrous mats for human motion monitoring. Carbon 126: 360-371.

Wei, D., L. Grande, V. Chundi, R. White, C. Bower, P. Andrew, et al. 2012. Graphene from electrochemical exfoliation and its direct applications in enhanced energy storage devices. Chem. Comm. 48(9): 1239-1241.

Wu, J., M. Agrawal, H.A. Becerril, Z. Bao, Z. Liu, Y. Chen, et al. 2010. Organic light-emitting diodes on solution-processed graphene transparent electrodes. ACS Nano 4(1): 43-48.

Xia, Y., J. Wang, J.-L. Xu, X. Li, D. Xie, L. Xiang, et al. 2016. Confined formation of ultrathin ZnO nanorods/reduced graphene oxide mesoporous nanocomposites for high-performance roomtemperature NO_2 sensors. ACS Appl. Mater. Interfaces. 8(51): 35454-35463.

Xiao, Y., Q. Yang, Z. Wang, R. Zhang, Y. Gao, P. Sun, et al. 2016. Improvement of NO_2 gas sensing performance based on discoid tin oxide modified by reduced graphene oxide. Sens. Actuator B. Chem. 227: 419-426.

Yan, C., J. Wang, W. Kang, M. Cui, X. Wang, C. Y. Foo, et al. 2014. Highly stretchable piezoresistive graphene-nanocellulose nanopaper for strain sensors. Adv. Mater. 26(13): 2022-2027.

Ye, S., H. Huang, C. Yuan, F. Liu, M. Zhai, X. Shi, et al. 2014. Thickness-dependent strain effect on the deformation of the graphene-encapsulated Au nanoparticles. J. Nanomater. 2014.

Yoo, E., T. Okata, T. Akita, M. Kohyama, J. Nakamura and I. Honma. 2009. Enhanced electrocatalytic activity of Pt subnanoclusters on graphene nanosheet surface. Nano Lett. 9(6): 2255-2259.

Yu, X. and S. Ye. 2007. Recent advances in activity and durability enhancement of Pt/C catalytic cathode in PEMFC: Part II: Degradation mechanism and durability enhancement of carbon supported platinum catalyst. J. Power Sources 172(1): 145-154.

Zhang, L., Z. Ding, T. Tong and J. Liu. 2017. Tuning the work functions of graphene quantum dot-modified electrodes for polymer solar cell applications. Nanoscale 9(10): 3524-3529.

Zhang, W., Y. Zeng, C. Xu, N. Xiao, Y. Gao, L.-J. Li, et al. 2012. A facile approach to nanoarchitectured three-dimensional graphene-based Li–Mn–O composite as high-power cathodes for Li-ion batteries. Beilstein J. Nanotechnol. 3(1): 513-523.

Zhang, Y. and C. Pan. 2012. Measurements of mechanical properties and number of layers of graphene from nano-indentation. Diam. Relat. Mater. 24: 1-5.

Zhou, S.Y., G.-H. Gweon, A. Fedorov, P.N. First, W. De Heer, D.-H. Lee, et al. 2007. Substrate-induced bandgap opening in epitaxial graphene. Nat. Mater. 6(10): 770-775.

Zhou, X., F. Wang, Y. Zhu and Z. Liu. 2011. Graphene modified $LiFePO_4$ cathode materials for high power lithium ion batteries. J. Mater. Chem. 21(10): 3353-3358.

Zhu, Y., S. Murali, W. Cai, X. Li, J.W. Suk, J.R. Potts, et al. 2010. Graphene and graphene oxide: Synthesis, properties, and applications. Adv. Mater. 22(35): 3906-3924.

8

Advances in Ceramic PZT/PA6 Matrix Composite Materials for Energy Harvesting Applications: Structural, Dielectric, Piezoelectric and Mechanical Study

Rida Farhan[1,2,*], Nabil Chakhchaoui[3,4], Adil Eddiai[2], Mounir Meddad[5], Mohamed Rguiti[6], M'hammed Mazroui[2] and Omar Cherkaoui[4]

8.1 INTRODUCTION

In the last 30 years, technological advances have prompted major changes in communications, consumer electronics, automotive, and the leisure industry. During that time, the sensor technology qualified substantial increase, both in use and in development, globally thanks to the large-scale implementation and commercialization of microelectromechanical systems (Bowen et al. 2014, Suzuki 2008, Lee et al. 2016, Kansal et al. 2004, Farhan et al. 2019a).

The demand for portable electronic devices and wireless detectors has increased with the growth in the production of microelectronic and mechanical systems (MEMS) in the past decade (Meddad et al. 2019, Chakhchaoui et al. 2020a, b, Oumghar et al. 2020, Eddiai et al. 2019). In recent times, the development of intelligent materials has been recently used for numerous applications such as electrochromic systems (Yildirim et al. 2008), sensors (Rault et al. 2014, Meddad et al. 2016), and batteries (Ryu et al. 2000) for perfectly improved performance. Ensuring the energy sovereignty of mobile devices and leveraging the ambient energy like mechanical vibrations (Chakhchaoui et al. 2018, Eddiai et al. 2016), wind (Li et al. 2011), thermal (Benahdouga et al. 2016), and light (Paradiso et al. 2005), and transforming these resources is becoming a problem for the scientific community. Studies indicate that piezoelectric composites are dependable energy conversion harvesters due to their processability, flexibility, high productivity, and optical efficiency even in great thicknesses. Piezoelectric composites are formed using particulate piezoelectric ceramic phases that are dispersed in a polymer matrix. Nanocomposites made in this way are also possible candidates for sensor devices (Wenger et al. 1996). The composite materials are an interesting technology because the use of nanoscale fillers enables major structural enhancements compared

[1] Laboratoire Nanotechnologies et Nanosystèmes, LN2, CNRS, 3IT (Institut Interdisciplinaire d'Innovations Technologiques), Université de Sherbrooke, Sherbrooke, QC, Canada.
[2] Laboratory of Physics of Condensed Matter (LPMC), Faculty of Sciences Ben M'Sick Hassan II University, Casablanca, Morocco.
[3] BGIM Laboratory, Higher Normal School (ENS), Hassan II University, Casablanca, Morocco.
[4] REMTEX Laboratory, Higher School of Textile and Clothing Industries (ESITH), Casablanca Morocco.
[5] LAS Laboratory of Setif, Mohamed el Bachir el Ibrahimi BBA University, Algeria.
[6] Laboratory of ceramic materials and associated processes, Univ. Valenciennes, EA 2443 - LMCPA, F-59313 Valenciennes, France.
* Corresponding author: Rida.Farhan@usherbrooke.ca

to polymeric micro-reinforcement materials that work well with industrial aims to produce lighter, thinner and firmer structures at reduced costs (Jagur-Grodzinski 2016). Electroactive polymers like poly(vinylidene fluoride)-co-hexafluoropropylene (PVDF-HFP), Poly(vinylidene fluoride) (PVDF), and polyamides (PA) are strong candidates for the energy advanced application because they possess many advantages such as high flexibility and good electromechanical coupling coefficients (Teyssèdre and Lacabanne 1995, Zheng et al. 2019, Chakhchaoui et al. 2020c, Farhan et al. 2020).

Polyamide (PA) is well known for its exceptional mechanical and thermal properties as a thermoplastic material that is commonly used in industrial applications (e.g. nanocomposite, film, and smart textiles, etc.). However, these benefits are accompanied by limitations such as moisture absorption, poor dimensional stability, and relatively low impact strength. Therefore, the improvement of the physical properties of PA by using new products has received a lot of interest (Chow et al. 2003, Liu et al. 2012). The dispersion of micrometre-scale reinforcement can significantly increase selected polyamide properties, such as dielectric properties and conductivity (Kim et al. 2001, Lonjon et al. 2013). Recently, reinforcement of the polyamide matrix by the use of a second polymer with the insertion of nanoparticles has been gaining interest. However, PZT nanoparticles are among the most utilized ferroelectric materials. It is commonly used in sensors due to its exceptional piezoelectric properties (Sodano et al. 2005, Chakhchaoui et al. 2017), though the vibratory harvesting application of the PZT is prevented due to a lack of flexibility, the vibratory harvesting application via PZT has been limited.

Polymer blends have been defined as a usually recognized category of materials with several characteristics that target specific applications and have been met by the best academic and technological interest. The properties of the mixture are strongly influenced by their components, the morphology, and the design formed during processing (Rhim et al. 2013). Though certain polymer mixtures are completely or partially miscible, more polymer blends are immiscible and have several step morphologies (Xiang et al. 2012). Mixing of different polymers has become a commonly recognized method for the development of new products with favorable properties.

There have been numerous studies on the behavior of piezoelectric materials and their application in piezoelectric device development. Eddiai et al. (2016) and Shinde et al. (2016) have established expressions to calculate the dielectric, piezoelectric, elastic constants, and modulus for binary systems such as poly(vinylidene-fluoride)/PZT. Thus, this material is defined as a highly dielectric compound suitable for the harvesting of electrical energy. Polyurethane (PU), Polyamide-11 (PA11), and PVDF are known as semi-crystalline polymers, with exceptional dielectric, mechanical and piezoelectric properties. Consequently, we have developed a novel series of piezoelectric 0-3 connectivity composites (PA6/PZT) by adding PZT into a matrix of polyamide-6 that is capable of realizing composites with high dielectric permittivity and piezoelectric coefficient.

8.2 POTENTIAL FOR AMBIENT ENERGY HARVESTING

Currently, electrochemical batteries are the main sources of electrical energy for powering energy-efficient systems, especially portable electronic devices. This type of source has the following characteristics: high energy density, low cost, and no moving parts. However, the life of an electrochemical battery is limited. While it is not used for discharge by electricity-consuming appliances, its power tends to slowly decrease due to the well-known self-discharge phenomenon. As a result, regular replacement of the batteries becomes an inevitable task. For certain applications (biomedical implant, corrosion sensor inserted in concrete, etc.), this operation is costly and delicate. Moreover, the miniaturization of electrochemical cells, compared to that of electronic circuits, which have greatly reduced in size in recent decades, is progressing slowly. Consequently, today, the relatively large size of the electrochemical battery is becoming an obstacle to the miniaturization of electronic devices. Moreover, for applications such as WSN (Wireless Sensor Network) composed of thousands of sensors distributed in the environment, it is unthinkable to envisage a power supply of electrochemical batteries which could be a major source of pollution.

Faced with these drawbacks, it is essential to develop new micro sources with a very long lifespan, small foot-print, and no environmental impact. The ideal approach allows the electronic device to harvest energy itself in this working environment for self-feeding (harvesting of ambient energy), which is an important current research topic. The primary sources of ambient energy are very varied, such as vibrational energy, solar radiation, thermal energy induced by temperature gradients, electromagnetic energy, the kinetic energy of rain, etc.

As shown in Fig. 8.1, the possible energy harvesting tools can be analyzed in or around mechanical systems to provide adequate power for self-powered WSNs. These sources can generally be divided into internal and external sources of energy representing the energy that comes from mechanical systems or their environment. They are then further categorized into light, electricity, heat, and motion depending on the energy forms.

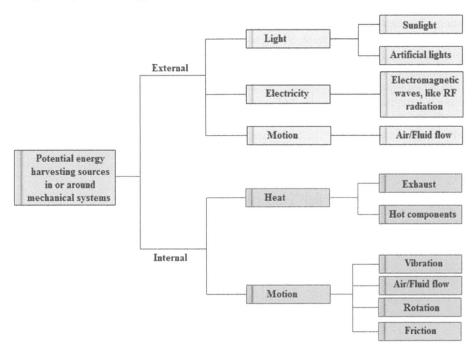

Fig. 8.1   Potential sources of energy extracted in mechanical devices

Solar power is a natural, sustainable, and effective source of energy in the environment where most mechanical systems are operated. Solar energy may be collected and processed at a large level. Sunlight is significantly affected by area, weather, and length of the day. Artificial light has no association with these variables, but its illumination degree is smaller than that of sunlight in many orders of magnitude. Because of the extensive norm of radio transmitters, electromagnetic waves (such as RF radiation) are another type of energy primarily contained in the surroundings. Both light and radiofrequency energy may be recovered for use in wireless body field networks (Ghamari et al. 2016). The force created by fluid flows (such as wind and tidal (Khaligh and Omnar 2009)) is an actual source of dynamic energy. External energy sources are very affected by climate, so the adjustment to condition control systems is becoming increasingly difficult for consistent generators. Their mutual applications include generating energy for power plants (Choi and Yeom 2018).

Conversely, the machine's internal energy is fairly efficient and steady. The performance of the powerful machines (generally motors) is less than 50%. The lost energy offers an incentive to be harvested and then applied to power the WSN to track the working conditions. The lost energy could be the source of power for maintenance-free wireless sensor nodes. The energy available for harvesting is energy dynamic and thermal.

In summary, external and internal energy resources can be working independently or simultaneously using several energy harvesting technologies (Yang and Daoud 2018). Owing to its abundance, solar energy may be scavenged to provide power for WSNs added in the monitoring of the outdoor machine condition. Owing to a large amount of excess thermal and dynamic energy in mechanical devices, most machines have ample power for harvesting to attain a WSN-based maintenance-free condition monitoring system.

8.3 RESEARCH PROBLEM

This work aims to study the potential of piezoelectric composite for sensing applications. Due to various their excellent properties, which are improved by the use of piezoelectric ceramic, polymeric composites are becoming widely attractive in smart materials research and energy harvesting applications. Therefore, flexible material composites with particular physical properties are the best choice for developing flexible generators. The use of these composites is expanding but a few energy harvesting structures exploit their potential. Many technical advances have not yet been made: through its reliable characterization, improvements have been made to materials used as generators.

A novel composite with a high dielectric constant is developed with the incorporation of the polyamide matrix through ceramic PZT microparticles. Structural, dielectric, and mechanical characterizations of PA6/PZT composite have been performed to estimate their potential for good application in harvesting energy. The found results show that the insertion of the PZT microparticles increases suggestively the intrinsic parameters of our composites. Consequently, these increases make these composites as potential candidates for energy harvesting.

8.4 ENHANCEMENT OF PHYSICAL PROPERTIES OF COMPOSITES PA6/PZT FOR ENERGY HARVESTING

The fundamental properties of composites (dielectric constant and dielectric loss factor, Young's modulus) play an essential role in their performance to develop the response of the density of power harvested in generator mode. The purpose of this section is to prepare the various PA6/PZT composites and to provide an overview of the structural, dielectric, piezoelectric, and mechanical behavior of these composites.

8.4.1 Processing of PA6 Based Composites

Many manufacturing methods were used for the manufacture of thermoplastic composites such as solution casting, *in situ* polymerization, and manufacturing of solutions. Melting of micro fillers with PA6 blends by using traditional processing techniques is especially desirable among these processes, as the process is quick, clear, solvent-free, and available in advanced applications. Polymer blend composites may result in some kind of high-performance material that combines the advantages of polymer blend and the strengths of polymer composites (Dayma and Satapathy 2012).

According to a study conducted by the ceramic materials and associated processes laboratory, particles dispersed by melting in a polymer matrix show a good increase in dielectric constant, and the volume of micro charges is low, thus Young's modulus is maintained. Polyamide is among the most used commercial polymers for manufacturing the composite; therefore, polyamide will be used to develop the composites in this work.

Polyamides are polymers that contain repeating amide, -CO-NH-, linkage (Fig. 8.2). The most famous synthetic polyamides are often referred to as nylon and they are aliphatic polyamides. Nevertheless, other synthetic polyamides are also important, including aromatic polyamide, Kevlar ® and plastics made from UREA. The term used to describe the linear aliphatic polyamide (NYLON) is based on the number of carbon atoms in the repeating unit.

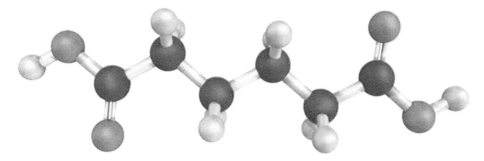

Fig. 8.2 Molecule of the polyamide-6

These polyamide-6 composites were prepared in the laboratory using thermoplastic polyamide 6 as the matrix. The pure PA6 films as well as their filled counterparts were prepared by solution casting (Shanmugaraj et al. 2019).

The development process presented below is described for the preparation of a composite using PA6. First, a quantity of 2 g of PA6 is dissolved in 20 ml of formic acid (AF) via mechanical stirring at a temperature of T = 120°C for 2 hours. To limit the evaporation of (AF) (solvent), use a clear 100 ml Duran glass bottle with a stopper. As the various percentages of PZT are prepared, this is the step which consists of dispersing the microparticles of PZT (powder) in 5 ml of the solvent. The dispersion is carried out using sonication for 3 min (Power 15 W and Amplitude 76%). Then, the two solutions are mixed, to have more homogenization; the mixture is left under sonication at the same amplitude (76%) for 3 min. To ensure good homogenization, stirring is carried out for 2 hours under the same temperature. Then, the mixture is poured and scraped onto a 34 cm × 38 cm glass plate using a special pouring system for this form of polymer which allows calibrated and very fine stripes to be produced: the K Hand Coater by ERICHSEN ® (Fig. 8.3).

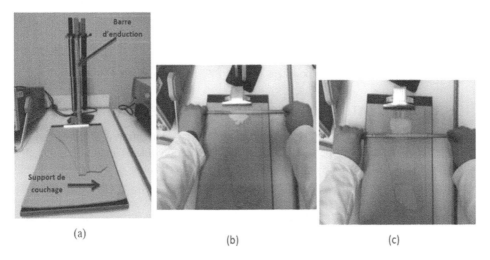

(a)

(b)

(c)

Fig. 8.3 The K Hand Coater device from ERICHSEN (a) before casting, (b) start of casting, (c) during casting

Finally, the film is subsequently placed in an oven at a temperature T = 40°C, for 2 hours, to ensure the drying of the composite film and the evaporation of the solvent (Fig. 8.4). Therefore, we have developed different polyamide-6 matrix composites filled by varying the different volumes of PZT and with a PZT powder particle size: ≈ 10 μm.

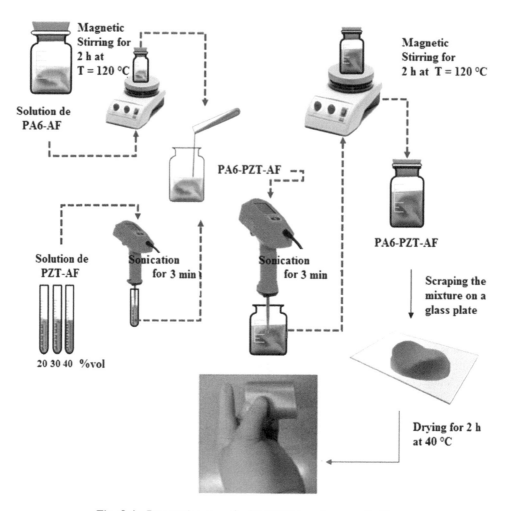

Fig. 8.4   Preparation steps for PA6/PZT-based composite films

8.5   CHARACTERIZATION OF PA6/PZT COMPOSITES

The objective of this section is to present the principle of characterization of structural, dielectric, piezoelectric and mechanical properties as well as the clarification of experimental results approved to analyze the role of PZT particles and their influence on the performance of composites for energy harvesting.

8.5.1   Structural Characterization

8.5.1.1   Principal

Scanning electron microscopy (SEM) is a non-destructive technique based on the principle of electron-matter interactions allowing high resolution images of the surface of a sample to be obtained. Indeed, under the impact of the beam of accelerated primary electrons (from 10 to 30 keV), the backscattered electrons and secondary electrons emitted by the sample are selectively collected by detectors which transmit a signal to a cathode ray screen whose scanning is synchronized with the scanning of the sample.

The morphology of the diverse materials with polyamide composites (PA6/PZT) was analyzed by scanning electron microscopy (SEM, Hitachi S-3500N). The films studied underwent a cryofracture with liquid nitrogen to study the dispersion of the particles in the volume of the polymer. The backscattered electron detector mode was used to promote the contrast between the polymer matrix and the PZT particles.

SEM observation is a means of studying the state of dispersion of PZT particles in the polymer matrix. The grain distribution in our composites can be observed on the surface of the film, but also in its thickness. Before switching to SEM, a slight deposit of a metallic layer (silver in our case) is necessary. The film is therefore metalized so that its surface is conductive to allow the flow of excess electrons. This prevents the film from loading when it is exposed in the SEM to the electronic probe.

8.5.1.2 Surface View of PA6/PZT Composite

In reference (Farhan et al. 2019b), there is the microscopic view of the surface of the composite film, based on PA6/PZT. The grains of PZT used for these pouring have a size less than 10 μm, and the composite is filled with 20, 30, and 40% by volume of PZT.

The surface microstructure of the composites loaded in 20, 30, and 40% by volume show that the PZT particles are distributed through the PA6 matrix. It was seen that the incorporation of the PZT ceramic homogeneously dispersed in the polymer matrix and PZT is collected in the form of nodules. The ultrasonic agitation and the rapid transfer of the solution containing the PA6 and the PZT at the doctor blade made it possible to obtain this good dispersion of the PZT throughout the polyamide matrix. The homogeneity of the dispersion of PZT particles in a PA6 matrix displays an important role as a filler and influences the dielectric, piezoelectric and mechanical properties of composites.

8.5.1.3 Thickness View of PA6/PZT Composite

The composite film is fractured after quenching in liquid nitrogen, which makes it possible to obtain fragile facies of rupture. A typical example of these facies is presented in Fig. 8.5. The band in thickness is uniform without segregation and sedimentation of the particle of PZT. In conclusion, this protocol allows us to obtain homogeneous PA6/PZT composites, with a volume fraction of charge 20, 30, and 40%.

Fig. 8.5 SEM image of the thickness view of PA6/PZT composite films (a) PA6/20% PZT, (b) PA6/30% PZT, (c) PA6/40% PZT

8.5.2 Dielectric Characterization

The dielectric characterizations at low frequency will be focused on the study of the following variables: the relative permittivity and the dielectric losses of the PA6/PZT composite. The principle of the dielectric characterization is based on the measurement of the capacitance of the capacitor realized with the material to be studied as a dielectric element. The dielectric constant of the composites filled with fillers was calculated at room temperature using an HP 4284A LCR meter across a large range of frequencies (from 12 Hz to 1 MHz) to estimate the contribution of space charge.

8.5.2.1 Experimental Results and Discussions

Previous studies in our laboratory have notably shown that the energy harvesting capacity of composite materials can be considerably improved by the incorporation of PZT in the matrix of electrostrictive polymers of polyurethane (PU) type (Krause et al. 2009). Consequently, the dielectric properties which are considered as a crucial parameter for energy harvesting will be presented below. It should be noted that the range of thicknesses of the samples tested is between 50 and 70 μm.

8.5.2.2 Dielectric Constant and Loss Index

The variation of the dielectric constant for the various materials studied with the composites based on polyamide matrice (PA6) as a function of the frequency is illustrated in Fig. 8.6(a). According to this variation, we can distinguish two zones of interesting functioning, one less than 40 Hz and the other between 40 Hz and 100 Hz. For low frequencies (below 40 Hz), an important increase in dielectric constant is observed for the different materials. This increase can be explained by the theory of dielectric behavior (Grannan et al. 1981, Dang et al. 2002, Mezzenga et al. 2003). In this frequency band, all polarization phenomena intervene, in addition to the interfacial polarization or the Maxwell (Farhan et al. 2019b).

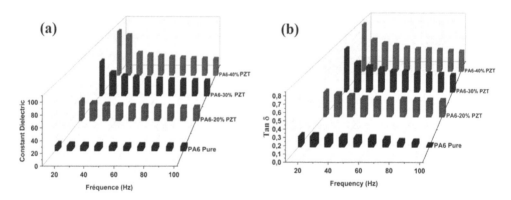

Fig. 8.6 (a) The dielectric constant, (b) The dielctric losses as a function of frequency for PA6/PZT composites

For frequencies between 40 Hz and 100 Hz, the relative permittivity of the materials is almost constant. This corresponds to the plateau of the orientation polarization. We can explain that by the orientation of the molecules which have a permanent dipole moment permanent. The structure of these molecules is asymmetrical: the center of gravity resulting from all the negative charges of such a molecule does not coincide with that of all its positive charges. As shown in Fig. 8.6(a), the permittivity relative of the filled composites is always higher compared to those of the pure composite. This improvement of the dielectric parameters is due to the addition of the microparticles of PZT which makes it possible to create new dipoles. On the other side, it also generates an imbalance in the structure of the polymer, hence a favoring of the orientation polarization translated by an increase in the permittivity. As predicted, the difference between the dielectric constant values for pure composites and micro filled composites at higher frequencies is not that large, which suggests that the filler content is small relative to the threshold value. Also, Carponcin et al. (2014) elaborated a composite composed of polyamide 11 (PA11), PZT, and CNT. The authors stated that the incorporation of 0.2 vol percent of CNTs resulted in improved dielectric permittivity for composite PA 11/PZT. The addition of 30 vol percent of PZT particles with PA11 increases its permittivity from 3 to 6.6 at 1 kHz where a dielectric permittivity value of 32 is achieved when 0.2 vol percent of CNT is introduced into the composite.

Moreover, the variation of dielectric loss for a pure PA6 and filled composites versus frequency is shown in Fig. 8.6(b). From this figure, dielectric losses are still slightly increased by the incorporation of PZT particles into polymer matrices, but the value of these losses is still relatively low and constant in the frequency range considered. Despite the disadvantage that the permittivity increases parallel to the conduction and therefore to the losses, encouraging results propose that this charging method holds great promise for increasing the electromechanical properties of composite material. The main effect of the existence of charges in a dielectric material is the change of the internal distribution of the electric field; this phenomenon is greatly improved when particles are dispersed in the dielectric matrix because they are equivalent to dipoles.

8.5.3 Piezoelectric Characterization

To achieve piezoelectric comportment on a macroscopic scale, it is necessary to go through the polarization process, which consists of applying a strong electrical field higher than the coercive field of the material, to orient the electrical dipoles in the preferred direction.

8.5.3.1 The Polarization Protocol

Titano zirconate is a ferroelectric, which means that it owns spontaneous electrical polarization, which can be overturned by the application of an electric field. For it to develop macroscopic piezoelectricity, it must be polarized beforehand. This polarization is governed by three parameters: the value of the applied electric field, the time, and the polarization temperature. To attain this polarization, a very fine electrode is deposited on each surface of the sample by vacuum evaporation (Fig. 8.7).

Fig. 8.7 Metallic composite film

The polarization is done in a silicone oil bath to avoid breakdown at excessively strong electrical voltages. The silicone oil is heated and maintained at 60°C, in a way to facilitate the movement of the domain walls and an electrical voltage of 2 kV for 15 min is applied.

The sample is placed in the polarization device so that one of the faces is connected at the positive terminal of the generator, and the other to mass. For some samples, we have previewed that sometimes some electrodes did not support the polarization cycle in the oil bath at 60°C (this has been observed especially for composites with large grains of PZT).

8.5.3.2   Piezoelectric Coefficient

To measure the piezoelectric coefficient d_{31}, we used a piezometer from Piezotest Ltd ©. The system works by clamping the sample and subjecting it to a low-frequency force (110 Hz). From the electrical signals generated by the sample, the piezoelectric coefficient d_{31} is immediately provided by the system. The piezoelectric coefficient increases with the volume fraction of PZT for different composites of PA6/PZT. This coefficient reaches a saturation value for 15 minutes of polarization: it is the time necessary to align the maximum of dipoles in the case of different composites. However, the polarization time of a massive ceramic depends on the nucleation time of the ferroelectric domains as well as on the kinetics of their walls (Hua et al. 2017, Kamel et al. 2007). The value of the piezoelectric coefficient is directly related to the amount of electroactive phase present in the composites. These results are confirmed by the comparison of the hysteresis cycles obtained for the different composites. The evolution of the coefficient piezoelectric with the rate of particles is more marked at high concentrations of PZT.

At room temperature, the coefficient d_{31} increases with the volume fraction of PZT: the composite filled at 40% of PZT induces a maximum coefficient piezoelectric increase of 7 pC/N. The same, the composite filled at 20% vol, is characterized by a coefficient of 4 pC/N; at 30% by volume, the coefficient d_{31} is equal to 5 pC/N (Farhan et al. 2019b). Moreover, Hua et al. (2017), Kamel et al. (2007) have observed maximum values of coefficient piezoelectric of 6.6 pC/N for polyamide 11/ Titanate de barium films (at 50 vol% of $BaTiO_3$) at room temperature.

8.5.3.3   Hysteresis Cycle

To measure the piezoelectric properties, gold electrodes were deposited on the samples, and then the polarization process was carried out in silicone oil at 166.6 kV/cm for a direct current electric field. The hysteresis cycles were recovered by a Radiant Precision Workstation system of test ferroelectric (Park et al. 2004). The appearance of the hysteresis cycle confirms the ferroelectric nature of the PA6/PZT composites prepared. Measuring polarization switching is a common method for assessing the polarization and coercivity of a ferroelectric material.

During the evolution of the PA6/PZT composite hysteresis cycle, we have found that the hysteresis cycle area increases with the concentration of particles (Farhan et al. 2019b). Indeed, under an electric field of 1 kV, the remanent polarization passes from 0.17 to 0.4 $\mu C.m^2$ when the percentage increases from 20 to 40% of the PZT particles. These marks can be attributed to the influence of PZT which has important piezoelectric properties. Therefore, the composites elaborated in this chapter combine the advantages of the polymer matrix with those of the ceramic phase, which has high piezoelectric properties.

In composites material, the electrical field applied to the ceramic phase is determined by the electrical conductivity and the dielectric constant of the ceramic and polymer phases. If the polarization time is greater than the dielectric relaxation time constant of each of the two phases, it is estimated that the distribution of the electric field in the PZT and polymer phases will be well controlled by their electrical resistivity rather than by their dielectric properties. Therefore, we expect to obtain a significantly higher quality of polarization of the PZT phase in polarized composites than what emerges from the hysteresis measurements, which have been realized over a relatively short period.

8.5.4   Mechanical Characterization

The analysis of the mechanical properties of composite proves is complicated due to the wide range of behaviors observed as a function of deformation, temperature, and stress strength. Our interest is generally focused on the deformation. Depending on the equipment, different stress modes can be used (traction, shear, simple bending, etc.). Piloting can be done either by fixing the applied force and by measuring the deformation, or by working with imposed deformation and by measuring the force. In our case, the tensile tests were chosen for the ease of set up. In traction at room temperature, Young's modulus is a characteristic of the stiffness of the material, the lower this parameter the more flexible the composite. The proposed approach consists in measuring Young's modulus for the different composites studied based on PA6 and seeing the influence of the incorporation of PZT in the starting matrix on the variation of the elastic modulus (Regrain et al. 2009).

8.5.4.1 Principle of Performance

Tensile tests were carried out on different materials. The rectangular-shaped samples are placed between the jaws of a tensile machine, which pulls the sample until it ruptures. The tensile tests give the initial information for modeling the behavior of our composites: linear or non-linear relationship between stresses and strains, apparent Young's modulus, elasticity limit, and rupture stress.

8.5.4.2 Experimental Results and Discussions

The tensile tests on the composites were carried out to define the parameters necessary for modeling the behavior of the composite. Figure 8.8 represents the stress-strain curve of PA6/PZT composites. From this figure, it can be seen that the resulting strain limit decreases sharply with the increase in the volume fraction of PZT leading to an increase in Young's modulus. The results obtained clearly show that adding particles in the starting matrix influences Young's modulus. This slight increase still allows the flexibility of the elastomers to be maintained for a good actuation performance and takes advantage of the increase in relative permittivity by the incorporation of particles for energy harvesting. In general, we see that the maximum stress for the different composites is close to the pure PA6 one for the volume fraction of 20% and 30% (between 9 MPa and 16 MPa) and the ruptures arrives above 1.5% of strain with speed traction 2 mm.min^{-1}.

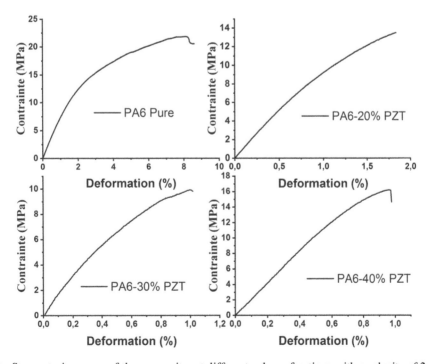

Fig. 8.8 Stress-strain curves of the composites at different volume fractions with a velocity of 2 mm/min

Young's modulus values obtained are presented in Table 8.1. We notice that the various materials have a Young modulus higher than that of the starting matrix indicating that the Young modulus increases with the increase in PZT. However, the strengthening of the permittivity of the materials generates an increase in the performance of the piezoelectric polymers. It is therefore essential to develop methods capable of increasing the dielectric properties while guaranteeing a low Young's modulus to maintain the flexible character of polymers.

TABLE 8.1   Tensile testing of different materials

Matériaux	Maximal Force (N)	Strain (%)	Stress (MPa)	Module d'Young (MPa)
PA6 pure	21.5	8.6	20.5	675.20
PA6-20%PZT	16.1	1.8	13.4	990.53
PA6-30%PZT	11.8	1.01	9.7	1356.23
PA6-40%PZT	19.4	0.9	14.6	2091.47

Several studies are being conducted to increase the efficacy and life span of polyamide. For example, further studies on polyamide have been conducted to improve the mechanical properties of the prepared composites by adding layered silicates for their vibration energy harvesting performance (Leveque et al. 2017). The elaborated composites (PA11/Cloisite Na+) displayed a lower energy harvesting performance as compared to PA 11 because of the mechanical rigidity of the composite films increased by the incorporation of nano-clays. For this reason, we propose our filled composites (PA6/PZT) as a candidate material for energy harvesting technology.

8.6   SUMMARY

The subject addressed in this chapter is about electrical energy generation based on the harvesting of human mechanical energy to power portable electronics. Composite materials have been studied as a transducer for the conversion of mechanical energy into electrical energy. After a state of the art on the vast array of promising energy harvesting technologies and systems to achieve self-powered WSNs in machine condition monitoring, different aims were defined.

The first section of this chapter has been dedicated to the preparation of composite material with the solution casting method by adding microparticles (PZT) into a semi-crystalline thermoplastic matrix (PA6) to increase the dielectric and piezoelectric properties of these composites for advanced applications. These properties were the objective of the second part. In this framework, it has been shown that the dielectric constant of the composites with different volumes depended closely on the functioning frequency. In particular, the effect of PZT on the enhancement of dielectric parameters has been shown to increase the relative permittivity and simultaneously increase Young's modulus. For the time being, the piezoelectric coefficient increases with the volume fraction of the PZT microparticle. In this context, the ceramic injection disperses uniformly in the polymer matrix and no large agglomerations of ceramic powder exist.

Finally, we conclude that the micro fillers of PZT have a good effect on the dielectric, piezoelectric, and mechanical properties for energy harvesting.

8.7   REFERENCES

Benahdouga, S., R. Khenfer, M. Meddad, A. Eddiai and K. Benkhouja. 2016. New material connected with Matlab for physicals characteristics tracer of a thermogenerator. Mole. Crystals and Liq. Crystals. 628: 41-48.

Bowen, C.R., J. Taylor, E. LeBoulbar, D. Zabek, A. Chaulchan and R. Vaish. 2014. Pyroelectric materials and devices for energy harvesting applications. Energy. Environ. Sci. 712: 3836-3856.

Carponcin, D., E. Dantras, J. Dandurand, G. Aridon, F. Levallois, L. Cadiergues, et al. 2014. Electrical and piezoelectric behavior of polyamide/PZT/CNT multifunctional nanocomposites. Adv. Eng. Mater. 16: 1018.

Chakhchaoui, N., H. Ennamiri, A. Hajjaji, A. Eddiai, M. Meddad and Y. Boughaleb. 2017. Theoretical modeling of piezoelectric energy harvesting in the system using technical textile as a support. Polym. Adv. Technol. 28: 1170.

Chakhchaoui, N., H. Jaouani, H. Ennamiri, A. Eddiai, A. Hajjaji, M. Meddad, et al. 2018. Modeling and analysis of the effect of substrate on the flexible piezoelectric films for kinetic energy harvesting from textiles. J. Compos. Mater. 53: 3349.

Chakhchaoui, N., H. Jaouani, R. Farhan, A. Eddiai, M. Meddad, O. Cherkaoui, et al. 2020a. An enhanced power harvesting from woven textile using piezoelectric materials. IOP Conference Series: Materials Science and Engineering 827: 012046.

Chakhchaoui, N., R. Farhan, A. Eddiai, M. Meddad, O. Cherkaoui, Y. Boughaleb, et al. 2020b. Improved piezoelectric properties of PLA/PZT hybrid composite films. IOP Conference Series: Materials Science and Engineering 827: 012012.

Chakhchaoui, N., R. Farhan, M. Boutaldat, M. Rouway, A. Eddiai, M. Meddad, et al. 2020c. Piezoelectric β-polymorph formation of new textiles by surface-modification with coating process based on interfacial interaction on the conformational variation of poly (vinylidene fluoride) (PVDF) chains. Eur. Phys. J. Appl. Phys. 91: 31301.

Choi, K. and Yeom, J. 2018. Modeling of management system for hydroelectric power generation from water flow. In Proceedings of the 2018 Tenth International Conference on Ubiquitous and Future Networks (ICUFN), Prague, Czech Republic, 3–6 July 2018: 229-233.

Chow, W.S., Z.A. Mohd Ishak, J. Karger-Kocsis, A.A. Apostolov and U.S. Ishiaku. 2003. Compatibilizing effect of maleated polypropylene on the mechanical properties and morphology of injection molded polyamide 6/polypropylene/organoclay nanocomposites. Polymer 44: 7427-7440.

Dang, Z.-M., Y. Shen and C.-W. Nan. 2002. Dielectric behavior of three phase percolative Ni TBaTiO$_3$/ polyvinylidene fluoride composite. Appl. Phys. Lett. 81: 44814.

Dayma, N. and B.K. Satapathy. 2012. Microstructural correlations to micromechanical properties of polyamide-6/low density polyethylene-grafted-maleic anhydride/nanoclay ternary nanocomposites. Mater. Des. 33: 510-522.

Eddiai, A., M. Meddad, M. Mazroui, Y. Boughaleb, M. Idiri, R. Khanfer, et al. 2016. Strain effects on an electrostrictive polymer composite for power harvesting: Experiments and modeling. Poly. for Advan. and Technol. 27: 677-684.

Eddiai, A., M. Meddad, R. Farhan, M. Mazroui, M. Rguiti and D. Guyomar 2019. Using PVDF piezoelectric polymers to maximize power harvested by mechanical structure. Superl. and Microst. 127: 20-26.

Farhan, R., A. Eddiai, M. Meddad, M. Mazroui and D. Guyomar. 2019a. Electromechanical losses evaluation by energy-efficient method using the electrostrictive composites: Experiments and modeling. Smart Mater. Struct. 28: 035024.

Farhan, R., M. Rguiti, A. Eddiai, M. Mazroui, M. Meddad and C. Courtois. 2019b. Evaluation of performance of polyamide/lead zirconate titanate composite for energy harvesters and actuators. J. Compos. Mater. 53: 345-352.

Farhan, R., A. Eddiai, M. Meddad, N. Chakhchaoui, M. Rguiti and M. Mazroui. 2020. Improvement in energy conversion of electrostrictive composite materials by new approach via piezoelectric effect: Modeling and experiments. Polym. Adv. Technol. Doi: https://doi.org/10.1002/pat.5066.

Ghamari, M., B. Janko, R. Sherratt, W. Harwin, R. Piechockic, C. Soltanpur, et al. 2016. A survey on wireless body area networks for eHealthcare systems in residential environments. Sensors 16: 831.

Grannan, D.M., J.C. Garland and D.B. Tanner. 1981. Critical behavior of the dielectric constant of a random composite near the percolation threshold. Phys. Rev. Lett. 46: 375-378.

Hua, Z., X. Shi and Y. Chen. 2017. Preparation, structure, and property of highly filled polyamide 11/BaTiO$_3$ piezoelectric composites prepared through solid-state mechanochemical method. Polym. Compos. 40: E177-E185.

Jagur-Grodzinski, J. 2016. Polymers for tissue engineering, medical devices, and regenerative medicine. Concise general review of recent studies. Polym. Adv. Technol. 17: 395-418.

Kamel, T.M., F.X.N.M. Koolsand and G. De With. 2007. Poling of soft piezoceramic PZT. J. Eur. Ceram. Soc. 27: 2471-2479.

Kansal, A., D. Potter and M.B. Srivastava. 2004. Performance aware tasking for environmentally powered sensor networks. In: Proceedings of the Measurement and Modeling of Computer Systems p. 223-234.

Khaligh, A. and O.C. Omnar. 2009. Energy harvesting: solar, wind, and ocean energy conversion systems. CRC Press: Boca Raton, FL, USA, ISBN 978-1-4398-1509-0.

Kim, G.-M., D.-H. Lee, B. Hoffmann, J. Kressler and G. Stöppelmann. 2001. Influence of nanofillers on the deformation process in layered silicate/polyamide-12 nanocomposites. Polymer 42: 1095-1100.

Krause, B., P. Pötschke and L. Häubler. 2009. Influence of small scale melt mixing conditions on electrical resistivity of carbon nanotube-polyamide composites. Compos. Sci. Technol. 69: 1505-1515.

Lee, J.H., J. Kim, T. Yun Kim, M.S. Al Hossain, S.W. Kim and J. Ho Kim. 2016. All-in-one energy harvesting and storage devices. J. Mater. Chem. A. 4: 7983-7999.

Leveque, P.K.M., C. Douchain, M. Rguiti, K. Prashantha, C. Courtois and M.-F. Lacrampe. 2017. Vibrational energy-harvesting performance of bio-sourced flexible polyamide 11/layered silicate nanocomposite films. Int. J. Polym. Anal. Charact. 22: 72-82.

Li, S., J. Yuan and H. Lipson. 2011. Ambient wind energy harvesting using cross-flow fluttering. J. Appl. Phys. 109: 026104.

Liu, X-Q., W. Yang, B.-H. Xie and M.-B. Yang. 2012. Influence of multiwall carbon nanotubes on the morphology, melting, crystallization and mechanical properties of polyamide 6/acrylonitrile–butadiene–styrene blends. Mater. Des. 34: 355-362.

Lonjon, A., I. Caffrey, D. Carponcin, E. Dantras and C. Lacabanne. 2013. High electrically conductive composites of Polyamide 11 filled with silver nanowires: Nanocomposites processing, mechanical and electrical analysis. J. Non Cryst. Solids. 376: 199-204.

Meddad, M., A. Eddiai, A. Cherif, D. Guyomar and A. Hajjaji. 2016. Enhancement of electrostrictive polymer power harvesting using new technique SSHI-Max. Opt. and Quant. Electron. 48: 94.

Meddad, M., A. Eddiai, R. Farhan, S. Benahadouga, M. Mazroui and M. Rguiti. 2019. Design hybridization system of TEG/PZT for power generation: Modelling and experiments. Superl. and Microst. 127: 86-92.

Mezzenga, R., J. Ruokolainnen, G.H. Fredrickson, E.J. Kramer, D. Moses, A.J. Heeger, et al. 2003. Templating organic semiconductors *via* self-assembly of polymer colloids. Science 299: 1872.

Oumghar, K., N. Chakhchaoui, R. Farhan, A. Eddiai, M. Meddad, O. Cherkaoui, et al. 2020. Enhanced piezoelectric properties of PVdF-HFP/PZT nanocomposite for energy harvesting application. IOP Conference Series: Materials Science and Engineering 827: 012034.

Paradiso, J.A. and T. Starner. 2005. Energy scavenging for mobile and wireless electronics. IEEE Pervasive Comput. 4: 18-27.

Park, G.-T., J.-J. Choi, C.-S. Park, J.-W. Lee and H.-E. Kim. 2004. Piezoelectric and ferroelectric properties of 1 µm-thick PZT film fabricated by a double spin coating process. Appl. Phys. Lett. 85: 2322.

Rault, T., A. Bouabdallah and Y. Challal. 2014. Energy efficiency in wireless sensor networks: A top-down survey. Comput. Networks. 67: 104-122.

Regrain, C., L. Laiarinandrasana and S. Toillon. 2009. Experimental and numerical study of creep and creep rupture behavior of PA6. Eng. Fract. Mech. 76: 2656-2665.

Rhim, J.-W., H.-M. Park and C.-S. Ha. 2013. Bio-nanocomposites for food packaging applications. Prog. Polym. Sci. 38: 1629-1652.

Ryu, K.S., K.M. Kim, S.G. Kang, G.J. Lee, J. Joo and S.H. Chang. 2000. Electrochemical and physical characterization of lithium ionic salt doped polyaniline as a polymer electrode of lithium secondary battery. Synth. Met. 110: 213-217.

Shanmugaraj, P., A. Swaminathan, R. Kumar Ravi, M. Dasaiah, P. Senthil Kumar, A. Sakunthala, et al. 2019. Preparation and characterization of porous PVdF-HFP/graphene oxide composite membranes by solution casting technique. J. Mater. Sci.: Mater. Electron. 30: 20079-20087.

Shinde, A.V., S.A. Pande, S.S. Joshi and S.A. Acharya. 2016. Novel ceramic-polyamide nanocomposites approach to make flexible film of PZT ceramics: Structural and dielectric study. Ferroelectrics 502: 187-196.

Sodano, H.A., D.J. Inman and G. Park. 2005. Comparison of piezoelectric energy harvesting devices for recharging batteries. J. Intell. Mater. Syst. Struct. 16: 67-75.

Suzuki, Y. 2008. International Symposium on Micro-NanoMechatronics and Human Science, Nagoya, Japan.

Teyssèdre, G. and C. Lacabanne. 1995. Study of the thermal and dielectric behavior of P(VDF-TrFE) copolymers in relation with their electroactive properties. Ferroelectrics 171: 125-144.

Wenger, M.F., P. Blanas, R.J. Shuford and D.K. Das-Gupta. 1996. Acoustic emission signal detection by ceramic/polymer composite piezoelectrets embedded in glass-epoxy laminates. Polym. Eng. Sci. 36: 2945-2954.

Xiang, F., Y. Shi, X. Li, T. Huang, C. Chen, Y. Peng, et al. 2012. Cocontinuous morphology of immiscible high density polyethylene/polyamide 6 blend induced by multiwalled carbon nanotubes network. European Poly. J. 48: 350-361.

Yang, X. and W.A. Daoud. 2018. Design parameters impact on output characteristics of flexible hybrid energy harvesting generator: Experimental and theoretical simulation based on a parallel hybrid model. Nano Energy 50: 794-806.

Yildirim, A., S. Tarkuc, M. Ak and L. Toppare. 2008. Syntheses of electroactive layers based on functionalized anthracene for electrochromic applications. Electrochim. Acta 53: 4875.

Zheng, R.R., X. Zhang, Z.P. Zhang, H. Jun Niu, C. Wang and W. Wang. 2019. Preparation and multifunction of electrochromic polyamides containing flexible backbone chains with electrochemical, fluorescence and memory properties. App. Surf. Science. 478: 906-915.

9

Role of Nitrides in Hydrogen Production and CO$_2$ Reduction

Gurvinder Singh[1, #], Anupma Thakur[1, 2, #],
Praveen Kumar[3] and Pooja Devi[1, 2, *]

9.1 INTRODUCTION

Climate change has emerged as one of the biggest global problems, necessitating universal governments and scientists to divulge in sustainable policies and solutions. The increasing global population has been putting a tremendous burden on conventional energy sources (Lesnikowski et al. 2019, Zhou et al. 2013, Di et al. 2017). Thus, limited availability of fossil fuels and corresponding threats to the environment has motivated scientists to explore new energy technologies (Hoffmann 2019). Sun is the epitome energy source amongst all available energies, which can be harvested in forms of gaseous and chemical fuels. Hydrogen energy has been realized as a promising future renewable energy (RE) fuel owing to its high combustion energy, abundance, and environment-friendly combustion by-products. It unlocks the opportunities for mitigation of greenhouse gas emission, which otherwise is a major bottleneck of conventional fuels (Hoffmann 2019). Thus, the need of the hour is to tap in solar energy for hydrogen production as well as greenhouse gases fixation. Nitride materials can harvest the majority of visible energy in solar hydrogen technologies and can thus play a vital role in CO$_2$ reduction.

9.1.1 Global Energy Status

Energy plays the lead role behind the development of all segments of life (economic, social, and political) required for a safer and cleaner environment. Since the looming industrialization in 1760s, the demand for energy has increased many folds. Worldwide, our conventional resources' reserve contains 1047.7 billion barrels of oil, 5501.5 trillion standard cubic foot of natural gas (NG), and 984 billion tons of coal, which might be sufficient to satisfy our energy needs for 50.7, 52.2, and 114 years. Currently, 11 billion tonnes of oil produced from fossil fuels is being consumed every year, but these reserves are depleting at a rate of 4 billion tonnes per year. At present, fossil fuel accounts for 86.4% of the present energy consumption and the rest of the energy is being supplied by renewable energy sources (Rabiu et al. 2019). With the tremendous growth in population, the energy production from fossil fuels has increased many times which may have catastrophic societal effects. According to U.S. Energy Information Administration (EIA) report, the world energy consumption will grow by 50% between 2018 and 2050.

---

[1] Central Scientific Instruments Organisation, Sector 30 C, Chandigarh, India.
[2] Academy of Scientific and Innovative Research (AcSIR), Ghaziabad-201002, India.
[3] School of Material Science, Indian Association for Cultivation of Sciences, Kolkata, India.
[#] Equally Contributed
[*] Corresponding author: poojaiitr@csio.res.in

9.1.2   Climate Change and Energy Challenge

Climate change can have catastrophic effects in terms of increased number of floods, droughts, cold-related deaths, etc., and conventional fuels have significantly contributed to this. This is because of the emission of greenhouse gases (GHG, which mainly includes carbon dioxide, water vapor, methane) on their combustion (Eveloy 2019). Burning of hydrocarbons produce huge CO_2 emissions and other pollutants in the environment. According to the latest report, the current concentration level of CO_2 in the environment is around 410 ppm, which scientifically should have been less than 350 ppm (Wuebbles et al. 2016). The escalated level of GHG is predicted to cause an increase in the Earth's temperature by ~2°C and elevation in sea level at the end of the 21st century (Brown et al. 2019). According to U.S. EIA, the CO_2 emissions in the environment will increase by 0.6% per year between 2018 and 2040 (Fig. 9.1).

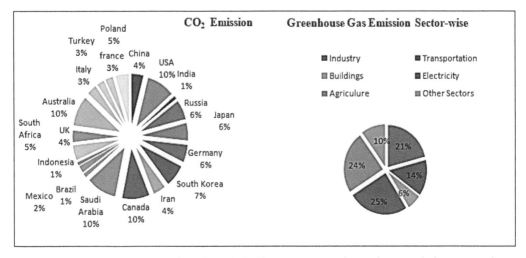

Fig. 9.1   Country-wise per capita CO_2 emission in percentage and greenhouse emission sector-wise

Thus, exhaustion of natural resources accelerated demand for energy, and a negative impact on the environment has forced planners and policymakers to look for alternative energy sources (Bond et al. 2007). But at present, only 14% of the total energy requirement is being achieved from renewable energy sources in which biofuel and biomass contribute to 10%. Since 2004, there is a six-fold increase in the production of energy from renewable energy sources; still, fossil fuel usage continues to increase. Nearly 120,000 TW solar radiation is shining on the Earth's surface (Kabir et al. 2018). Therefore, it is clear that solar energy can be exploited depending upon the global needs without affecting the non-renewable resource and without producing greenhouse gases (Kabir et al. 2018).

Solar energy can be either used directly or in storage mode in the form of batteries, chemical fuels, hydrogen fuels, etc. Hydrogen fuel is of vital interest in the transport sector, which consumes around 57.7% of the fossil fuel accounting for about 14% of the global emissions as shown in Fig. 9.1. Hydrogen, biofuels, and electrification are the most suitable alternatives for the replacement of conventional fuels (Kobayashi et al. 2009).

9.1.3   Hydrogen and its Production

Hydrogen is the lightest element with the symbol H and atomic number 1 in the periodic table and the third most abundant element in the universe after carbon and oxygen. It can be perceived as a clean future energy carrier, owing to its high energy density and H_2O vapors as a combustion byproduct (Badwal et al. 2014). Hydrogen is thus deliberated as the cleanest fuel and energy carrier owing to its skillful comparative properties (in terms of molecular weight, density, ignition energy, diffusion coefficient, energy density and fuel efficiency) over gasoline, natural gas, diesel, LPG propane, CNG and LPG butane, as presented in Table 9.1 (Yip et al. 2019).

TABLE 9.1 Comparative qualities (in terms of molecular weight, density, ignition energy, diffusion coefficient, energy density and fuel efficiency) of hydrogen with other common fuels

Fuel	Molecular Weight	Density (g/L)	Ignition Energy (mJ)	Diffusion Coefficient (cm^3/s)	Energy Density (kJ/kg)	Fuel Efficiency
Hydrogen	2	0.0893	0.017	0.61	142.00	60%
LPG propane	44.09	118	0.24	3.04	25.3	–
CNG	16	0.7-0.9	0.28	1.90	45.30	–
LPG butane	58.14	121	0.26	3.17	27.7	–
Gasoline	100-105	710-770	0.24	21.34	48.6	22%
Natural gas	19	70	0.30	1.1	0.03	–
Diesel	200-300	832	–	–	33.8	45%

Thus, hydrogen has many potential applications, including fuel cells, domestic heating systems, and aircraft. But the majority of hydrogen production is from fossil fuels, biomass, and water as shown in Fig. 9.2(a). In thermochemical processes, heat and chemical reactions are used to release hydrogen from fossil fuels/biomass. In other processes like coal gasification, heat, in combination with closed-chemical cycles, produces hydrogen from feedstocks such as coal, etc. Most of these technologies rely on conventional fuels, which makes it unsuitable to realize the objective of clean and sustainable technology. Electrolysis is one of the highly sought sustainable options for hydrogen production from water. Hydrogen and electricity are substitutable via electrolyzer and fuel cell (FC); the electrolyzer makes hydrogen from electricity, and the FC makes electricity from hydrogen. One of the utmost fascinating feature of hydrogen as a fuel is that it could be proficiently transformed to electricity and produced from the water in a reversible reaction (Zeng and Zhang 2010):

$$H_2 + \frac{1}{2}O_2 \xleftrightarrow{\text{Fuel cell/Electrolyzers}} \text{Electricity} + \text{Water}$$

In recent years, solar technology has been developed considerably with solar photovoltaics achieving efficiencies greater than 30% (Zeng and Zhang 2010, Glover 2017). However, direct photolysis is not a viable option as it requires temperatures of 4000°C, which is not practically achievable with concentrated systems. To overcome this temperature limitation, multistep thermo-chemical cycles have been investigated to reduce the required temperature. In photoelectrochemical (PEC) systems, the photocatalyst must be prepared on the conductive substrate in the form of electrodes, so H_2 and O_2 gases are produced on different electrodes, and therefore, there is no need for separation of gases as required in PC route for hydrogen production. For photoelectrochemical hydrogen to succeed, the number of requirements have to be satisfied. Firstly, these semiconductor materials are required to have the correct band alignment to effectively split water. It has to withstand the chemical solutions implemented over an adequate working timescale. PEC systems currently lack inefficient light-absorbing capabilities, and corrosion of the semiconductor and the difficulties of matching the band-edge energies to redox potentials are the major challenges (Zeng and Zhang 2010, Bahadar and Khan 2013). Ongoing research and development of PEC materials, devices, and systems is making important strides, benefiting from strong synergies with contemporary research efforts in photovoltaics, nanotechnologies, and computational materials. Thus, the continued improvements in efficiency, durability, and cost are still needed for market viability. The PEC efficiencies are being improved through enhanced sunlight absorption and better surface catalysis. The durability and lifetime of PEC materials are being improved with protective surface coatings. Hydrogen production costs are being lowered through reduced materials and materials processing costs.

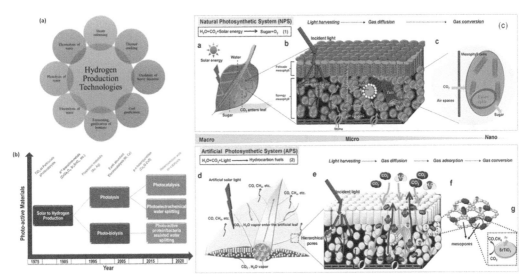

Fig. 9.2 (a) Key industrial hydrogen production technologies, (b) Current progress of cutting-edge photocatalysts for photocatalytic, photoelectrochemical and photobiological water splitting, (c) The schematic diagram for photosynthesis, artificial photosynthesis, and photocatalysis (Reprinted with permission from Han Zhou, *Leaf-architectured 3D hierarchical artificial photosynthetic system of pervoskite titanates towards CO$_2$ ptotoreduction into hydrocarbon fuels.* Scientific reports 3(2013), 1667)

9.1.4   CO$_2$ Pollution and its Fixation

Carbon dioxide is one of the major greenhouse gas contributing tremendously to global warming. Thus, the reduction of CO$_2$ in the environment and prevention of its increase is one of the important tasks for the researchers (25). Worldwide increasing demand for energy has added volumes to the emission level of CO$_2$. Excessive emissions of carbon dioxide in the environment are responsible for the drastic climate change. To save the environment, reduction of the CO$_2$ and prevention of further increase is required which can be done with action plan/time-based strategies, namely, electrochemical, photoelectrochemical, photocatalytic, thermochemical and hydrothermal methods to overcome CO$_2$ reduction. In the electrochemical method, an electrochemical reactor, such as electrolyzer, is powered by electricity generated from renewable sources to reduce CO$_2$ and water into hydro-fuels. Generally, this process involves electrolysis of solution containing dissolved CO$_2$ and voltage greater than the required potential for CO$_2$ splitting which is of the order of 1.47 V. However, due to ohmic losses and electrode resistance, a cell voltage of 2 V is required (Jitaru 2007). The reduction of CO$_2$ under hydrothermal conditions has attracted more attention in recent years. In the Earth's outermost shell and hydrothermal vents in deep-sea, hydrothermal reaction resulted in fossil fuel formation. The hydrothermal reaction does not involve solar radiation and carry out reactions under hydrothermal conditions (temperature and pressure). Thermochemical method, instead of using solar energy directly, uses two-step water splitting thermochemical cycles resulting in integrated solar hydrogen production and CO$_2$ reduction integrated technology.

Out of all these methods, PEC method is the most explored and potential method for hydrogen evolution and CO$_2$ reduction because if we tied CO$_2$ mitigation with the renewable photoelectrochemical process, then it would be a win-win situation from the environmental and economic point of view. To resolve this issue of CO_2 reduction, PEC and photocatalytic routes, and mimicking artificial photosynthesis processes are the most viable options by converting CO_2 into useful industrial chemicals like methanol, carbon monoxide, formic acid, methane, and even higher hydrocarbons depending upon the type of reaction conditions. The redox reactions involving CO_2 reductions are listed below (Chouhan et al. 2017):

$$CO_2 + 2e^- + 2H^+ \rightarrow CO + H_2O \qquad E = -0.76 \text{ V}$$
$$CO_2 + 2e^- + 2H^+ \rightarrow HCOOH \qquad E = -0.85 \text{ V}$$
$$CO_2 + 2e^- \rightarrow CO + CO_3^{2-} \qquad E = -0.79 \text{ V}$$
$$CO_2 + 4e^- + 4H^+ \rightarrow HCHO \qquad E = -0.62 \text{ V}$$
$$CO_2 + 6e^- + 8H^+ \rightarrow CH_3OH \qquad E = -0.72 \text{ V}$$

9.1.5   Solar Hydrogen Production and CO_2 Fixation

In the case of the hydrogen energy economy, electrical energy is used to electrolyze water to hydrogen, producing around 4% of the world's total hydrogen (Dincer 2012). Lately, there is an augmented emphasis on the usage of photonic energy towards solar-driven hydrogen production. The abundant solar energy could be tapped for the production of hydrogen from naturally available water resources by solar-assisted water splitting. Thermodynamically, a water-splitting reaction is a type of uphill reaction that necessitates external energy to drive. Water splitting can be achieved by (1) solar photovoltaics and electrolyzer (2) concentrated solar thermal and electrolyzer (3) thermochemical (4) biological processes (5) photo-catalysis (PC) and photo-electrochemical (PEC) and (6) photo-biochemical energy (Uyar and Beşikci 2017). The photolysis process of water encompasses the decomposition of water molecules into hydrogen and oxygen molecules by the consequence of photonic energy.

To achieve this, photo-catalyst has a major role to play. Figure 9.2(b) shows the evolution of a class of material possessing photo activity for solar water splitting through photocatalytic, photoelectrochemical and photobiological routes. This figure represents the research trends in PEC materials investigated in the years for solar water splitting assisted hydrogen production. Likewise, the major challenge associated with the reduction of CO_2 is the requirement of a significant amount of energy/or extremely reactive reagents as CO_2 is extremely stable in molecular form. However, with artificial photosynthesis, using PEC, as shown in Fig. 9.2(c), can convert CO_2 into important chemical products with the absorption of solar energy. Initial step in the photocatalytic reduction of CO_2 involves the absorption of solar energy by the photosensitizer (p) resulting in excitation of a photosensitizer (p*). Sacrificial donor (d) quenched the excited state by reducing and simultaneously oxidizing photosensitizer (p⁻) and donor (d·⁺), respectively. The reduced photosensitizer transfers electron to catalyst species (cat), thus generting reduced catalyst species (cat⁻). Finally, binding of cat⁻ to CO_2 takes place resulting in intended products and regenerating the catalyst. The mechanism has been shown in following reactions (Kumar et al. 2012).

$$p + h\nu \rightarrow p^*$$
$$p^* + d \rightarrow p^- + d^\circ$$
$$p^- + \text{cat} \rightarrow p + \text{cat}^-$$
$$\text{cat}^- + CO_2 \rightarrow \text{cat} + \text{products}$$

Although a large number of photocatalytic materials have been reported for CO_2 reduction, the production rates are low for a photocatalytic method to be economically viable (Morris et al. 2009). In photoelectrochemical (PEC) reduction of CO_2, electrons (e⁻) and holes (h⁺) are generated in semiconductors as a result of the absorption of solar radiation. These charge carriers carry out the reduction of CO_2 resulting in the production of hydrocarbons. Inoue and co-workers studied semiconductors for CO_2 reduction and suggested following reactions for conversion of CO_2 to various hydrocarbons like formic acid, formaldehyde, methanol, and methane (Inoue et al. 1979).

$$H_2O + 2h^+ \rightarrow 1/2O_2 + 2H^+$$
$$CO_{2\,aq.} + 2H^+ + 2e^- \rightarrow HCOOH$$
$$HCOOH + 2H^+ + 2e^- \rightarrow HCHO + H_2O$$
$$HCHO + 2H^+ + 2e^- \rightarrow CH_3OH + H_2O$$
$$CH_3OH + 2H^+ + 2e^- \rightarrow CH_4 + H_2O$$

9.2 MATERIAL PROPERTIES FOR SOLAR HYDROGEN PRODUCTION AND CO$_2$ REDUCTION

Semiconductor materials are considered as the most potential materials in the progress of the PEC hydrogen production and CO$_2$ reduction. This is due to their ability to absorb incident photons and convert them into free charge carriers. The most explored material has been TiO$_2$ semiconductor, but a bandgap of 3.2eV limits its light absorption, i.e. only 5% of the visible solar spectrum (Inoue et al. 1979).

The most studied materials for PECs are metal oxides, sulfides, and nitrides materials containing either transition metal cations of d^0 configuration or d^{10} configuration (Li and Zhang 2010). These materials should satisfy certain electrochemical and semiconducting properties such as (a) band gap, (b) flat band potential, and (c) Schottky barrier, which are discussed below:

9.2.1 Band Gap

Band gap (E_g) of semiconducting material is defined as the smallest difference between the valence band top and the conduction band bottom. This gap between the bands is an important parameter for the selection of material for solar water splitting and CO$_2$ reduction. The optical band gap value for the efficient performance of photo-electrode must lie between 1.6 eV (1.23 eV + over potential) and 2.5 eV, so that photo-electrode can harvest the visible part of the sunlight (Yin et al. 2010). The other important point for photoelectrochemical water splitting is the position of conduction and valence band of semiconductors and the potential required for oxidation and reduction of water. To split water, conduction band energy value of semiconductor must be larger than that of the redox energy required for water reduction and, similarly, valence band energy value must be lower than the water oxidation energy. Another method of increasing the efficiency of the PEC water splitting is to design a hybrid/tandem photo-electrode with a variety of materials having different bandgaps.

9.2.2 Flat – Band Potentials

Band structures of both electrodes, involving the photoelectrode of semiconductor and metallic counter electrode, illustrate various parameters like work function and band bending. As two electrodes come in contact, transfer of electrons takes place from the electrode with lower work function to the electrode with higher work function, until the Fermi level of both electrodes reaches the same value (Butler and Ginley 1978). In case the Fermi level of the metal is lower than the semiconductor, electrons would flow from semiconductor to metal, which results in the accumulation of positive charge near the surface of the semiconductor, moving the Fermi level close to the valence band and away from the conduction band. This leads to a modification in the surface potential of semiconductor leading to upward band bending. The distance near the surface where the concentration change is significant is called the depletion layer or space charge region. This concept is important in deciding the extent of external bias to generate the voltage sufficient for water decomposition (Gelderman et al. 2007).

9.2.3 Schottky Barrier

The Schottky barrier exibhits a key role in preventing the recombination of charges produced during the process of photo-ionization. Under the equilibrium, the Fermi level would be flat throughout the semiconductor, and the relative movement of the Fermi level to the band edges would be accompanied by bending of the band edges position. The bending of the band produces a built-in field to prevent the further flow of electrons from semiconductors to metal. This barrier is known as a Schottky barrier. If the semiconductor is said to be of n-type, then the electron concentration change is negligible as electrons are majority charge carriers; however, the relative concentration of holes changes significantly. This change in concentration can be expressed as (Yang and Wang 2017):

$$n = n_0 + \Delta n$$
$$p = p_0 + \Delta p$$

where n_0 and p_0 represents the electrons and holes concentration in equillibrium and Δn and Δp represent the additional charges generated by illumination. The variation between the equilibrium Fermi level and quasi-fermi level of electrons is negligibly small, but the quasi-fermi level of holes differs significantly from the equilibrium Fermi level of holes. The difference between the quasi-fermi level of electrons and holes defines the ideal upper limit of the achievable photovoltage. Higher the difference between the quasi-Fermi levels, higher the photovoltage produced.

9.3 NITRIDES IN SOLAR HYDROGEN PRODUCTION

Nitrides are semiconductor nitrides that have excellent material, optical and electrical properties. Most of these materials are non-toxic, chemically stable in aqueous solutions. Research in this area mainly focuses on materials of high quality with defined tailored properties. Binary nitrides group, which involves GaN, InN, GaN and their heterostructures, are most reported for solar water splitting, and are discussed in this section in detail. Ternary alloys like InGaN, AlGaN and their heterostructures, which have been reported for PEC activity, are also discussed. These nitrides also possess great physicochemical characteristics such as high electrical conductivity, corrosion resistance, chemical stability and unique electronic structure, thus enhancing their remarkable performance for solar hydrogen production. In this section, the nitrides are categorized on the basis of interstitial compounds, comprising the parent metals with nitrogen atoms integrated into the interstitial sites and are further discussed for their application in solar hydrogen production as tabulated in Table 9.2.

9.3.1 III-Nitrides

The III-Nitrides mainly consist of Al, Ga and In nitrides having ternary and quaternary structures. They are generally direct band semiconductors with a variable band gap ranging from 0.65 eV-6.2 eV, depending upon the concentration. Thus, they could cover the major portion of the solar spectrum. Nitride semiconductors are mainly grown by metal-organic vapor phase epitaxy (MOVPE), molecular beam epitaxy (MBE) and high-pressure bulk growth in growing single crystal and as thin films. The binary group nitrides can crystallize in three phases: (1) wurtzite (2) zinc blende (3) rock salt (NaCl). Wurtzite and zinc-blended structures have many similarities in terms of structure, as shown in Fig. 9.3(a) (Yadav et al. 2016). In both structures, a group III element is bounded by four Group-V elements and Group-V element is bounded by four group III elements (Biefeld et al. 2015). The fabrication of these nitride materials is based upon the values of band gap and valence band positions for binary alloys and shifts in these values for ternary alloys (Kim and Park 2005). The wurtzite structure of GaN material band gap is found to be 3.5 eV; however, this slight band gap modification takes place for other structures like zinc blende due to symmetry differences. Band gap of the AlN and InN was found to be 6.2 eV and 0.65 eV. For ternary alloys like InGaN, the variation of band gap with composition is still not established and found to vary non-linearly with composition (86).

TABLE 9.2 Summary of nitride materials in Solar Hydrogen Production

Order	Material	Electrolyte	Illumination Intensity	Performance Parameters
		III-Nitrides		
1	(a) p-GaN (b) n-GaN (Fujii and Ohkawa 2005)	0.5 M H_2SO_4	150 W Xe	50 μA/cm^2 700 μA/cm^2
2	n-GaN (Fujii and Ohkawa 2006)	1 M HCL	150 W Xe	0.8 mA/cm^2
3	GaN nanowire Si-doped GaN nanowire (AlOtaibi et al. 2013)	1 M KBr 1 M KBr	13.2 mW/cm^2	10 mA/cm^2 38 μmol/h H$_2$ IPCE = 15% 14 mA/cm^2 IPCE = 18.7% 38 μmol/h H$_2$
4	NanoporousGaN (Kim et al. 2014)	1 M NaCl	200 W Hg	Photocurrent = 7 mA/cm^2
5	GaN/NiO (Kang et al. 2014)	1 M NaCl	500 mW/cm^2 300 W Xe	3.14 mA/cm^2 Overall water splitting efficiency of 0.66%
6	ZnS/GaN (Hassan et al. 2018)	1 M NaOH	300 W Xe 500 mW/cm^2	2.8 mA/cm^2
7	MoS$_2$ decorated p-GaN (Ghosh et al. 2020)	0.5 M H_2SO_4	100 mW/cm^2	2.15 mA/cm^2 ABPE = 3.18% 89.5 μmol/h H$_2$
8	InN (Lindgren et al. 2002)	0.1 M KI	1000 W/m^2	2.2 μA/ cm^2
9	InN film (Lindgren et al. 2006)	0.1 NaoH	1000 W/m^2	Quantum efficiency of 2%
10	InN nanowire + Hole scavanger (Kamimura et al. 2016)	1 M NaOH	1000 mW/m^2	4 mA/cm^2
11	ZnO/InN (Liu et al. 2018)	0.1 M/L	100 mW/cm^2	0.56 μmol/h/cm^2 H$_2$ 150 μA/cm^2
12	InN/InGaN (Nötzel 2017)	0.5 M H_2SO_4	98 mW/cm^2 1000 W Xe	IPCE = 77% 81 μmol/h/cm^2 5 mA/cm^2
13	InGaN (Aryal et al. 2010)	1 mol/L HBr	132 mW/cm^2	1.2 mA/cm^2
14	InN/InGaN quantum dot (Rodriguez et al. 2015)	0.5 M Na$_2$SO$_4$	1000 W Xe 100 mW/cm^2	IPCE = 56% H$_2$ = 133 μmol/h/cm^2
15	InGaN/ GaN Rh/Cr$_2$O$_3$ (Kibria et al. 2013)	CH$_3$OH	300 W Xe	Internal quantum efficiency of 13% at 440 nm
16	IN$_{0.25}$Ga$_{.75}$N/Si (Kumar et al. 2019)	0.1 M Na$_2$SO$_4$	100 mW/cm^2	75 μmol/h/cm^2 H$_2$ Photo-to-current efficiency of 44% ABPE = 4.1%
17	InGaN/ GaN multiquantum well (Ebaid et al. 2015)	1 M HCL	100 mW/cm^2	IPCE = 8.6% at 350 nm 1.5 mA/cm^2
18	Monolithic p-In$_{0.42}$Ga$_{0.58}$N n-Si (Wang et al. 2019)	0.5 M H$_2$SO$_4$	100 mW/cm^2	1.23 mA/cm^2 ABPE = 4%
19	InGaN/Si double band photoanode (Fan et al. 2017)	1 M HBr	100 mW/cm^2	16.3 mA/cm^2 ABPE = 8.3%
		Carbon Nitrides		
20	Polymeric CN (Zhang and Antonietti 2010)	0.1 M KCl	50 mW/cm^2, Xe	IPCE = 3% at 420 nm 150 mA/cm^2
21	CN (Su et al. 2015)	1 M NaOH	100 mW/cm^2	IPCE = 7.3% at 400 nm 1.98 mA/cm^2, 36.6 μmol/h H$_2$ evolution

Contd.

TABLE 9.2 Contd.

22	CN (Liu et al. 2015)	1 M Na_2SO_4	500 W Xe	IPCE 6.6%, 30.2 $\mu A/cm^2$
23	Porous CN Film (Peng et al. 2018)	1 M KOH	100 mW/cm^2	12 $\mu A/cm^2$
24	Compact (C-rich) CN film (+ hole scavenger) (Bian et al. 2016)	0.1 M Na_2SO_4 + 0.1 M Na_2SO_3 + 0.01 M Na_3S	100 mW/cm^2	100 $\mu A/cm^2$ IPCE = 16%
25	Boron-doped CN	0.1 M Na_2SO_4	100 mW/cm^2, AM 1.5	103.2 $\mu A/cm^2$
26	CN-rGO film (Hou et al. 2016)	1 M KOH	100 mW/cm^2, AM 1.5	Quantum efficiency = 60% at 400 nm, 72 $\mu A/cm^2$ 0.8 mol/h/g of H_2
27	TiO_2 nanorod/CN (Kang et al. 2018)	0.5 M Na_2SO_4	300 W Xe, $\lambda > 420$ nm and AM 1.5 filtes	290 $\mu A/cm^2$
28	ZnO nanorods/thin CN hybrid (Park et al. 2016)	0.5 M Na_2SO_4	100 mW/cm^2, AM 1.5	120 $\mu A/cm^2$
29	$BiVO_4$/CN (Wang et al. 2017)	0.5 M KH_2PO_4, pH 7 phosphate buffer + 1 M Na_2SO_3	300 W Xe, $\lambda > 420$ nm	200 $\mu A/cm^2$
	Titanium Nitrides			
30	n-TiN/P-Si heterojunction (Solovan et al. 2013)		800 mW/cm^2	1.38 mA/cm^2
31	TiN/TiO_xN_y (Mohamed et al. 2020)	0.1 M NaOH	N.A	1.6 mA/cm^2
32	TiN/TiO_2 composite (Solovan et al. 2014)	0.5 M K_2SO_4	35 mW/cm^2	= 0.12 mA/cm^2, 7.27% power conversion efficiency
	Boron Nitrides			
33	Photocatalytic Au/TiO_2/BN (Ide et al. 2014)	25 ml, 50 ppm Au/TiO_2/BN	150 W Xe	10.3 $\mu mol/h$ H_2
34	CCTO/h-BN (Uosaki et al. 2016)	1 M KOH	150 W halogen	8.1 $\mu mol/h$, 0.97 mA/cm^2
	Tantalum Nitrides			
35	Ta_3N_5 nanorod (Li et al. 2013b)	0.5 M aq. Na_2SO_4	100 mW/cm^2	IPCE = 41.3% at 440 nm, 3.8 mA/cm^2
36	IrO_2/Ta_3N_5 (Yokoyama et al. 2011)	0.1 M aq. Na_2SO_4	300 W Lamp	3 mA/Cm^2
37	Ta_5N_5 film95	0.1 mM $Fe(CN)_6^{3}$/0.1 M $Fe(CN)_6^{4}$	300 mW/cm^2Xe	4.5 mA/cm^2
38	Co-Pi/Ba-Ta_3N_5 (Li et al. 2013c)	0.1 M aq. Na_2SO_4	–	6.7 mA/cm^2, IPCE = 86% for 400 nm, 130 $\mu molcm^{-2}$ H_2 evolution, solar conversion efficiency of 1.5%
39	Multilayer Ta_3N_5 film (Feng et al. 2010)	0.5 M Na_2SO_4	100 mW/cm^2	5.2 mA/cm^2 at 1.21 V, IPCE = 43.5% at 470 nm
40	Co_3O_4/Fh/Ta_3N_5 (Liu et al. 2014)	1 M NaOH	100 mW/cm^2	5.2 mA/cm^2
41	$Ta_8O_{1-y}N_x$ (Cong et al. 2012)	0.1 M Na_2SO_4	100 mW/cm^2	1.5 mA/cm^2 at 0.7 V, AgCl

Fujii et al. compared the PEC properties in *p*-type and *n*-type GaN, and current in *n*-type GaN was found to be higher, while the photocorossion is lower in *p*-type GaN because holes in *p*-type GaN oxidize the semiconductor itself (Fujii and Ohkawa 2005). ALOtaibi et al. demonstrated the photoelectrochemical properties of both Si-doped and GaN nanowires with IPCE of 15% to 18%, respectively, under 350 nm light illumination (AlOtaibi et al. 2013). Ryu et al. reported heavily doped and nanoporous GaN with a greater photocurrent density of 7 mA/cm^2 in comparison to planer GaN, which was only 14% (Ryu et al. 2011). The efficiency and stability of GaN photoanodes are further enhanced by decorating with electroctalyst such as NiO-Co and achieved a current density of 3.14 mA/cm^2. GaN, in combination with other materials to form heterostructures, are also investigated for PEC water splitting. ZnS/GaN heterostructure photoanode exhibited 1.75 fold photocurrent in comparison to that of reference GaN photoanode (Hassan et al. 2018). Similarly, dichalcogenide (MoS$_2$) decoration onto nitrides is found to enhance PEC applied bias photon-to-current conversion efficiency to 38% with high hydrogen evolution rate of 89.56 μmolh^{-1}cm^{-2} at 0.3 V vs RHE (Ghosh et al. 2020). InN is a representative group III-V nitride semiconductor known for its narrow band gap of 0.7 eV at 300 K and high electron mobility of 14000 cm^{2}V^{-1}s^{-1} (Polyakov and Schwierz 2006). Kamimura et al. developed the InN nanowire on Si substrates by using plasma-assisted molecular beam epitaxy with H$_2$O$_2$ in the electrolyte as hole scavenger resulting in high photocurrent of 4 mA/cm^2 (Kamimura et al. 2016). Liu et al. developed InN nanopyramid arrays as an active catalyst on a single crystal ZnO substrate via conventional chemical vapor deposition with 0.56 μmol/h/cm^2 hydrogen evolution rate found to be 30 times higher than the bare ZnO (Liu et al. 2018).

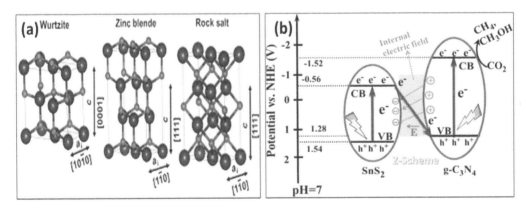

Fig. 9.3 (a) Atomic structure of different crystal structures (wurtzite, zinc blende, rock salt) (Reprinted with permission from Satyesh Kumar Yadav *Ab initio modeling of zinc blende AlN layer in Al-AlN-TiN multilayers* (Journal of apllied physics, 2016), 22430, (b) Z-scheme of photoelectrochemical water splitting (Reprinted with permission from *Di Tingmin, A direct Z-scheme g-C3N4/SnS2 photocatalyst with superior visible-light CO$_2$ reduction performance* (Journal of catalysis, 2017), 532-541)

Ternary metal nitrides such as InGaN/AlGaN materials are also the most explored materials for artificial photosynthesis. Aryal and coworkers reported *p*-type InGaN in aqueous HBr for hydrogen generation from solar water splitting and it has been found that *p*-type InGaN alloys possess much higher conversion efficiency of 1.2 mA/cm^2 *p*-type InGaN and 0.4 mA/cm^2 for *p*-type *p*-GaN (Aryal et al. 2010). Kibira and co-workers reported the multiband InGaN/GaN nanowire heterostructures, decorated with rhodium/chromium-oxide core-shell nanoparticles. The internal quantum efficiency, estimated to be 13% with 3μmolh^{-1}cm^{-2} of H$_2$ and 10 μmolh^{-1}cm^{-2} of O$_2$, was produced (Kibria et al. 2013). Kumar et al. reported the quantum dot activated In$_{0.25}$Ga$_{0.75}$N layer activated by InN quantum dot photoanode with IPCE of 44% at 550 nm and 0.4 V bias voltage with 75 μmolh^{-1}cm^{-2} and 33 μmolh^{-1}cm^{-2} of hydrogen and oxygen production at 0.2 V under 100 mWcm^{-2} illumination (Kumar et al. 2019). Ebaid et al. developed coaxial InGaN/GaNmulti-quantum well nanowires with IPCE of

8.6% at 350 nm, the photocurrent of 1.5 mA/cm^2 and 0.21% solar to hydrogen conversion efficiency (Ebaid et al. 2015). Al$_x$Ga$_{1-x}$N alloys are known for their optoelectronic applications because of their tenability of direct bandgap energy ranging from 2.4 eV to 6.1 eV. Nepal et al. studied the effects of temperature and composition on the energy band gap of AlGaN alloys. For temperature range of 10 K to 800 K, photoluminescence spectra shows decrease in band gap from 4.72 eV at 10 K to 3.92 K at 800 K with $x = 0.38$ and 5.77 eV to 5.40 eV at $x = 0.9$ (Nepal et al. 2005). Medvedev et al. developed GaN/AlGaN p-n structure by hydride vapor phase epitaxy with chloride p-n structure. The AlGaN film of 100 mm thickness is sandwiched between p and n Al$_x$Ga$_{1-x}$N barrier with hydrogen evolution rate of 0.56 ml/cm^2/h by 100 mm thick active AlGaN film sandwiched between p and n Al$_x$Ga$_{1-x}$N barrier with hydrogen evolution rate of 0.56 ml/cm^2/h (Medvedev et al. 2017). To achieve enhanced PEC efficiency, Z-scheme (tandem or two steps photoexcitation system) is well explored with III nitrides as presented in Figure 9.3(b) is another well explored with III-nitrides. In this approach, a variety of small band gap materials are used and coupled via a redox shuttle. Although high efficiencies have been reported, still very few reports are available on Z-scheme.

9.3.2 Carbon Nitrides

Another important class of nitride materials are graphitic carbon nitride with a large variation of bandgap from 1.6 eV to 2.7 eV, depending upon the method of synthesis and carbon to nitrogen concentration ratio, i.e polymeric carbon nitride, also called melon and triazine based graphitic carbon nitride, generally called graphitic carbon nitride (g C$_3$N$_4$) (Algara-Siller et al. 2014). The triazine based graphitic carbon nitride is stable and has a direct band gap. Carbon nitrides show excellent photoanodic activity in PEC cells, making investigations for conversion of solar energy and energy storage applications due to bandgap from 1.6 eV to 2.7 eV. Zhargyuanjian and co-workers reported carbon nitride photoanode with IPCE of 3% at 420 nm and low photocurrent of about 150 μAcm^{-2} (Zhang and Antonietti 2010). Later, Su et al. showed carbon nitride/TiO$_2$ nanotube array heterojunction with better IPCE of 7.3% at 400 nm. Doping of graphitic-carbon nitride has further increased the current density to 1.98 mAcm^{-2} at 0 V vs Ag/AgCl, when used in combination with TiO$_2$ nanostructures (Su et al. 2015). Liu et al. reported the first direct access of graphitic carbon nitride films by anodic aluminum oxide membrane assisted growth with 30.2 μA/cm^2 photocurrent (Liu et al. 2015). Again, Bian et al. reported the enhancement in PEC efficiency of carbon nitride films with high carbon content with 100 μA/cm^2 and 16% quantum efficiency (Bian et al. 2016). Effect of Boron doping in carbon nitride resulted in high IPCE of 10% with photocurrent density of 103.2 μA/cm^2. Peng et al. reported the carbon nitride/reduced graphene oxide film for photoelectrochemical activity with a quantum efficiency of 60%, 72 μA/cm^2 photocurrent density and 0.8 mol/h hydrogen evolution rate (Hou et al. 2016). Also, the photoelectrochemical study has been performed for directly coated carbon nitride on one-dimensional TiO$_2$ nanorod with 290 μA/cm^2 photocurrent (Kang et al. 2018).

9.3.3 Titanium Nitrides

Titanium nitrides (TiN), also called tinite, is most commonly used as a coating to improve substrate's surface properties. TiN is a ceramic material that is non-stoichiometric and has a wide band gap of the order of 3.42 eV-3.62 eV (Solovan et al. 2014). These materials possess high corrosion resistance, high chemical stability, low thermal conductivity, and high electrical conductivity (Solovan et al. 2014). TiN materials exhibit small interband losses in the visible spectrum and negative permittivity. Thus, they exhibit plasmonic applications that result in the enhancement of absorption of light incident on it. Solovan et al. studied the electrical and optical properties of the thin films and found the band gap of 3.2 eV and carrier concentration of the order of semiconductors (Solovan et al. 2014). Solovan et al. studied the photoelectric properties of n-TiN/p-Si photoanode by depositing TiN film on polished Si-wafer by reactive magnetron sputtering of pure titanium with photocurrent of 1.36

mA/cm^2 (Solovan et al. 2013). Li and co-workers studied titanium nitride/titanium oxide composite photoanodes for dye-sensitized solar cells and water splitting with 7.27% power conversion efficiency and the current density of 0.12 mA/cm^2 (Li et al. 2015).

9.3.4   Boron Nitrides

Boron nitrides (BN) are transition metal-free nitrides. Boron nitrides are isoelectric and isostructural to their carbon counterparts possessing many forms, mainly including (a) soft hexagonal (h-BN) (b) hard cubic (c-BN) (c) amorphous BN (Haubner et al. 2002). c-BN is the second hardest known material so far. h-BN band gap has a wide variation from 5.5 eV to 6.4eV and exhibits high thermal conductivity and negligible electrical conductivity (Haubner et al. 2002). They have emerged as a strong candidate for 2-D nanostructure because of similar planer geometry to graphene but in contrast to graphene, its band gap decreases with hydrogenation (Haubner et al. 2002). Li et al. reported the semi-hydrogenated BN sheet as visible light driven efficient photocatalyst for water splitting. The redox potentials for H$^+$ to H$_2$ and H$_2$O to O$_2$ were 4.4 eV and 5.67 eV, which showed that the reduction potential is slightly below the conduction band and the oxidation potential lies in between the band gap (Li et al. 2013a). Ide et al. studied the hybridized Au nanoparticle-loaded TiO$_2$ with BN nanosheets and studied photocatalytic activity with 10.3 μmol/h hydrogen production rate (Ide et al. 2014). Uosaki et al. reported a hydrogen evolution at insulating boron nitride nanosheet on an inert gold substrate with 15 mA/cm^2 photocurrent (Uosaki et al. 2016). Kawrani et al. reported the enhancement in photoelectrochemical performance of calcium copper titanium oxide with the addition of 3% h-BN nanosheet leading to 16 times enhancement in photocurrent (Kawrani et al. 2020).

9.3.5   Tantalum Nitrides

Tantalum nitride (Ta$_3$N$_5$) belongs to the most promising class of nitride materials because of its band gap. It has a band gap of 2.1 eV, which is most appropriate for solar water splitting using photoelectrochemical cell and is also stable in aqueous solutions. Thus, it has the potential to split water even without the application of external bias (Chun et al. 2003). In PEC systems, Ta$_3$N$_5$ is used as photoanode producing oxygen. Li et al. reported the vertically aligned crystalline Ta$_3$N$_5$ nanorod arrays for water splitting using solar radiation with 3.8 mA/cm^2 of photocurrent with 41.3% IPCE at 440 nm and solar conversion efficiency of 0.5% (Li et al. 2013b). Yoloyama et al. studied the properties of Ta$_3$N$_5$ photoanode prepared by sputtering and its modification with IrO$_2$ for suppression of self-oxidation of photoanode. Ishikawa et al. studied the electrochemical behavior of Ta$_3$N$_5$ using cyclic voltammograms and current-time curves for depicting the conduction band of photoanode was found to be 0.8 V vs Ag/AgCl and current density of 4.5 mA/cm^2 (Yokoyama et al. 2011). Li et al. reported the highly efficient cobalt phosphate-modified barium-doped tantalum nitride photoanode with solar conversion efficiency 1.5% greater than that reported for vertically doped Ta$_3$N$_5$ photoanode (Li et al. 2013c). Multilayered tantalum nitride hollow sphere-nanofilms were synthesized using the combination of oil-water interfacial self-assisted strategy with reported photocurrent of 5.2 mA/cm^2 and IPCE of 43.5% at 470 nm (Feng et al. 2010). Liu et al. reported the first highly stable modified tantalum nitride with ferrihydrate layer, which resulted in sustainable water oxidation for at least 6 h with photocurrent over 5 mA/cm^2 (Liu et al. 2014).

9.4   NITRIDES IN CO$_2$ REDUCTION

In view of past decades, the extensive upsurge in the carbon dioxide (CO$_2$) emissions and the alarming situation about the substantial energy supply have become the major challenges of the 21st century. The CO$_2$ conversion to renewable chemical fuels by the process of artificial photosynthesis has been viewed as the supreme strategy to overcome these challenges. Of late, the first discovery of photo-assisted reduction of CO$_2$ in the presence of a semiconductor photocatalyst has been a significant effort in investigating photocatalysts for CO_2 reduction. Currently, many semiconductor metal oxides, sulfides, and nitrides-based photocatalysts have been employed towards photo-assisted CO_2 reduction under UV light or 1-sun illumination, yet they show significantly low efficiency for practical applications. Research in this area mainly focuses on nitride-based materials of high quality with defined tailored properties (Table 9.3).

TABLE 9.3 Summary of nitride materials in CO_2 reduction

Order	Material	Experimental Characteristics	Products in CO_2 Reduction	Performance Metrics and Process
1	GaN nanowire/silicon (Wang et al. 2016)	Copper: co-catalyst	CH_4 at 1.4 V vs Ag/AgCl	19% Faradaic efficiency; photoelectrocatalytic
2	AlGaN/GaN (Lee et al. 2019)	NaCl aqueous solution and seawater: electrolyte	HCOOH	Photoelectrocatalytic
3	InGaN/GaN (Sheu et al. 2017)	NaCl electrolyte	HCOOH	Energy conversion efficiency 1.09%; photoelectrocatalytic
4	AlGaN/GaN device (Yotsuhashi et al. 2012)	n⁺-GaNaselectrical-conduction layer	HCOOH	Photoelectrocatalytic
5	InGaN/two Si p-n junctions (Sekimoto et al. 2015)	–	HCOOH	Energy conversion efficiency 0.97%; photoelectrocatalytic
6	Mg-Al-LDH/C_3N_4 (Hong et al. 2014)	Pd: cocatalyst	CH_4	6.5 mmol in 72 h, turnover number (TON) is 3.8; photocatalytic
7	Graphene oxide (rGO)/ protonated g-C_3N_4 (pCN) (Ong et al. 2015a)	2D/2D morphology	CH_4	13.93 mmol/g$_{catalyst,}$ photochemical quantum yield of 0.560%; photocatalytic
8	Pt NPs/g-C_3N_4 (Ong et al. 2015b)	2 wt% Pt NPs loading	CH_4	13.03 mmol/g$_{catalyst}$ after 10 h; photocatalytic
9	Pd/g-C_3N_4 and Pt/g-C_3N_4 (Gao et al. 2016)	Single layer palladium/ platinum (Pd/Pt)	HCOOH and CH_4	Photocatalytic
10	AgCl/carbon nitride (CN) (Putri et al. 2016)	–	CH_4	2.5-fold enrichment; photocatalytic
11	C_3N_4/Ru(II) complexes (Kuriki et al. 2018)	Time-resolved emission spectroscopic study	CH_4	Photocatalytic
12	Cu_2O/g-C_3N_4 (Chang and Tseng 2018)	8 hour-illumination	CO	Photocatalytic
13	g-C_3N_4 (Maeda et al. 2018)	Modified with AgNPs; RuP cocatalyst	HCOO⁻	Photocatalytic
14	TiO_2/C_3N_4 (Tseng et al. 2019)	–	CO and CH_3OH	Photocatalytic
15	CoAg/CN (Nazir et al. 2019)	0.5 M $KHCO_3$ as electrolyte	CH_4	Electrocatalytic conversion
16	Fe/BN, Co/BN, and Pt/BN (Tan et al. 2019)	Theoretical computational study	CH_4	Very low onset potentials of −0.52, −0.68 and −0.60; electrocatalytic conversion
17	O/BN nanosheets (Cao et al. 2020)	–	CO	12.5 μmol/g/h
18	TiN (Back and Jung 2017)	Theoretical computational study	CH_4	Electrochemical CO_2 reduction

There are continuing research interests in the pursuit of investigating robust, efficient and stable photocatalysts for practical applications. To address these, numerous research efforts have been made for the investigation of nitrides towards the CO_2 reduction as shown in Fig. 9.4. GaN semiconductor is a viable semiconductor for CO_2 reduction as its valence band and conduction band straddles the potential for CO_2 reduction and water splitting (Halmann 1978). Bare GaN nanowires are preferred for CO production over CH_4 from CO_2. The enhancement in product selectivity for methane has been observed with platinum co-catalyst on GaN nanowires (Halmann 1978). Alloying with indium offers tunning of band gap from 0.65 to 3.4 eV, leading to the absorption of the nearly entire solar spectrum. With 50% concentration of indium, bandgap of InGaN is 1.7 eV which can straddle CO_2 as well as water oxidation potential. Similarly, GaN nanowires in an aqueous PEC cell can produce syn gas (synthesis gas, CO_2 + H_2 mixtures), which is another source of energy that can produce variety of liquid fuels like methanol, kerosene, and diesel at large scale while making a healthier environment (Halmann 1978). Wang and co-workers demonstrated GaN/Si solar cell photocathode that exhibits exceptional merits for the photoreduction of CO_2, such as enhanced absorption of solar spectrum and highly efficient photogenerated electron transfer. This photocathode, by means of copper as the co-catalyst, shows 19% Faradaic efficiency for the photo-assisted reduction of CO_2 to CH_4 at 1.4 V vs Ag/AgCl (Wang et al. 2016).

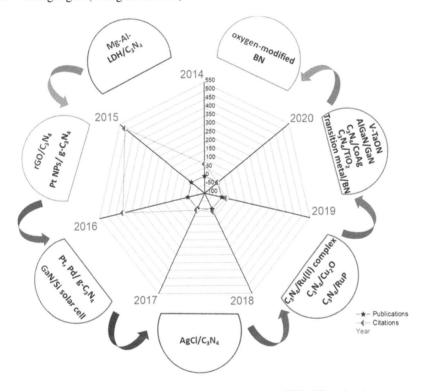

Fig. 9.4 Investigated nitride materials towards PEC CO_2 reduction

Also, Lee and co-workers presented the unique aluminum gallium nitride/gallium nitride hybrid heterostructures as the photoanodes for the production of formic acid by photoelectrocatalytic carbon dioxide reduction (Lee et al. 2019). In the photoelectrocatalytic conversion of CO_2, the aluminum gallium nitride/gallium nitride hybrid heterostructures were evaluated in sodium chloride aqueous solution and seawater as supporting electrolytes. Their studies revealed that formic acid is the only product yielded during the photoelectrocatalytic carbon dioxide reduction reactions (Lee et al. 2019). Sheu and co-workers presented InGaN/GaN epitaxial wafers for direct photoelectrolysis of water and CO_2 reduction by using CO_2-containing NaCl electrolyte. The energy conversion efficiency of CO_2 to HCOOH and H_2 generated were measured as 1.09% and 5.48%, respectively (Sheu et al. 2017). Thereafter, Yotsuhashi and co-workers reported an AlGaN/GaN device that consists of an unintentionally doped AlGaN photoabsorption layer and an n^+-GaN electrical-conduction layer for the production of formic acid by photoelectrocatalytic carbon dioxide reduction. The production rate of formic acid by AlGaN/n^+-GaN photoelectrode is about double as compared to –GaN/n^+-GaN device (Yotsuhashi et al. 2012). Also, Sekimoto and co-workers demonstrated a tandem photo-electrode (TPE) of InGaN and two Si p-n junctions for photoelectrocatalytic carbon dioxide reduction. The energy conversion efficiency for HCOOH by carbon dioxide reduction production reached 0.97% (Sekimoto et al. 2015). In 2014, Hong et al. reported the self-assembly of carbon nitride (C_3N_4) and layered double hydroxide (LDH), i.e. Mg-Al-LDH/C_3N_4 for the photocatalytic CO_2 reduction (Hong et al. 2014). The reactive carbonate anions supplemented in the LDH/C_3N_4 photocenters with Pd as a cocatalyst revealed remarkably high reduction CO_2 efficiency to CH_4. The yielded product is 6.5 mmol of CH_4 in 72 h, and the measured turnover number (TON) is 3.8 (Hong et al. 2014). In 2015, Ong and co-workers reported the hybrid reduced graphene oxide (rGO) and protonated g-C_3N_4 (pCN) heterojunction photocatalyst for the photocatalytic CO_2 reduction (Ong et al. 2015a). Correlated with pure g-C_3N_4 and rGO/CN, the rGO/pCN photocatalysts demonstrated a remarkable enhancement on the photocatalytic reduction of CO_2 to CH_4. The better performance of rGO/pCN photocatalysts was attributed to their extraordinary 2D/2D morphology allied with interfacial coupling, thus successfully harnessing light absorption to aid the photogenerated charge transfer. The rGO/pCN photocatalysts exhibited the boosted CH_4 evolution of 13.93 mmol/$g_{catalyst}$ with a photochemical quantum yield of 0.560% (Ong et al. 2015a). Similarly, Ong and co-workers also reported the usage of noble-metal Pt nanoparticles decorated on graphitic carbon nitride (g-C_3N_4) for the photocatalytic CO_2 reduction (Ong et al. 2015b). In comparison with pure g-C_3N_4, the Pt NPs/g-C_3N_4 photocatalysts demonstrated a remarkable enhancement on the photocatalytic reduction of CO_2 to CH_4. The better performance of 2 wt% Pt NPs/g-C_3N_4 photocatalysts was attributed to their enhanced visible light absorption and the efficient photogenerated charge transfer from g-C_3N_4 to Pt due to the lower Fermi level of Pt. The Pt NPs/g-C_3N_4 photocatalysts exhibited the boosted CH_4 evolution of 13.03 mmol/$g_{catalyst}$ after 10 h of light irradiation (Ong et al. 2015b).

Later in 2016, Gao and co-workers investigated by density function theory (DFT) calculations that single palladium/platinum (Pd/Pt) deposited on g-C_3N_4, i.e. Pd/g-C_3N_4 and Pt/g-C_3N_4 photocatalysts, can efficiently opt for the photocatalytic CO_2 reduction (Gao et al. 2016). The ideal yielded products are HCOOH and CH_4 of the photocatalytic CO_2 reduction on the Pd/g-C_3N_4 and Pt/g-C_3N_4 photocatalysts by means of 0.66 eV and 1.16 eV rate-determining barrier, respectively (Gao et al. 2016). Likewise, Putri and co-workers investigated an effective photocatalyst silver chloride (AgCl)/carbon nitride (CN) hybrid for the photocatalytic reduction of CO_2 (Putri et al. 2016). The as-synthesized AgCl/CN photocatalyst revealed a boosted visible-light absorption due to SPR effect of Ag, showing about 2.5-fold enrichment in the performance of AgCl/CN as compared to pure CN. This improved photocatalytic activity for the reduction of CO_2 is also ascribed to the hierarchical AgCl/CN heterojunction hybrid scheme which impedes photogenerated electron-hole recombination (Putri et al. 2016).

Similarly, Chang and co-workers showed the effect of several morphologies of cuprous oxide (Cu_2O) crystals coated onto graphitic carbon nitride (g-C_3N_4) to assess their photoactivity for the photoreduction of CO_2 under solar illumination. Only CO was the chief gaseous product yielded from the photocatalytic reduction of CO_2 after 8 hour-illumination by employing Cu_2O/gC_3N_4 photocatalyts (Chang and Tseng 2018). Equally, Maeda and co-workers reported the unique synthesis protocol for graphitic carbon nitride (g-C_3N_4) by simply heating urea, as a sole precursor, in air at diverse temperatures (773-923 K), and the as-synthesized g-C_3N_4 photocatalyst as employed for photocatalytic CO_2 reduction (Maeda et al. 2018). The g-C_3N_4 was additionally modified with nanoparticles of Ag, which was skilled in reducing CO_2 into formate under solar light, with the assistance of a RuP cocatalyst (Maeda et al. 2018). In the same year, Tseng and co-workers presented the *in-situ* deposition of anatase TiO_2 nanoparticles on urea derived C_3N_4 for the photoreduction of CO_2 under solar illumination (Tseng et al. 2019). The common yielded products were carbon monoxide and methanol; however, the Ti/uC_3N_4 samples with higher amine functional groups produced more carbon monoxide, while the samples with hydroxyl functional groups produced methanol (Tseng et al. 2019). Similarly, Nazir and co-workers stated the applicability of an exceptional bimetallic alloy of CoAg nanoparticles coated on carbon nitride for the electrocatalytic conversion of carbon dioxide to chemical fuels. In the electrocatalytic conversion of CO_2, the CoAg/CN samples were evaluated in 0.5 M $KHCO_3$ as supporting electrolyte. The studies reveal that CoAg/CN show better ECR of CO_2 as compared to Co/CN, as observed by the decreased cathodic current at and below −0.9 V versus Ag/AgCl (Nazir et al. 2019).

Later in 2019, Tan and co-6workers deliberated a systematic first principle study to testify the ability of single transition metal such as Sc, Zn, Mo, Rh, Ru, Pd, Ag, Pt, and Au, coated on single-layered defective boron nitride to accomplish electrocatalytic CO_2 reduction (ECR) process utilizing efficient electrocatalysts (Tan et al. 2019). Their theoretical computational study reveals that Fe/BN, Co/BN, and Pt/BN nanosheets show boosted ECR activities with very low onset potentials of −0.52, −0.68, and −0.60 V, respectively. The enhanced ECR activities of these TM/BN nanosheets are ascribed to their high selectivity of CO_2 ECR to CH_4 as the only yielded product (Tan et al. 2019). Recently in 2020, Cao and co-workers reported the fabrication of ultrathin oxygen-modified-bismuth nitride (O/BN) nanosheets encompassing B−O bonds (Cao et al. 2020). Upon these ultrathin O/BN nanosheets, the CO_2 can be seized and gets bonded with the B−O group. Consequently, the CO_2 gets converted into carbon active species onto O/BN surface and O/BN nanosheets showcase enhanced photocatalytic CO_2 conversion efficiency with the CO production rate of 12.5 µmol/g/h (Cao et al. 2020). Recently, Back and Jung reported the theoretical density functional theory calculations to investigate the catalytic properties of TiC, TiN, and single-atom catalysts supported on them for the electrochemical CO_2 reduction. This work demonstrates the great potential of TiN catalysts as active and selective CO_2 reduction catalysts (Back and Jung 2017).

9.5 FUTURE PROSPECTIVE NITRIDE MATERIALS

9.5.1 Tantalum (oxy)nitride

Lately, numerous oxy-nitride materials have been reconnoitered as outstanding photocatalyst candidates for utilizing the maximum/complete solar spectrum. Admist, TaON is a moderate band gap based oxy-nitride photocatalysts. It is known as an idyllic solar-driven photocatalyst for various photo-assisted applications (Cristea et al. 2019). Unsuitably, pristine TaON reveals low photocatalytic behavior because of its less quantum yield, self-deactivation of TaON when irradiated with light, very-poor photo-stability due to translation of TaON back to Ta_2O_5 and upsurge recombination rate charge carriers generated by solar radiation. No metal doping of photocatalysts is expected to improve the quantum yield of photocatalysts and diminish the recombination rate of charge carriers using light radiaton. Henceforth, various non-noble metals, namely, V, Cu and Co, as dopants are investigated to boost the photocatalytic activity of TaON, which has continually been of prodigious significance in this research domain (Cristea et al. 2019, Zhen et al. 2016). Recently, Chi and co-workers investigated for the first time the use of vanadium (V) as a dopant to boost the photoactivity of TaON photocatalysts for efficient photocatalytic reduction of CO_2 into chemical fuels (Le Chi et al. 2019). Their work studied the effect of the replacement of several Ta atoms by V dopants leading to the straddling of the band edges, i.e. decrease in the conduction band minimum and increase in the valence band maximum of the as-synthesized V-TaON heterostructure. This band alignment of V-TaON well absorbed the significant amount of solar spectrum for the production of photogenerated charge carriers, which participate in the photocatalytic reduction reactions of CO_2 to produce CH_4 and CO. The production rates of the yielded products, CH_4 and CO, are 673 μmol/g/h and 206 μmol/g/h, respectively (Le Chi et al. 2019). However, the work in this domain is very limited. In the future, various other metals as dopants can be examined to enhance the PC/PEC activity of TaON, and in so doing the innovative research opportunities can be explored.

9.5.2   Oxynitride Semiconductors

Oxynitride based semiconductor family encompasses lanthanum titanium oxynitride, Ti-Pd mixed oxynitride, $BaTaO_2N$, $(ZrSn)Ti_xN_y$, etc. Thin films of these materials possess conduction and valence band edges for H_2 as well as O_2 production, and narrow band gap in the visible region, thus harvesting a major portion of the solar illumination. Le Paven-Thivet et al. investigated the photo-electrochemical properties of crystalline pervoskite lanthanum titanium oxynitride film of band-gap 2.1 eV, and observed remarkable enhancement in photoactivity on surface modification with IrO_2 colloidal onto the film (Le Paven-Thivet et al. 2009). Allam and co-workers studied Ti-Pd oxynitride nanotube and observed 5 fold upsurge in photoconversion efficiency than pure TiO_2 (Allam et al. 2011). Shaheen et al. and Higashi and co-workers reported the synthesis of niobium oxynitride for light-driven water splitting and found 1000% enhancement inefficiency as compared to mesoporous niobium oxynitride (Shaheen et al. 2016). The impressive performance of oxynitride semiconductors' materials was reported with IPCE of about 76% at 400 nm illumination and photocurrents of the order of 4-5 mAcm^{-2} for water oxidation can be obtained for heterojunctions (Abe et al. 2010). Ueda et al. studied the $BaTaO_2N$ photoanodes and found the photocurrent of 4.2 mAcm^{-2} at 1.2 V RHE (Ueda et al. 2015). The effect of Mo^{6+} dopant is studied on $BaTaO_2N$ photoanode, which resulted in higher photocurrent (Higashi et al. 2015). It has been observed that the material performance also depends upon the synthesis condition, humidity, and concentration. The oxynitride materials have been found to show the best performance when synthesized at a higher temperature of the order of 850°C, which makes this material the less preferred choice for transparent glass substrate. Thus, the future research on oxynitride based materials will be directed towards various synthesis routes to achieve high performance and also avoid harsh conditions.

9.6   SUMMARY

In summary, we have provided an in-depth overview of the present status of material aspect of nitride semiconductors for solar driven artificial photosynthesis including both solar photo-electrochemical water splitting and CO_2 reduction. Initially, these materials suffer from challenges like low quantum efficiency and poor stability in the electrolyte. But, in recent years, with new materials and synthesis methods, these are becoming potential materials for water splitting and CO_2 reduction. Nitride materials are well known for their tunable bandgap and stability. Therefore, the authors discussed various pros and cons of nitride materials as solar light harvesting electrodes and CO_2 reduction. The realization of high performance nitride materials for practical application is still facing a number of challenges: (i) large-scale and reproducible metal nitride synthesis with controlled morphology, (ii) stability and durability issues in aqueous electrolytes, and (iii) the solar-to-hydrogen conversion efficiency of most metal nitride-based semiconductors is still very low. Although great progress has been made in this field, the fascinating properties of nitride semiconductors are yet to be fully explored. The growth of efficient photoelectrode with high light-harvesting and low overpotential for water splitting for hydrogen production and CO_2 reduction with minimum coupling losses could make artificial photosynthesis devices a reality.

9.7 REFERENCES

Abe, Ryu, Masanobu Higashi and Kazunari Domen. 2010. Facile fabrication of an efficient oxynitride TaON photoanode for overall water splitting into H$_2$ and O$_2$ under visible light irradiation. JACS. 132(34): 11828-11829.

Algara-Siller, Gerardo, Nikolai Severin, Samantha Y. Chong, et al. 2014. Triazine-based graphitic carbon nitride: A two-dimensional semiconductor. Angew. Chem. Int. 53(29): 7450-7455.

Allam, Nageh K., Adam J. Poncheri and Mostafa A. El-Sayed. 2011. Vertically oriented Ti–Pd mixed oxynitride nanotube arrays for enhanced photoelectrochemical water splitting. ACS Nano. 5(6): 5056-5066.

AlOtaibi, B., M. Harati, S. Fan, et al. 2013. High efficiency photoelectrochemical water splitting and hydrogen generation using GaN nanowire photoelectrode. Nanotechnology 24(17): 175401.

Aryal, K., B.N. Pantha, J. Li, J.Y. Lin and H.X. Jiang. 2010. Hydrogen generation by solar water splitting using p-InGaN photoelectrochemical cells. Appl. Phys. Lett. 96(5): 052110.

Back, Seoin and Yousung Jung. 2017. TiC-and TiN-supported single-atom catalysts for dramatic improvements in CO$_2$ electrochemical reduction to CH$_4$. ACS Energy Lett. 2(5): 969-975.

Badwal, Sukhvinder P.S., Sarbjit S. Giddey, Christopher Munnings, Anand I. Bhatt and Anthony F. Hollenkamp. 2014. Emerging electrochemical energy conversion and storage technologies. Front. Chem. 2: 79.

Bahadar, Ali and M. Bilal Khan. 2013. Progress in energy from microalgae: A review. Renewable Sustainable Energy Rev. 27: 128-148.

Bian, Juncao, Lifei Xi, Chao Huang, Kathrin M. Lange, Rui-Qin Zhang and Menny Shalom. 2016. Efficiency enhancement of carbon nitride photoelectrochemical cells via tailored monomers design. Adv. Energy Mater. 6(12): 1600263.

Biefeld, Robert M., Daniel D. Koleske and Jeffrey G. Cederberg. 2015. The Science and Practice of Metal-Organic Vapor Phase Epitaxy (MOVPE). pp. 95-160. *In*: T.F. Kuech [ed.]. Handbook of Crystal Growth, 2nd Edition Volume IIIA (Basic Techniques). Elsevier.

Bond, Tami C., Ekta Bhardwaj, Rong Dong, et al. 2007. Historical emissions of black and organic carbon aerosol from energy-related combustion. Global Biogeochem. Cycles. 21(2): 1850-2000.

Brown, B.E., R.P. Dunne, P.J. Somerfield, et al. 2019. Long-term impacts of rising sea temperature and sea level on shallow water coral communities over a ~40 year period. Sci. Rep. 9(1): 1-12.

Butler, M.A. and D.S. Ginley. 1978. Prediction of flatband potentials at semiconductor-electrolyte interfaces from atomic electronegativities. J. Electrochem. Soc. 125(2): 228-232.

Cao, Yuehan, Ruiyang Zhang, Tianli Zhou, et al. 2020. B-O bonds in ultrathin boron nitride nanosheets to promote photocatalytic carbon dioxide conversion. ACS Appl. Mater. Interfaces. 12(8): 9935-9943.

Chang, Po-Ya and I-Hsiang Tseng. 2018. Photocatalytic conversion of gas phase carbon dioxide by graphitic carbon nitride decorated with cuprous oxide with various morphologies. J. CO$_2$ Util. 26: 511-521.

Chouhan, Neelu, Ru-Shi Liu and Jiujun Zhang. 2017. Photochemical Water Splitting: Materials and Applications, CRC Press.

Chun, Wang-Jae, Akio Ishikawa, Hideki Fujisawa, et al. 2003. Conduction and valence band positions of Ta$_2$O$_5$, TaON, and Ta$_3$N$_5$ by UPS and electrochemical methods. J. Phys. Chem. B. 107(8): 1798-1803.

Cong, Yanqing, Hyun S. Park, Hoang X. Dang, Fu-Ren F. Fan, Allen J. Bard and C. Buddie Mullins. 2012. Tantalum cobalt nitride photocatalysts for water oxidation under visible light. Chem. Mater. 24(3): 579-586.

Cristea, Daniel, Luis Cunha, Camelia Gabor, et al. 2019. Tantalum oxynitride thin films: Assessment of the photocatalytic efficiency and antimicrobial capacity. Nanomaterials 9(3): 476.

Di, Tingmin, Bicheng Zhu, Bei Cheng, Jiaguo Yu and Jingsan Xu. 2017. A direct Z-scheme g-C3N$_4$/SnS$_2$ photocatalyst with superior visible-light CO$_2$ reduction performance. J. Catal. 352: 532-541.

Dincer, Ibrahim. 2012. Green methods for hydrogen production. Int. J. Hydrog. Energy. 37(2): 1954-1971.

Ebaid, Mohamed, Jin-Ho Kang, Seung-Hyuk Lim, et al. 2015. Enhanced solar hydrogen generation of high density, high aspect ratio, coaxial InGaN/GaN multi-quantum well nanowires. Nano Energy 12: 215-223.

Eveloy, Valerie. 2019. Hybridization of solid oxide electrolysis-based power-to-methane with oxyfuel combustion and carbon dioxide utilization for energy storage. Renewable Sustainable Energy Rev. 108: 550-571.

Fan, Shizhao, Ishiang Shih and Zetian Mi. 2017. A monolithically integrated InGaN nanowire/Si tandem photoanode approaching the ideal bandgap configuration of 1.75/1.13 eV. Adv. Energy Mater. 7(2): 1600952.

Feng, Xinjian, Thomas J. LaTempa, James I. Basham, Gopal K. Mor, Oomman K. Varghese and Craig A. Grimes. 2010. Ta_3N_5 nanotube arrays for visible light water photoelectrolysis. Nano Letters 10(3): 948-952.

Fujii, Katsushi and Kazuhiro Ohkawa. 2005. Photoelectrochemical properties of *p*-type GaN in comparison with *n*-type GaN. Jpn. J. Appl. Phys. 44(7L): L909.

Fujii, Katsushi and Kazuhiro Ohkawa. 2006. Bias-assisted H_2 gas generation in HCl and KOH solutions using *n*-type GaN photoelectrode. J. Electrochem. Soc. 153(3): A468.

Gao, Guoping, Yan Jiao, Eric R. Waclawik and Aijun Du. 2016. Single atom (Pd/Pt) supported on graphitic carbon nitride as an efficient photocatalyst for visible-light reduction of carbon dioxide. JACS. 138(19): 6292-6297.

Gelderman, K., L. Lee and S.W. Donne. 2007. Flat-band potential of a semiconductor: Using the Mott-Schottky equation. J. Chem. Educ. 84(4): 685.

Ghosh, Dibyendu, Pooja Devi and Praveen Kumar. 2020. Modified *p*-GaN micro-wells with vertically aligned 2D-MoS_2 for enhanced photoelectrochemical water-splitting. ACS Appl. Mater. Interfaces. 13797-13804.

Glover, E.N. 2017. Design, synthesis and characterisation of novel materials for photocatalytic water splitting, UCL (University College London).

Halmann, M. 1978. Photoelectrochemical reduction of aqueous carbon dioxide on *p*-type gallium phosphide in liquid junction solar cells. Nature 275(5676): 115-116.

Hassan, Mostafa Afifi, Jin-Ho Kang, Muhammad Ali Johar, Jun-Seok Ha and Sang-Wan Ryu. 2018. High-performance ZnS/GaN heterostructure photoanode for photoelectrochemical water splitting applications. Acta Mater. 146: 171-175.

Haubner, R., M. Wilhelm, R. Weissenbacher and B. Lux. 2002. Boron nitrides – properties, synthesis and applications. pp. 1-45. *In*: M. Jansen [ed.]. High Performance Non-Oxide Ceramics II, Springer, Berlin, Heidelberg.

Higashi, Masanobu, Yuta Yamanaka, Osamu Tomita and Ryu Abe. 2015. Fabrication of cation-doped $BaTaO_2N$ photoanodes for efficient photoelectrochemical water splitting under visible light irradiation. APL Mater. 3(10): 104418.

Hoffmann, Peter. 2019. The Forever Fuel: The Story of Hydrogen: Routledge.

Hong, Jindui, Wei Zhang, Yabo Wang, Tianhua Zhou and Rong Xu. 2014. Photocatalytic reduction of carbon dioxide over self-assembled carbon nitride and layered double hydroxide: The role of carbon dioxide enrichment. Chem. Cat. Chem. 6(8): 2315-2321.

Hou, Yang, Zhenhai Wen, Shumao Cui, Xinliang Feng and Junhong Chen. 2016. Strongly coupled ternary hybrid aerogels of N-deficient porous graphitic-C_3N_4 nanosheets/N-doped graphene/NiFe-layered double hydroxide for solar-driven photoelectrochemical water oxidation. Nano Letters 16(4): 2268-2277.

Ide, Yusuke, Fei Liu, Jun Zhang, et al. 2014. Hybridization of Au nanoparticle-loaded TiO_2 with BN nanosheets for efficient solar-driven photocatalysis. J. Mater. Chem. A. 2(12): 4150-4156.

Inoue, Tooru, Akira Fujishima, Satoshi Konishi and Kenichi Honda. 1979. Photoelectrocatalytic reduction of carbon dioxide in aqueous suspensions of semiconductor powders. Nature 277(5698): 637-638.

Jitaru, Maria. 2007. Electrochemical carbon dioxide reduction-fundamental and applied topics. JUCTM. 42(4): 333-344.

Kabir, Ehsanul, Pawan Kumar, Sandeep Kumar, Adedeji A. Adelodun and Ki-Hyun Kim. 2018. Solar energy: Potential and future prospects. Renewable Sustainable Energy Rev. 82: 894-900.

Kamimura, J., P. Bogdanoff, M. Ramsteiner, L. Geelhaar and H. Riechert. 2016. Photoelectrochemical properties of InN nanowire photoelectrodes for solar water splitting. Semicond. Sci. Technol. 31(7): 074001.

Kang, Jin-Ho, Soo Hee Kim, Mohamed Ebaid, June Key Lee and Sang-Wan Ryu. 2014. Efficient photoelectrochemical water splitting by a doping-controlled GaN photoanode coated with NiO cocatalyst. Acta Mater. 79: 188-193.

Kang, Suhee, Joonyoung Jang, Rajendra C. Pawar, Sunghoon Ahn and Caroline Sunyong Lee. 2018. Direct coating of a gC_3N_4 layer onto one-dimensional TiO_2 nanocluster/nanorod films for photoactive applications. Dalton Trans. 47(21): 7237-7244.

Kawrani, Sara, Amr A. Nada, Maged F. Bekheet, et al. 2020. Enhancement of calcium copper titanium oxide photoelectrochemical performance using boron nitride nanosheets. CEJ. 389: 124326.

Kibria, Md G., Hieu P.T. Nguyen, Kai Cui, et al. 2013. One-step overall water splitting under visible light using multiband InGaN/GaN nanowire heterostructures. ACS Nano. 7(9): 7886-7893.

Kim, Doek Kyu and Choon Bae Park. 2005. Photoluminescence studies of GaN films on Si (111) substrate by using an AlN buffer control. J. Korean Phys. Soc. 47(6): 1006-1009.

Kim, Soo Hee, Mohamed Ebaid, Jin-Ho Kang and Sang-Wan Ryu. 2014. Improved efficiency and stability of GaN photoanode in photoelectrochemical water splitting by NiO cocatalyst. Appl. Surf. Sci. 305: 638-641.

Kobayashi, Shigeki, Steven Plotkin and Suzana Kahn Ribeiro. 2009. Energy efficiency technologies for road vehicles. Energy Effic. 2(2): 125-137.

Kumar, Bhupendra, Mark Llorente, Jesse Froehlich, Tram Dang, Aaron Sathrum and Clifford P. Kubiak. 2012. Photochemical and photoelectrochemical reduction of CO_2. Annu. Rev. Phys. Chem. 63: 541-569.

Kumar, Praveen, Pooja Devi and Rishabh Jain, et al. 2019. Quantum dot activated indium gallium nitride on silicon as photoanode for solar hydrogen generation. Commun. Chem. 2(1): 1-7.

Kuriki, Ryo, Chandana Sampath Kumara Ranasinghe, Yasuomi Yamazaki, Akira Yamakata, Osamu Ishitani and Kazuhiko Maeda. 2018. Excited-state dynamics of graphitic carbon nitride photocatalyst and ultrafast electron injection to a Ru(II) mononuclear complex for carbon dioxide reduction. J. Phys. Chem. C. 122(29): 16795-16802.

Le Chi, Nguyen Thi Phuong, Nguyen Thi Dieu Cam and Doan Van Thuan, et al. 2019. Synthesis of vanadium doped tantalum oxy-nitride for photocatalytic reduction of carbon dioxide under visible light. Appl. Surf. Sci. 467: 1249-1255.

Lee, Ming-Lun, Po-Hsun Liao, Guan-Lun Li, Hung-Wei Chang, Chi-Wing Lee and Jinn-Kong Sheu. 2019. Enhanced production rates of hydrogen generation and carbon dioxide reduction using aluminum gallium nitride/gallium nitride heteroepitaxial films as photoelectrodes in seawater. Sol. Energy Mater. Sol. Cells. 202: 110153.

Le Paven-Thivet, Claire, Akio Ishikawa, Ahmed Ziani, et al. 2009. Photoelectrochemical properties of crystalline perovskite lanthanum titanium oxynitride films under visible light. J. Phys. C. 113(15): 6156-6162.

Lesnikowski, Alexandra, James D. Ford, Robbert Biesbroek and Lea Berrang-Ford. 2019. A policy mixes approach to conceptualizing and measuring climate change adaptation policy. Clim. Change. 156(4): 447-469.

Li, Chun-Ting, Sie-Rong Li, Ling-Yu Chang, et al. 2015. Efficient titanium nitride/titanium oxide composite photoanodes for dye-sensitized solar cells and water splitting. J. Mater. Chem. A. 3(8): 4695-4705.

Li, Xingxing, Jin Zhao and Jinlong Yang. 2013a. Semihydrogenated BN sheet: A promising visible-light driven photocatalyst for water splitting. Sci. Rep. 3: 1858.

Li, Yanbo, Tsuyoshi Takata, Dongkyu Cha, et al. 2013b. Photoelectrodes: Vertically aligned Ta_3N_5 nanorod arrays for solar-driven photoelectrochemical water splitting (Adv. Mater. 1/2013). Adv. Mater. 25(1): 152-152.

Li, Yanbo, Li Zhang, Almudena Torres-Pardo, et al. 2013c. Cobalt phosphate-modified barium-doped tantalum nitride nanorod photoanode with 1.5% solar energy conversion efficiency. Nat. Commun. 4(1): 1-7.

Li, Yat and Jin Zhong Zhang. 2010. Hydrogen generation from photoelectrochemical water splitting based on nanomaterials. Laser Photonics Rev. 4(4): 517-528.

Lindgren, Torbjörn, Magnus Larsson and Sten-Eric Lindquist. 2002. Photoelectrochemical characterisation of indium nitride and tin nitride in aqueous solution. Sol. Energy Mater. Sol. Cells. 73(4): 377-389.

Lindgren, Torbjörn, Gemma Romualdo Torres, Jesper Ederth, Richard Karmhag, Claes-Göran Granqvist and Sten-Eric Lindquist. 2006. DC magnetron reactive sputtered InN thin film electrodes as photoanodes in aqueous solution. A study of as prepared and nitrogen annealed electrodes. Thin Solid Films 510(1-2): 6-14.

Liu, Guiji, Jingying Shi, Fuxiang Zhang, et al. 2014. A tantalum nitride photoanode modified with a hole-storage layer for highly stable solar water splitting. Angew. Chem. Int. 53(28): 7295-7299.

Liu, Huiqiang, Xinzhou Ma, Zuxin Chen, et al. 2018. Controllable synthesis of (11−2−2) faceted InN nanopyramids on ZnO for photoelectrochemical water splitting. Small 14(17): 1703623.

Liu, Jian, Hongqiang Wang, Zu Peng Chen, et al. 2015. Microcontact-printing-assisted access of graphitic carbon nitride films with favorable textures toward photoelectrochemical application. Adv. Mater. 27(4): 712-718.

Maeda, Kazuhiko, Daehyeon An, Ryo Kuriki, Daling Lu and Osamu Ishitani. 2018. Graphitic carbon nitride prepared from urea as a photocatalyst for visible-light carbon dioxide reduction with the aid of a mononuclear ruthenium (II) complex. Beilstein J. Org. Chem. 14(1): 1806-1812.

Medvedev, O.S., M.V. Puzyk, A.S. Usikov, H. Helava and Yu N. Makarov. 2017. Solar water splitting with III-N nanocolumn structures. Paper read at Journal of Physics: Conference Series.

Mohamed, S.H., Huaping Zhao, Henry Romanus, et al. 2020. Optical, water splitting and wettability of titanium nitride/titanium oxynitride bilayer films for hydrogen generation and solar cells applications. Mat. Sci. Semicon. Proc. 105: 104704.

Morris, Amanda J., Gerald J. Meyer and Etsuko Fujita. 2009. Molecular approaches to the photocatalytic reduction of carbon dioxide for solar fuels. Acc. Chem. Res. 42(12): 1983-1994.

Nazir, Roshan, Anand Kumar, Sardar Ali, Mohammed Ali Saleh Saad and Mohammed J Al-Marri. 2019. Galvanic exchange as a novel method for carbon nitride supported CoAg catalyst synthesis for oxygen reduction and carbon dioxide conversion. Catalysts 9(10): 860.

Nepal, N., J. Li, M.L. Nakarmi, J.Y. Lin and H.X. Jiang. 2005. Temperature and compositional dependence of the energy band gap of AlGaN alloys. Appl. Phys. Lett. 87(24): 242104.

Nötzel, Richard. 2017. InN/InGaN quantum dot electrochemical devices: New solutions for energy and health. Natl. Sci. Rev. 4(2): 184-195.

Ong, Wee-Jun, Lling-Lling Tan, Siang-Piao Chai, Siek-Ting Yong and Abdul Rahman Mohamed. 2015a. Surface charge modification via protonation of graphitic carbon nitride (g-C_3N_4) for electrostatic self-assembly construction of 2D/2D reduced graphene oxide (rGO)/g-C_3N_4 nanostructures toward enhanced photocatalytic reduction of carbon dioxide to methane. Nano Energy 13: 757-770.

Ong, Wee-Jun, Lling-Lling Tan, Siang-Piao Chai and Siek-Ting Yong. 2015b. Heterojunction engineering of graphitic carbon nitride (gC_3N_4) via Pt loading with improved daylight-induced photocatalytic reduction of carbon dioxide to methane. Dalton Trans. 44(3): 1249-1257.

Park, Tae Joon, Rajendra C. Pawar, Suhee Kang and Caroline Sunyong Lee. 2016. Ultra-thin coating of gC_3N_4 on an aligned ZnO nanorod film for rapid charge separation and improved photodegradation performance. RSC Adv. 6(92): 89944-89952.

Peng, Guiming, Lidan Xing, Jesús Barrio, Michael Volokh and Menny Shalom. 2018. A general synthesis of porous carbon nitride films with tunable surface area and photophysical properties. Angew. Chem. Int. 57(5): 1186-1192.

Polyakov, Vladimir M. and Frank Schwierz. 2006. Low-field electron mobility in wurtzite InN. Appl. Phys. Lett. 88(3): 032101.

Putri, Lutfi Kurnianditia, Wee-Jun Ong, Wei Sea Chang and Siang-Piao Chai. 2016. Enhancement in the photocatalytic activity of carbon nitride through hybridization with light-sensitive AgCl for carbon dioxide reduction to methane. Catal. Sci. Technol. 6(3): 744-754.

Rabiu, K.O., L.K. Abidoye and A.T. Oriaje. 2019. The Potential of bioethanol as a viable alternative to fossil fuels. AZOJETE. 15(3): 554-559.

Rodriguez, Paul Eduardo David Soto, Pavel Aseev, Víctor Jesús Gómez, Waheed ul Hassan, Magnus Willander, and Richard Nötzel 2015. In N/InGaN quantum dot photoelectrode: Efficient hydrogen generation by water splitting at zero voltage. Nano Energy 13: 291-297.

Ryu, Sang-Wan, Yu Zhang, Benjamin Leung, Christopher Yerino and Jung Han. 2011. Improved photoelectrochemical water splitting efficiency of nanoporous GaN photoanode. Semicond. Sci. Technol. 27(1): 015014.

Sekimoto, Takeyuki, Shuichi Shinagawa, Yusuke Uetake, et al. 2015. Tandem photo-electrode of InGaN with two Si pn junctions for CO_2 conversion to HCOOH with the efficiency greater than biological photosynthesis. Appl. Phys. Lett. 106(7): 073902.

Shaheen, Basamat S., Ahmed M. Hafez, Banavoth Murali, et al. 2016. 10-fold enhancement in light-driven water splitting using niobium oxynitride microcone array films. Sol. Energy Mater. Sol. Cells. 151: 149-153.

Sheu, Jinn-Kong, P.H. Liao, T.C. Huang, K.J. Chiang, Wei-Chi Lai and ML Lee. 2017. In GaN-based epitaxial films as photoelectrodes for hydrogen generation through water photoelectrolysis and CO_2 reduction to formic acid. Sol. Energy Mater. Sol. Cells. 166: 86-90.

Solovan, M.M., V.V. Brus and P.D. Maryanchuk. 2013. Electrical and photoelectric properties of anisotype n-TiN/p-Si heterojunctions. Semiconductors 47(9): 1174-1179.

Solovan, M.N., V.V. Brus, E.V. Maistruk and P.D. Maryanchuk. 2014. Electrical and optical properties of TiN thin films. Inorg. Mater. 50(1): 40-45.

Su, Jingyang, Ping Geng, Xinyong Li, Qidong Zhao, Xie Quan and Guohua Chen. 2015. Novel phosphorus doped carbon nitride modified TiO$_2$ nanotube arrays with improved photoelectrochemical performance. Nanoscale 7(39): 16282-16289.

Tan, Xin, Hassan A. Tahini, Hamidreza Arandiyan and Sean C. Smith. 2019. Electrocatalytic reduction of carbon dioxide to methane on single transition metal atoms supported on a defective boron nitride monolayer: First principle study. Adv. Theory Simul. 2(3): 1800094.

Tseng, I, Yu-Min Sung, Po-Ya Chang and Chin-Yi Chen. 2019. Anatase TiO$_2$-decorated graphitic carbon nitride for photocatalytic conversion of carbon dioxide. Polymers 11(1): 146.

Ueda, Koichiro, Tsutomu Minegishi, Justin Clune, et al. 2015. Photoelectrochemical oxidation of water using BaTaO$_2$N photoanodes prepared by particle transfer method. JACS. 137(6): 2227-2230.

Uosaki, Kohei, Ganesan Elumalai, Hung Cuong Dinh, Andrey Lyalin, Tetsuya Taketsugu and Hidenori Noguchi. 2016. Highly efficient electrochemical hydrogen evolution reaction at insulating boron nitride nanosheet on inert gold substrate. Sci. Rep. 6: 32217.

Uyar, Tanay Sıdkı and Doğancan Beşikci. 2017. Integration of hydrogen energy systems into renewable energy systems for better design of 100% renewable energy communities. Int. J. Hydrog. Energy 42(4): 2453-2456.

Wang, Yan, Jianyang Sun, Jiang Li and Xu Zhao. 2017. Electrospinning preparation of nanostructured g-C$_3$N$_4$/BiVO$_4$ composite films with an enhanced photoelectrochemical performance. Langmuir 33(19): 4694-4701.

Wang, Yichen, Shizhao Fan, Bandar AlOtaibi, Yongjie Wang, Lu Li and Zetian Mi. 2016. A monolithically integrated gallium nitride nanowire/silicon solar cell photocathode for selective carbon dioxide reduction to methane. Chem. Eur. J. 22(26): 8809-8813.

Wang, Yongjie, Srinivas Vanka, Jiseok Gim, et al. 2019. An In0. 42Ga0. 58N tunnel junction nanowire photocathode monolithically integrated on a nonplanar Si wafer. Nano Energy 57: 405-413.

Wuebbles, Donald, David R. Easterling, Katharine Hayhoe, Thomas Knutson, Robert Kopp, James Kossin, et al. 2016. Our globally changing climate. pp. 38-97. In: D.J. Wuebbles, D.W. Fahey, K.A. Hibbard, D.J. Dokken, B.C. Stewart and T.K. Maycock [eds.]. Climate Science Special Report: A Sustained Assessment Activity of the U.S. Global Change Research Program, Washington, DC, USA.

Yadav, Satyesh Kumar, Jian Wang and X-Y Liu. 2016. Ab initio modeling of zincblende AlN layer in Al-AlN-TiN multilayers. J. Appl. Phys. 119(22): 224304.

Yang, Xiaogang and Dunwei Wang. 2017. Photophysics and photochemistry at the semiconductor/electrolyte interface for solar water splitting. pp. 47-80. In: Z. Mi, L. Wang, C. Jagadish [eds.]. Semiconductors and Semimetals, vol. 97, Elsevier.

Yin, Wan-Jian, Houwen Tang, Su-Huai Wei, Mowafak M. Al-Jassim, John Turner and Yanfa Yan. 2010. Band structure engineering of semiconductors for enhanced photoelectrochemical water splitting: The case of TiO$_2$. Phys. Rev. B. 82(4): 045106.

Yip, Ho Lung, Aleš Srna, Anthony Chun Yin Yuen, et al. 2019. A Review of hydrogen direct injection for internal combustion engines: Towards carbon-free combustion. Appl. Sci. 9(22): 4842.

Yokoyama, Daisuke, Hiroshi Hashiguchi, Kazuhiko Maeda, et al. 2011. Ta$_3$N$_5$ photoanodes for water splitting prepared by sputtering. Thin Solid Films 519(7): 2087-2092.

Yotsuhashi, Satoshi, Masahiro Deguchi, Hiroshi Hashiba, et al. 2012. Enhanced CO$_2$ reduction capability in an AlGaN/GaN photoelectrode. Appl. Phys. Lett. 100(24): 243904.

Zeng, Kai and Dongke Zhang. 2010. Recent progress in alkaline water electrolysis for hydrogen production and applications. Prog. Energy Combust. Sci. 36(3): 307-326.

Zhang, Yuanjian and Markus Antonietti. 2010. Photocurrent generation by polymeric carbon nitride solids: An initial step towards a novel photovoltaic system. Chem. Asian J. 5(6): 1307-1311.

Zhen, Chao, Runze Chen, Lianzhou Wang, Gang Liu and Hui-Ming Cheng. 2016. Tantalum (oxy) nitride based photoanodes for solar-driven water oxidation. J. Mater. Chem. A. 4(8): 2783-2800.

Zhou, Han, Jianjun Guo, Peng Li, Tongxiang Fan, Di Zhang and Jinhua Ye. 2013. Leaf-architectured 3D hierarchical artificial photosynthetic system of perovskite titanates towards CO$_2$ photoreduction into hydrocarbon fuels. Sci. Rep. 3: 1667.

10

Oxide Perovskites and Their Derivatives for Photovoltaics Applications

Sanjay Sahare[1,2], Manjeet Kumar[3], Prachi Ghoderao[4], Radhamanohar Aepuru[5], Shern-Long Lee[1] and Ju-Hyung Yun[3,*]

10.1 INTRODUCTION

Development of sustainable renewable energy is an important task to fulfill energy needs of ~7.8K million inhabitants since the conservative energy sources, viz. coal, petroleum, and fossil fuels, will be over one day. Solar energy is a valuable source of energy to convert into electricity (photovoltaics) or fuels (chemical), which implies the forefront of renewable energy research. To achieve such an imperative goal, searching of novel functional materials is essential. Especially in photovoltaics (PV), semiconductor materials were used to absorb solar energy for direct electricity conversion. So far, silicon has dominated tremendously in the arena of PVs compared to other semiconductor materials such as GaAs, CdTe, CIGS, CZTS, dye-sensitized TiO_2, organic polymer, material, perovskite, etc. (Polman et al. 2016, Ferguson et al. 2019, Sinke 2019). In the last decade, perovskite material has emerged as the exotic material for the futuristic PVs owing to the plethora of functionalities including semiconducting (Stoumpos et al. 2013), conducting (Loi and Hummelen 2013), ferroelectric (Loi and Hummelen 2013, Frost et al. 2014), thermoelectric (Takahashi et al. 2011), dielectric (Juarez-Perez et al. 2014), magnetoresistive (Loi and Hummelen 2013), electro-optic (Bhalla et al. 2000), and super-conduction (Kim et al. 2014). A perovskite is any compound that has the same structure as that of perovskite mineral in the form of oxides, nitrides, or halides (Nuraje and Su 2013). The journey of perovskite materials toward PVs' application began with Methylammonium (MA) lead iodide perovskite ($CH_3NH_3PbI_3$/$MAPbI_3$), and then most of the research focused on substitution of halide ions with different stoichiometric ratio to achieve better optoelectronic properties and the PCE augmentation (McMeekin et al. 2016, Wang et al. 2017, Cho et al. 2018, Balakrishna et al. 2018). However, these halide perovskites showed poor stability under open atmospheric conditions (Jing et al. 2017, Chen et al. 2019). Recently, the oxide perovskites and their derivative structures have been investigated comprehensively as a class of adaptable materials owing to their unique structural and compositional flexibility, along with better material stability for numerous

[1] Instiute for Advanced Study, Shenzhen University, Shenzhen, Guangdong 518060, China.
[2] Laboratory of Optoelectronic Devices and Systems of Ministry of Education and Guangdong Provence, College of Optoelectronics Engineering, Shenzhen University, Shenzhen, Guangdong 518060, China.
[3] Department of Electrical Engineering, Incheon National University, Incheon 406772, South Korea.
[4] Department of Applied Physics, Defence Institute of Advanced Technology, Pune 411025, India.
[5] Departamento de Ingeniería Mecánica, Facultad de Ingeniería, Universidad Tecnológica Metropolitana, Santiago, Chile.
[*] Corresponding author: juhyungyun@inu.ac.kr

applications such as photocatalysis and photovoltaics (Liu et al. 2007, Yang et al. 2009). Materials possess extraordinary characteristics, which are due to the involvement of transition metals' outer d-orbital electrons. The cations possess multiple oxidation states with variation in metal-oxygen bonding from ionic to metallic, resulting in a unique property in oxide perovskite. Their potential has been proven and showed enormous promises in terms of photocatalytic activity and stability of the devices (Nechache et al. 2015, Liu et al. 2014). The oxide perovskite, for example, $Pb(Zr,Ti)O_3$, $BaTiO_3$, $BiFeO_3$, etc. has been used as a photo absorber or charge transport layer in PV devices (Won et al. 2011, Zheng et al. 2014, Yang et al. 2009, Fan et al. 2015, Huang et al. 2017). Loureiro et al. have discussed another important perovskite oxide material that is proton conducting material, typically $BaCeO_3$-$BaZrO_3$, which has been developed using sintering technology and used as protonic ceramic fuel cells (Loureiro et al. 2019). These perovskite materials possess ferroelectric properties, which offer an effective driving force for charge generation and quick separations because of strong intrinsic dipoles. Numerous theories propose the foundation of this intriguing phenomenon. However, specifically in the PVs effects, it is quite challenging to identify a single approach, which can be explained universally in different forms of oxide perovskites.

Perovskites have some serious shortcomings, the foremost is perovskites contain materials such as lead, gallium, tellurium, cadmium, arsenic, which are toxic in nature (Ke and Kanatzidis 2019). Perovskite materials show very low stability, e.g. lead in perovskite oxide easily makes iodine volatility, and is easy to decompose when crystals are in the wet form. If we use the perovskite battery power, it is likely to decompose the seepage flow to the roof or in the soil. In addition, this material breaks down or decomposes easily due to heat, moisture, snow etc. (Han et al. 2015). Perovskite oxide shows unique structural and compositional flexibility with high material stability and therefore oxide perovskites and their derivatives are being extensively explored. Hence, perovskite oxide is being used as a versatile material for various applications in electrocatalysis (EC), photocatalysis (PC) and photovoltaics (PV), with great promises in catalytic activity and device stability (Yin et al. 2019). In addition, perovskite oxide material also shows properties like perfect thermal stability, ideal band gap "3-4 eV," and difference in size between the cations of A- and B-sites, which allows different dopants' addition for controlling semiconducting properties and their catalytic properties (Chilvery et al. 2019). In this chapter, the structural and optoelectronic properties of oxide perovskite will be discussed along with the energy conversion process by PVs phenomenon.

10.2   PEROVSKITES STRUCTURE AND THEIR DERIVATIVES

Generally, ABO_3 perovskite exhibits a six-coordinated B-site with O by forming an octahedral structure surrounded by an A-site. The oxide perovskites have ABO_3 composition due to their elemental characteristics of the metal-oxygen bonds (for instance, A-O and B-O bonds). In the ABO_3 formulation, each anion and cation own unique advantages, which tailor optoelectronic properties of the material. Subsequently, modify the device stability and photo-conversion efficiency. The alteration of anion and cation in ABO_3 decides the wide range of structural and compositional flexibility in oxide perovskites. The inherent flexibility could be due to different types of distortion, the ideal structure, which further reduces the symmetry. The distortion includes tilting of the octahedral, cation displacements at the centers of their coordination polyhedral. As per the arrangement of A-site and octahedron, obtained perovskites can be categorized into four different types: 0-D, 1-D, 2-D and 3-D perovskites. Because of the distortion, many physical properties can be altered, for instance, dielectric, electronic, and magnetic (Fu and Itoh 2011). Such material engineering produces single, double perovskites, and their derivative structures (Anderson et al. 1993). The compound perovskites and their derivative structures have been shown in Figure 10.1. The idyllic single crystal structure of oxide perovskite ABO_3 has a high symmetry cubic structure, which can be observed at high temperatures beyond 1000 K (Fig. 10.1, center). As soon as the temperature reduces, the cubic perovskites undergo octahedral rotations along with their axis symmetry, such as (100), (010), and (001) that subsequently transforms to subordinate symmetry phases (Lufaso et al. 2006). The formation of double perovskites can be achieved upon A or B site replacement with two different types of cations, following $A`A``B_2O_6$ (double A-site) or $A_2B`B``O_6$ (double B-site) formulae. Such modifications produce rock salt, random and layered structures, which have been shown in Figure 10.1. The crystallographic information for these common structures has been tabulated in Table 10.1. The perovskite derivatives can be obtained by large distortion, where few of B-O bonds may break. As a result, lower-dimensional perovskite derivatives such as 1D and 2-D perovskite are determined by breaking 3-D framework as depicted in Figure 10.1. These reduced structures of perovskite derivatives have exhibited a larger surface to bulk ratios over their 3-D structure. Hence, these materials are more suitable for PVs and catalytic applications.

TABLE 10.1   Crystallographic information for common double perovskites (Fu and Itoh 2011)

S. No.	Sub-lattice Type	Cell Size	Space Group	Crystal System
1	Random	$1a_p \times 1a_p \times 1a_p$ $\sqrt{2}a_p \times \sqrt{2}a_p \times \sqrt{2}a_p$	$Pm\bar{3}m$ $Pbnm$	Cubic Orthorhombic
2	Rock-salt	$2a_p \times 2a_p \times 2a_p$ $\sqrt{2}a_p \times \sqrt{2}a_p \times \sqrt{2}a_p$	$Fm\bar{3}m$ $P2_1/n$	Cubic Monoclinic
3	Layered	$2a_p \times 2a_p \times 2a_p$	$P2_1/m$	Monoclinic

$CaTiO_3$ is the originally named perovskite; however, similar crystal structure has been found with diverse oxidation states and steric sizes of different cations. The oxide perovskite has compositions of multivalent metal cations including Ca^{2+}, Ti^{4+}, Fe^{3+}, etc., where the larger radius cations occupy the A-site while smaller ions occupy the B-site, and O^{2-}. In the perfect perovskite structure of ABO_3, the radii of all these cations and anions should meet the tolerance factor (t) within the range of 0.87-1.0, and octahedral factor $\mu = 0.44$-0.90. Failing to achieve the standard value 't', the cubic/cubic-like crystal structure will be distorted or destroyed (Zheng et al. 2014). If value of 't' is smaller, the result would be lower symmetrical tetragonal or orthorhombic structures. Similarly, 'μ' value governs the stability of the octahedron and subsequently affects the stability of the perovskite structure (Pérez-Tomás et al. 2018). There are several elements found at A and B site with wide oxidation states in the range of 1^+ to 7^+ and diverse, demonstrating huge flexibility of compositions for the formation of the perovskite crystal structure. Theoretically, ABO_3 (single perovskite) can be segregated into five groups as per the various possible combinations of oxidation states of cations A and B, for example, $A^{1+}B^{5+}O_3$, $A^{2+}B^{4+}O_3$, $A^{3+}B^{3+}O_3$, $A^{4+}B^{2+}O_3$, and $A^{5+}B^{1+}O_3$. Still, there is a long list of single perovskites that have not been synthesized experimentally (Nechache et al. 2015). Similar to single perovskite, double perovskite, i.e. $A_2B`B``O_6$ also shows compositional flexibility since it has substantial tolerance to different elemental combinations of (B`, B``). Few of the flexible compositions applicable to PVs have been tabulated in Table 10.2. These kinds of flexibility are essential to design PVs devices to improve their performance and stability.

TABLE 10.2   Compositional flexibility of oxide perovskite along with its tolerance

S. No	Perovskite Compositions	Tolerance	Ref.
1	$Pb(ZrTi)O_3$	0.991	(Zheng et al. 2014)
2	$(PbLa)TiO_3$	0.996	(Pérez-Tomás et al. 2018)
3	$(PbLa)(ZrTi)O_3$	0.970	(Pérez-Tomás et al. 2018)
4	Bi_2FeCrO_6	0.948	(Nechache et al. 2015)
5	$(KNbO_3)_{1-x}(BaNi_{0.5}Nb_{0.5}O_{3-\delta})_x$	1.055 ($x = 0.1$)	(Grinberg et al. 2013)

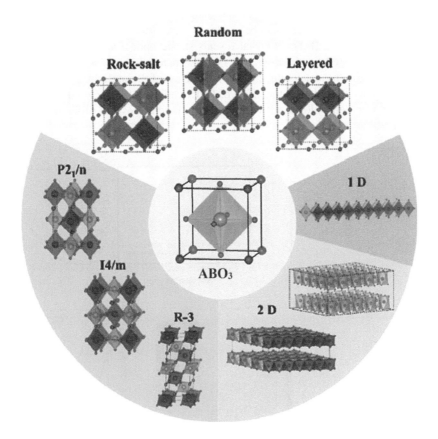

Fig. 10.1 The structural flexibility of oxide perovskite that shows single cubic perovskite (center) to double cubic perovskite with the B-site ordering of rock salt, random, and layered structures. Through (BO6) octahedral tilt, cubic phases can transfer to different tilted phases, with three dominating ones showed on the bottom left. Under large distortion of (BO6) octahedral, the three-dimensional (3-D) connection of octahedra can be broken, leading to 2-D and 1-D perovskite derivatives (Yin et al. 2019)

10.3 OPTOELECTRONIC PROPERTIES OF THE OXIDE PEROVSKITES

Optoelectronic properties of oxide perovskite material are crucial for PV device application, which provide fundamental information on chemical bonding, bandgap, hybridization, absorption coefficient, and many others. Especially, bandgap that shows huge variation from 1.1 to 3.8 eV suggests a wide range of solar spectrum absorption (Grinberg et al. 2013). Among oxide perovskite, $BiFeO_3$ is the most studied material for PVs due to a flexible range of bandgap, i.e. ~2.2-2.8 eV, and exhibited appreciable photoconductivity in the visible spectral region (Yang et al. 2010, Choi et al. 2009). Chen et al. have demonstrated a comparative optical study of 16 nm nanoparticle and epitaxially grown thin film of $BiFeO_3$ through absorption and ellipsometry spectroscopy, which depicts direct bandgap with large absorption coefficient (10^5 cm^{-1}) (Chen et al. 2010). In inset of Figure 10.2(a), the triangle marks indicate a 2.67 eV bandgap of the rhombohedral film, whereas the square marks depict the 2.43 eV bandgap of 16 nm $BiFeO_3$ nanoparticles. The optical response described a small bandgap and matched with the solar, which translates into a broader active wavelength range in a PV device. However, $BiFeO_3$ has a limitation that does not absorb more than 80% of solar spectrum owing to its bandgap. This constraint prompts toward the investigation and development of new material with better semiconducting properties. Double perovskites, i.e. Bi_2FeCrO_6 material, which can be constructed from $BiFeO_3$, offers tremendous potential in PVs applications (Nechache et al.

2012). The bandgap of Bi_2FeCrO_6 is a bit smaller than $BiFeO_3$ due to electron-electron interaction. This signifies that more photocurrent can be produced under illumination. Recently, Wu et al. have studied optical properties of Bi_2FeCrO_6 that reveals strong optical absorption near 600 nm (Wu et al. 2020). The calculated band gap (1.5 eV) of the material shows direct-allowed nature and illustrating their absorption capacity under the main portion of the sunlight spectrum (Figure 10.2b, c). Nechache et al. have investigated the effect of Fe/Cr cationic ordering on the Bi_2FeCrO_6 optical properties by thin film deposition at different growth temperatures and different depositions time (Nechache et al. 2015).

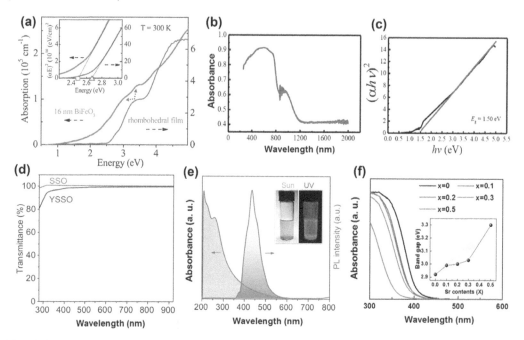

Fig. 10.2   (a) The 300 K absorption coefficient r(E) of 16 nm $BiFeO_3$ nanoparticles as compared with that of an epitaxially grown rhombohedral thin film. Square dot arrows denote the charge transfer excitation split by 0.5 eV. Inset: Direct bandgap analysis of thin films and nanoparticles. The triangle marks the 2.67 eV charge gap of the rhombohedral film, whereas the square marks the 2.43 eV gap of 16 nm $BiFeO_3$ (Chen et al. 2010), (b) Optical absorption of Bi_2FeCrO_6, (c) the corresponding plot of $(\alpha h v)^2$ vs hv curve. b, c from ref (Wu et al. 2020), (d) Transmission spectra of the SSO and YSSO NPs thin films deposited on a quartz substrate, (e) Optical absorption and PL emission spectra of the annealed YSSO NPs (at 900°C for 30 min) in solution, inset: photograph of the corresponding YSSO NPs solutions under visible illumination and UV lamp excitation at 365 nm. (d, e) from ref. (Guo et al. 2019), (f) Absorbance spectra of synthesized $Ba_{1-x}Sr_xSnO_3$ solid solution (0 r × r 0.5) (Shin et al. 2019)

Transparency of the thin layer is also as important as the absorption since many of the oxide perovskites can be utilized as an electron transport layer in the PV device. To pass incident light towards the active layer with the approximately same intensity is essential for the electron transport layer (Zhu et al. 2017). The $SrSnO_3$ perovskite and yttrium-doped $SrSnO_3$ have been used as an ETL in solar cell applications owing to their transparent property under the visible spectrum (Guo et al. 2019). The thin film of $SrSnO_3$ and yttrium doped $SrSnO_3$ on quartz substrate show high transparency (more than 97%) with the minimum optical loss in the visible region (Fig. 10.2d). Further, the yttrium-doped $SrSnO_3$ colloidal solution appeared transparent under visible light (inset of Fig. 10.2e). The absorption peak of this colloidal solution was observed at 257 nm, whereas an emission peak centered at 435 nm. This absorption and emission spectra reveal that one could be achieved at a large Stokes shift. This has been confirmed from the inset image of Figure 10.2(e) upon illumination of UV light (365 nm). One can believe that there is huge potential for a mass-production of ETL thin films for hybrid solar cells. Similar to doping, Sr has substituted in the BaSnO$_3$ perovskite compounds, and altered bandgap of Ba$_{1-x}$Sr$_x$SnO$_3$ was investigated by a UV-Vis spectroscopy. Figure 10.2(f) shows an analysis of UV-Vis spectroscopy, where a gradual blue shift was observed at the absorption edge.

As the concentration of Sr increases from 0 to 0.3, in the perovskite compounds, this leads to widening an optical bandgap from 2.9 to 3.1 eV. Therefore, the band structure also changed, which is attributed to the Sn-O bond hybridization. Especially at $x = 0.5$, a large increment in bandgap was observed, which indicates the abrupt change in structure from a cubic to tetragonal perovskite due to a large amount of Sr substitution (Shin et al. 2019). In short, the Sn-O bond lengths are reduced due to the increment of Sr content (Zhang et al. 2007). Further, it was leading to an increase in the Sn-O bond hybridization and changed the energy band structure.

10.4 PHOTOVOLTAIC APPLICATIONS

The working mechanism of the oxide perovskite is entirely different from p-n junction (conventional) solar cells since these materials show ferroelectric properties. The generated photocurrent can flow without forming an interface or junction that will further drive by the ferroelectric polarization-induced internal electric field (Yuan et al. 2014). When single crystal oxide perovskite is exposed to the illumination, a short-circuit current is generated. This is parallel in the direction of the ferroelectric polarization axis. The photogenerated electrons and holes are further driven toward the electrode (cathode/anode) by the polarization-induced internal electric field that extends over the entire volume. In these materials, the spontaneous electric polarization breaks the strong inversion symmetry, which leads to the separation of photo-excited carriers and allows voltages higher than the bandgap. This might be enabled to enhance device efficiency than a conventional p-n junction solar cell (Yang et al. 2010, Kreisel et al. 2012, Young and Rappe 2012). Oxide perovskites exhibit a strong PV effect with numerous advantages compared to traditional solar energy conversion materials (Alexe and Hesse 2011). The photo-voltage is not limited by bandgap like semiconductor PVs. In addition, oxide perovskite can be abundant, stable, and cheap with tunable optoelectronic properties, making them promising candidates for thin-film PVs. The PV properties were originally investigated over 30 years ago in various oxide perovskite viz. LiNbO$_3$ (Glass et al. 1975), BaTiO$_3$ (Glass et al. 1995), and Pb(Zr,Ti)O$_3$ (Brody and Crowne 1975). Yang et al. showed PV activity in BiFeO$_3$ with strong self-polarization to enhance the PV efficiency (Yang et al. 2009). The I-V measurement was taken at dark and under white light illumination (285 mWcm^{-2}) depicting diode like behavior and PV effect (Figure 10.3a). The open-circuit voltage obtained approximately 0.8-0.9 V, whereas photocurrent density is low, which is around 1.5 mAcm^{-2}. The probable cause behind this has been determined through external quantum efficiency measurements that quantify the short circuit current density as a function of incident wavelength. Figure 10.3(b) shows that the maximum conversion efficiency is around 10% for light energy is more than the BiTiO$_3$ bandgap. Under UV light exposure, it shows maximum current conversion efficiency. However, due to larger bandgap, this material is unable to achieve optimum efficiency as other PV material (Green 2003, Würfel 2005). Similarly, Tiwari et al. have also examined highly pure phase of BiFeO$_3$ synthesized by the solution process method for PV application (Tiwari et al. 2015). Figure 10.3(c) showed BiFeO$_3$/ZnO heterojunction solar cell, which delivered 3.98% PCE with 0.64 V_{oc} and 12.47 mA/cm^2 J_{sc}. This is the highest photoconversion efficiency for BiFeO$_3$ material reported so far in the literature. The external quantum efficiency (EQE) illustrated that the wavelength-dependent efficiency was obtained around 70% between 350 and 560 nm. Further, Figure 10.3(d) shows a long tail at longer wavelengths owing to either high recombination rates or short minority carrier lifetimes. To obtain better performance, further engineering has been performed in the perovskite octahedral sites of the BaTiO$_3$ structure. These efforts in the material are effective in making a low bandgap without affecting their ferroelectric properties. With this suggestion, many researchers have synthesized Bi$_2$FeCrO$_6$ thin films and utilized them successfully in the area of PVs (Nechache et al. 2015).

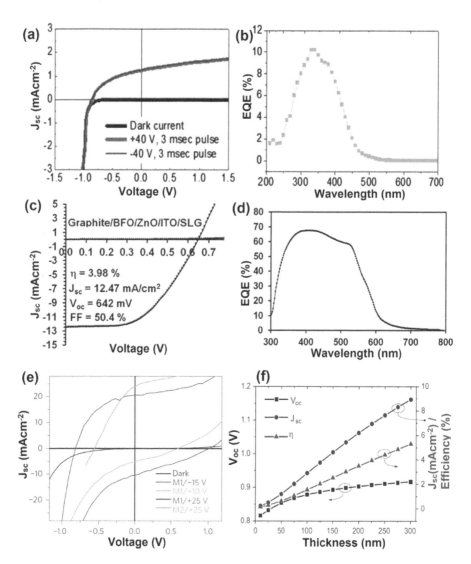

Fig. 10.3 (a) Light and dark J-V measurements completed at 2.85 suns intensity reveals PV effects in BiFeO$_3$ based device. There is no observed change in the light J-V response upon application of an electric field, (b) Average EQE measurements for five different contacts on a single sample reveals efficiencies 10% under illumination with band gap light (a, b from Ref. (Yang et al. 2009)), (c) Room temperature J-V characteristics of BiFeO$_3$ all-oxide and all-solution based solar cell, (d) EQE of BiFeO$_3$ solar cell (c,d from ref. (Tiwari et al. 2015)), (e) J-V characteristics of Bi$_2$FeCrO$_6$ multilayer devices under AM1.5 G illumination (Nechache et al. 2015), (f) Evolution of the V_{oc}, J_{sc} and efficiency simulated with a p-LVO/n-ZnO junction (Jellite et al. 2018)

Recently, Rosei's group demonstrated considerable PVs response in the thin film of Bi$_2$FeCrO$_6$ (Nechache et al. 2011). The power conversion efficiency was obtained over 6% under the illumination (Intensity = 1.5 mW/cm^2) of monochromatic light (635 nm). However, the efficiency in a single junction solar cell is still lower in such Bi$_2$FeCrO$_6$ material/thin film. Further, they demonstrated a new move toward bandgap tuning of double perovskite by engineering the cationic (Cr/Fe) ordering of Bi$_2$FeCrO$_6$ (Nechache et al. 2015). Using the multilayered approach, 8.1% power conversion efficiency of the device has reported under 1.5 AM illumination for Bi$_2$FeCrO$_6$ thin-film solar cells. For a multilayered device, the sequences of growth rates have chosen to deposit the high/ low bandgap material. They grew the largest bandgap layer as a 'front-cell material', and the layer possessing low bandgap was used as 'rear-cell material', for improved efficiency. Figure 10.3(e) shows J-V curves for two Bi_2FeCrO_6 multilayer devices, which were grown at 580°C (device M1) and 720°C (M2) temperatures. These layers possess band gap variation, i.e. 1.6 eV for M1 and 1.3 eV for M2, and hence dissimilar absorption coefficients leading to variation in PCE. The device M1 yielded unprecedented PCE of 8.1% with $J_{sc} = 20.6$ mAcm^{-2}, $V_{oc} = 0.84$, whereas the M2 device depicts 4.3% PCE. The cationic modification in Bi_2FeCrO_6 shows an enormous potential of PVs application.

Other than bismuth based oxide perovskite, $BiMnO_3$, $LaMnO_3$, $SrMnO_3$, and $LaRhO_3$ have been studied for PV application; however, the observed photocurrent was very low in the devices (Chakrabartty et al. 2013, Nakmura et al. 2010, 2015). Recently, Jellite et al. synthesized $LaVO_3$ perovskite using sputtering deposition followed by annealing (Jellite et al. 2018). Prior to the fabrication of the actual device, PV performance has been analyzed theoretically with the variation in thickness of the photo absorbing layer, which is shown in Figure 10.3f. During these studies, other electronic parameters such as electrical mobility, carrier concentration, and electron affinity were kept constant. One can observe from the figure that the V_{OC} does not show much variation with the absorbing layer thickness, whereas the corresponding I_{SC} and efficiency increases. This might be due to the absorption of more photon as thickness increases, especially in the higher wavelength range. As per this simulation, the PCE should reach 5% for 300 nm of the $LaVO_3$ layer. Further, the device was fabricated experimentally, however it showed poor PCE than theoretically calculated. Therefore, they suggested that the material requires some alteration for the PV applications.

Similar to the multi-dimensional $BiFeO_3$, the 1D $BiFeO_3$ nanostructures such as nanowires, nanofibers, nanorod, and nanotubes have drawn significant attention due to enhanced multiferroic and photocatalytic activity (Zhang et al. 2005, Wang et al. 2010, Liu et al. 2011, Gao et al. 2006, Dutta et al. 2010). These structures have been used for various applications including photocatalysis and photovoltaics. In general, such 1D nanostructures are synthesized through different synthesis techniques viz. sol-gel, hydrothermal, electrospinning, etc. (Wang et al. 2010, Liu et al. 2011, Gao et al. 2006, Dutta et al. 2010, Baji et al. 2011, Xie et al. 2008). Recently, Mohan and Subramanian have synthesized the free-standing $BiFeO_3$ nanotubes from electrospinning nanofiber (Mohan and Subramanian 2019). The obtained nanotubes have demonstrated relatively improved photocatalytic activity as compared to the nanofibers and build particle of $BiFeO_3$. Further, Li et al. have synthesized La and Mn co-doped $BiFeO_3$ nanofibers by sol-gel and electrospinning methods (Li et al. 2013). In a while, Fei et al. have explored $BiFeO_3$ nanofibers as a potential candidate for photovoltaic applications (Fei et al. 2015). To measure the photovoltaic response, metal-insulator-metal "sandwich" structures have been designed. Two types of configuration were involved in the measurement system, as shown in Figure 10.4. The left portion in Figure 10.4(a) shows the parallel capacitor configuration, which could be offered a relatively large photocurrent but low voltage whereas, in another approach, laterally aligned inter-digital electrodes (right portion, Figure 10.4a) demonstrate exactly reverse effect, i.e. lower photocurrent but high voltage. In order to achieve hassle-free measurement, they have developed high-quality $BiFeO_3$ nanofibers, which is quite tricky and critical. The schematic set up of $BiFeO_3$ nanofiber-based photovoltaic device measurements have been shown in Figure 10.4(b). The calculated efficiency of the device is quite low (0.2%) due to very little short circuit current density (3.1×10^{-4} mAcm^{-2}), as shown in Figure 10.4(c). A couple of reasons may be associated with these poor results, and one of them is a low output voltage due to a single "sandwich" configuration, which is proportional to film thickness (typically < 1 μm). Another cause may be that ferroelectric polarization gets suppressed by clampic effect. The superior thing out of work is that they could develop and propose the $BiFeO_3$-nanofiber photovoltaic devices with a new configuration of nanofibers/IDE/substrate. They demonstrate the possibility of making a new type of photovoltaic device with laterally aligned BFO nanofibers electrically connected with interdigital electrodes.

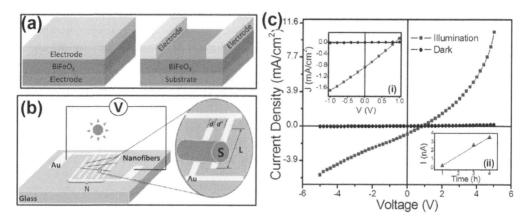

Fig. 10.4 (a) Schematic setup for the measurement of thin film-based photovoltaic devices: left-device with a parallel capacitor configuration, and right-device with laterally aligned interdigital electrodes, (b) Schematic setup for the measurement of random BiFeO$_3$ nanofiber-based photovoltaic devices, (c) The photo-response profile of the nanofibers. The inset (i) shows expanded view of current density behavior around zero-bias. The inset (ii) shows averaged photocurrent after several measurements for different deposition time from 1 to 4 h (a-c from the ref. (Fei et al. 2015))

10.5 METAL DOPED OXIDE PEROVSKITE

Metal doping plays a crucial role in oxide perovskite to enhance their electronic and optical properties. To date, several metal-doped systems have been used in oxide perovskite for different applications (Ghosh et al. 2000, Zhu et al. 2016). Especially, Lanthanide ions viz. Er^{3+}, Yb^{3+}, Pr^{3+}, etc., have been used for up-conversion, and to improve the NIR absorption intensity as a sensitizer of photoanode layer (Zhou et al. 2015, Yao et al. 2016, Lian et al. 2013). Among them, only a few systems have been utilized for PV applications including La-BaSnO$_3$, yttrium-doped SrSnO$_3$, co-doped (Er^{3+}/Yb^{3+}) BaSnO$_3$, etc. as photon absorption layer and electron transport layer (Guo et al. 2019, Zhu et al. 2017, Kumar et al. 2018). Kumar et al. have made an attempt to synthesize pristine and co-doped (Er^{3+}/Yb^{3+}) BaSnO$_3$ nanocrystallites using sol-gel route (wet chemical) (Kumar et al. 2018). The pristine and co-doped (Er^{3+}/Yb^{3+}) BaSnO$_3$ samples were prepared in order to evaluate optical properties. Particularly, stoichiometry of the synthesized co-doped (Er^{3+}/Yb^{3+}) Ba$_{1-x-y}$Er$_x$Yb$_y$SnO$_3$ samples were maintained with the variation of $x + y = 0.10$, and $x/y = 0.11, 0.18$ and 0.25, respectively. The PL behavior of pristine and co-doped Ba$_{1-x-y}$Er$_x$Yb$_y$SnO$_3$ was measured with an excitation wavelength of 350 nm, as shown in Figure 10.5(a). The samples with a different stoichiometry of dopants exhibited a broad peak between 380 and 400 nm that indicates bandgap around 3.2 eV. The observed excitation energy is a bit less than the band edge transition at 3.64-3.68 eV, depicting the presence of extra trap levels just below the conduction band. One can see a small but consistent blue shift in the case of dopant samples for the x/y ratio of 0.11, 0.18, and 0.25, respectively, which reveals band-edge transitions O2p $\rightarrow$ Sn5s.

The up-conversion spectra of the Er^{3+}/Yb^{3+} co-doped Ba$_{1-x-y}$Er$_x$Yb$_y$SnO$_3$ have been evaluated with 980 nm exciting wavelength and 200 mW power. The graphical representations have been shown in Figure 10.5(b). The pristine BaSnO$_3$ has worked as a host matrix, where Yb^{3+} and Er^{3+} ions occupy the Ba^{2+} sites. The photon energy (980 nm) is enough to move to its first excitation level $^2F_{5/2}$ from the ground state, i.e. $^2F_{7/2}$. As soon as the electron reached the $^2F_{5/2}$ level, it is quite possible to get relaxed to the ground state of Er^{3+} ions' adjacent site. The major peaks of emission spectra obtained at 525, 548-559, and 654-679 nm corresponds to $^2H_{11/2} \rightarrow ^4I_{15/2}$ (520 nm), $^4S_{3/2} \rightarrow ^4I_{15/2}$ (548-459 nm) and $^4F_{3/2} \rightarrow ^4I_{15/2}$ (654-679 nm), respectively, which have been shown in Figure 10.5(b). Further, upon careful observation, the intensity of Er^{3+}/Yb^{3+} co-doped BaSnO$_3$ was found quite higher at the

Er^{3+}/Yb^{3+} ratio of 0.25 that infers an optimum emission. These optoelectronic properties have been utilized to fabricate Ba(Er/Yb)SnO$_3$ based DSSC to achieve better PV performance. The schematic representation of various components of DSSC has been shown in Figure 10.5(c). TiO$_2$/Ba(Er/Yb)SnO$_3$ photoanode based DSSC depicts 4.2% efficiency as compared to the pristine BaSnO$_3$ (PCE = 3%) with the enhanced short circuit current density. About 40% more efficiency than the pristine oxide perovskite has been observed (Fig. 10.5, d). The improved electrical conductivity of co-doped BaSnO$_3$ is also an important aspect, which could be responsible for achieving such a huge improvement in the TiO$_2$/Ba(Er/Yb)SnO$_3$ devices.

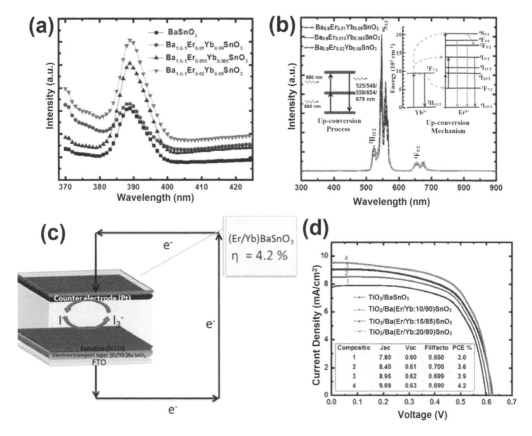

Fig. 10.5 (a) Photo-luminescence spectra, (b) Up-conversion spectra of pristine and Er^{3+}/Yb^{3+} co-doped BaSnO$_3$ nano-crystallites, (c) Schematic representation of the device, (d) J-V curves of pristine and Er^{3+}/Yb^{3+} co-doped BaSnO$_3$ nanocrystallites based DSSC (a-d, from ref. (Kumar et al. 2018))

In a very recent study, Guo et al. have demonstrated the utility of un-doped and Yttrium-doped SrSnO$_3$ as ETL in planner perovskite solar cell (Guo et al. 2019). The fabricated device with a schematic of the stack is shown in Figure 10.6 along with a cross-sectional SEM view. The analogous energy level diagram of each functional layer in the fabricated device has been shown in Figure 10.6. SrSnO$_3$ and Y-SrSnO$_3$ have been used as an ETL, whereas Spiro-OMe TAD acts as a hole transport layer. One can see in the figure that the current density-voltage has been measured in the reverse and forward condition. SrSnO$_3$ based device has shown the device efficiency of 16.9% in reversed condition and 14.2% in forward measurement condition. Similarly, Y-SrSnO$_3$ based device has shown enhanced device performance up to 19% in reverse scanned condition. Under forward scanning, device does not reduce efficiency much, as it was observed around 18.6%. In addition, the devices demonstrated the output, including steady current and good stability for

1000 hrs. Therefore, researchers have illustrated that the Y-SrSnO$_3$ oxide perovskite can be a potential alternative to fabricate economical, wide-scale, and high-performing planar perovskite solar cells in the near future.

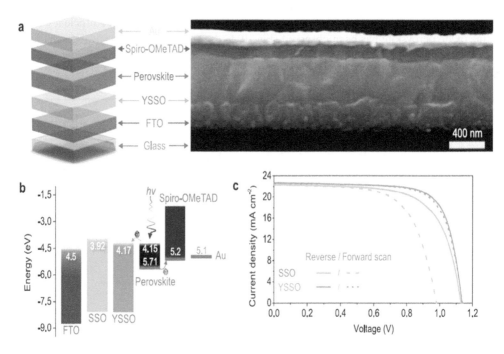

Fig. 10.6 (a) Schematic of the device architecture (left) and a colored cross-section SEM image (right) of the completed device (Stack: glass/FTO/YSSO/Perovskite/Spiro-OMeTAD/Au), (b) The corresponding energy band diagram relative to vacuum level. (c) Current-density (J-V) curves of the best-performing devices based on SSO and YSSO ETLs measured at: forward scan (0-1.2 V) and reverse scan (1.2-0 V) at a scan rate 0.1 Vs^{-1} (a-c, from ref. (Guo et al. 2019))

10.6 CONCLUSION AND OUTLOOK

The novel functional material is an immediate need in the direction of research and development of renewable energy. The perovskite material has become a star in the photovoltaic's era, owing to their unique but compositional flexibility. Stability of oxide perovskites and their derivative structures have been investigated as a class of multipurpose materials for different applications such as photocatalysis and photovoltaics. An alteration in perovskite structure causes the change in optoelectronic properties and subsequent applications. Here, we attempt to discuss how the oxide perovskite properties are tuned for specific applications. To enhance photovoltaic device efficiency, structural and compositional modifications have been discussed towards clean energy utilization. The high absorption coefficient in the visible region and transparent nature in UV and NIR regions make them perfect to utilize as a photo absorber and electron transport layer, respectively, in the solar cell devices. The structural properties and their application relationship of perovskite will help in the future to design novel functional material that may overcome the current challenges. Oxide perovskites behave differently than the conventional p-n junction solar cell due to spontaneous electric polarization. This effect separates photogenerated charge carriers quickly and produces high open-circuit voltage. Double perovskite Bi$_2$FeCrO$_6$ also shows effective progress, since it can tune the bandgap by varying order of cations. The wide ranges of bandgap have attracted more attention in multilayer solar cells to absorb a specific wavelength of photons. Similarly, metal-doped oxide perovskite has engineered the optoelectronic properties such as absorption range, photoemission, and their up-conversion etc. to achieve optimum photoconversion efficiency. For instance, the rare earth metal such as Er^{3+}/Yb^{3+} co-doped barium stannate demonstrated approximately 40% improvement in the solar cell as compared to the pristine barium stannate. Another example is of Yttrium-doped $SrSnO_3$ as an electron transport layer. The device stability has been improved up to 1000 hrs along with the steady current. Subsequently, the overall performance of the device has been increased significantly.

However, a few challenges are still coming into consideration, which are needed to be addressed as soon as possible for large scale productions. In the oxide perovskite, oxygen vacancies/deficiencies play a significant role in electrical conductivity, though their effect is not fully understood. Under such circumstances, it is quite challenging to identify the appropriate applications of oxide perovskites.

10.7 ACKNOWLEDGMENTS

Sanjay Sahare and Manjeet Kumar contributed equally to this work. The authors would like to acknowledge the financial supports of the post-doctoral research program (2018) through the Incheon National University.

10.8 REFERENCES

Alexe, M. and D. Hesse. 2011. Tip-enhanced photovoltaic effects in bismuth ferrite. Nat. Commun. 2: 1.

Anderson, M.T., K.B. Greenwood, G.A. Taylor and K.R. Poeppelmeier. 1993. B-cation arrangements in double perovskites. Prog. Solid State Chem. 22: 197.

Baji, A., Y.W. Mai, Q. Li, S.C. Wong, Y. Liu and Q.W. Yao. 2011. One-dimensional multiferroic bismuth ferrite fibers obtained by electrospinning techniques. Nanotechnology 22: 235702.

Balakrishna, R.G., S.M. Kobosko and P.V. Kamat. 2018. Mixed halide perovskite solar cells. Consequence of iodide treatment on phase segregation recovery. ACS Energy Lett. 3: 2267.

Bhalla, A.S., R. Guo and R. Roy. 2000. The perovskite structure-a review of its role in ceramic science and technology. Mater. Res. Innovations. 4: 3.

Brody, P.S. and F. Crowne. 1975. Mechanism for the high voltage photovoltaic effect in ceramic ferroelectrics. J. Electron. Mater. 4: 955.

Chakrabartty, J.P., R. Nechache, C. Harnagea and F. Rosei. 2013. Photovoltaic effect in multiphase Bi-Mn-O thin films. Opt. Express. 22: A80.

Chen, L., Y.Y. Tan, Z.X. Chen, T. Wang, S. Hu, Z.A. Nan, et al. 2010. Size-dependent infrared phonon modes and ferroelectric phase transition in $BiFeO_3$ nanoparticles. Nano Lett. 10: 4526.

Chen, L., Y.Y. Tan, Z.X. Chen, T. Wang, S. Hu, Z.A. Nan, et al. 2019. Toward long-term stability: single-crystal alloys of cesium-containing mixed cation and mixed halide perovskite. J. Am. Chem. Soc. 141: 1665.

Chilvery, A., S. Palwai, P. Guggilla, K. Wren and D. Edinburgh. 2019. Perovskite materials: recent advancements and challenges. In perovskite materials, devices and integration. IntechOpen.

Cho, Y., A.M. Soufiani, J.S. Yun, J. Kim, D.S. Lee, J. Seidel, et al. 2018. Mixed 3D-2d passivation treatment for mixed-cation lead mixed-halide perovskite solar cells for higher efficiency and better stability. Adv. Energy Mater. 8: 1703392.

Choi, T., S. Lee, Y. Choi, V. Kiryukhin and S.-W. Cheong. 2009. Switchable ferroelectric diode and photovoltaic effect in $BiFeO_3$. Science 324: 63.

Dutta, D.P., O.D. Jayakumar, A.K. Tyagi, K.G. Girija, C.G.S. Pillai and G. Sharma. 2010. Effect of doping on the morphology and multiferroic properties of $BiFeO_3$ nanorods. Nanoscale 2: 1149.

Fan, Z., J. Xiao, K. Yao, K. Zeng and J. Wang. 2015. Ferroelectric polarization relaxation in $Au/Cu_2O/ZnO/BiFeO_3/Pt$ heterostructure. Appl. Phys. Lett. 106: 102902.

Fei, L., Y. Hu, X. Li, R. Song, L. Sun, H. Huang, et al. 2015. Electrospun bismuth ferrite nanofibers for potential applications in ferroelectric photovoltaic devices. ACS Appl. Mater. Interfaces. 7: 3665.

Ferguson, V., S.R.P. Silva and W. Zhang. 2019. Carbon materials in perovskite solar cells: Prospects and future challenges. Energy Environ. Mater. 2: 107.

Frost, J.M., K.T. Butler, F. Brivio, C.H. Hendon, M.V. Schilfgaarde and A. Walsh. 2014. Atomistic origins of high-performance in hybrid halide perovskite solar cells. Nano Lett. 14: 2584.

Fu, D. and M. Itoh. 2011. Ferroelectricity in silver perovskite oxides. ferroelectrics – material aspects.

Gao, F., Y. Yuan, K.F. Wang, X.Y. Chen, F. Chen, J.M. Liu, et al. 2006. Preparation and photoabsorption characterization of $BiFeO_3$ nanowires. Appl. Phys. Lett. 89: 102506.

Ghosh, V.J., B. Nielsen and T. Friessnegg. 2000. Identifying open-volume defects in doped and undoped perovskite-type $LaCoO_3$, $PbTiO_3$, and $BaTiO_3$. Phys. Rev. B. 61: 207.

Glass, A.M., D. Von der Linde, D.H. Auston and T.J. Negran. 1975. Excited state polarization, bulk photovoltaic effect and the photorefractive effect in electrically polarized media. J. Electron. Mater. 4: 915.

Glass, A.M., D.V.D. Linde and T.J. Negran. 1995. High-voltage bulk photovoltaic effect and the photorefractive process in $LiNbO_3$. pp. 371-373. *In*: Y. Pochi and G. Claire [eds.]. Landmark Papers On Photorefractive Nonlinear Optics, World Scientific.

Green, M.A. 2003. Third Generation Photovoltaics. Advanced Solar Energy Conversion. Springer-Verlag, Berlin Heidelberg.

Grinberg, I., D.V. West, M. Torres, G. Gou, D.M. Stein, L. Wu, et al. 2013. Perovskite oxides for visible-light-absorbing ferroelectric and photovoltaic materials. Nature 503: 509.

Guo, H., H. Chen, H. Zhang, X. Huang, J. Yang, B. Wang, et al. 2019. Low-temperature processed yttrium-doped $SrSnO_3$ perovskite electron transport layer for planar heterojunction perovskite solar cells with high efficiency. Nano Energy 59: 1.

Han, Y., S. Meyer, Y. Dkhissi, K. Weber, J.M. Pringle, U. Bach, et al. 2015. Degradation observations of encapsulated planar $CH_3NH_3PbI_3$ perovskite solar cells at high temperatures and humidity. J. Mater. Chem. A. 3: 8139.

Huang, W., C. Harnagea, D. Benetti, M. Chaker, F. Rosei and R. Nechache. 2017. Multiferroic Bi_2FeCrO_6 based *p-i-n* heterojunction photovoltaic devices. J. Mater. Chem. A. 5: 10355.

Jellite, M., J.L. Rehspringer, M.A. Fazio, D. Muller, G. Schmerber, G. Ferblantier, et al. 2018. Investigation of $LaVO_3$ based compounds as a photovoltaic absorber. Solar Energy. 162: 1.

Jing, Q., M. Zhang, X. Huang, X. Ren, P. Wang and Z. Lu. 2017. Surface passivation of mixed-halide perovskite $CsPb$ $(Br_xI_{1-x})_3$ nanocrystals by selective etching for improved stability. Nanoscale 9: 7391.

Juarez-Perez, E.J., R.S. Sanchez, L. Badia, G.G.-Belmonte, Y.S. Kang, I.M.-Sero, et al. 2014. Photoinduced giant dielectric constant in lead halide perovskite solar cells. J. Phys. Chem. lett. 5: 2390.

Ke, W. and M.G. Kanatzidis. 2019. Prospects for low-toxicity lead-free perovskite solar cells. Nat. Commun. 10: 965.

Kim, H.S., S.H. Im and N.G. Park. 2014. Organolead halide perovskite: New horizons in solar cell research. J. Phys. Chem. C. 118: 5615.

Kreisel, J., M. Alexe and P.A. Thomas. 2012. A photoferroelectric material is more than the sum of its parts. Nat. Mater. 11: 260.

Kumar, A.A., J. Singh, D.S. Rajput, A. Placke, A. Kumar and J. Kumar. 2018. Facile wet chemical synthesis of Er^{3+}/Yb^{3+} co-doped $BaSnO_3$ nanocrystallites for dye-sensitized solar cell application. Mater. Sci. Semiconductor Proc. 83: 83.

Li, B., C. Wang, W. Liu, M. Ye and N. Wang. 2013. Multiferroic properties of La and Mn co-doped $BiFeO_3$ nanofibers by sol-gel and electrospinning technique. Mater. Lett. 90: 45.

Lian, H., Z. Hou, M. Shang, D. Geng, Y. Zhang and J. Lin. 2013. Rare earth ions doped phosphors for improving efficiencies of solar cells. Energy 57: 270.

Liu, B., B. Hu and Z. Du. 2011. Hydrothermal synthesis and magnetic properties of single-crystalline $BiFeO_3$ nanowires. Chem. Commun. 47: 8166.

Liu, B., C.H. Wu, J. Miao and P. Yang. 2014. All inorganic semiconductor nanowire mesh for direct solar water splitting. ACS Nano 8: 11739.

Liu, J.W., G. Chen, Z.H. Li and Z.G. Zhang. 2007. Hydrothermal synthesis and photocatalytic properties of $ATaO_3$ and $ANbO_3$ (A = Na and K). Int. J. Hydrogen Energy. 32: 2269.

Loi, M.A. and J.C. Hummelen. 2013. Hybrid solar cells: Perovskites under the sun. Nat. Mater. 12: 1087.

Loureiro, F.J., N. Nasani, G.S. Reddy, N.R. Munirathnam and D.P. Fagg. 2019. A review on sintering technology of proton conducting $BaCeO_3$-$BaZrO_3$ perovskite oxide materials for protonic ceramic fuel cells. J. Power Sources. 438: 226991.

Lufaso, M.W., P.W. Barnes and P.M. Woodward. 2006. Structure prediction of ordered and disordered multiple octahedral cation perovskites using SPuDS. Acta Crysta. Sect. B: Struct. Sci. 62: 397.

McMeekin, D.P., G. Sadoughi, W. Rehman, G.E. Eperon, M. Saliba, M.T. Hörantner, et al. 2016. A mixed-cation lead mixed-halide perovskite absorber for tandem solar cells. Science 351(6269): 151-155.

Mohan, S. and B. Subramanian. 2019. A strategy to fabricate bismuth ferrite (BiFeO$_3$) nanotubes from electrospun nanofibers and their solar light-driven photocatalytic properties. RSC Adv. 3: 23737.

Nakamura, M., A. Sawa, J. Fujioka, M. Kawasaki and Y. Tokura. 2010. Interface band profiles of Mott-insulator/Nb: SrTiO$_3$ heterojunctions as investigated by optical spectroscopy. Phys. Rev. B. 82: 201101(R).

Nakamura, M., Y. Krockenberger, J. Fujioka, M. Kawasaki and Y. Tokura. 2015. Perovskite LaRhO$_3$ as a p-type active layer in oxide photovoltaics. Appl. Phys. Lett. 106: 072103.

Nechache, R., C. Harnagea, S. Licoccia, E. Traversa, A. Ruediger, A. Pignolet, et al. 2011. Photovoltaic properties of Bi$_2$FeCrO$_6$ epitaxial thin films. App. Phys. Lett. 98: 202902.

Nechache, R., C. Harnagea and A. Pignolet. 2012. Multiferroic properties-structure relationships in epitaxial Bi$_2$FeCrO$_6$ thin films: Recent developments. J. Phys.: Cond. Matter. 24: 096001.

Nechache, R., C. Harnagea, S. Li, L. Cardenas, W. Huang, J. Chakrabartty, et al. 2015. Bandgap tuning of multiferroic oxide solar cells. Nat. Photonics. 9: 61.

Nuraje, N. and K. Su. 2013. Perovskite ferroelectric nanomaterials. Nanoscale 5: 8752.

Pérez-Tomás, A., A. Mingorance, D. Tanenbaum and M. Lira-Cantú. 2018. Metal oxides in photovoltaics: all-oxide, ferroic, and perovskite solar cells. pp. 267-356. *In*: M. Lira-Cantu [ed.]. The future of semiconductor oxides in next-generation solar cells. Elsevier.

Polman, A., M. Knight, E.C. Garnett, B. Ehrler and W.C. Sinke. 2016. Photovoltaic materials: Present efficiencies and future challenges. Science 352: aad4424.

Shin, S.S., J.H. Suk, B.J. Kang, W. Yin, S.J. Lee, J.H. Noh, et al. 2019. Energy-level engineering of the electron transporting layer for improving open-circuit voltage in dye and perovskite-based solar cells. Energy Environ. Sci. 12: 958.

Sinke, W.C. 2019. Development of photovoltaic technologies for global impact. Renewable Energy 138: 911.

Stoumpos, C.C., C.D. Malliakas and M.G. Kanatzidis. 2013. Semiconducting tin and lead iodide perovskites with organic cations: Phase transitions, high mobilities, and near-infrared photoluminescent properties. Inorg. Chem. 52: 9019.

Takahashi, Y., R. Obara, Z.Z. Lin, Y. Takahashi, T. Naito, T. Inabe, et al. 2011. Charge-transport in tin-iodide perovskite CH$_3$NH$_3$SnI$_3$: Origin of high conductivity. Dalton Trans. 40: 5563.

Tiwari, D., D.J. Fermin, T.K. Chaudhuri and A. Ray. 2015. Solution processed bismuth ferrite thin films for all-oxide solar photovoltaics. J. Phys. Chem. C. 119: 5872.

Wang, J., M. Li, X. Liu, L. Pei, J. Liu, B. Yu, et al. 2010. Synthesis and ferroelectric properties of Nd doped multiferroic BiFeO$_3$ nanotubes. Chin. Sci. Bull. 55: 1594.

Wang, Z., D.P. McMeekin, N. Sakai, S. van Reenen, K. Wojciechowski, J.B. Patel, et al. 2017. Efficient and air-stable mixed-cation lead mixed-halide perovskite solar cells with n-doped organic electron extraction layers. Adv. Mater. 29: 1604186.

Won, C.J., Y.A. Park, K.D. Lee, H.Y. Ryu and N. Hur. 2011. Diode and photocurrent effect in ferroelectric BaTiO$_{3-\delta}$. J. Appl. Phys. 109: 084108.

Wu, H., Z. Pei, W. Xia, Y. Lu, K. Leng and X. Zhu. 2020. Structural, magnetic, dielectric and optical properties of double-perovskite Bi$_2$FeCrO$_6$ ceramics synthesized under high pressure. J. Alloys Compounds. 819: 153007.

Würfel, P. 2005. Physics of solar cells. From principles to new concepts. WILEY VCH Verlag GmbH and Co, KGaA.

Xie, S.H., J.Y. Li, R. Proksch, Y.M. Liu, Y.C. Zhou, Y.Y. Liu, et al. 2008. Nanocrystalline multiferroic BiFeO$_3$ ultrafine fibers by sol-gel based electrospinning. Appl. Phys. Lett. 93: 222904.

Yang, S.Y., L.W. Martin, S.J. Byrnes, T.E. Conry, S.R. Basu, D. Paran, et al. 2009. Photovoltaic effects in BiFeO$_3$. Appl. Phys. Lett. 95: 062909.

Yang, S.Y., J. Seidel, S.J. Byrnes, P. Shafer, C.H. Yang, M.D. Rossell, et al. 2010. Above-bandgap voltages from ferroelectric photovoltaic devices. Nature Nanotechnol. 5: 143.

Yao, N., J. Huang, K. Fu, X. Deng, M. Ding and X. Xu. 2016. Rare earth ion doped phosphors for dye-sensitized solar cells applications. RSC Adv. 6: 17546.

Yin, W.J., B. Weng, J. Ge, Q. Sun, Z. Li and Y. Yan. 2019. Oxide perovskites, double perovskites and derivatives for electrocatalysis, photocatalysis, and photovoltaics. Energy Environ. Sci. 12: 442.

Young, S.M. and A.M. Rappe. 2012. First principles calculation of the shift current photovoltaic effect in ferroelectrics. Phys. Rev. Lett. 109: 16601.

Yuan, Y., Z. Xiao, B. Yang and J. Huang. 2014. Arising applications of ferroelectric materials in photovoltaic devices. J. Mater. Chem. A. 2: 6027.

Zhang, S.T., M.H. Lu, D. Wu, Y.F. Chen and N.B. Ming. 2005. Larger polarization and weak ferromagnetism in quenched $BiFeO_3$ ceramics with a distorted rhombohedral crystal structure. Appl. Phys. Lett. 87: 262907.

Zhang, W., J. Tang and J. Ye. 2007. Structural, photocatalytic, and photophysical properties of perovskite $MSnO_3$ (M = Ca, Sr, and Ba) photocatalysts. J. Mater. Res. 22: 1859.

Zheng, F., Y. Xin, W. Huang, J. Zhang, X. Wang, M. Shen, et al. 2014. Above 1% efficiency of a ferroelectric solar cell based on the $Pb(Zr,Ti)O_3$ film. J. Mater. Chem. A. 2: 1363.

Zhou, B., B. Shi, D. Jin and X. Liu. 2015. Controlling up-conversion nano-crystals for emerging applications, Nat. Nanotechnol. 10: 924.

Zhu, L., J. Ye, X. Zhang, H. Zheng, G. Liu, X. Pan, et al. 2017. Performance enhancement of perovskite solar cells using a La-doped $BaSnO_3$ electron transport layer. J. Mater. Chem. A 5: 3675.

Zhu, Y., W. Zhou, J. Sunarso, Y. Zhong and Z. Shao. 2016. Phosphorus-doped perovskite oxide as highly efficient water oxidation electrocatalyst in alkaline solution. Adv. Funct. Mater. 26: 5862.

11

Nanostructured Li_2MSiO_4 (M=Fe, Mn) Cathode Material for Li-ion Batteries

A.L. Sharma[1], Shweta Tanwar[1], Nirbhay Singh[1,3],
Vijay Kumar[4] and Anil Arya[2,1,*]

11.1 INTRODUCTION

Due to the crucial demand for energy resources across the globe, it becomes important to develop clean and renewable resources over the traditional ones. The rechargeable battery seems to be a more appropriate device for energy storage and convenient use (transport and portable electronic sectors) as per demand. We are fully aware of the importance of rechargeable batteries for daily life use. In some of these battery or energy systems, ceramic is playing an important role in their performance. It is the rechargeable battery for electrochemical battery, energy storage device, and conversion of chemical energy into electrical energy. The energy storage device can be any rechargeable battery such as charged electrical energy, i.e. charge can store also convert chemical energy into electrical energy and vice versa. (Zhang et al. 2015a, Arya and Sharma 2020). These batteries are only one-time use, i.e. there is only one discharge process in the battery. So basically rechargeable batteries, lithium-ion batteries have made significant progress in the rechargeable battery for safety, capacity, cycling stability, non-toxic, and high energy density as compared to the other rechargeable batteries (Bensalah and Dawood 2016, Arya and Sharma 2019). The lithium-ion battery has since become part of the portable electronics and its demands for the coming future. The lithium-ion batteries would also replace electric vehicles such as the trains, buses, and cars to fight against pollution, as they have the highest energy density in comparison to all other secondary batteries. The secondary batteries are environmentally friendly and safe than other sources that can be used for a long time. Present-day battery technologies are being outpaced by the ever-increasing power demands from new applications. It is inherently safe and has to integrate the concept of environmental sustainability. The present chapter will shine a light on the cathode materials and characteristic parameters of cathode material for the battery. Then advantages of orthosilicates materials will be discussed over other cathode materials. Different synthesis methods and important characterization techniques are also summarized. Overall, the chapter provides brief information and recent developments in orthosilicates based cathode material.

[1] Department of Physics, Central University of Punjab, Bathinda-151001, Punjab, India.
[2] Department of Physics, National Institute of Technology, Kurukshetra-136119, Haryana, India.
[3] Department of Physics, Babasaheb Bhimrao Ambedkar University, Lucknow.
[4] Department of Physics, Institute of Integrated and Honors Studies, Kurukshetra-136119, Haryana, India.
* Corresponding author: aniljaglan0581@yahoo.com

11.1.1 Fundamentals of Battery

The first electrochemical battery was built by A. Volta in 1800 and named as the voltaic pile which consists of a stack of Cu and Zn plates. Its ability to produce only a small current motivated Daniell to develop the first practical source of electricity in 1836, which comprises of a copper pot filled with a copper sulfate solution (Boulabiar 2004). After that progress in the primary battery was significant (Nagaura and Tozawa 1990), the lead-acid battery was invented followed by Ni-Cd battery, and then Ni-MH battery (Fig. 11.1). The first commercialized lithium-ion battery was made by Sony in 1991, it exchanged Li^+ ion between layered oxide cathode and graphite (Linden and Reddy 2002), with the transition metal. The average voltage is 3.8 V, which is much better than older lead-acid batteries. But this is not according to Moore's law in Electronics (memory capacity doubles every 18 months). So, a lot of work is to be done in the development of material for energy devices (Armand and Tarascon 2008).

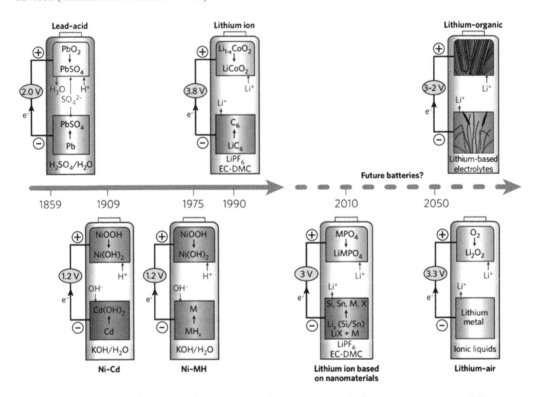

Fig. 11.1 Battery chemistry over the years (Reprinted with permission from Armand and Tarascon, Nature, Springer Publishing, 2008, 451: 652)

A secondary battery is a group of cells for the electrical energy in which the cell, after discharging, is restored to the original charged state, the electric current flowing in the opposite direction to flow of current, when the cell was discharged. The lithium-ion battery comprises of the key component anode, cathode, electrolyte, and separator. The separator plays the role of a physical separation for electrodes while the electrolyte acts as the medium for ion transportation. At present, the liquid electrolyte is used in the commercial system. The cathode material plays a very crucial role in determining the lithium-ion battery performance parameters, such as potential, thermal stability, and capacity. These parameters, directly and indirectly, contribute to the electrochemical reaction (Cheng and Chen 2012, Yi et al. 2015). Thus, the properties of cathode material need to be improved to achieve an enhancement in lithium-ion battery performance (Wu and Yushin 2017).

11.1.2 Principle for Li-ion Battery

The commercial lithium-ion battery was based on the lithium metal oxide cathode and carbon anode (Armand and Tarascon 2008). In the charging process, lithium-ion is extracted from the cathode and moves to anode via the electrolyte. In the discharge, process lithium-ion is released from anode toward the cathode. In both cases, the electron moves through the external circuit as the separator blocks the path of the electron (Fig. 11.2). The general battery configuration comprises three components: cathode, anode, and electrolyte cum separator. The cathode (generally LiCoO$_2$) is coated on aluminum foil, and the anode (graphite) is coated on copper foil. The electrolyte is a mixture of lithium salt in organic solvents. Besides, a separator is used to separate the electrodes and is made up of polyethylene or propylene (Tarascon and Armand 2011).

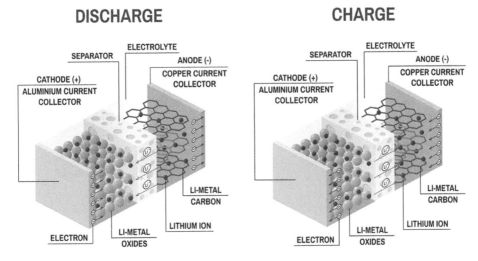

Fig. 11.2 Discharge and charge of Li-ion battery (Source: *sivVector/Shutterstock.com*)

Important configurations are shown in Fig. 11.3i. Figure 11.3ii shows the various structures of the commercial battery system.

11.1.3 Characteristics of Cathode Materials

A cathode is a very crucial component of LIB and decides the overall battery performance. The overall cell voltage of the battery is determined from the redox energies' difference of the anode and cathode. For cell operation, the cathode energy should be lower and anode energy should be higher (Manthiram 2020). The key requirements for a material to be successfully used as a cathode in a rechargeable lithium battery are as follows (Whittingham 2004):

1. The material contains a readily reducible/oxidizable ion, for example, a transition metal.
2. The material reversibly reacts with lithium.
   (a) This dictates an intercalation-type reaction in which the host structure essentially does not change as lithium is added.
3. The material reacts with lithium with the high free energy of the reaction.
   (a) High capacity, preferably at least one lithium per transition metal.
   (b) High voltage, preferably around 4 V (as limited by the stability of electrolyte).
   (c) This leads to high-energy storage.

4. The material reacts with lithium very rapidly both on insertion and removal.
    (a) This leads to high power density, which is needed to replace the Ni/Cd battery or for batteries that can be recharged using HEV regenerative braking.
5. The material is a good electronic conductor, preferably a metal.
    (a) This allows for the easy addition or removal of electrons during the electrochemical reaction.
    (b) This allows for a reaction at all contact points between the cathode active material and the electrolyte rather than at ternary contact points between the cathode active material, the electrolyte, and the electronic conductor (such as carbon black).
    (c) This minimizes the need for inactive conductive diluents, which takes away from the overall energy density.
6. The material is stable, i.e. no structure change during over-discharge, and overcharge.
7. The material is low cost and environmentally benign.

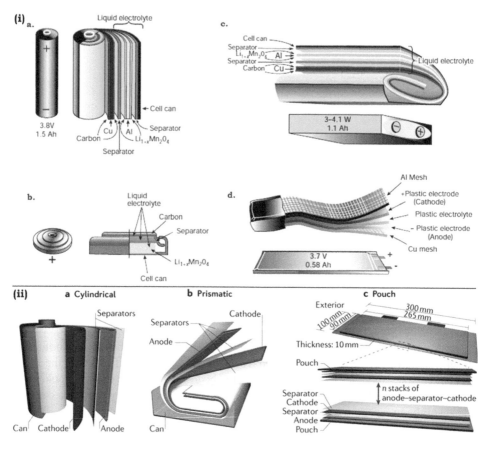

Fig. 11.3 (i) Schematic drawing showing the shape and components of various Li-ion battery configurations. a) Cylindrical; b) coin; c) prismatic; and d) thin and flat. Note the unique flexibility of the thin and flat plastic LiION configuration; in contrast to the other configurations, the PLiION technology does not contain free electrolyte (Reprinted with permission from Armand and Tarascon, *Nature, Springer Publishing, 2011, 414: 359-367*), (ii) Three representative commercial cell structures. a | Cylindrical-type cell. b | Prismatic-type cell. c | Pouch-type cell. The pouch dimensions are denoted, along with the internal configuration for *n* anode-separator-cathode stacks. (Reprinted with permission from Choi and Aurbach, *Nature, Springer Publishing, 2016, 1: 16013*)

The important characteristics for a good battery are shown in Fig. 11.4a, and characteristics of cathode materials are summarized in Fig. 11.4b.

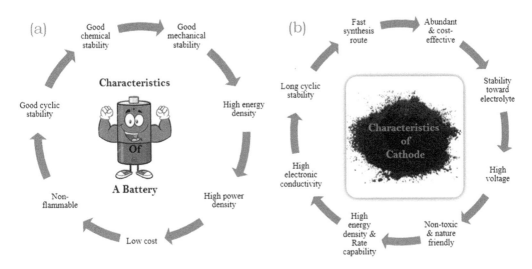

Fig. 11.4 (a) Characteristics of a battery, (b) Key characteristics of cathode material

Before the selection of any material for the electrode, its theoretical capacity (C_t) need to be interpreted. Faraday's first law of electrochemistry states that '1 gram equivalent weight of the material will deliver 96487 coulombs (or 26.8 Ah)'. It is associated with the number of active electrons (number of Li ions accommodated by host lattice) and the molar weight of the material. The smaller molecular weight and the accommodation (insertion/de-insertion) of more number of electrons is the key to achieve high theoretical capacity (Lee et al. 2015, Singh and Setiawan 2013). The calculation of theoretical capacity facilitates a quick comparison with the existing electrodes. The C_t is expressed as

$$C_t = \frac{n \times F}{3.6 \times M} \tag{1}$$

Here, n is the number of reactive electrons per formula unit, M is the molar weight of materials, and F is the Faraday constant.

11.1.4 Structure of Cathode Material

The cathode crystal structure is classified into three types of lithium insertion compounds as shown in Fig. 11.5. The classification is done based on ion diffusion and activation energy that promotes lithium-ion transport in the material (Goodenough 1994).

11.1.4.1 Olivine Compounds

The olivine family lithium iron phosphate crystallized is an orthorhombic group. It has a distorted hexagonal close-packed (hcp) structure. The HCP oxygen-containing framework has lithium (Li) and iron (Fe) in half of the octahedral sites, and one-eight of the tetrahedral site have Phosphorus (P) ion. Figure 11.5 shows the structure of olivine and sites for Li-ion removal. The FeO$_6$ octahedral having corner-shape are linked in the bc-plane, and LiO$_6$ forms an edge-sharing chain with a b-axis. The PO$_4$ tetrahedral group bridges the nearby layers of FeO$_6$ octahedral by sharing the common edge with one of FeO$_6$ octahedral and two edges of LiO$_6$ octahedral (Julien et al. 2014, Zaghib et al. 2007).

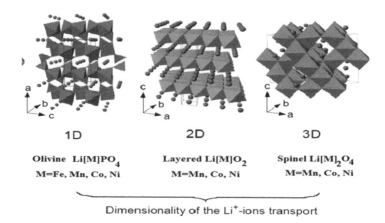

1D   **2D**   **3D**

Olivine Li[M]PO$_4$ **Layered Li[M]O$_2$**   **Spinel Li[M]$_2$O$_4$**
M=Fe, Mn, Co, Ni   **M=Mn, Co, Ni**   **M=Mn, Co, Ni**

Dimensionality of the Li$^+$-ions transport

Fig. 11.5 Crystal structure of the three lithium-insertion compounds in which the Li$^+$ ions are mobile through the 1-D (olivine), 2-D (layered) and 3-D (spinel) frameworks (Reprinted with permission from Julien et al., *Inorganics, MDPI Publishing, 2014, 2: 132-154*)

11.1.4.2 Layered Compounds

The layered compounds are two dimensional: Li(M)O$_2$ where (M = Co, Ni, Mn) and chemical structure is the layered α-NaFeO$_2$ type with the oxygen ion closed packed in the arrangement of cubic site. Lithium-ion and transition metal (e.g. Co) occupies the octahedral site, with the layers ABCABCABC....... this sequence is known as O$_3$ type crystal structure in Fig. 11.5. At high-temperature, LiCoO$_2$ possesses symmetric rhombohedral structure, in which Li is located at 3a, O is in 6c site, and Ni in 3b site (Thackeray 1995, Julien 2000). For maximum charge delocalization, the equivalent environment for cobalt ion is favorable and achieved by the ordering of Li and stacking sequence. One issue with the LiCoO$_2$ is that its Li-ion gets dissolved in the electrolyte and oxygen is released, which threatens the battery safety. So, to prevent the electrode breakdown, surface modification approach (using ZrO$_2$, Al$_2$O$_2$, TiO$_2$, etc) is used by coating the cathode (Julien 2000, Li et al. 2013, Li et al. 2014a, Li et al. 2014b).

11.1.4.3 Spinel Compound

The first report on the spinel oxides (chemical formula: LiMn$_2$O$_4$) was given by Michael Thackeray in the early 1980s (Thackeray et al. 1983). These compounds have various advantages over others such as eco-friendly, cost-effective, and better safety. The Li-ion diffusion occurs via the 3D path provided by the MnO$_2$ framework in the spinel structure (Fig. 11.5). Table 11.1 compares the specific capacity and average potential for spinel, olivine, and layered framework.

TABLE 11.1 Electrochemical characteristics of the three classes of insertion compounds

Framework	Compound	Specific Capacity^a (mAh g^{-1})	Average Potential (V vs. Li^0/Li$^+$)
Layered	LiCoO$_2$	272 (140)	4.2
	LiNi$_{1/3}$Mn$_{1/3}$Co$_{1/3}$O$_2$	272 (200)	4.0
Spinel	LiMn$_2$O$_4$	148 (120)	4.1
	LiMn$_{3/2}$Ni$_{1/2}$O$_4$	148 (120)	4.7
Olivine	LiFePO$_4$	170 (160)	3.45
	LiFe$_{1/2}$Mn$_{1/2}$PO$_4$	170 (160)	3.4/4.1

^{a} Value in parenthesis indicates the practical specific capacity of electrode.

(Reprinted with permission from Julien et al., *Inorganics, MDPI Publishing, 2014, 2: 132-154*)

11.1.5   Crystal Structure of Li$_2$FeSiO$_4$

The first report on Li$_2$FeSiO$_4$ was by Nytén et al. (Nytén et al. 2005). It shows an orthorhombic structure with space group Pmn21 and is based on β-Li$_3$PO$_4$ (Fig. 11.6). In the β-structure, a direction parallel to the chain of LiO$_4$ tetrahedral along with the alternating chain FeO$_4$ and SiO$_4$ tetrahedral is present. Nishimura et al. (Nishimura et al. 2008) reported the structure of Li$_2$FeSiO$_4$ prepared at 800°C and have γ structure (γ_s) with space group P21. (Sirisopanaporn et al. 2010) explored the crystal structure of γII polymorph of Li$_2$FeSiO$_4$ at quenching from 900°C and differs from the γ_s obtained structure by quenching from 800°C. In LFP (LiFePO$_4$), only one lithium-ion is extracted from the host theoretically and in LVP (Li$_3$V$_2$(PO$_4$)$_3$), three lithium-ion is extracted and has a high capacity. But toxic vanadium prevents its use. So, the orthosilicates based cathode material Li$_2$MSiO$_4$ (M = Fe, Ni, Co, Mn) has gained attention of researchers (Karthikeyan et al. 2010, Lyness et al. 2007, He and Manthiram 2014). The Li$_2$MSiO$_4$ has a high theoretical capacity of 333 mAhg^{-1} (Fan et al. 2010, Li et al. 2014a, Zaghib et al. 2006, Chung et al. 2002). Figure 11.6 shows the two lithium-ion extractions and insertion mechanism in Li$_2$FeSiO$_4$.

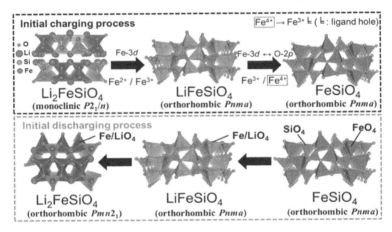

Fig. 11.6   Schematic summary of the two lithium-ion extractions and insertion mechanism in Li$_2$FeSiO$_4$ (Reprinted with permission from Masese et al., *The Journal of Physical Chemistry C, ACS Publishing, 2015, 119: 10206-10211*)

The advantage is high abundance, better thermal stability due to Si-O strong bonding, and low toxicity. Table 11.2 compares the capacity, average potential, and energy density of different cathodes and shines a light on the superiority of orthosilicates due to their high reversible capacity and energy density. It shows a high operating voltage of about 4.8 V for Li$^+$/Li electrode with better stability (Muraliganth et al. 2010, Islam et al. 2011).

TABLE 11.2   Electrochemical characteristics of some cathodes for lithium-ion batteries

Cathode	Reversible Capacity (mAh g^{-1})	Average Potential (vs Li$^+$/Li)	Energy Density (Wh kg^{-1})
LiCoO$_2$	150	3.9	580
LiMn$_2$O$_4$	120	4.1	490
LiFePO$_4$	165	3.4	560
LiMnPO$_4$	145	4.0	580
LiCoPO$_4$	130	4.8	62
Li$_2$FeSiO$_4$[a]	331	3.4	1120
[a] Theoretical values are adopted for Li$_2$FeSiO$_4$.			

(Reprinted with permission from Ni et al., *ACS Energy Letters, ACS Publishing, 2017, 2: 1771-1781*)

One important point that impacts the cell performance parameters is the content of the active material, binder, and conductive material (Andre et al. 2015). A drastic comparison of the silicates based materials is summarized in Table 11.3; silicate-based materials give improved performance at low active material loading.

TABLE 11.3 The ratio of active material, conducting agent, and binder used, for each cathode family, in the cell's energy calculation

Cathode Class	Ratio of Active Material	Ratio of Conducting Agent	Ratio of Binder
Oxides	90%	5%	5%
Conversions	75%	15%	10%
Phosphates	84%	5%	11%
Silicates	66%	29%	5%
Borate	81%	10%	9%

(Reprinted with permission from Andre et al., *Journal of Materials Chemistry A*, RSC Publishing, 2015, 3: 6709-6732)

11.1.6 Approaches to Modify Li_2FeSiO_4

From the above discussion, it may be concluded that the Li_2FeSiO_4 may be adapted as the alternative to replace the existing cathode materials. The key characteristics that strengthen its candidature are higher chemical and thermal stability due to Si-O covalent bond, safety and cost-effectiveness. Two important parameters that need to be improved to achieve optimum performance are electronic conductivity and ionic diffusivity. The existing cathode suffers from poor electronic conductivity and low Li diffusion. So, this limitation needs to be overcome by developing new strategies to tailor or modify the structure. It can be done in two ways: one by modifying the cathode material externally, and another by altering the structure internally. Some of the important strategies are discussed below.

11.1.6.1 Carbon Coating

The coating of conducting carbon around individual LFS nanoparticle is an effective strategy to enhance electrochemical performance. It is the most adapted technique that plays three roles: (i) improves electronic conductivity effectively, (ii) acts as a buffer later which prevents the growth of particle, and (iii) suppresses the tendency of agglomeration during calcination at high temperature. The coating can be done by two methods: (i) *in-situ* method: precursors are mixed with carbon source followed by annealing at high temperature. It results in the formation of a thin homogenous layer, and (ii) *ex-situ* method: pure materials mixed with carbon source are ball-milled followed by annealing at high temperature (Sun et al. 2013). One drawback that forces the researchers to develop an alternative to carbon coating is that ionic conductivity is smaller (10^{-11}-10^{-10} S cm^{-1}) than the electronic conductivity ($>10^{-9}$ S cm^{-1}) at room temperature.

11.1.6.2 Reduction of Particle Size and Tune of Morphology

Reduction of particle size is one of the important strategies along with morphology control (sphere, flower, rod, wire) to enhance the electrochemical properties of the cathode. The high surface area, which is the inherent characteristic of the nanoparticle, motivates the researchers to think it as an alternative to the coating. It is known that the smaller cathode size will facilitate faster Li-ion diffusion owing to the shortening of the diffusion length (Ding et al. 2014, Gaberscek et al. 2007). The diffusion coefficient (D) of ion (Li$^+$) is linked with the diffusion path length (L) and characteristic charge/discharge time (t) by the following relation;

$$\tau = \frac{L^2}{4\pi D} \qquad (2)$$

So, it suggests that the reduction of particle size or transition from micro to nanoparticles can effectively enhance the stability, rate capability, and interfacial properties along with energy density.

11.1.6.3 Metal Doping

This is an effective strategy to enhance the electrochemical properties and is superior as compared to carbon coating. Carbon coating enhances the electronic conductivity only at the surface that contributes to enhanced electrochemical properties, but cyclic stability remains an issue as it is inherent. The metal doping is dominating as it alters the crystal structure and will be more feasible to achieve the desirable properties. The doping results in enhanced specific capacity and cyclic stability (Zou et al. 2019). Some of the important metal ions are Mn, Cr, Cu, Cd, Mg, Ni, V, etc.

11.2 MATERIALS AND METHODOLOGY

11.2.1 Sol-Gel Method

It is a cost-effective technique and comes in the category of wet chemical techniques. Here, precursors are mixed in a solvent and stirred until a gel is obtained (Fig. 11.7). Then the gel is dried to obtain a fine powder. After this, the girding is done followed by calcination at high temperature. The main reaction for the sol-gel process involves chemistry, based on hydrolysis and condensation (Baccile et al. 2009, Muruganantham et al. 2017).

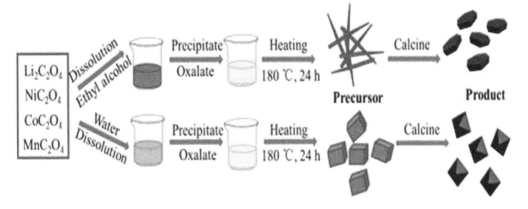

Fig. 11.7 The possible schematic illustration of the formation mechanism of the precursor and final product with various shapes and sizes (Reprinted with permission from Fu et al., *Journal of Alloys and Compounds, Elsevier Publishing, 2015, 618: 673-378*)

11.2.2 Hydrothermal Method

The advantage of hydrothermal over the sol-gel method is low cost, crystallinity with high purity, simplicity, short-time reaction, and low energy consumption (Song et al. 2010). Here, the stoichiometric amount of inorganic material is mixed in distilled water and is to be kept on stirring until the precursors get dissolved (Fig. 11.8). Then the solution is transferred into the Teflon-lined stainless steel sealed autoclave with optimized temperature and time. The autoclave is then cooled at room temperature. The final reacted product is in the form of brown precipitate which is filtrated and washed with distilled water to obtain cathode powder (Lu et al. 2015). One unique feature of the HT method is that it allows controlling of the morphology with different shapes, e.g. spherical, cubic, and plate-like.

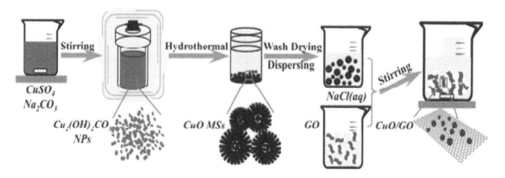

Fig. 11.8 Schematic illustration of the synthesis process for CuO MSs and CuO/GO hybrid (Reprinted with permission from Shi et al., *RSC Advances, RSC Publishing, 2015, 5: 85179-85186*)

11.2.3 Microwave Synthesis

The cathode material may be synthesized by the microwave method. In the microwave process, as microwave energy is absorbed by reactant, high temperatures required for a chemical reaction can be obtained easily. Also, heating is uniform in this method (Fig. 11.9). The need for low temperature and shorter time during synthesis reduces energy consumption than other methods.

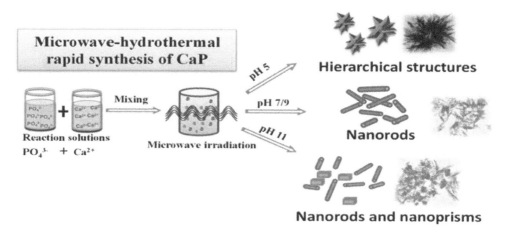

Fig. 11.9 Schematic illustration for the synthesis of calcium phosphate nanostructures with the microwave-assisted hydrothermal method under different pH values (Reprinted with permission from Cai et al., *Nanomaterials*, MDPI *Publishing, 2015, 5: 1284-1296*)

The microwave heating time plays an important role in controlling the size of the particle. If the heating time is longer, then the particle size is larger. Shorter heating time results in an incomplete crystalline structure, which decreases the charging and discharging capacity (Satyavani et al. 2016). The key advantage of the microwave method is that it heats the reactant to high temperature in a short time by providing a platform for direct interaction between different components (Qin et al. 2015, Peng et al. 2009). Unlike the sol-gel method, it also avoids the agglomeration of the prepared material (Liu et al. 2014).

11.2.4 Co-precipitation Method

Hydrothermal and sol-gel processes need optimization of temperature/time parameters and are time-consuming. Another drawback is the need for high pressure in hydrothermal method and long reaction time (depending on desirable morphology). So, the co-precipitation method seems to be better as compared to both and has advantages such as simplicity, and cost-effectiveness (Fig. 11.10). This method allows the preparation of material with spherical morphology and element mixing at the atomic level. Here, the pH value of the solution, coprecipitation temperature, and stirring intensity need to be optimized for better role of the precursor. The increase of pH value decreases particle size while an increase in stirring speed results in quasi-spherical morphology (Liang et al. 2014). Figure 11.10 shows the preparation method for Li_2FeSiO_4 materials. Here, Fe^{3+} salt is used as an iron source and polyethylene glycol (PEG200) as a dispersant and carbon source. This carbon coating enhances the electronic conduction as well as interfacial properties (Du et al. 2016).

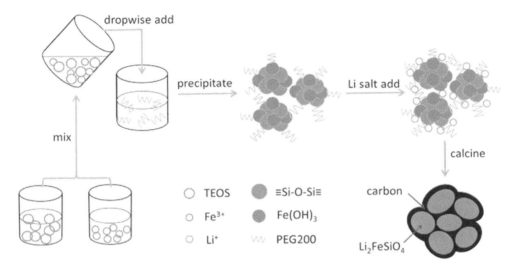

Fig. 11.10 Schematic illustration of the preparation process of Li_2FeSiO_4/C via co-precipitation method (Reprinted with permission from Du et al., *Electrochimica Acta, Elsevier Publishing, 2016, 188: 744-751*)

11.2.5 Solid State Reaction (SSR) Method

This is a cost-effective method, where a measured metal salt and lithium salt precursor are mixed in a mortar-paster or ball-milling. Then the material is heated at high temperatures to achieve uniform particle distribution. This is effective in large scale production. One drawback of the SSR is that there is a possibility of non-uniform particle formation and the impurity formation. Another key issue is the agglomeration that results in poor performance (Gong et al. 2014).

11.2.6 Solvothermal Method

Solvothermal is a method of producing suitable chemical compounds. It resembles in function to the hydrothermal process, the difference is of the precursor solution (non-aqueous). It includes the advantages of both sol-gel and hydrothermal routes. High temperature and high pressure are also essential conditions for this method. The chemical reaction that takes place in solvothermal is in the solvent phase and is suitable for water-sensitive compounds. The morphology and structure of the materials synthesized are better in solvothermal than the hydrothermal method (Fig. 11.11). This method provides an upper edge over the size, shape distribution, and crystallinity of the prepared products. The properties, morphologies, size, and structure of the synthesized nanostructure can be tailored and altered by varying the different reaction parameters like reaction time, pH, the concentration of the reactant, and filled volume of autoclave (Deng et al. 2011). Lee et al. synthesized the nanospheres and nanorods of Li_2FeSiO_4 material via solvothermal method (Lee et al. 2015).

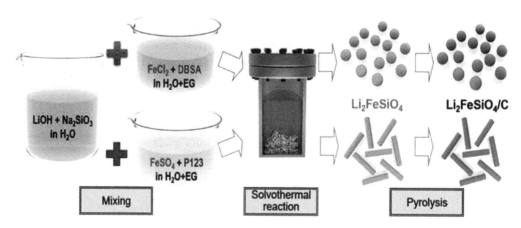

Fig. 11.11   Schematic diagram depicting the solvothermal process (Reprinted with permission from Lee et al., *Materials Letters*, *Elsevier Publishing, 2015, 160: 507-510)*

11.2.7   Chemical Vapor Deposition (CVD)

CVD is a vacuum deposition method whose application is to produce high quality, high-performance, solid materials. It is a process used to coat almost any metallic or ceramic compound, including elements, metals and their alloys and intermetallic compounds. The CVD method involves, firstly, the exposure of the wafer (substrate) surface to one or more volatile precursors, which react or decompose on the substrate surface to produce the appropriate deposit. Mostly, volatile products are also produced, which are removed by gas flow through the reaction chamber. The process is thermally driven but the photo and plasma-assisted methods are also used. The deposition of the film is controlled by the chemical reaction. The method is hence more versatile than many traditional methods. Other advantages of CVD are the potential for both conformal and large area growth, the possibility of achieving, very high level of purity in the prepared material. By varying experimental conditions like substrate material, substrate temperature, the composition of reaction gas mixture, total pressure gas flows, etc. material with a wide range of physical and chemical properties can be synthesized (Deokar et al. 2016). Zhang et al. prepared $Li_2FeSiO_4/C/G$ composite which was synthesized by a CVD assisted two-step solid-state reaction. $Li_2C_2O_4$, $FeC_2O_4\cdot2H_2O$, and tetraethyl orthosilicate (TEOS) were used as the starting materials (Zhang et al. 2015b).

11.2.8   Combustion Method

In this method, the metal salts are mixed initially with nitric acid/urea followed by heating at ignition temperature. The heat released during the reaction results in the formation of the cathode material. This method is cost-effective and no separate instrument is required. There are some limitations like large particle size and poor control overreaction (Xu et al. 2018).

11.3   CHARACTERIZATION TECHNIQUES

11.3.1   X-ray Diffraction (XRD)

XRD is an important and non-destructive technique. The structure of the cathode material is examined by the XRD and also allows the investigation of phase, grain size, and the lattice strain/defect in the material. In case of the possibility of the presence of various phases in the material, the XRD spectra are analyzed by the Rietveld analysis.

11.3.2   Field Emission Scanning Electron Microscopy (FESEM)

The FESEM is an important technique for obtaining information about the topography of the prepared material. The morphology of the cathode material can be confirmed. Along with this, the identification of various elements in the prepared material can be performed by the Energy Dispersive X-ray Analysis (EDAX). The elemental mapping of the different elements can be done to obtain information. It allows us to check whether the material is uniformly dispersed or agglomerated. With FESEM only the surface morphology can be investigated.

11.3.3   Transmission Electron Microscopy (TEM)

TEM is used to examine the structure, composition, and properties of specimens in submicron detail. It uses a particle beam of electrons to visualize samples and generate a highly-magnified image (Ensling et al. 2009).

11.3.4   Electrochemical Characterization of the Electrode Material

The electrochemical analysis of the cathode material in Li-ion batteries helps to determine various parameters of the device like specific capacitances, energy density, power density, Coulombic efficiency, and conductivity. The three-electrode and two-electrode systems are two popular systems to make electrochemical measurements. The electrochemical performances of the cathode material were analyzed via EIS (Electrochemical Impedance Spectroscopy), CV (Cyclic Voltammetry), and GCD (Galvanostatic charge/discharge) techniques.

11.3.4.1   EIS (Electrochemical Impedance Spectroscopy)

Impedance spectroscopy is used to obtain the transport properties and dielectric properties. The real part of impedance (Z') is linked to R_s and R_{ct} as

$$Z' = R_s + R_{ct} + \sigma\omega^{-1/2} \tag{3}$$

Here, R_s is Ohmic resistance, R_{ct} is charge transfer resistance, and σ is the Warburg factor which is related to the diffusion coefficient of lithium-ion. The value of Ohmic resistance (R_s), charge transfer resistance (R_{ct}), and low R_{ct} value are evaluated, and low R_s and R_{ct} suggests the high specific capacity and electronic conductivity. The diffusion coefficient of Li^+ is estimated using the equation

$$D = \frac{R^2 T^2}{2n^4 A^2 F^4 C^2 \sigma^2} \tag{4}$$

where R is the gas constant, T is the absolute temperature, A is the surface area of the cathode, n is the number of electrons per molecule during oxidization, F is the Faraday constant, and C is the concentration of lithium-ion for active electrode materials.

11.3.4.2   CV (Cyclic Voltammetry)

CV is a tool that is employed to investigate the electrochemistry of the cathode material. The shape of CV curve, peak potential, and peak currents resembles the electrochemical properties of the electrode and discloses the phase transition that occurs during charge-discharge experiments, which strongly affect the capacity fading during the cycle. Mostly, when a cathode material encounters phase transformation, a peak appears in CV curve due to the coexistence of two phases. The Li diffusion coefficients for an electrode are calculated using the Randles Sevcik equation, which describes the effect of the scan rate on the peak current. In a linear potential sweep voltammogram, the relation between the peak current and the scan rate (for low scan rates) is given by,

$$i_p = 0.4463 F \left(\frac{F}{RT} \right)^{1/2} C^* S^{1/2} A D^{1/2} \tag{5}$$

where i_p is the peak current, F is the Faraday's constant (96500 $Cmol^{-1}$), R is the gas constant (8.32 $JK^{-1}mol^{-1}$), T is the temperature (298.15 K), C^* is the initial Li-ion concentration for active electrode material, A is the electrode area, S is the scan rate, and D is the lithium diffusion coefficient (Hong and Zhang 2013).

11.3.4.3 GCD (Galvanostatic Charge/Discharge)

GCD is a technique that represents the graph for the potential range versus period for each charge/discharge cycle. The specific capacity, energy density, coulombic efficiency and capacity retention are estimated. Table 11.4 shows the essential parameters for testing the performance of a lithium ion cell.

TABLE 11.4 Essential parameters for testing the performance of a lithium ion cell

Parameters	Measuring Unit	Measuring Formula	Information
Operating voltage	Volts (V)	Instrumental	Energy density and safety
Current density	mAg^{-1}	Instrumental	For testing rate capabilities
Theoretical capacity	$mAhg^{-1}$	$TC = \dfrac{F \times x}{3.6 \times M.M \times y}$	Lithium ion storage capability
Gravimetric capacity	$mAhg^{-1}$	$C = \dfrac{I(mA) \times t(h)}{m(g)}$	Li+ storage capability measured per unit mass
Areal capacity	$mAhcm^{-2}$	$C = \dfrac{I(mA) \times t(h)}{A(cm^2)}$	Li+ storage capability measured per unit area
Volumetric capacity	$mAhcm^{-2}$	$C = \dfrac{I(mA) \times t(h)}{V(cm^3)}$	Li+ storage capability measured per unit volume
Energy density	Whg^{-1} or $Whcm^{-2}$ or $Whcm^{-3}$	$E = C \times V$	How much energy can be extracted
Power density	Wg^{-1} or Wcm^{-2} or Wcm^{-3}	$P = I \times V$	How fast the energy can be extracted
C rate	h^{-1}	$C_{rate} = \dfrac{J(mAg^{-1})}{C(mAhg^{-1})}$	Rate of charging/discharging
Coulombic efficiency	N/A	$\%E = \dfrac{C_{charging}}{C_{dishcarging}} \times 100$	Reversible capacity

(Reprinted with permission from Gulzar et al., *Journal of Materials Chemistry A*, RSC Publishing, 2016, 4: 16771-16800)

11.4 RECENT PROGRESS IN THE SILICATE-BASED CATHODE MATERIAL

The development of the Li_2MSiO_4 (M: Fe, Ni etc) based cathode materials has gained speed. So, for enhancing the electrochemical performances of the LMS, various strategies have been performed by researchers. The reduction of crystallite size (micro to nano), carbon coating, morphological variations, and the doping of various elements are important strategies. So, this section shines a light on some important work done concerning this.

Recently, Shen et al. (Shen et al. 2019) reported the preparation of the Li_2FeSiO_4/C hollow nanospheres by combining the hydrothermal method with annealing. The formation is shown in Fig. 11.12a. HRTEM analysis evidenced that approximately 6 nm amorphous carbon layers are anchored on surfaces of Li$_2$FeSiO$_4$ crystal matrix. Figure 11.12b depicts the XRD spectra, and monoclinic structure with a space group of P$_{21/n}$ is formed, which indicates the formation of single-phase Li$_2$FeSiO$_4$. The absence of any peak corresponding to carbon indicates the amorphous nature of carbon. Raman spectra show two bands. D band of sp^2-type carbon (at 1,330 cm^{-1}), and G band of sp^3-type carbon (at 1,592 cm^{-1}) and confirms the amorphous nature of carbon materials (Fig. 11.12c). The high value of peak intensity ratio (0.83) between the D and G bands (ID/IG) indicates the high degree of graphitization, which suggests high electronic conductivity. The electrochemical testing shows the discharge capacity (for 1st cycle) of about 168.1 mAhg^{-1}, and reduces to 155.6 mAhg^{-1} after the 100th cycle. The capacity retention of about 92.5% is obtained after 100 cycles and demonstrates excellent electrochemical stability. The high value of specific capacity is attributed to (i) a large amount of interior void space, (ii) 10 nm thick walls of the hollow nanosphere cathode.

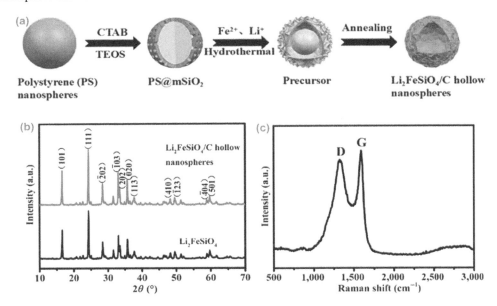

Fig. 11.12 (a) Schematic illustration of controlled synthesis of highly uniform Li$_2$FeSiO$_4$/C hollow nanospheres, XRD pattern, (b) and Raman spectra (c) of Li$_2$FeSiO$_4$/C hollow nanospheres (Reprinted with permission from Shen et al., *Nano Research, Springer Publishing, 2019, 12:357-363*)

To enhance the electronic conductivity of the Li$_2$FeSiO$_4$ (LFS), Singh et al. (Singh et al. 2017) tuned the intrinsic properties of LFS by replacing the O with Cl (less electronegative). They combined the DFT and experimental approach and correlated the results. The XRD (simulated and experimental) analysis confirmed the increase in the lattice parameter after Cl doping and volume expansion. It was also found that the Cl doping leads to the lowering of the deinsertion voltage, and indicates the use of this cathode with carbonate-based commercial electrolyte. The DOS (density of state) analysis evidenced the increase in electronic conductivity. Figure 11.13a shows the cycling profile before and after doping. For the doped LFS, the first discharge capacity was 198 mAhg^{-1}, and after 50th cycle decreases to 182.5 mAhg^{-1}. Figure 11.13b shows the cyclic stability for different current rates, and doped LFS shows high capacity as compared to undoped. This enhancement is attributed to the faster kinetics in doped LFS. From the impedance analysis, the charge transfer resistance (R_{ct}) for the doped sample was 32 Ω and is less than the pure sample (55 Ω). This low resistance value suggests the high capacity retention for the doped system (Fig. 11.13c). It was concluded that the doping approach is beneficial for the enhancement of the performance of pure LFS material.

Another report by Qiu et al. (Qiu et al. 2017) prepared the Li$_2$Fe$_{1-x}$Ti$_x$SiO$_4$/C cathode materials ($x = 0$, $x = 0.02$, $x = 0.04$, $x = 0.10$) by a facile sol-gel method, and examined the effect of Ti doping.

The corresponding sample code are T0, T2, T4, and T10, respectively. Ti was chosen for doping due to its ability to achieve a smaller radius (0.068 nm) than Fe^{2+} (0.076 nm), and the electrochemical inactive nature of Ti^{4+} in 1.5-4.5 V is beneficial for crystal structure stability. The average crystallite size was lowest for T4 and is 13.18 nm. The T2 electrode demonstrates excellent specific capacity and stability. The specific capacity was 115.6 mAhg^{-1} (at 5 C) and 97.1 mAhg^{-1} (at 10 C) and after the 1000th cycle reduces to 102.8 and 91.1 mAhg^{-1} (Fig. 11.13d). The charge transfer resistance for T2 was 140.8 Ω and is smaller than the T0 (315.4 Ω). It was concluded that the Ti doping does not affect the structure and microstructure, and enhancement in the electrochemical performance is attributed to the faster ion kinetics and structural stability.

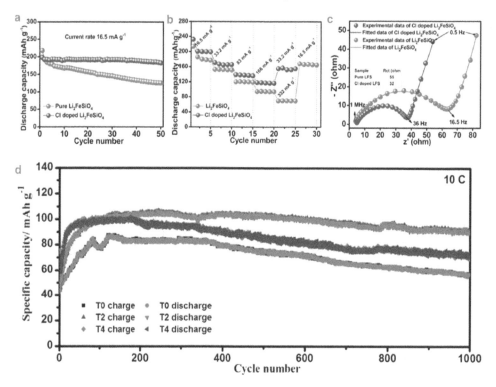

Fig. 11.13 (a) Cyclic and stability performance of pure ($x = 0$) and doped ($x = 0.1$) $Li_2FeSiO_{4-x}Cl_x$ at 20°C at 16.5 mA g^{-1} current rate, (b) Rate performance of pure ($x = 0$) and doped ($x = 0.1$) $Li_2FeSiO_{4-x}Cl_x$, (c) Nyquist plots of pure ($x = 0$) and doped ($x = 0.1$) $Li_2FeSiO_{4-x}Cl_x$ sample at OCP (Reprinted with permission from Singh et al., *ACS applied materials and interfaces*, ACS Publishing, *2017, 9: 26885-26896*, (d) Cycling performance of the T0, T2 and T4 at the rate of 10°C (Reprinted with permission from Qiu et al., *Journal of Alloys and Compounds, Elsevier Publishing, 2017, 725: 860-868*)

Recently, Peng et al. (Peng et al. 2019) synthesized the core-shell heterostructure CNT@ Li_2FeSiO_4@C cathode material (impurity-free and structurally stable). The inner layer of CNT and out carbon leads to high electronic conductivity. Figure 11.14a shows the detailed preparation process. XRD analysis confirmed the pure phase formation and XPS analysis confirms the formation of orthogonal silicate structure (SiO_4). The electrochemical testing of the cell was performed in the voltage window of 1.5-4.8 V (at 0.2°C). The first discharge capacity was 178 mAhg^{-1} and after 150th cycle, 89.3% capacity was retained. Figure 11.14b shows the CV profile and redox potentials corresponding to Fe^{2+}/Fe^{3+} reaction are 3.1 V (anodic) and 2.7 V (cathodic), respectively. The diffusion coefficient of Li ions for the anodic and cathodic reactions are 7.32×10^{-11} and 0.64×10^{-12} cm^2 s^{-1} and are within the desirable limit (Fig. 11.14c). Table 11.5 summarize the synthesis methods and electrochemical performance of various orthosilicate based cathode.

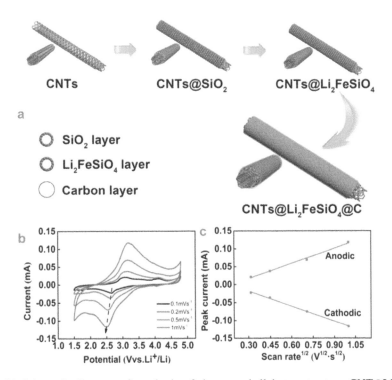

Fig. 11.14 (a) Schematic diagram of synthesis of the core-shell heterostructure CNT@Li$_2$FeSiO$_4$@C, Kinetic analysis of CNT@Li$_2$FeSiO$_4$@C using CV, (b) CV profiles at various scan rates, (c) peak current as a function of square root of scan rates (Reprinted with permission from Peng et al., *Nanoscale research letters*, Springer *Publishing, 2019, 14: 1-7*)

TABLE 11.5 Properties of some orthosilicate based cathode

Orthosilicate-based Compound	Synthetic Method	Structure	Initial Discharge Capacity	Reversible Capacity	Rate Capability	Ref.
Li$_2$FeSiO$_4$ nanorods (LFSNRs) anchored on graphene	Hydrothermal method	Orthorhombic Pmn21	300 mAhg^{-1} at 1.5-4.8 V	134 mAhg^{-1} at 12°C	95% over 240 cycles)	Yang et al. 2015
Li$_2$FeSiO$_4$/MWCNT	Sol gel synthesis	Orthorhombic with Pmn21 space group	240 mAhg^{-1} at 16.5 mAg^{-1}	–	166 mAhg^{-1} after 30 cycles	Singh and Mitra 2014
Li$_2$FeSiO$_4$/C	Co-precipitation method	monoclinic with space group P21/n	190 mAhg^{-1} at 0.1 C	190 mAhg^{-1} for 60 cycles	90% at 0.5 C after 400 cycles	Du et al. 2016
Li$_2$MnSiO$_4$/C nanopowders	Tartaric acid-assisted sol-gel process	Orthorhombic with Pmn21 space group	147 mAhg^{-1}	–	105 mAhg^{-1} after 30 cycles	Park et al. 2014b
M-Li$_2$MnSiO$_4$/C	Hydrothermal method	Pmn21	275 mAhg^{-1} at 8 mAg^{-1}	115 mAhg^{-1} after 50 cycles at 8 mAg^{-1}	59 mAhg^{-1} at 0.5C	Pei et al. 2016
Li$_2$MnSiO$_4$	Molten salt synthesis	Orthorhombic Pmn21	194 mAhg^{-1} at a low rate of 0.05C	–	165 mAhg^{-1} after 50 cycles at 0.1C.	Wang et al. 2014

Contd.

TABLE 11.5   Contd.

Li_2MnSiO_4/C	Ionothermal synthesis and solid-state reaction	Orthorhombic with Pmn21 space group	218.2 mAhg^{-1} at 0.1C	175.7 mAhg^{-1} after 50 cycles at 0.1C	62.6 mAhg^{-1} after 50 cycles at 5C	Li et al. 2014a
Li_2MnSiO_4/C nanocomposites	Spray pyrolysis with ball milling	Orthorhombic with Pmn21 space group	167 mAhg^{-1} at 0.1C	108 mAhg^{-1} after 20 cycles at 0.1C	115 mAhg^{-1} at 1C	Shao and Taniguchi 2014
Li_2MnSiO_4/C/graphene composite	PEO-600 assisted solid state reaction	Orthorhombic with Pmn21 space group	215.3 mAhg^{-1} at 0.05C	175 mAhg^{-1} after 40 cycles	–	Gong et al. 2014
Li_2MnSiO_4/C	Sol-gel and electrospinning technique	Orthorhombic nanofibers with the Pmn21 space group	108 mAhg^{-1} after 20 cycles at 0.1C	160 mAhg^{-1} after 20 cycles at 0.05C	80 mAhg^{-1} at 1C	Park et al. 2014a
Li_2MnSiO_4/C	Solid-state reaction	Orthorhombic with Pmn21 space group	204 mAhg^{-1} at 10 mAg^{-1}	105 mAhg^{-1} after 20 cycles at 10 mAg^{-1}	128 mAhg^{-1} after 5 cycles at 200 mAg^{-1}	Peng et al. 2013
Li_2MnSiO_4/C	Sol-gel method	Orthorhombic with Pmn21 space group	250 mAhg^{-1} at 0.05C	48 mAhg^{-1} after 16 cycles at 0.05C	136 mAhg^{-1} at 0.2C	Devraj et al. 2013
Li_2MnSiO_4/C	Sol-gel method	Orthorhombic with Pmn21 space group	240 mAhg^{-1} at 8 mAg^{-1}	109 mAhg^{-1} after 30 cycles at 8 mAg^{-1}	125 mAhg^{-1} at 160 mAg^{-1}	Qu et al. 2014
Li_2MnSiO_4/C	Sol-gel method	Orthorhombic	216.3 mAhg^{-1} at 80 mAg^{-1}	44% after 20 cycles at 80 mAg^{-1}	149.9 mAhg^{-1} at 320 mAg^{-1}	Liu et al. 2013
ZnO-coated $Li_2MnSiO_4/$ C composite	Sol-gel method and wet chemical	Orthorhombic with Pmn21 space group	134 mAhg^{-1} at 1C	64 mAhg^{-1} after 25 cycles at 1C	70 mAhg^{-1} at 4C	Zhu et al. 2015
TiO_2-coated Li_2MnSiO_4/C	Sol-gel method and wet chemical	Orthorhombic with Pmn21 space group	151 mAhg^{-1} at 0.5C	50 mAhg^{-1} after 50 cycles at 0.5C	131 mAhg^{-1} at 1C	Zhu et al. 2016
$Li_2Fe_{1/3}Mn_{1/3}Ni_{1/3}SiO_4/C$	Feasible solution process in ternary systems	Orthorhombic with the Pmn21 space group	181.4 mAhg^{-1} at 0.05C	172.9 mAhg^{-1} after 20 cycles at 0.05C	106.1 mAhg^{-1} at 1C	Yang et al. 2016
$Li_{2.05}MnSi_{0.95}Al_{0.05}O_4/C$	Mixed solvothermal process	Orthorhombic phase	290 mAhg^{-1} at 0.05C	204 mAhg^{-1} after 50 cycles at 0.05C	About 180 mAhg^{-1} at 2C	Zhang et al. 2014
$Li_{1.8}MnSi_{0.8}P_{0.2}O_4/C$	Sol-gel method	Pmn21 polymorph	153 mAhg^{-1} at 10 mAg^{-1}	80 mAhg^{-1} after 20 cycles at 10 mAg^{-1}	150 mAhg^{-1} at 40 mAg^{-1}	Gummow et al. 2014
$Li_2MnSi_{0.75}V_{0.25}O_4/C$	Sol-gel method	Pmn21 space group	132 mAhg^{-1} at 0.5C	25 mAhg^{-1} after 100 cycles at 0.5C	About 110 mAhg^{-1} at 1C	Wagner et al. 2016

11.5 SUMMARY

A low cost, safe, and high energy storage device is a Li-ion battery and has gained the attention of researchers. Recently, orthosilicates based cathodes (Li$_2$MSiO$_4$) have emerged as an attractive candidate for the LIBs. This material provides better structural stability as compared to the oxide materials as a cathode. The presence of Si-O covalent bonding in the structure and availability of two Li$^+$ for transportation provides high capacity and energy density. Still, the voltage degradation issue and efficiency need to be overcome before use in commercial batteries. So, different strategies need to be developed to enhance the energy density to meet the requirement in portable electronics. Along with this, the selection of electrolyte and anode need to be taken care of before the fabrication of cell. The electrolyte needs to be safe and flexible for fulfilling future device needs. Also, the full solid-state battery needs to be explored by examining the interface, mechanism, and low conductivity issue.

11.6 REFERENCES

Andre, D., S.J. Kim, P. Lamp, S.F. Lux, F. Maglia, O. Paschos, et al. 2015. Future generations of cathode materials: An automotive industry perspective. J. Mater. Chem. A. 3: 6709-6732.

Armand, M. and J.M. Tarascon. 2008. Building better batteries. Nature 451: 652.

Arya, A. and A.L. Sharma. 2019. Electrolyte for energy storage/conversion (Li$^+$, Na$^+$, Mg^{2+}) devices based on PVC and their associated polymer: A comprehensive review. J. Solid State Electrochem. 23: 997-1059.

Arya, A. and A.L. Sharma. 2020. A glimpse on all-solid-state Li-ion battery (ASSLIBs) performance based on novel solid polymer electrolytes: A topical review. J. Mater. Sci. 55: 6242-6304.

Baccile, N., F. Babonneau, B. Thomas and T. Coradin. 2009. Introducing ecodesign in silica sol-gel materials. J. Mater. Chem. 19: 8537-8559.

Bensalah, N. and H. Dawood. 2016. Review on synthesis, characterizations, and electrochemical properties of cathode materials for lithium ion batteries. J. Material Sci. Eng. 5: 1000258.

Boulabiar, A., K. Bouraoui, M. Chastrette and M. Abderrabba. 2004. A historical analysis of the Daniell cell and electrochemistry teaching in French and Tunisian textbooks. J. Chem. Educ. 81(5): 754.

Cai, Z.Y., F. Peng, Y.P. Zi, F. Chen and Q.R. Qian. 2015. Microwave-assisted hydrothermal rapid synthesis of calcium phosphates: Structural control and application in protein adsorption. Nanomaterials 5: 1284-1296.

Cheng, F. and J. Chen. 2012. Metal–air batteries: From oxygen reduction electrochemistry to cathode catalysts. Chem. Soc. Rev. 41: 2172-2192.

Choi, J.W. and D. Aurbach. 2016. Promise and reality of post-lithium-ion batteries with high energy densities. Nat. Rev. Mater. 1: 16013.

Chung, S.Y., J.T. Bloking and Y.M. Chiang. 2002. Electronically conductive phospho-olivines as lithium storage electrodes. Nat. Mater. 1: 123-128.

Deng, D., X. Pan, L. Yu, Y. Cui, Y. Jiang, J. Qi, et al. 2011. Toward N-doped graphene *via* solvothermal synthesis. Chem. Mater. 23: 1188-1193.

Deokar, G., D. Vignaud, R. Arenal, P. Louette and J.F. Colomer. 2016. Synthesis and characterization of MoS$_2$ nanosheets. Nanotechnology 27: 075604.

Devaraj, S., M. Kuezma, C.T. Ng and P. Balaya. 2013. Sol-gel derived nanostructured Li$_2$MnSiO$_4$/C cathode with high storage capacity. Electrochim. Acta. 102: 290-298.

Ding, K., H. Gu, C. Zheng, L. Liu, L. Liu. X. Yan, et al. 2014. Octagonal prism shaped lithium iron phosphate composite particles as positive electrode materials for rechargeable lithium-ion battery. Electrochim. Acta. 146: 585-590.

Du, X., H. Zhao, Y. Lu, C. Gao, Q. Xia and Z. Zhang. 2016. Electrochemical properties of nanostructured Li$_2$FeSiO$_4$/C synthesized by a simple co-precipitation method. Electrochim. Acta. 188: 744-751.

Ensling, D., M. Stjerndahl, A. Nytén, T. Gustafsson and J.O. Thomas. 2009. A comparative XPS surface study of Li$_2$ FeSiO$_4$/C cycled with LiTFSI-and LiPF$_6$-based electrolytes. J. Mater. Chem. 19: 82-88.

Fan, X.Y., Y. Li, J.J. Wang, L. Gou, P. Zhao, D.L. Li, et al. 2010. Synthesis and electrochemical performance of porous Li$_2$FeSiO$_4$/C cathode material for long-life lithium-ion batteries. J. Alloy. Compd. 493: 77-80.

Fu, F., Y. Huang, P. Wu, Y. Bu, Y. Wang and J. Yao. 2015. Controlled synthesis of lithium-rich layered Li$_{1.2}$Mn$_{0.56}$Ni$_{0.12}$Co$_{0.12}$O$_2$ oxide with tunable morphology and structure as cathode material for lithium-ion batteries by solvo/hydrothermal methods. J. Alloy. Compd. 618: 673-678.

Gaberscek, M., R. Dominko and J. Jamnik. 2007. Is small particle size more important than carbon coating? An example study on $LiFePO_4$ cathodes. Electrochem. Commun. 9: 2778-2783.

Gao, H., Z. Hu, K. Zhang, F. Cheng, Z. Tao and J. Chen. 2014. Hydrothermal synthesis of spindle-like Li_2FeSiO_4-C composite as cathode materials for lithium-ion batteries. J. Energy Chem. 23: 274-281.

Gong, H., Y. Zhu, L. Wang, D. Wei, J. Liang and Y. Qian. 2014. Solid-state synthesis of uniform Li_2MnSiO_4/C/graphene composites and their performance in lithium-ion batteries. J. Power Sources. 246: 192-197.

Goodenough, J.B. 1994. Design considerations. Solid State Ion. 69: 184-198.

Gulzar, U., S. Goriparti, E. Miele, T. Li, G. Maidecchi, A. Toma, et al. 2016. Next-generation textiles: From embedded supercapacitors to lithium ion batteries. J. Mater. Chem. A. 4: 16771-16800.

Gummow, R.J., G. Han, N. Sharma and Y. He. 2014. Li_2MnSiO_4 cathodes modified by phosphorous substitution and the structural consequences. Solid State Ion. 259: 29-39.

He, G. and A. Manthiram. 2014. Nanostructured $Li2MnSiO_4$/C Cathodes with hierarchical macro-/mesoporosity for lithium-ion batteries. Adv. Funct. Mater. 24: 5277-5283.

Hong, L. and Z. Zhang. 2013. Effect of carbon sources on the electrochemical performance of Li_2FeSiO_4 cathode materials for lithium ion batteries. Russ. J. Electrochem. 49: 386-390.

Islam, M.S., R. Dominko, C. Masquelier, C. Sirisopanaporn, A.R. Armstrong and P.G. Bruce. 2011. Silicate cathodes for lithium batteries: Alternatives to phosphates. J. Mater. Chem. 21: 9811-9818.

Julien, C. 2000. Local cationic environment in lithium nickel-cobalt oxides used as cathode materials for lithium batteries. Solid State Ion. 136-137: 887-896.

Julien, C.M., A. Mauger, K. Zaghib and H. Groult. 2014. Comparative issues of cathode materials for Li-ion batteries. Inorganics 2: 132-154.

Karthikeyan, K., V. Aravindan, S.B. Lee, I.C. Jang, H.H. Lim, G.J. Park, et al. 2010. A novel asymmetric hybrid supercapacitor based on Li_2FeSiO_4 and activated carbon electrodes. J. Alloy. Compd. 504: 224-227.

Lee, J.J., H.C. Dinh, S.I. Mho, I.H. Yeo, W.I. Cho and D.W. Kim. 2015. Morphology-controlled solvothermal synthesis of Li_2FeSiO_4 nanoparticles for Li-ion battery cathodes. Mater. Lett. 160: 507-510.

Li, X., J. Liu, X. Meng, Y. Tang, M.N. Banis, J. Yang, et al. 2013. Significant impact on cathode performance of lithium-ion batteries by precisely controlled metal oxide nanocoatings via atomic layer deposition. J. Power Sources. 247: 57-69.

Li, X., Y. Liu, Z. Xiao, W. Guo and R. Zhang. 2014a. Ionothermal synthesis and characterization of Li_2MnSiO_4/C composites as cathode materials for lithium-ion batteries. Ceram. Int. 40: 289-296.

Li, X., J. Liu, M.N. Banis, A. Lushington, R. Li, M. Caib, et al. 2014b. Atomic layer deposition of solid-state electrolyte coated cathode materials with superior high voltage cycling behavior for lithium ion battery application. Energy Environ. Sci. 7: 768-778.

Liang, L., K. Du, Z. Peng, Y. Cao, J. Duan, J. Jiang, et al. 2014. Co-precipitation synthesis of $Ni_{0.6}Co_{0.2}Mn_{0.2}(OH)_2$ precursor and characterization of $Ni_{0.6}Co_{0.2}Mn_{0.2}O_2$ cathode material for secondary lithium batteries. Electrochim. Acta. 130: 82-89.

Linden, D. and T.B. Reddy [eds.]. 2002. Handbook of Batteries, 3rd edn.

Liu, N., X. Wang, W. Xu, H. Hu, J. Liang and J. Qiu. 2014. Microwave-assisted synthesis of MoS_2/graphene nanocomposites for efficient hydrodesulfurization. Fuel 119: 163-169.

Liu, S., J. Xu, D. Li, Y. Hu, X. Liu and K. Xie. 2013. High capacity Li_2MnSiO_4/C nanocomposite prepared by sol-gel method for lithium-ion batteries. J. Power Sources. 232: 258-263.

Lu, X., H. Wei, H.C. Chiu, R. Gauvin, P. Hovington, A. Guerfi, et al. 2015. Rate-dependent phase transitions in Li_2FeSiO_4 cathode nanocrystals. Sci. Rep. 5: 8599.

Lyness, C., B. Delobel, A.R. Armstrong and P.G. Bruce. 2007. The lithium intercalation compound Li_2CoSiO_4 and its behaviour as a positive electrode for lithium batteries. Chem. Comm. 46: 4890-4892.

Manthiram, A. 2020. A reflection on lithium-ion battery cathode chemistry. Nat. Commun. 11: 1-9.

Masese, T., C. Tassel, Y. Orikasa, Y. Koyama, H. Arai, N. Hayashi, et al. 2015. Crystal structural changes and charge compensation mechanism during two lithium extraction/insertion between Li_2FeSiO_4 and $FeSiO_4$. J. Phys. Chem. C. 119: 10206-10211.

Muraliganth, T., K.R. Stroukoff and A. Manthiram. 2010. Microwave-solvothermal synthesis of nanostructured Li_2MSiO_4/C (M = Mn and Fe) cathodes for lithium-ion batteries. Chem. Mater. 22: 5754-5761.

Muruganantham, R., Y.T. Chiu, C.C. Yang, C.W. Wang and W.R. Liu. 2017. An efficient evaluation of F-doped polyanion cathode materials with long cycle life for Na-ion batteries applications. Sci. Rep. 7: 14808.

Nagaura, T. and K. Tozawa. 1990. Progress in batteries and solar cells. JEC Press 9: 209.

Ni, J., Y. Jiang, X. Bi, L. Li and J. Lu. 2017. Lithium iron orthosilicate cathode: Progress and perspectives. ACS Energy Lett. 2: 1771-1781.

Nishimura, S.I., S. Hayase, R. Kanno, M. Yashima, N. Nakayama and A. Yamada. 2008. Structure of Li$_2$FeSiO$_4$. J. Am. Chem. Soc. 130: 13212-13213.

Nytén, A., A. Abouimrane, M. Armand, T. Gustafsson and J.O. Thomas. 2005. Electrochemical performance of Li$_2$FeSiO$_4$ as a new Li-battery cathode material. Electrochem. Commun. 7: 156-160.

Park, H., T. Song, R. Tripathi, L.F. Nazar and U. Paik. 2014a. Li$_2$MnSiO$_4$/carbon nanofiber cathodes for Li-ion batteries. Ionics 20: 1351-1359.

Park, K.S., Y.H. Jin, L.S. Kang, G.H. Lee, N.H. Lee, D.W. Kim, et al. 2014b. Enhanced electrochemical performance of carbon-coated Li$_2$MnSiO$_4$ nanoparticles synthesized by tartaric acid-assisted sol-gel process. Ceram. Int. 40: 9413-9418.

Pei, Y., Q. Chen, C.Y. Xu, H.X. Wang, H.T. Fang, C. Zhou, et al. 2016. Chelate-induced formation of Li$_2$MnSiO$_4$ nanorods as a high capacity cathode material for Li-ion batteries. J. Mater. Chem. A. 4: 9447-9454.

Peng, T., W. Guo, Y. Zhang, Y. Wang, K. Zhu, Y. Guo, et al. 2019. The core-shell heterostructure CNT@Li$_2$FeSiO$_4$@C as a highly stable cathode material for lithium-ion batteries. Nanoscale Res. Lett. 14: 1-7.

Peng, Z., H. Miao, H. Yin, C. Xu and W.G. Wang. 2013. PEG-assisted solid state synthesis and characterization of carbon-coated Li$_2$MnSiO$_4$ cathode materials for lithium ion battery. Int. J. Electrochem. Sci. 8: 903-913.

Peng, Z.D., Y.B. Cao, G.R. Hu, K. Du, X.G. Gao and Z.W. Xiao. 2009. Microwave synthesis of Li$_2$FeSiO$_4$ cathode materials for lithium-ion batteries. Chin. Chem. Lett. 20: 1000-1004.

Qin, W., T. Chen, L. Pan, L. Niu, B. Hu, D. Li, et al. 2015. MoS$_2$-reduced graphene oxide composites via microwave assisted synthesis for sodium ion battery anode with improved capacity and cycling performance. Electrochim. Acta. 153: 55-61.

Qiu, H., H. Yue, X. Wang, T. Zhang, M. Zhang, Z. Fang, et al. 2017. Titanium-doped Li$_2$FeSiO$_4$/C composite as the cathode material for lithium-ion batteries with excellent rate capability and long cycle life. J. Alloy. Compd. 725: 860-868.

Qu, L., S. Fang, L. Yang and S.I. Hirano. 2014. Synthesis and characterization of high capacity Li$_2$MnSiO$_4$/C cathode material for lithium-ion battery. J. Power Sources. 252: 169-175.

Satyavani, T.V.S.L., A.S. Kumar and P.S. Rao. 2016. Methods of synthesis and performance improvement of lithium iron phosphate for high rate Li-ion batteries: A review. Int. J. Eng. Sci. Technol. 19: 178-188.

Shao, B. and I. Taniguchi. 2014. Synthesis of Li$_2$MnSiO$_4$/C nanocomposites for lithium battery cathode employing sucrose as carbon source. Electrochim. Acta. 128: 156-162.

Shen, S., Y. Zhang, G. Wei, W. Zhang, X. Yan, G. Xia, et al. 2019. Li$_2$FeSiO$_4$/C hollow nanospheres as cathode materials for lithium-ion batteries. Nano Res. 12: 357-363.

Shi, L., X. Fu, C. Fan, S. Yu, G. Qian and Z. Wang. 2015. Carbonate-assisted hydrothermal synthesis of porous, hierarchical CuO microspheres and CuO/GO for high-performance lithium-ion battery anodes. RSC Adv. 5: 85179-85186.

Singh, R. and A.D. Setiawan. 2013. Biomass energy policies and strategies: Harvesting potential in India and Indonesia. Renewable Sustainable Energy Rev. 22: 332-345.

Singh, S. and S. Mitra. 2014. Improved electrochemical activity of nanostructured Li$_2$FeSiO$_4$/MWCNTs composite cathode. Electrochim. Acta. 123: 378-386.

Singh, S., A.K. Raj, R. Sen, P. Johari and S. Mitra. 2017. Impact of Cl doping on electrochemical performance in orthosilicate (Li$_2$FeSiO$_4$): A density functional theory supported experimental approach. ACS Appl. Mater. Interfaces. 9: 26885-26896.

Sirisopanaporn, C., A. Boulineau, D. Hanzel, R. Dominko, B. Budic, A.R. Armstrong, et al. 2010. Crystal structure of a new polymorph of Li$_2$FeSiO$_4$. Inorg. Chem. 49: 7446-7451.

Song, H.K., K.T. Lee, M.G. Kim, L.F. Nazar and J. Cho. 2010. Recent progress in nanostructured cathode materials for lithium secondary batteries. Adv. Funct. Mater. 20: 3818-3834.

Sun, D., H. Wang, P. Ding, N. Zhou, X. Huang, S. Tan, et al. 2013. In-situ synthesis of carbon coated Li$_2$MnSiO$_4$ nanoparticles with high rate performance. J. Power Sources. 242: 865-871.

Tarascon, J.M. and M. Armand. 2011. Issues and challenges facing rechargeable lithium batteries. Nature 414: 359-367.

Thackeray, M.M., W.I.F. David, P.G. Bruce and J.B. Goodenough. 1983. Lithium insertion into manganese spinels. Mater. Res. Bull. 18: 461-472.

Thackeray, M.M. 1995. Structural considerations of layered and spinel lithiated oxides for lithium ion batteries. J. Electrochem. Soc. 142: 2558-2563.

Wagner, N.P., P.E. Vullum, M.K. Nord, A.M. Svensson and F. Vullum-Bruer. 2016. Vanadium substitution in Li_2MnSiO_4/C as positive electrode for Li ion batteries. J. Phys. Chem. C. 120: 11359-11371.

Wang, F., Y. Wang, D. Sun, L. Wang, J. Yang and H. Jia. 2014. High performance Li_2MnSiO_4 prepared in molten KCl-NaCl for rechargeable lithium ion batteries. Electrochim. Acta. 119: 131-137.

Whittingham, M.S. 2004. Lithium batteries and cathode materials. Chem. Rev. 104: 4271-4302.

Wu, F. and G. Yushin. 2017. Conversion cathodes for rechargeable lithium and lithium-ion batteries. Energy Environ. Sci. 10: 435-459.

Xu, L., F. Zhou, B. Liu, H. Zhou, Q. Zhang, J. Kong, et al. 2018. Progress in preparation and modification of $LiNi_{0.6}Mn_{0.2}Co_{0.2}O_2$ cathode material for high energy density Li-ion batteries. Int. J. Electrochem. 2018: 1-12.

Yang, J., L. Hu, J. Zheng, D. He, L. Tian, S. Mu, et al. 2015. Li_2FeSiO_4 nanorods bonded with graphene for high performance batteries. J. Mater. Chem. A. 3: 9601-9608.

Yang, R., L. Wang, K. Deng, M. Lv and Y. Xu. 2016. A facile synthesis of $Li_2Fe_{1/3}Mn_{1/3}Ni_{1/3}SiO_4/C$ composites as cathode materials for lithium-ion batteries. J. Alloys Compd. 676: 260-264.

Yi, T.F., S.Y. Yang and Y. Xie. 2015. Recent advances of $Li_4Ti_5O_{12}$ as a promising next generation anode material for high power lithium-ion batteries. J. Mater. Chem. A. 3: 5750-5777.

Zaghib, K., A.A. Salah, N. Ravet, A. Mauger, F. Gendron and C.M. Julien. 2006. Structural, magnetic and electrochemical properties of lithium iron orthosilicate. J. Power Sources. 160: 1381-1386.

Zaghib, K., A. Mauger, J.B. Goodenough, F. Gendron and C.M. Julien. 2007. Design and properties of $LiFePO_4$ positive electrode materials for Li-ion batteries. pp. 115-149. *In*: S.S. Zhang [ed.]. Advanced Materials and Methods for Lithium-ion Batteries. Transworld Research Network, Kerala, India.

Zhang, K., X. Han, Z. Hu, X. Zhang, Z. Tao and J. Chen. 2015a. Nanostructured Mn-based oxides for electrochemical energy storage and conversion. Chem. Soc. Rev. 44: 699-728.

Zhang, M., S. Zhao, Q. Chen and G. Yan. 2014. $Li_{2+x}MnSi_{1-x}Al_xO_4/C$ nanoparticles for high capacity lithium-ion battery cathode applications. RSC Adv. 4: 30876-30880.

Zhang, Z., X. Liu, Y. Wu and H. Zhao. 2015b. Graphene modified Li_2FeSiO_4/C composite as a high performance cathode material for lithium-ion batteries. J. Solid State Electrochem. 19: 469-475.

Zhu, J., H. Tang, Z. Tang and C. Ma. 2015. Improved electrochemical performance of zinc oxide coated lithium manganese silicate electrode for lithium-ion batteries. J. Alloy. Compd. 633: 194-200.

Zhu, J., Z. Tang, H. Tang, Q. Xu and X. Zhang. 2016. Titanium dioxide and carbon co-modified lithium manganese silicate cathode materials with improved electrochemical performance for lithium ion batteries. J. Electroanal. Chem. 761: 37-45.

Zou, L., J. Li, Z. Liu, G. Wang, A. Manthiram and C. Wang. 2019. Lattice doping regulated interfacial reactions in cathode for enhanced cycling stability. Nat. Commun. 10: 1-11.

12

Nanostructured Ceramics:
Role in Water Remediation

Kriti Bijalwan[1], Aditi Kainthola[1], Smriti Negi[1],
Himani Sharma[2] and Charu Dwivedi[1,*]

12.1 INTRODUCTION

Water is vital for sustaining all forms of life on earth. Over the years, there has been a decline in the quality of water due to the entry of unwanted material or pollutants into the water bodies (Warren 1971, Goel 2006). This form of environmental degradation can be attributed to the growth in human population in the last 200 years and rapid industrialization. Indirect or direct discharge of pollutants into the water bodies without treatment leads to the degradation in the quality of water (Schweitzer and Noblet 2018). Presence of detrimental matter in large quantities in water makes it inappropriate for its designated usage and for the aquatic ecosystems, this is known as water pollution (Tesh and Scott 2014). The regional and seasonal availability of water and the quality of surface and groundwater influence the economic growth and development to a significant extent (Abraham and Rosencranz 1986, Maria 2003, Rajaram and Das 2008). Therefore, polluting this limited natural resource adversely impacts the growth and development of mankind. Chemical pollution of water has become a considerable concern all around the world (Sartor and Boyd 1972, Helmer et al. 1997). The quality of water has been menaced by raw sewage, industrial waste and oil spills. The natural purification of polluted water is a slow process; therefore, human effort is required to undo the pollution that is of our own creation (Schwarzenbach et al. 2010).

Anthropogenic sources of water contamination consist of organic as well as inorganic substances. Organic sources of contamination include pesticides, insecticides, herbicides, organohalides, pharmaceutical waste, degreasers, adhesives, gasoline, fuel additives, volatile organic chemicals (VOCs) and other organic compounds (Maria 2003, Qu et al. 2013). Bacteria from sewage and livestock farming, food processing waste, pathogens, etc. fall into the broad category of bio-organic waste (Warren 1971, Schwarzenbach et al. 2010). Inorganic contaminants include heavy metals like arsenic, lead, copper, chromium, mercury, antimony, cadmium, etc., which are highly toxic and tend to interfere with the normal functioning of plant and animal bodies. Contaminated water can be remediated for its reuse by traditional approach (filtration using a column of bed and bank material) and modern methods (Goel 2006).

Riverbank filtration systems are used to remove organic matter. Subsurface processes are also available naturally on the Earth's crust for water remediation (e.g. hydrolysis, leaching, oxidation, reduction and precipitation) (Bhattacharya et al. 2018). Complex molecular structures of the organic contaminants prohibit their detection and subsequent removal by the majority of wastewater

[1] Department of Chemistry, Doon University, Dehradun-248001, India.
[2] Department of Physics, Doon University, Dehradun-248001, India.
[*] Corresponding author: cdwivedi.ch@doonuniversity.ac.in

treatment plants. Therefore, a lot of research work is being carried out to develop the methods that are suitable for the estimation and elimination of these contaminants. (Anju et al. 2010, Halder and Islam 2015). Techniques such as ultrafiltration, photocatalysis, adsorption, etc. have been found to be effective in removing organic contaminants from wastewater (Sartor and Boyd 1972, Helmer et al. 1997, Goel 2006, Rajaram and Das 2008).

These techniques are being developed to make them useful in large-scale industrial effluent treatment plants. Photocatalysis and bio-electrochemical systems are known to be energy-efficient and sustainable methods for the treatment of wastewater (Halder and Islam 2015, Schweitzer and Noblet 2018, Baghbanzadeh et al. 2016). Evolution and upgradation of the water purification technologies is required to cope up with the new contaminants being discovered on regular basis (Wang 2007, Theron et al. 2008). The entry of nanotechnology in wastewater remediation has proved to be a noteworthy advancement in this domain. With the progress of nanotechnology, new nanomaterials have been introduced for wastewater treatment. Nano remediation decreases the time required for the cleanup of the pollutants and leads to the *in situ* removal of the contaminant concentration to a great extent (Lu and Astruc 2018).

Nanoparticles (NPs) have been extensively explored for their potential in remediation applications. Living systems have low levels of identifiability for NPs as foreign bodies which may lead to the bioaccumulation of NPs in the living organisms, this is a major challenge associated with the use of NPs. Metals such as silver (Ag) in the form of nanoparticles have been used as biocides (Maldonado et al. 2018). Porous nanocomposites offering nano pockets for containing nanoparticles have also been exploited heavily for their surface area (Theron et al. 2008, Prabhakar et al. 2013, Halder and Islam 2015, Schweitzer and Noblet 2018). Carbon based nanomaterials are also considered cost effective and efficient for water remediation. Carbon nanotubes act like nano sorbents and have been reported to show antimicrobial activity, high porosity, ease of tunability and functionalization (Chen et al. 2015). Srivastava et al. have demonstrated the method for synthesis of radially aligned carbon nanotube filters. It has been reported that the cylindrical membranes removed *Escherichia coli, Staphylococcus aureus* and Poliovirus Sabin 1 from contaminated water resources (Srivastava et al. 2004).

Also, novel membranes consisting of poly(dimethoxysiloxane)-crossed microfluidic channels separated by a nano porous polycarbonate track-etched membrane having surface functionality have potential use in the future water purification technologies (Prabhakar et al. 2013). Nanotechnology can be used both as absorptive technology (sequestration) and reactive technology (degradation). Also, it can be used for the treatment of contaminants at the very source (*in situ*) as well as for the treatment of contaminants after removal (*ex situ*) (Helmer et al. 1997, Prabhakar et al. 2013). Self-assembled monolayers on mesoporous supports (SAMMS) having high selectivity for contaminants is one of the examples of application of nanotechnology for contaminant remediation by adsorption. Other types of nanostructured materials include dendritic polymers that can remove metals such as Cu (II) and Pb (II) from wastewater (Theron et al. 2008).

A ceramic is an inorganic solid made up of either metal or non-metal compounds fabricated at high temperatures. Compared to other materials such as metals and polymers, ceramics possess properties such as strength and brittleness, superior wear resistance, and high thermal, mechanical and chemical stability, which makes ceramic materials useful in a variety of areas (Kiani et al. 2014, Vunain et al. 2017). Revolution has occurred in the use of ceramics by reducing their size to the nanometer levels and enabling them to be applied to biosystems. Advancement in nanotechnology has enabled the manufacturers to develop ceramic materials with unique properties such as transparency, ductility etc. (Chronakis 2005). Advanced ceramics are generally based on oxides or non-oxides or their composites. Typical oxides include alumina (Al_2O_3) and zirconia (ZrO_2), titania (TiO_2), ZnO, SnO_2, etc. Non-oxides include borides, carbides, silicide and nitrides, e.g. silicon carbide (SiC), boron carbide (B_4C), molybdenum disilicide ($MoSi_2$), etc. Newer varieties of ceramics such as those based on (Ca, Sr) (Zr, Ti) O_3, $MgOTiO_2$, $BaO-TiO_2$ and $BaO-TiO_2$ have been developed over the years for their application as dielectric materials in communication and computer technology (Beall and Pinckney 1999).

Advanced ceramic materials and their composites have become a part of everyday life due to their application in a wide range of industries such as electronics, automotive, aerospace etc. Bio-ceramics or advanced ceramics such as zirconia and alumina, having chemical inertness and high mechanical strength, are being used in modern medicine as dental implants and prosthetics (Sigmund et al. 2000, Naslain and Langlais 1986). Ceramic composite materials (CMCs) have emerged as valuable structural components in classic reinforcement from microscale materials to new age materials (Das-Gupta and Doughty 1988). CMCs include ceramics having at least one dimension in the nanoscale range; they can be nanosized or nanostructured particles, nanofibers, thin films or even nanostructured surfaces (Prabhakar et al. 2013, Kiani et al. 2014).

CMCs possess unique mechanical and morphological characteristics such as super plasticity, strength and bioactivity due to their fine grain size, controllable crystallinity and abundant grain boundaries. Mats composed of electro spun ceramic nanofibers, such as ZrO_2, TiO_2 and SnO_2 have been proposed as efficient and durable substrates for noble-metal nanostructures to be used in a variety of catalytic applications (Abellán et al. 2007, Khataee et al. 2011). Xia et al. have synthesized metal nanoparticles (Pd, Pt and Rh) loaded ZrO_2 and TiO_2 nanofiber mats as catalyst. These mats showed good efficiency in cross-coupling reactions as well as in hydrogenation of azo bonds (Formo et al. 2008). Electro spun ceramic nanofibers of hollow TiO_2 are also reported as photocatalysts. The porosity in this system was introduced by using Pluronic (P123) as pore generating agent (Zhan et al. 2006). In this chapter, a brief account of the major ways in which the oxide, carbide, nitride and glass nanoceramics have been reported for the treatment of wastewater and removal of both organic and inorganic water pollutants has been presented. The methods include photocatalysis, adsorption, absorption, Fenton and other treatments like membrane techniques and bioremediation.

12.2 PHOTOCATALYSIS

Solar energy in the form of radiant light and heat energy from the sun has been the most abundant and clean source of energy available. Literature suggests that the solar energy striking Earth in an hour is relatively higher than the yearly energy consumption by humans. Thus, extensive research is being carried out at various levels for the development of materials that are capable of harvesting solar energy efficiently and can be used for wastewater treatment (Lee et al. 2016). Photocatalysis has attracted the interest of many researchers as a sustainable way to solve the problem of eliminating residual dyes pollutants from wastewater (Khataee et al. 2009, Lee et al. 2016). A wide variety of organic compounds discharged into the water bodies from textile industries, paper and pulp industries and pharmaceutical industries are major sources of organic contaminants. These pollutants pose a serious threat to the organisms living in water due to their complex structures and presence of multiple bonds. These organic contaminants, due to their non-biodegradable nature, persist in the environment. When they come in contact with living organisms, they have adverse effects on their well-being. Therefore, the removal of these pollutants prior to the discharge into the natural water resources is essential (Tan et al. 2015).

12.2.1 Oxides

Various techniques have been employed to degrade the residual dyes and other organic contaminants; among these, advanced heterogeneous photocatalysis has proved to be inexpensive and the most effective. Several types of photocatalysts, such as zinc oxide (ZnO), titanium dioxide (TiO_2) and zirconia (ZrO_2) have been actively used in environmental waste management systems (Vinu and Madras 2010). There have been a number of successful attempts of treating textile wastewater using nano oxide ceramics. Nano oxide ceramic composites of zirconium oxide and activated carbon known as ZrO_2-activated carbon assembly (ZrSAC) can be synthesized by microwave synthesis method. These materials have shown high photocatalytic activity and adsorptivity for dye degradation.

The HR-SEM micrographs of ZrSAC samples (Fig. 12.1) reveal that these nanoparticles are spherical and agglomerated. The size of pure ZrO_2 ranged from 20-50 nm, while the size of ZrSAC samples ranged from 20-120 nm. SEM images reveal ZrO_2 loading on carbon support (Suresh et al. 2014). This is due to the fact that activated carbon support acts as a semiconductor which favours the photocatalytic activity. Activated carbon facilitates the promotion of electrons from valence band to conduction band resulting in electron hole pair formation. Thus, the generated electrons interact with the absorbed species of oxygen molecule through the sp^2 hybridized layers of activated carbon. The superoxide radical formed during the interaction further undergoes a reaction with H_2O, initiating the generation of oxidizing radical group. These radicals further bring out the degradation of organic matter (Suresh et al. 2014).

Fig. 12.1 HR-SEM images of zirconium oxide nanoparticles-activated carbon assemblies (left) pure zirconium oxide nanoparticles (right). (Reprinted with permission from Suresh et al., copyright (2014) Elsevier)

Among the various oxide nanoceramics, TiO_2 is the most extensively explored catalyst in photocatalytic reactions because of its chemically and biologically inert nature. Its outstanding optical and electronic properties, high chemical stability, photoactivity, non-toxicity, low cost, reusability and eco-friendliness have made it a promising candidate for its application in photocatalysis. TiO_2 exists in three phases, namely, anatase, rutile, and brookite (Ramasundaram et al. 2013). Anatase nanoparticles are stable at lower temperatures and show better photocatalytic properties because of high surface area and elevated solid-solid interactions. On the contrary, at higher temperatures nano-TiO_2 does not show significant photocatalytic activity (Vinu and Madras 2010, Lee et al. 2013).

Discovery of photocatalytic water splitting on titanium (IV) dioxide (TiO_2) electrodes has also piqued the interest of the researchers in application of TiO_2 as photocatalyst for the wastewater remediation (Fujishima and Honda 1972). The effective photoexcitation of TiO_2 photocatalysts demands that the energy of the incident light should be higher than its band-gap energy for creating electrons in the conduction band and holes in the valence band. Hydroxyl ions and water can act like traps for holes by forming hydroxyl radicals ($OH^•$) and the electrons can combine with the adsorbed oxygen species, forming unstable superoxide species ($O_2^•$) (shown in Fig. 12.2) (Pawar et al. 2018). The reactive oxygen species generated ($OH^•$, $O_2^•$) react with molecules adsorbed on the photocatalyst surface, resulting in its hydroxylation, oxidation and finally mineralization to carbon dioxide and water (Linsebigler et al. 1995, Mills and Le Hunte 1997).

Modifications in the TiO_2 photocatalytic system have been successfully employed in case of organic dye degradation, aimed at improving the efficiency, activity in the visible light radiation and reusability of the catalyst, etc. The main barriers in case of TiO_2 utilization are its activation by high energy ultraviolet light (only 3% of the solar spectrum) because of the wide band gap of TiO_2 (~3.2 eV for anatase and ~3.0 eV for rutile). Photogenerated electron-hole recombination is another major drawback of TiO_2 which leads to deterioration of the photocatalytic activity. Therefore, modifications are done with the aim of reducing the bandgap of TiO_2 to make it visible light activated and to enhance the electron-hole separation process, adopting different strategies such as doping, co-doping, sensitization and coupling with metals (Au, Ag, Ce, Cd, Se) and non-metals (C, S, N, F) (Subramanian et al. 2004). In different studies, 99% dye decomposition was achieved using Ag-TiO₂/UV system over a wide pH range (Özkan et al. 2004, Suwarnkar et al. 2014).

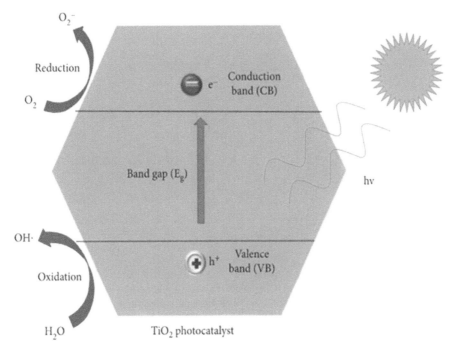

Fig. 12.2 Mechanism of titanium oxide photocatalysis. (Reprinted with permission from Pawar et al., copyright (2018) Pawar et al.)

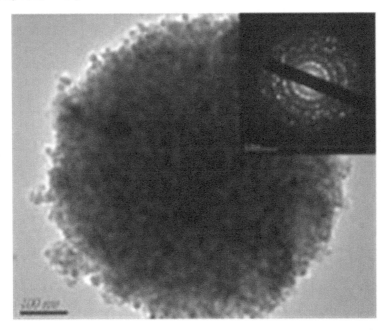

Fig. 12.3 TEM image of a single cerium doped titanium dioxide microsphere. Inset shows the corresponding SAED pattern. (Reprinted with permission from Xie et al., copyright (2010) Elsevier)

Figure 12.3 shows a TEM image of a synthesized polycrystalline single Ce-TiO$_2$ microsphere (~10 nm). When cerium is introduced in the TiO$_2$ matrix, there is a significant increase in MB degradation. This is because TiO$_2$ microspheres have Ce^{4+} dopants, which due to varied valencies and special 4f level serve as electron traps in the reaction (Xie et al. 2010). The authors have also reported metal-loaded TiO$_2$, prepared by loading Pd on TiO$_2$ by using photo deposition method, which is used in Methyl Orange (MO) degradation experiments. Figure 12.4 shows the time dependent degradation of MB and rhodamine B (rhB), respectively, when Ce-TiO$_2$ and Pd-TiO$_2$ photocatalysts were used (Xie et al. 2010).

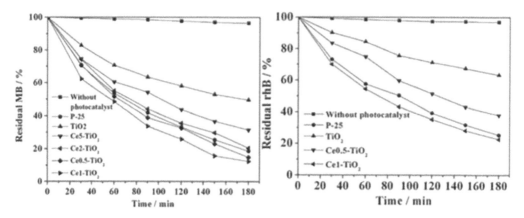

Fig. 12.4 Photocatalytic degradation of methylene blue (left) and rhodamine B (right) as a function of irradiation time over pure titanium dioxide nanoparticles and different cerium and palladium doped titanium dioxide microspheres. (Reprinted with permission from Xie et al., copyright (2010) Elsevier)

For the degradation of MO by Pd-TiO$_2$ nanocomposites, the catalytic efficiency was found to be dependent on the concentration of palladium. This system has been reported to be capable of absorbing light over a wide wavelength range of the solar spectra (Anas et al. 2016). The photocatalytic degradation of different commercial textile dyes was investigated by Khataee et al., using TiO$_2$ nanoparticles immobilized on glass plate under UV light irradiation. SEM images (Fig. 12.5) reveal that the crystallite mean size was 5-10 nm. Absorbance peak decrease indicated the degradation of the aromatic part of the dyes. The attack on the -N=N- azo group by the hydroxyl radical leads to the decolorization of the dye. Ammonium ion is obtained as one of the products of the azo dye degradation process. Hydroxyl radical is generated when a light of wavelength less than 390 nm is illuminated on TiO$_2$ nanoparticles. The electron-hole pair generation leads to the formation of hydroxyl radical which acts as a powerful oxidizing agent (Khataee et al. 2009).

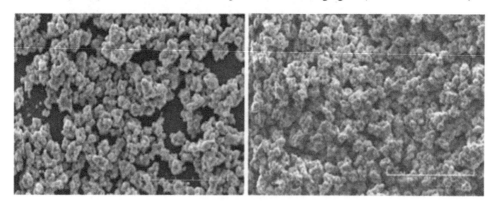

Fig. 12.5 Scanning electron microscopy images of TiO$_2$ nanoparticles deposited on glass plate after initial (left) and final (right) coat. (Reprinted with permission from Khataee et al., copyright (2009) Elsevier)

TiO$_2$ nanowires (NW) were prepared via an alkaline hydrothermal process by Safajou et al. (2017). The composites of Graphene-Palladium-Titanium dioxide nanoparticles (Gr/Pd/TiO$_2$-NPs) and Graphene-Palladium-Titanium dioxide nanowires (Gr/Pd/TiO$_2$-NWs) were synthesized by a combination of hydrothermal and photo deposition methods. To stimulate TiO$_2$ under visible light, transition metals are incorporated into TiO$_2$ in different proportions. Noble metals deposited on a TiO$_2$ surface increase its photocatalytic activity and the transition metals act as electron trapping sites. This leads to interfacial charge transfer in the composites. The nanowires have palladium which is one of the active metals for interacting with the surface of the oxide supports. The results described confirm that the Gr/Pd/TiO$_2$-NWs composites show the highest photocatalytic activity among the synthesized and studied systems due to more available surface area of TiO$_2$ substrate. The process of degradation was studied on rhB as a water pollutant. Pd acts as an electron acceptor by collecting the photoinduced electrons and Gr acts as a high electrically conductive path for photogenerated electrons, resulting in producing active sites for dye degradation. In contrast to nanoparticles, nanowires have much higher surface area and hence numerous active sites leading to higher catalytic efficiency. Modification of the nanowire TiO$_2$ photocatalyst by Gr and clusters resulted in enhanced catalytic activity in comparison with other samples (Safajou et al. 2017).

In another study, photocatalytic degradation of an anthraquinone dye was carried out using TiO$_2$ nanoparticles under UV light irradiation in a rectangular photoreactor. The anthraquinone group is subjected to attack by the hydroxyl radicals. The phenolic compounds formed are spontaneously oxidized to butanedioic acid, benzoic acid and other short chain acids that are ultimately converted into CO$_2$ (Khataee et al. 2011). Pure and La doped TiO$_2$ nanoparticles with different La content prepared by a sol-gel process using Ti (OC$_4$H$_9$)$_4$ as a raw material have been employed in the degradation of phenol in water. La dopant had a great inhibition on TiO$_2$ phase transformation and enhanced the PL intensity. The stronger PL intensity is related to larger content of oxygen vacancy and defects and higher photocatalytic activity (Liqiang et al. 2004, Aware and Jadhav 2016). Shape controlled photo-decomposition efficiency of TiO$_2$ is explained by Yun et al. (2009). The authors have compared the photocatalytic activities of the ellipsoidal and spherical TiO$_2$ for formic acid (HCOOH) degradation. Their studies concluded that the high current density at high aspect ratio of the ellipsoidal TiO$_2$ nanoparticles show better photocatalytic efficiency. Therefore, TiO$_2$ in nano form can be used to treat acidic water with considerably low pH values (Yun et al. 2009).

The presence of drugs in the aquatic media has also been identified in the last decade as a major environmental risk. Pharmaceuticals are generally stable organic compounds and can be eliminated from the body either after being partially or completely converted to water soluble metabolites or in some cases, without being metabolized (Abellán et al. 2007). It is important to eliminate these harmful substances from water resources in order to prevent antibiotic resistance. To prevent the accumulation of pharmaceutical compounds in the aquatic environment, nanosized TiO$_2$ has been investigated as an alternative to degrade these organic pollutants in water. A photocatalytically active stainless-steel filter (P-SSF) was prepared by Ramasundaram et al. (2013), by integrating electro spun TiO$_2$ nano fibers (NFs) on the SSF surface through a hot-press process using a poly (vinylidene fluoride) (PVDF) nanofibers interlayer as a binder. The photocatalytic efficacies of P-SSFs were evaluated based on the efficiencies for photocatalyzed oxidation of selected pharmaceuticals (acetaminophen, cimetidine, sulfamethoxazole and propranolol) in the bench top dead-end filtration set-up. They observed that when the thickness of TiO$_2$ NFs was increased from 10 to 29 mm, the photocatalytic oxidation of cimetidine increased from 42% to 90%. Cimetidine degradation can be attributed to TiO$_2$ mediated production of hydroxyl ion or valence band holes (Ramasundaram et al. 2013). Figure 12.6 shows the SEM images of bare SSF and PVDF/SSF and the schematic of the process followed to integrate TiO$_2$ NFs on PVDF/SSF.

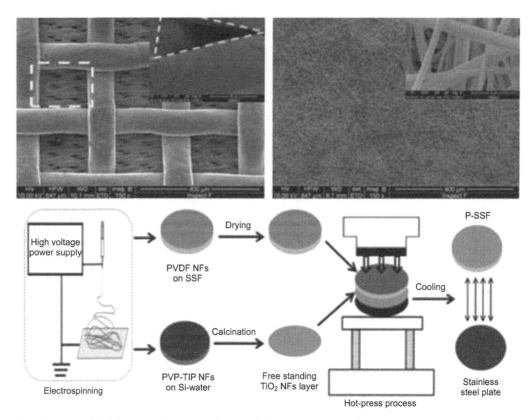

Fig. 12.6 FE-SEM images of bare stainless-steel filter (left), poly (vinylidene fluoride) - stainless-steel filter (right) and the process to integrate TiO$_2$ nano fibers on stainless-steel filter using poly (vinylidene fluoride) binder layer. (Reprinted with permission from Ramasundaram et al., copyright (2013) Elsevier)

12.2.2   Carbides

The dyes generally used as colorants have complex structure and high molecular weight, they are water soluble, difficult to degrade and carcinogenic in nature. Carbide nano-ceramics have also been used in the photocatalytic degradation of the dyes from wastewater, making the process eco-friendly and less tedious in comparison to the other methods (Noll 1991). Cationic MB, rhB, fuchsine and anionic Congo red dyes have been successfully degraded using a ternary nanocomposite system as a catalyst under visible light illumination. The system was developed by suspension drying, using graphitic carbon nitride based on which silicon carbide and photoluminescent carbon dots were anchored (CN/SiC/C-Dots). The photoactivity of the system was found to be much higher than single, binary or bulk systems due to the coupling of graphitic carbon nitride with SiC and carbon dots. Being versatile, this system can be used for the removal of different pollutants that persist in water bodies. Figure 12.7 shows the HRTEM and TEM images of CN/SiC/C-Dots (Asadzadeh-Khaneghah et al. 2020).

It has been reported that 3C-SiC nanoparticles can be used to make g-C$_3$N$_4$ a visible light active catalyst with reduced charge recombination. Xu et al. demonstrated the photocatalytic activity of these 3C-SiC/g-C$_3$N$_4$ nanocomposites for MO dye degradation under visible light (Xu et al. 2017). Nanostructured boron carbide (B$_4$C) synthesized by solvothermal method has been reported to possess unique characteristics for industrial wastewater treatment. Photocatalytic degradation of MB and industrial Synazol yellow dye has been reported to use nanostructured B$_4$C. It has shown remarkable efficiency for dye degradation owing to the defect states or shallow states present in the material which lead to the enhancement of the photocatalytic activity of nanostructured B_4C. Increase in concentration of B_4C catalyst corresponds to a larger number of surface-active sites and hence higher dye degradation rate. The reusability of B_4C exhibited its cost-effectiveness and high capability as a metal free visible light harvesting photocatalyst. These nanoparticles were found to be non-cytotoxic; this makes them environmentally benign material for water remediation (Singh et al. 2018a).

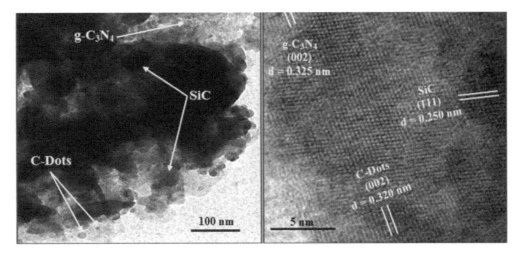

Fig. 12.7 TEM (left) and HRTEM (right) images of the CN/SiC/C-Dots nanocomposite. (Reprinted with permission from Xu et al., copyright (2017) The Royal Society of Chemistry)

B_4C/SnO_2 nanocomposite have also been reported to be prepared by reflux method. The addition of tin oxide in the base matrix leads to the increase in the defect states in boron carbide as indicated by the calculated texture coefficient and Nelson-Riley factor. Also, the composite was used for the catalytic degradation of industrially used dyes such as novacron red huntsman (NRH) and MB. The degradation analysis of industrial pollutants established the potential of the nanocomposites as efficient catalyst. Degradation study shows that this catalyst can degrade MB and NRH dye significantly under sunlight irradiation. Figure 12.8 shows the TEM image of the B_4C/SnO_2 catalyst along with its catalytic efficiency for the degradation of MB (Singh et al. 2018b).

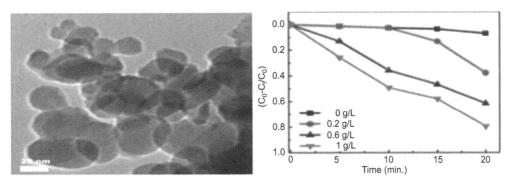

Fig. 12.8 TEM image of B_4C/SnO_2 nanostructures (left) and degradation efficiency of different amounts of B_4C/SnO_2 catalyst with aqueous dispersion of methylene blue dye as function of irradiation time (right). (Reprinted with permission from Singh et al., copyright (2018) Springer Nature)

12.2.3 Nitrides

The high degree of aromaticity and extensively conjugated chromophores impart stability to synthetic dyes such as MO towards degradation. Green photoreduction of MO has been reported by using Au nanoparticles loaded graphitic carbon nitride (g-C$_3$N$_4$) nanosheets. These nanosheets were prepared by ultrasonication-assisted exfoliation of bulk g-C$_3$N$_4$ via green photoreduction of Au (III) under visible light. In comparison with the bulk carbon nitride, carbon nitride nanosheets and AuNP/bulk carbon nitride hybrids, the Au nanoparticles loaded graphitic carbon nitride (g-C$_3$N$_4$) nanosheets showed superior photocatalytic activities for MO dye decomposition under visible light. AuNPs check the recombination of photogenerated electron-hole pairs and the excitation of the surface plasmon resonance in AuNPs enhances the photocatalytic activity. Au/(g-C$_3$N$_4$) nanosheets are, therefore, potential functional material for degradation of organic pollutants and other photocatalytic applications (Cheng et al. 2013).

Ternary composites obtained by depositing both Ag nanoparticles (AgNPs) and carbon dots (CDs) onto the surface of g-C$_3$N$_4$ nanosheets (CNNS) were used for the sunlight facilitated photodegradation of 4-Nitrophenol and MO, which are persistent organic pollutants present in industrial wastewater. Figure 12.9 shows the FESEM images of AgNPs/CDs/CNNS (Dadigala et al. 2017). This system showed higher light harvesting capability and better charge carrier separation than the individual, binary and other systems. The photodegradation of these pollutants was mainly due to O$_2$· species and photogenerated holes. Also, due to high reusability and excellent stability, the system has potential application in treatment of other organic pollutants present in wastewater (Dadigala et al. 2017).

Fig. 12.9 FESEM images of graphitic carbon nitride nanosheets deposited with carbon dots and silver nanoparticles. Reprinted with permission from Dadigala et al., copyright (2017) Elsevier

Sulphur-doped carbon nitride porous rods are also reported for photocatalytic degradation of rhB under visible light (Fan et al. 2017). The surface area of the doped nanorods was found to increase with increase in the heating temperature. The sulphur doping made the nanorods exhibit broader light adsorption range and narrower band gaps than that of bulk g-C$_3$N$_4$. The sulphur-doped carbon nitride porous rods' samples exhibit superior physical adsorption and photocatalytic activity compared to bulk g-C$_3$N$_4$; this is because of the synergetic effects of sulphur doping. Due to the efficient photodegradation of rhB by the sulphur-doped carbon nitride rods, these nanorods have potential use in the treatment of organic pollutants present in wastewater which are otherwise difficult to remove (Fan et al. 2017).

Magnetically separable g-C$_3$N$_4$/Fe$_3$O$_4$/Ag$_2$CrO$_4$ nanocomposites have also been reported to be employed as visible-light-driven photocatalysts for the removal of organic pollutants such as rhB in aqueous solution. The results showed that the g-C$_3$N$_4$/Fe$_3$O$_4$/Ag$_2$CrO$_4$ (20%) nanocomposite exhibited superior activity, higher than that exhibited by g-C$_3$N$_4$ and g-C$_3$N$_4$/Fe$_3$O$_4$ samples. Figure

12.10 shows the mechanism of degradation of rhB over the g-C$_3$N$_4$/Fe$_3$O$_4$/Ag$_2$CrO$_4$ nanocomposites (Habibi-Yangjeh and Akhundi 2016). The efficient charge separation and visible light driven activity leads to higher catalytic efficiency of the system. Due to easy magnetic separation and high recyclability of the nanocomposites from the reaction medium, an efficient visible-light-driven photocatalyst based on g-C$_3$N$_4$ can prove to be of importance in removal of pollutants from wastewater (Habibi-Yangjeh and Akhundi 2016).

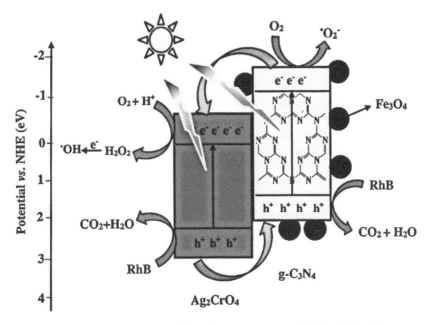

Fig. 12.10   The degradation mechanism of RhB in the presence of g-C$_3$N$_4$/Fe$_3$O$_4$/AgCrO$_4$ nanocomposites. Reprinted with permission from Habibi-Yangjeh and Akhundi, copyright (2016) Elsevier

Ternary photocatalysts of g-C$_3$N$_4$/CDs/AgBr nanocomposites have been synthesized by allowing *in situ* growth of AgBr nanoparticles on CD modified g-C$_3$N$_4$ nanosheets. These g-C$_3$N$_4$/CDs/AgBr nanocomposites are said to exhibit excellent photocatalytic efficiency for the degradation of organic pollutants like rhB, as compared to the individuals or their binary combinations. The introduction of CDs onto g-C$_3$N$_4$/AgBr can improve the photocatalytic activity, as CDs function as the light absorber and electron mediator. Moreover, it has been reported that the catalytic system was reusable even after four photodegradation cycles (Miao et al. 2017). Boron nitride and titanium oxide nano-composites' (BN-TiO$_2$) photocatalysts synthesized by the ice bath method have been reported for efficient dye degradation. The high specific surface area offered by the material was beneficial for catalytic oxidation. The MO photodegradation rate of BN-TiO$_2$ is reported to be 79% in 200 min, higher than that with TiO$_2$ (only 32% in the visible region). BN-TiO$_2$ can be used in water pollutant removal and water purification (Singh et al. 2017).

Hexavalent chromium Cr (VI) is a toxic metal ion and is introduced in the environment by many anthropogenic activities like leather tanning, metal-electroplating, wood preservation, textile manufacturing, etc. Generally, Cr is available in two stable oxidation states in aqueous environment, i.e. trivalent (III) and hexavalent (VI), having distinct chemical and bio-chemical properties. Among Cr (VI) and Cr (III), the former is identified as a mutagen, is carcinogenic and is a much more hazardous species even in trace amounts. Therefore, the reductive transformation of Cr (VI) to Cr (III) in wastewater is essential. Highly efficient Pd/CeO$_2$/g-C$_3$N$_4$ photocatalyst synthesized by a simple hydrothermal and ultrasonic deposition process has been reported for the reduction reaction of Cr (VI) to Cr (III). This ternary system showed enhanced visible light driven photoactivity due to reduced charge recombination (Saravanakumar et al. 2017).

Photocatalytic reduction of Cr (VI) by sulphur doped carbon nitride microsphere has been reported. The sulphur species doped into the lattice of carbon nitride is identified as formation of C-S and N-S in the hybrid framework. Sulphur doping improves conjugated structure, expands visible light harvesting and elevates conduction band reduction potential. Also, the reduction rate for the sulphur doped carbon nitride microsphere was found to be higher than the un-doped sample. This work provides a method for removal of toxic heavy metals in neutral condition (Cui et al. 2018).

12.2.4   Glass Ceramics

TiO_2 and ZnO are the most active photocatalysts under UV irradiation and they show best performance in nanosized particles (Hashimoto et al. 2005, Chen and Mao 2007). Therefore, TiO_2 and ZnO usually need to be coated on appropriate substrates, such as on the surface of a glass carrier or a ceramic for practical applications (Fujishima et al. 2000, Ollis 2000). However, it was seen that the surface was damaged as a result of easy peeling off of the films from the substrate over time, which in turn leads to short service life. Glass-ceramics have been reported to overcome this problem due to their precipitation in a glassy matrix. Margha et al. have reported the synthesis of a nanocrystalline transparent glass-ceramic containing two photoactive components (ZnO and TiO_2) having low melting temperature with SiO_2, TiO_2, ZnO, B_2O_3, Na_2O, K_2O, P_2O_5, Li_2O and BaO as their components (Margha et al. 2015). The photocatalytic activity of the prepared glass-ceramic materials was investigated based on its ability to degrade humic acid (HA) which is the major disinfection by-product (DBP) in water.

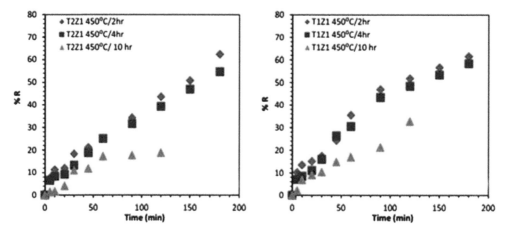

Fig. 12.11 HA removal by the glass ceramic catalysts prepared by using titanium dioxide (left) and zinc oxide (right) in different ratios. Reprinted with permission from Margha et al. copyright (2015) Elsevier

Figure 12.11 shows the variation in HA removal with the increase in irradiation time using the samples. The samples contained TiO_2 and ZnO in different ratios, heat treated for different time duration and showed different photoactivity. Heat treatment for 4 hrs at 450°C was reported to be the best condition for the development of TiO_2 and ZnO crystals and higher photocatalytic activity. The prepared materials were efficient for the degradation of HA, and therefore decreased the HA content in the water. An improvement of the glass ceramic microhardness was caused by the high content of TiO_2 or ZnO. The assembly was reported to be efficient in the dye degradation process. Therefore, the transparent nanocrystalline glass-ceramics have potential for photocatalytic wastewater treatment (Margha et al. 2015).

Many researchers have used TiO_2 powder for the degradation of organic pollutants present in wastewater under UV irradiation. TiO_2 coating on the glass substrate is an upcoming area of interest for photocatalytic degradation of pollutants and photocatalytic antibacterial activity. Kumar et al. have reported the antibacterial properties of TiO_2 (anatase) micro crystallized BaO-TiO_2-B_2O_3 glass ceramic (Xu et al. 1999, Sayilkan et al. 2007). They have reportedly achieved a 98.3% of bacterial inhibition under the solar irradiation. Singh et al. developed stable rutile phase of TiO_2 nanocrystals in the CaO-BaO-B_2O_3-Al_2O_3-TiO_2-ZnO (CBBATZ) glass by the process of heat treatment. The photocatalytic activity of these glass nanocomposite was investigated and the effect of different heat-treated glass nanocomposites on the photocatalysis and transparency was studied. The photocatalytic activity of the material is attributed to the TiO_2 crystallization and band gap reduction. Electron hole pairs played a significant role in the generation of the Reactive Oxygen Species (ROS). Electron hole pair generation on UV irradiation led to the formation of reactive OH species. These react with the pollutants (dyes) and ultimately produce the degradation products (Singh et al. 2019a).

CaF$_2$ nano crystals in glasses have recently gained attention due to their antibacterial and photocatalytic behavior (Bocker et al. 2015). Shinozaki et al. have reported to fabricate nanocrystals of CaF_2 (~10 to 70 nm) in $25CaF_2$-$5CaO$-$20Al_2O_3$-$50SiO_2$ glass composition (Shinozaki et al. 2011). The melt quenching method followed by heat treatment at a temperature of 650°C was used for the processing of glass ceramics containing CaF_2 nanoparticles. Singh et al. fabricated transparent calcium borate glass ceramics with surface crystallized CaF_2 at room temperature (Singh et al. 2019b). Crystallization of CaF_2 on the surface made the glass ceramic a photocatalytic and antibacterial surface ceramic. The glass-ceramic sample containing the highest fraction of CaF_2 crystals was found to have the best photocatalytic and antibacterial activity as compared to the other samples. Around 68% of MO dye was degraded using the assembly. It was noticed that the presence of CaF_2 particles on the surface of the glass can also help in the bacterial depletion by rupturing the flagella and cell walls of bacterial cells and thus helping in the removal of bacteria from wastewater. Fluoride ions from CaF_2 change the membrane activity of the bacterial cells which ultimately leads to their death (Vakifahmetoglu 2011).

12.3 SORPTION

Sorption is an alternate technique preferred for waste removal due to the low cost, flexibility, simple design and operation, and high efficiency. The type of sorbent used in the sorption process is an important factor for a successful purification process. Clays, carbon nanotubes (CNTs), metal oxides, tree barks, sawdust, cotton seed hulls, fly ash, charcoal, polymers and glass-ceramics are some of the popular absorbent materials (Vakifahmetoglu 2011, Ahmetoglu et al. 2017). Removal of pollutants by adsorption is a cost-effective alternative to other water purifying procedures. Water purifying assemblies involving adsorption require low capital investments, low running costs and have the capability of treating large volumes of water at one go which makes them highly effective. To make these assemblies more economically viable, the demand for low cost adsorbent material is increasing at a tremendous rate. Researchers all over the world are working towards exploring materials having suitable structural, compositional and chemical characteristics, making them good sorbents having high sorption efficiency and high retention potential.

12.3.1 Oxides

Oxides nanoceramics like aluminium oxide, zirconium oxide, zinc oxide etc. have been utilized by many researchers as effective low-cost adsorbents. A variety of pollutants have been removed by adsorption using oxide nanoceramics including phosphates, organic dyes, persistent organic pollutants (POPs), heavy metals, di and polyprotic inorganic acids and radioactive material etc. (Tan et al. 2015). Adsorption is commonly used as the final step to remove organic and inorganic contaminants in water and wastewater treatment. Drawbacks of conventional adsorbents are their low selectivity and specificity, slow adsorption kinetics and relatively small surface area.

Nano-adsorbents offer significant improvement with their associated sorption sites and extremely high surface area, short intraparticle diffusion barrier, and tunable surface morphology (Qu et al. 2013). It is possible that nanomaterials will become more economical than conventional sorbents for sorption applications. The adsorption capacities of different nanomaterials (such as CeO, ZnO) for various heavy metals (Ni (II), Cu (II) and Cr (VI)) have been reviewed intensively by Hua et al. (Hua et al. 2011). The high adsorption capacities exhibited by these nanomaterials suggest that they can be used effectively in wastewater treatment (Tan et al. 2015).

Aluminium and zirconium oxides are attractive adsorbent choices for water purification. Aluminium oxide is abundant and has low cost while zirconium oxide has good resistance to oxidizing agents, high thermal stability, low water solubility and is non-toxic (Xie et al. 2010, Su et al. 2013). According to studies, aluminium oxide and zirconium oxides have been found to be good phosphate adsorbents. Leaching of phosphates from soils into waterways can cause significant damage to fresh water bodies that are at a risk of eutrophication. WHO has suggested the maximum discharge limit of phosphorus 0.5-1 mg/L. The removal of phosphates is important to prevent the eutrophication of aquifers and freshwater. It has been reported that aluminium oxides strongly adsorb the phosphates and organophosphate on their surfaces. Di-phenyl phosphates, phenyl phosphate or orthophosphate are adsorbed on the surface of alumina (Johnson et al. 2002). Genz et al. have studied phosphate removal by adsorption for its suitability as a post treatment step for membrane bioreactor effluents low in phosphorus concentration and particle content. In this report, activated aluminium oxide has been used as an adsorbent in fixed bed adsorbers for the removal of phosphate from wastewater. Adsorption mechanism suggested in the study involves the formation of inner or outer sphere complexes or the surface precipitation of $AlPO_4$ on the surface at low pH (Genz et al. 2004).

Zirconium oxides have also shown high adsorption capacity due to highly hydrated structure and porous nature. It is reported that zirconium oxide has one of the highest values of adsorption capacity for phosphate removal which was 99.01 mg/g at pH 6.2 even in the presence of high concentration of competing ions. The prepared amorphous zirconium oxide nanoparticles exhibited 80-90% phosphate removal capability even after desorption. This suggests that the reported ceramic nano adsorbents could be reused several times, which can lower the operation costs of many phosphate removal systems (Su et al. 2013). Arsenic is a toxic metalloid even at a very low concentration. A lot of research work has been carried out for effective removal of this metalloid from wastewater. Aluminium oxide nanoparticles supported on cryogels have been reported to efficiently adsorb 69% of the amount of the arsenic present in wastewater (Peng and Di 1994, Ren et al. 2011). Chromium (Cr) in the form of Cr (VI) is toxic and known mutagen for most organisms. Chromium electroplating is identified as a major source of Cr (VI) emissions in the environment. The permissible limit for Cr (VI) for industrial wastewater lies in the range of 0.1 and 0.5 mg/L (Ying et al. 2015).

Aluminium oxide was used as an absorbent for the removal of Cr (VI) from electroplating wastewater, the maximum adsorption capacity reported was 78.8 mg/g. The adsorption of Cr (VI) takes place by an outer sphere complex. The formation of outer sphere complex is found to be decreased at higher ionic strength of the solution, establishing that the outer sphere complex formation between adsorbed Cr (VI) and Al_2O_3 is the mode of removal of the chromium ion (Álvarez-Ayuso et al. 2007). USEPA and WHO along with most national environment protection agencies have identified Pb as one of the most toxic and carcinogenic pollutants. The permissible limit of Pb in drinkable water is between 10-15 µg per/L. Several inorganic nano oxides have been employed for the effective removal of lead from water through the formation of inner-sphere complexes with the contaminants. Hybrid nanomaterial made up of zirconium oxide nanoparticles encapsulated in polystyrene beads have been employed in the sorption of Pb (II) (Trivedi et al. 2001). The polymer matrix entrapped ZrO_2 nanoparticles are also proposed as specific adsorbents for Pb^{2+}. The specificity of this system relies on the inner-sphere complex formed by Pb ions with the nanoparticles of ZrO_2.

Negatively charged functional groups immobilized onto the polymer matrix enhance the diffusion of positively charged Pb (II) inside the adsorbent due to electrostatic interactions. Zr-MPS nanocomposites consist of two different sorption sites for the removal of Pb (II) by two routes. One takes place by nonspecific interaction of sulphonate group with Pb (II) and other by strong specific coordination of Pb (II) with ZrO_2 (Zhao et al. 2011). Copper in water bodies can enter the food chain and subsequently lead to bioaccumulation leading to severe health problems. Hence, the water from the electroplating industries needs to be properly treated before its discharge into water bodies. Zinc oxide and magnesium oxide, both ceramic nano-adsorbents, have been employed for the removal of copper from industrial wastewater. In this study, it was observed that MgO was a better adsorbent than ZnO owing to its high adsorption capacity and spongy nature. The SEM analysis of the nanoparticles (shown in Fig. 12.12) reveals porous structure of MgO and rod-like structure of ZnO and shows that these materials have larger surface area (Rafiq et al. 2014).

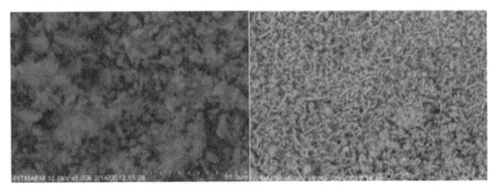

Fig. 12.12 SEM micrographs of sponge-like magnesium oxide (left) and rod-like zinc oxide (right). Reprinted with permission from Rafiq et al., copyright (2014) Elsevier

Another ceramic, namely, MgO in the form of nanoflowers has been employed in the removal of several ions such as Co (II), Cd (II), Zn (II), Cu (II), Mn (II), etc., from synthetic wastewater. These nanoflowers can be used to treat the metal rich industrial effluents. Flower-like nanostructure of MgO nanoparticles can be obtained by using Acacia gum as a shape directing agent. In the Acacia gum solution, the surface of nanosized MgO acquired a positive charge. The negatively charged carboxylate group of gum Acacia forms electrostatic and H-bonding interactions with the positively charged surface sites on MgO and leads to nanoflower formation. Further, the particle size was found to be greater than bare MgO nanoparticles. The surface of MgO nanoflowers have the tendency to attract positively charged metallic species. Figure 12.13 shows the SEM images of the MgO nanoflowers (Srivastava et al. 2015).

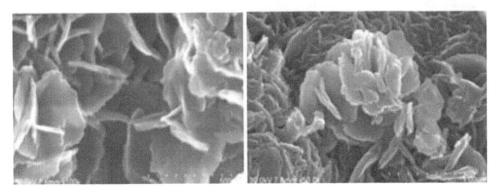

Fig. 12.13 SEM micrographs of nanoflowers of magnesium oxide. Reprinted with permission from Srivastava et al., copyright (2015) Elsevier

MgO has been employed for the purpose of removing of U (VI) from aqueous solution. Three types of MgO nanoparticles Mg600-N, Mg600S and Mg600Cl were reported which exhibited excellent adsorption capacities. Chemical adsorption plays a leading role in the removal process by the MgO nanoparticles which are strongly pH dependent. Hence, the use of MgO nanoparticles is one of the better alternatives for the removal of radioactive uranium from the wastewater of various nuclear facilities (Wang et al. 2017, Nagpal and Kakkar 2019). Nanosized aluminium oxide has been reported for the adsorptive removal of industrial dyes. The efficiency of γ-Al$_2$O$_3$ as an adsorbent was found to be dependent on its shape, size and reaction temperature. The aluminium source used for the preparation of γ-Al$_2$O$_3$ and the structure directing agent also affected the adsorption performance towards dye removal (Liu et al. 2015a).

Porous zinc oxide nanospheres derived from zinc hydroxide carbonate precursor have also been used in the removal of Congo red dye. The maximum adsorption amount reported was 334 mg/g. Porous zinc oxide nanoparticles (shown in Fig. 12.14) exhibited better adsorption efficiency as compared to commercial zinc oxide and have proved to be an efficient adsorbent for the removal of anionic organic dyes from wastewater. The enhanced adsorption capacity of both mesoporous γ-Al$_2$O$_3$ and hierarchical porous ZnO nanospheres can be attributed mainly to the large specific surface area which provides more active sites for the adsorption process (Cai et al. 2012, Lei et al. 2017).

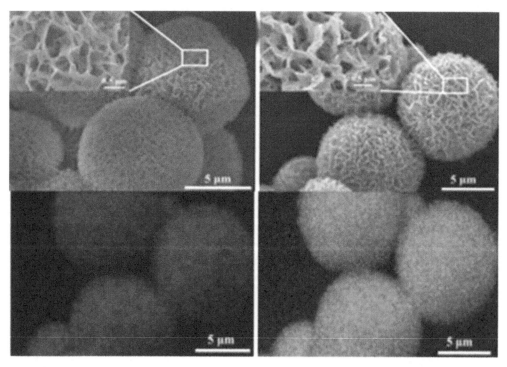

Fig. 12.14   SEM images of the zinc hydroxide carbonate precursor of porous ZnO (left) and porous ZnO sample (right) along with the element mapping images of the porous ZnO. Reprinted with permission from Lei et al., copyright (2017) Elsevier

Organic compounds that are resistant to environmental degradation are termed as persistent organic pollutants (POP), which are a matter of global concern due to their higher bioaccumulation and long-range transportation. POPs include organo halogens and furans, polychlorinated biphenyls, hexachlorobenzene, cyclic aromatic compounds and organotin compounds. POPs are hazardous and toxic, posing severe negative effects on ecosystems and the health of living beings. POPs can also act as endocrine disruptors by disrupting the reproductive cycle of animals and humans alike.

Many developed countries have banned the usage of POPs, yet in certain developing nations these are still being used due to their low cost. Triphenyltin chloride (TPT) is one such POP used as biocide, particularly to prevent fouling of boats and ships. Hence, there is an alarming need for the removal of this pollutant from dockyard and industrial water before it's discharged into the environment. ZnO nanoparticles have been reported as adsorbents for the efficient removal of TPT from dockyard waste. 0.50 g of ZnO was reported to be able to remove 95.4% of TPT from the contaminated dockyard wastewater (Ayanda et al. 2014).

Manganese dioxide nanoparticles have been reported to be anchored on reduced graphene oxide to form composites (RGO-MnO_2). The nanocomposites were further immobilized on river sand with the help of chitosan. These composites formed by redox like reaction between the metal precursor and RGO were used in removing Hg (II) from water. They were reported to possess a high distribution coefficient for Hg (II) uptake. The reported distribution coefficient values for the developed composites were higher than those for the parent RGO and GO for the Hg (II) uptake (Sreeprasad et al. 2011). Cellulose and nanoscale manganese oxide composites (C-NMOC) have also been reported for the efficient and fast elimination of lead from the aqueous solutions by physisorption over a wide pH range. The lead adsorption capacity of C-NMOC was reported to be a multiple times higher than the commercially available manganese oxide and at least twice larger than the nanoscale manganese oxide (Maliyekkal et al. 2010).

12.3.2 Carbides

The removal of MO and Congo red from wastewater was studied using SiC nanoparticles activated carbon composite as adsorbent. The adsorption capacity was found to be higher for the system than that for activated carbon. SiC modifies the surface of activated carbon which enhances the activation process, making it a promising and new adsorbent for the removal of acidic dyes during wastewater treatment. The thermodynamic adsorption parameters showed that the adsorption of both dyes onto the adsorbent process was spontaneous under experimental conditions (Ghasemian and Palizban 2016). Direct or sewer disposal of waste drugs like Aflatoxin B1 (AFB1) is hazardous and causes serious water pollution. It is not only toxic but also carcinogenic, and therefore its removal from waste water is important. It has been reported that SiC nanoparticles can be used as nano adsorbent for the adsorption of AFB1 in basic medium. The results show that the adsorption amount of AFB1 by SiC nanoparticles was higher than that obtained by using silicon nanoparticles as adsorbent (Gupta et al. 2017).

Endocrine disrupting chemicals (EDC) such as Bisphenol A (BPA) have potential environmental and human health hazards: they can cause complications like infertility, deformities and mutations in wildlife. BPA is used to synthesize polycarbonates and epoxy resins etc., and can remain in water for decades after their release into the environment. Therefore, it is necessary to search for effective methods and advanced catalysts that are highly efficient for the elimination of BPA. Jin et al. have reported zero-valent iron/iron carbide nanoparticles embedded within the N-doped carbon matrix (N-doped Fe^0/Fe_3C@C), and the system showed good adsorption capacity for BPA under a wide pH operation range for BPA removal. The main adsorption mechanism of BPA (shown in Fig. 12.15) was the π-π interactions that take place between the π orbital on the basal carbon planes and the electronic density present in the aromatic rings of BPA. The improved degradation was reported to be a result of generation and evolution of the radicals in phenol oxidation. N-doped Fe^0/Fe_3C@C can not only efficiently adsorb and degrade BPA but it has also shown potential of such systems for the removal of organic pollutants (Jin et al. 2018).

The extremely stable magnetic titanium carbide nanocomposite has been reported for treatment of wastewater and removal of pollutants. It was reported to be successfully synthesized via hydrothermal reaction and was employed for the adsorptive removal of mercury (Hg) ions. Hg is one of the most toxic heavy metals with comparatively high solubility in water and can cause deleterious effects on living organisms such as central nervous system damage, chromosomes alteration, birth defects, chest pain and kidney illnesses in living organisms. The techniques being previously used have disadvantages because they require high energy to operate and produce toxic sludge. The TiC nanocomposites exhibited high Hg (II) removal over a wide pH range and was reported to be stable and reusable for a multiple cycles (Shahzad et. al. 2018).

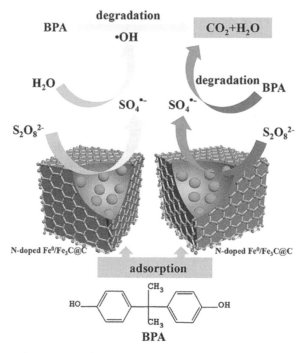

Fig. 12.15 Mechanism of adsorption of BPA by N-doped $Fe^0/Fe_3C@C$. Reprinted with permission from Jin et al., copyright (2018) Elsevier

Titanium carbide nanosheets are also reported for the removal of Cr (VI) from water. Ying et al. have synthesized highly stable TiC nanosheets by intercalation process using HF solution. Their system was able to remove Cr (VI) from water with the residual concentration of chromium less than 5 ppb. Along with the removal of Cr (VI), these nanosheets can also absorb the Cr (III) simultaneously (Fig. 12.16). These significantly broaden the range of application of titanium carbide nanosheets in water treatment (Ying et al. 2015).

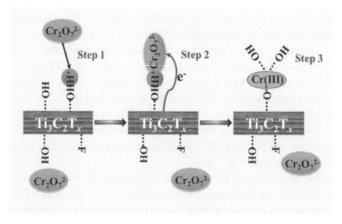

Fig. 12.16 Mechanism of Cr (VI) removal from wastewater by the titanium carbide nanosheets. Reprinted with permission from Ying et al., copyright (2015) American Chemical Society

12.3.3 Nitrides

Organic solvents and dyes discharged by the textile industries, oil spillage, paper and tannery industries are primary pollutants of water sources. Sorption is one of the most explored techniques for water remediation but the tedious regeneration and recycling of sorbent are the major challenges associated with the process. Layered boron nitride nanosheet (BNNS) structures, due to their high porosity and high uptake gravimetric capacity, have been reported to efficiently absorb these water pollutants. Due to high specific surface area, these nanosheets can absorb oils and organic compounds up to 33 times their own weight while repelling water as a result of super hydrophobicity. Because of their strong resistance to oxidation, the saturated BNNS can be prepared for the reuse by heating in air or burning (Lei et al. 2013).

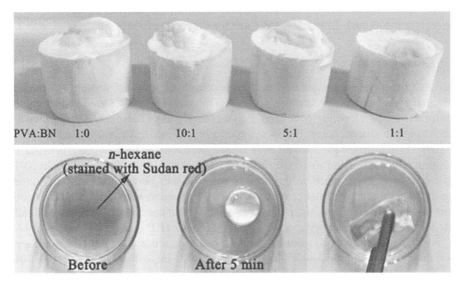

Fig. 12.17 Removal of *n*-hexane from water by BN-modified PVA aerogel. Reprinted with permission from Zhang et al., copyright (2017) Elsevier

Lian et al. synthesized boron nitride (BN) ultrathin fibrous networks (average diameter of ~8 nm) by one-step solvothermal process. The nanonets were reported to exhibit excellent performance for wastewater treatment due to the property of ultrafast adsorption, especially for synthetic dyes like MB and other pollutants through the filtration process (Lian et al. 2013). Reusable Oxygen-doped bundle-like porous boron nitride (OBPBN) has been reported for efficient adsorption of heavy metal ions such as Cu^{2+}, Pb^{2+}, Zn^{2+} and Cr^{3+}. The OBPBN have been synthesized through a reaction involving thermal evaporation of the solvent and high-temperature pyrolysis processes. They have interconnected mesoporous microstructures and facilitate monolayer adsorption of the heavy metal ions. The high surface area, porosity and oxygen doping induced lattice defects lead to high metal ion adsorption capacity (Liu et al. 2018).

Nanosheet-structured boron nitride spheres (NSBNSs) consisting of radially oriented ultrathin nanosheets with the surface-oriented sheet edges were synthesized by a thermal evaporation method. Formation of NSBNSs is a temperature-controlled reaction which yields high surface area and adsorptivity due to a unique structure. The high-density edge sites on the surface result in a high adsorption capacity for dyes, oil and heavy metal ions. It has been reported that the oil uptake by NSBNSs reaches 7.8 times its own weight. Also, the adsorption for malachite green and MB are reported to be 324 and 233 mg/g, while those for the toxic heavy metals Pb^{2+}, Cu^{2+} and Cd^{2+} are 536.7, 678.7 and 107.0 mg/g, respectively. The adsorption capacities of the NSBNSs for Pb^{2+} and Cu^{2+} are much higher than those of the previously reported adsorbents. The results advocate the potential of NSBNSs for water cleaning and treatment (Liu et al. 2015b).

Boron nitride (BN)-modified PVA aerogel and C_3N_4-modified PVA aerogel were prepared by freeze-drying method, and have been reported for the removal of organic pollutants. PVA acts as a scaffold of aerogel to support BN and C_3N_4 nanosheets, which can modulate the surface properties like wettability from hydrophilicity (0°) to hydrophobicity (94.9°-100.8°) of the PVA aerogel. Also, the modified PVA aerogel possesses low density (41.8-60.0 mg/cm³), favorable porous structure and good adsorption capacity (12-38 g/g), making it a promising wastewater treatment material. Figure 12.17 shows the removal of *n*-hexane from water by BN-modified PVA aerogel (Zhang et al. 2017).

Boron nitride (BN) nano carpets synthesized by a hot-press thermal method consist of twisted nanorods embedded on BN nanosheets. Besides the high-density structural defects, BN nano carpets also have a large surface area. Due to these characteristics, BN nano carpets have been reported to exhibit a high adsorption rate for MB. Furthermore, these BN nano carpets can also selectively adsorb MB. Therefore, the selectivity of BN nano carpets for organic molecules can be tapped for recovery of valuable organic compounds from wastewater and for water remediation (Zhang et al. 2012).

12.3.4 Glass Ceramics

It is noted that nano glass-ceramics have various applications including waste water management. The molecular composition of preceramic polymers (PPs) can be easily altered and processed to get desirable forms by plastic forming techniques. Polymer derived ceramics are nanostructured materials composed of carbon, silicon, boron, nitrogen and oxygen obtained by thermal decomposition of PPs. Among all the PPs, the most economical ones are polysiloxanes, as they can be processed and used under ambient temperature and are readily available (Haug et al. 1999, Vakifahmetoglu 2011, 2016). Zehdanyli et al. performed the pyrolysis of PPs polysiloxanes between 800-1400°C temperature resulting in the formation of silicon oxycarbide (SiOC) amorphous material. Carbonaceous SiOC can be obtained by HF etching of the pyrolyzed product. Both the samples were tested as adsorbent materials for the removal of heavy metals' ions (Pb (III), Cr (III) and Cd (II)) and dyes (MB, rhB and Crystal Violet (CV)) from aqueous solutions. It has been reported that high surface area SiOC samples treated with HF showed high adsorption affinity for these dyes (Zeydanli et al. 2018).

Arsenic is one of the major public health concerns, according to the World Health Organization, and the inorganic compounds of arsenic like arsenic (III) oxide, arsenic acid and arsenic (V) oxide are also listed on the hazardous waste list by United States Environment Protection Agency (Peng and Di 1994). In water, arsenic creates oxidized form of arsenate (As (V)) and reduced form of arsenite (As (III)), which are known as inorganic oxyanions (Wang et al. 2003). Metal oxide-based materials such as iron oxide and aluminium oxide have been reported for arsenic removal by adsorption. These materials are valued due to their low cost and environmentally benign nature (Ren et al. 2011). They have high affinity towards arsenic oxyanions due to which they are used as adsorbents in water purification. As (V) and As (III) tend to form surface complex with iron oxides leading to their removal (Su and Puls 2001). Several iron (III) oxide materials such as granular ferric hydroxide are popular iron adsorbents for arsenic removal. In 1982, Oscarson et al. investigated the amounts of As (III) and pure Fe oxide and Al oxide. In 2007, Jeong et al. investigated the adsorption of arsenate (As (V)) onto Al_2O_3 and Fe_2O_3. Though the adsorption capacities for Fe_2O_3 and Al_2O_3 are significantly lower, they remain one of the popular adsorbents for arsenic ion removal (Weidner and Ciesielczyk 2019).

Molybdenum is a transitional metal which exhibits interconvertible +II and +VI oxidation state (Halmi and Ahmad 2014). Molybdenum is toxic in high concentrations and can cause various health problems (Xu et al. 2006, Torres et al. 2014). Verbinnen et al. reported zeolite-supported magnetite for molybdenum removal (Verbinnen et al. 2012). The precipitation of magnetite onto the surface from Fe (II) and Fe (III) salts resulted in the desired complex capable of molybdenum removal through the formation of an inner-sphere $FeOMoO_2(OH) \cdot 2H_2O$ complex (Weidner and Ciesielczyk 2019). Porous ceramics are lightweight, having heat insulation as well as sound absorption properties. These materials can also be used for water cleanup as their pores can absorb the contaminants from water (Takami et al. 2004). Kato et al. have demonstrated the fabrication of porous nano ceramics from industrial wastes (glass, low-grade silica and waste alumina) (Kato et al. 2008).

12.4 FENTON PROCESS

The Fenton process was discovered by Fenton about a century ago, but only in the late 1960s was it used for the catalytic degradation of organic contaminants. Ferrous ions (Fe^{2+}) and hydrogen peroxide in aqueous acidic medium are employed in Fenton reaction. Hydroxyl radicals and ions are generated by the hydrogen peroxide decomposition, which leads to the oxidation of Fe^{2+} to Fe^{3+}. The Fe^{3+} ions catalyse hydrogen peroxide and generate hydroxyl radicals again (Feng et al. 2003, Zubir et al. 2014). Hydroxyl radicals abstract protons and produce organic radicals ($R^{\cdot}$), CO_2 and H_2O are then produced from the degradation of these organic radicals. Fenton's reagent has been reported for the remediation of diesel-contaminated soil, colour from textile industries and removal of organics and poly aromatic hydrocarbon from water. Large quantity of chemical reagents, slower ferrous ions' generation, acidic condition and production of ferric hydroxide sludge are the problems faced in the process (Babuponnusami and Muthukumar 2012). Another Fenton reaction utilizing UV irradiation for the generation of hydroxyl radical is known as photo-Fenton reaction (described by equations (1) and (2)). Regeneration of Fe^{2+} also takes place during the photolysis of the Fe^{3+} ions/complex (Feng et al. 2003, Shahwan et al. 2011, Wang et al. 2014).

$$H_2O_2 + UV \rightarrow HO\cdot + HO\cdot \tag{1}$$
$$Fe^{3+} + H_2O + UV \rightarrow HO\cdot + Fe^{2+} + H^+ \tag{2}$$

Nano-zero valent iron (NZVI) has been used as heterogenous photo electro Fenton-like (PEF-like) system for removal of phenol. The removal efficiency is observed to be positively affected by increase in NZVI dosage and negatively by increase in initial phenol concentration and initial pH, and the best results were obtained near neutral pH (Babuponnusami and Muthukumar 2012).

Stable nanocomposites of silicate and iron oxide have been reported to be synthesized by a reaction between the dispersion of Laponite clay and iron salt solution, for the catalysis of the photo Fenton degradation of Orange I azo-dye. The catalyst was found to be cost effective as compared to the Nafion-based catalysts. The results show that the catalyst significantly accelerates the Orange II degradation at $\lambda = 254$ nm. The mineralization of Orange II and 75% total organic carbon undergoes a slower kinetics than the discolouration of Orange II. Figure 12.18 shows the effect of dye concentration and the H_2O_2 concentration on the dye degradation under various conditions (Feng et al. 2003). Iron oxide nanoparticles immobilized in alumina coated mesoporous silica are reported to exhibit a good catalytic efficiency as heterogeneous Fenton catalyst due to synergistic effect of alumina coating on mesoporous silica. This occurs by facilitating the redox cycle of iron species and increasing the dispersion of the iron oxide nanoparticles (Lim et al. 2006).

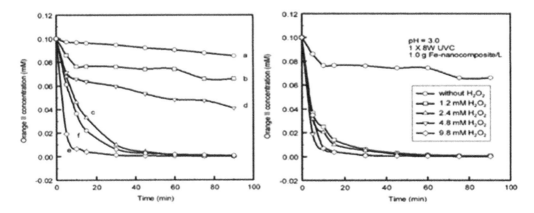

Fig. 12.18 Orange II concentration as a function of the reaction time under different conditions (left) and effect of the H_2O_2 molar concentration on the degradation of Orange II (right). Reprinted with permission from Feng et al., copyright (2003) American Chemical Society

12.4.1   Oxides

Iron oxide was reported to be deposited on reduced graphene and coated with zinc oxide to form magnetically-separable tertiary composites (ZnO-Fe$_3$O$_4$/rGO), for degradation of organic pollutants using photo Fenton heterogenous catalysis reaction. Magnetic Fe$_3$O$_4$ nanospheres offer superior stability to the catalyst and are convenient to reuse. In comparison to pristine ZnO, Fe$_3$O$_4$, their composites such as ZnO/Fe$_3$O$_4$ and Fe$_3$O$_4$/GO, the synthesized ZnO-Fe$_3$O$_4$/rGO composites have shown better catalytic activity towards dye degradation due to synergistic effects. These catalysts proved to be promising for the photocatalytic degradation of organic dyes (Ojha et al. 2017). Also, iron oxide nanoparticles produced from the extracts of tea leaves have been reported as a Fenton-like catalyst for MO and MB dyes degradation. These nanoparticles demonstrated high catalytic efficiency as compared to the borohydride reduced iron nanoparticles (Shahwan et al. 2011).

Shin et al. synthesised Fe$_3$O$_4$-poly(3,4-ethylene-dioxythiophene) (PEDOT) core-shell nanoparticles for heterogeneous Fenton catalysis by acid etching-mediated chemical oxidation polymerization. Compared to the commercialized iron oxide nano powder or pristine Fe$_3$O$_4$ nanoparticles, these core-shell nanoparticles show higher catalytic performance for the Orange II and reactive Black 5 (RB5) degradation. Also, the co-ordination of thin PEDOT shell with iron cation enhances the rate of the reaction and prevents the leaching of Fe (III) (Shin et al. 2008). Nanostructured iron oxide and magnetite were used as heterogeneous Fenton catalysts for the paracetamol mineralization. It is reported that the efficiency of this process is affected by temperature and oxidant dosage (Velichkova et al. 2013).

Nanoparticle gold/iron oxide aerogels are reported as a photocatalyst for the degradation of azo dye. In the mechanistic step, the azo dye gets absorbed on the iron oxide support and the hydroxyl radical generated near the catalyst surface brings out the oxidative degradation of the dye. The presence of Au nanoparticles increases the catalytic activity by band gap energy reduction of iron oxide. The results have shown the efficacy of aerogel powders as photocatalysts for dye removal from textile wastewater as shown in Fig. 12.19 (Wang 2007).

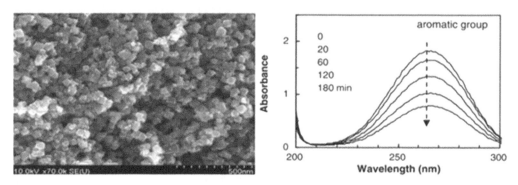

Fig. 12.19   FESEM image of 5 wt% Au/Fe$_2$O$_3$ aerogel (left) and UV-Vis absorption spectra for Blue 79 dye as a function of UV irradiation time (right). Reprinted with permission from Wang, copyright (2007) Elsevier

Fenton-like catalyst for the decolourization of acid black 194 in solution was reported to be synthesized using the extracts obtained from *Eucalyptus tereticornis*, *Melaleuca nesophila*, and *Rosmarinus officinaliI*. Iron-polyphenol nanoparticles (Fe-P NPs) thus formed show significant TOC removal dye degradation. The Fenton-like reaction with Fe-P NPs takes place more carefully and sustainably within 200 min as compared to the conventional Fenton reaction. Furthermore, there is no need for pH adjustment for the reaction (Wang et al. 2014). Wang et al. have reported the graphene oxide sheets supported dispersed iron oxide nanoparticles (i.e. GO-Fe$_3$O$_4$ nanocomposite) for Acid Orange 7 dye (AO7) degradation. Results demonstrated a fast 80% degradation in 20 min, while 98% of AO7 was successfully removed after 180 min of reaction time. The degradation kinetics of AO7 was most influenced by temperature and pH and can be described by a pseudo-first-order reaction. These findings have presented new insights into the influence of operational parameters in the heterogeneous Fenton like oxidation using GO-Fe_3O_4 nanocomposite (Zubir et al. 2014).

12.5 MEMBRANE TECHNIQUES

12.5.1 Carbides

Azamat et al. used molecular dynamics simulations to study the removal of Cd (II) from wastewater using armchair silicon carbide nanotubes (SiCNT) arranged in a manner to act like a membrane. On the basis of calculation of potential mean force, it has been found that armchair SiCNT under external electric fields are capable of separating Cd (II) from wastewater. This study can be used as the basis for development of ceramic nanotubes for heavy metals removal from industrial effluents (Azamat 2017). Nitrates contaminate water resources, give rise to eutrophication and are capable of causing potential human health hazards such as hypertension, methemoglobinemia in children, stomach cancer in adults and thyroid malfunctioning. Silicon nitride (SiN) membrane embedded with armchair (8,8) SiC nanotubes have been found to separate nitrate from wastewater under an applied external electric field by molecular dynamics simulations. The time required for nitrate removal was reduced on increasing the electric field. Therefore, fabricated membranes can potentially act as efficient nitrate separators (Khataee et al. 2016). Separation of extremely toxic cyanide, which is present in effluents from steel plant, pharmaceuticals, ore leaching etc., has been reported using molecular dynamics simulations. Armchair silicon carbide (SiC) nanotubes of different diameters were used to study the cyanide separation at different applied pressures. External pressure was applied to the system having (5,5), (6,6), (7,7) and (8,8) SiC nanotubes situated between two graphene sheets. The results showed that only the (5,5) SiC nanotube facilitated complete cyanide ion rejection (Khataee et al. 2017).

12.5.2 Nitrides

Ni (II) is a heavy metal present in industrial wastewater that causes not only respiratory problems but is carcinogenic and toxic as well. It is persistent, non-biodegradable and cannot be satisfactorily removed by general methods such as chemical reduction, ion-exchange, adsorption and reverse osmosis. Based on molecular dynamics simulations, boron nitride nanosheets (BNNS) can be used as a membrane having functionalized pore to exclusively remove Ni (II) from aqueous solutions under applied electric field. It has been reported that with the increase in the strength of the applied electric field, the number of ions passing through the membrane increased. Also, in comparison to graphene nanosheets, the retention time of Ni (II) in BNNS is less as the energy barrier of Ni (II) is low in BNNS (Azamat 2017). Pore functionalized BNNS are also known to act as a barrier for highly toxic mercury and copper ions on the basis of molecular dynamics simulations studies. Ion potential mean force showed that the heavy metals faced an energy barrier while permeating the BNNS; therefore, the separation of the toxic heavy metal from the wastewater can take place only under the influence of applied voltage as a source of additional energy (Azamat et al. 2016).

On the basis of a computer simulation, perforated BNNS can be used for separation of Zn (II) from wastewater. Zn (II) is discharged from pigments, fertilizers and steel processing industries, it is hazardous and causes human health disorders like other heavy metals which are persistent in the environment. More ions were reported to pass through the membranes and the residence time of the ions decreased on increasing applied voltage to the considered system, which was confirmed by calculating the radial distribution functions and density profile (Azamat et al. 2015). Highly selective Boron nitride nanotubes (BNNT) embedded in silicon nitride membrane can absorb toxic

Cd (II) from wastewater. Based on molecular simulations and calculation of potential mean force, it has been found that armchair BNNT under external electric fields are capable of separation of Cd (II) from wastewater (Azamat and Hazizadeh 2018). Cyanide, which is present in effluents from steel plant, pharmaceuticals, ore leaching etc., is extremely toxic in all forms. On the basis of molecular simulation analysis, BNNT with (6,6), (7,7) and (8,8) chirality can be used for removal of cyanide from aqueous solutions on application of pressure or external electric field. It was found that with the increase in applied pressure or electric field, more amount of cyanide ions could easily pass through the nanotubes (Azamat and Khataee 2017).

12.6 OXIDATION

Advanced oxidation processes (AOP) have emerged to be promising, efficient, reliable and economical methods for the removal of pollutants from wastewater. AOP techniques like H_2O_2 photolysis, ozonation, Fenton and photo-Fenton process, and heterogeneous photocatalysis can destroy the pollutants by chemical oxidation or reduction at ambient room temperature and atmospheric pressure (Rasalingam et al. 2014). Electro-oxidation is an emerging AOP, where pollutants can be oxidized directly by the exchange of electrons between the anode and the pollutants or indirectly by strong reactive oxygen species (ROS) like OH which is generated on the anode surface (Schulman et al. 2002).

Glass ceramic electrodes based on doped-TiO have been developed for the application in electrochemical wastewater treatment (Panizza et al. 2000). The TiO_2 belongs to a homologous series (Magnéli phases having the empirical formula Ti_nO_{2n-1}, ($n \geq 3$)); these materials have good corrosion resistance, high conductivity and chemical stability (Andersson et al. 1957). For example, Ti_4O_7 has electrical conductivity of 166 Ω^{-1} cm^{-1}, which is significantly greater than the conductivity of TiO_2 (Graves et al. 1992). It has been reported that the ceramic electrodes act as non-active anodes in radical generation. A few studies have shown application of sub-stoichiometric TiO_2 nanotube arrays-based electrodes in wastewater treatment (Ganiyu et al. 2015).

Degradation of water pollutants using catalytic properties of perovskites is becoming a significant area of interest in case of wastewater treatment (Anipsitakis and Dionysiou 2003). Janowska et al. synthesised cerium-doped strontium ferrite SrO.85CeO.15FeO$_3$-δ (SCF) perovskite oxides by solution combustion synthesis (SCS) for the degradation of Bisphenol A. The degradation occurs mainly due to OH radicals' generation which participates in abatement of organic pollutants. The catalytic efficiency was reported to be influenced by the variation in temperature conditions (Janowska et al. 2018).

12.7 BIOREMEDIATION

Several technologies exist for the removal of cadmium and lead, but they are generally complex and expensive treatments. Bioremediation is a more popular alternative that has gained attention in the past few years. To be used, the microbial biomass needs to be immobilized inside a bioreactor to increase their concentration and maintain a low growth rate. The ideal bedding material should be inexpensive and have a large surface per volume, does not collapse easily and is highly durable. Therefore, nano glass-ceramics are an interesting alternative as they include all the above characteristics. They can be regenerated by increasing the temperature and are highly stable, both chemically and thermodynamically (Ginebra et al. 1999, TenHuisen and Brown 1999). Also, they have been proved to adsorb the heavy metals present in water.

A glass-ceramic material has been reported to be made by SiO_2–P_2O_5–CaO–Al_2O_3–X_2O (where X is an alkaline element system) from industrial and urban wastes. This waste may contain animal bones, glass bottles, sludge from the pulp and paper industry and diatomite soil from the brewing industry (Pretto et al. 2003). To obtain good water circulation in the bioreactor, the porous material was shaped into Raschig rings so as to provide an optimal design. The microorganism *Escherichia Coli* was used and its growth was analysed in the presence of cadmium and lead, separately and jointly at different concentrations. Furthermore, they have established the degree of participation of the microorganisms in retaining heavy metals by using 10 mg/L Pb and Cd solutions, respectively. The tests were conducted with and without the use of *Escherichia Coli*. This mechanism can prove to be better for the elimination of heavy metals from wastewaters (García et al. 2003).

12.8 SUMMARY

From the past few decades, ceramics have attracted huge attention in the field of environmental protection. The pollutants have been removed by various processes that can be scaled up to study the technical feasibility and economic viability vis-a vis competing technologies. Nanoceramics can replace several processes in the industry which are quite unprofitable and less efficient. At nano meter scale, ceramics exhibit some special properties such as quantum effects, high surface areas, small size effect etc. These properties contribute to their excellent adsorption capacities and reactivities, both of which are favourable for the removal and treatment of a wide spectrum of contaminants.

Oxide nanoceramics possess unique properties in terms of their morphology, which make them quite effective in case of adsorption related processes. These properties have not only enabled the adsorption of heavy metals but also have facilitated the adsorption of other types of water pollutants like phosphates, dyes, and persistent organic pollutants. Oxides of zirconium and aluminum have been proved quite effective in the removal of phosphates, while the oxides of zinc and magnesium have been employed for the removal of several metal ions and organic contaminants like copper which is a major pollutant derived from copper electroplating industry. Oxides have been used in the adsorptive removal of pharmaceuticals from water to prevent the spread of multidrug resistant microbes. It may prevent future epidemics in the backdrop of a pandemic sweeping across the world.

Nano-oxide ceramics have also shown promising results in the field of photocatalysis. Photocatalysis is a highly energy efficient process which utilizes the energy of the sun. Oxide of TiO_2 is widely used as a photocatalyst, and the electron-hole recombination in TiO_2 generates OH radical; this can be increased by the addition of a variety of dopants which tend to reduce the band gap in TiO_2. These assemblies exhibit superior photocatalytic activity than pure TiO_2 and have been used to efficiently degrade a number of synthetic dyes like MB, Congo red, Rhodamine etc. 90-99% dye removal efficiency has been achieved by employing various modifications of TiO_2. Oxides of other metals like zirconium have also been used in photocatalysis. Water pollutants have also been removed using the Fenton degradation process. Oxide ceramics of Zn and Fe in the form of composites have been utilized as heterogeneous Fenton catalysts. These assemblies also have shown excellent catalytic performance. Novel ceramic carbides and nitride nano-assemblies have been reported as adsorbents as well as photocatalysts. Carbides of Si and Fe have shown high adsorption capacities as well as photocatalytic efficiencies. Carbides nano assemblies in various forms have been used to degrade dyes in general. They can be used to deal with the menace of drugs like Aflatoxin and endocrine disrupting chemicals like Bisphenol A in the waste. Titanium carbide in the form of nanosheets has shown excellent results as far as the adsorptive removal of Cr (VI) is concerned.

Boron carbide nanocomposites have exhibited encouraging results as far as photodegradation of dyes is concerned. Boron nitride has garnered considerable attention from researchers around the world. This ceramic in the form of nanotubes and nanosheets has been used in a variety of dye degradation processes, while in the form of ultrathin fibrous nets, it has been employed in adsorptive removal of a number of metals from water. Molecular dynamic simulation results suggest the use of carbides and nitrides in a variety of membrane techniques for the removal of various water contaminants especially cyanide. Nano glass-ceramics have been used extensively in the domain of water treatment and purification. In combination with certain oxides, they have shown impressive results as far as photodegradation and adsorption are concerned. They have also been employed in bioremediation which involves the removal of harmful pathogens from water. In addition to this, nano-glass ceramics also remove pollutants through oxidation. Nanoceramic materials have proved to be promising because of their idiosyncratic mechanical, physical, and chemical properties. They possess corrosion/oxidation resistance, excellent wear resistance, high thermal shock resistance, low densities, and optical, electrical and magnetic properties. These properties make them highly desirable in the field of water remediation. Currently, the focus should be directed towards exploring more types of nanoceramics which can be successfully utilized in water purification.

12.9 ACKNOWLEDGEMENTS

The authors are grateful to Doon University, Uttarakhand for providing the facilities for the successful completion of the above studies.

12.10 REFERENCES

Abellán, M.N., B. Bayarri, J. Giménez and J. Costa. 2007. Photocatalytic degradation of sulfamethoxazole in aqueous suspension of TiO_2. Appl. Catal. B: Environ. 74(3): 233-241.

Abraham, C. and A. Rosencranz. 1986. An evaluation of pollution control legislation in India. J. Inst. Eng. (India) 11: 101.

Ahmetoglu, C., D. Zeydanli, M. Innocentini, F. Ribeiro, P. Lasso and G. Soraru. 2017. Gradient-hierarchic-aligned porosity SiOC ceramics. Sci. Rep. 7: 41049.

Álvarez-Ayuso, E., A. García-Sánchez and X. Querol. 2007. Adsorption of Cr (VI) from synthetic solutions and electroplating wastewaters on amorphous aluminium oxide. J. Hazar. Mater. 142(1): 191-198.

Anas, M., D.S. Han, K. Mahmoud, H. Park and A. Abdel-Wahab. 2016. Photocatalytic degradation of organic dye using titanium dioxide modified with metal and non-metal deposition. Mater. Sci. Semicon. Proc. 41: 209-218.

Andersson, S., B. Collén, U. Kuylenstierna and A. Magnéli. 1957. Phase analysis studies on the titanium-oxygen system. Acta Chem. Scand. 11(10): 1641-1652.

Anipsitakis, G.P. and D.D. Dionysiou. 2003. Degradation of organic contaminants in water with sulfate radicals generated by the conjunction of peroxymonosulfate with cobalt. Environ. Sci. Technol. 37(20): 4790-4797.

Anju, A., R.S. Pandey and S. Bechan. 2010. Water pollution with special reference to pesticide contamination in India. J. Water Res. Protec. 2010(2): 432-448.

Asadzadeh-Khaneghah, S., A. Habibi-Yangjeh, M. Shahedi Asl, Z. Ahmadi and S. Ghosh. 2020. Synthesis of novel ternary g-C_3N_4/SiC/C-Dots photocatalysts and their visible-light-induced activities in removal of various contaminants. J. Photochem. Photobio. A. 392: 112431.

Aware, D.V. and S.S. Jadhav. 2016. Synthesis, characterization and photocatalytic applications of Zn-doped TiO_2 nanoparticles by sol-gel method. Appl. Nanosci. 6(7): 965-972.

Ayanda, D.O., O. Fatoki, F. Adekola, L. Petrik and B. Ximba. 2014. Application of nano zinc oxide (nZnO) for the removal of triphenyltin chloride (TPT) from dockyard wastewater. Water S. A. 40(4): 659-664.

Azamat, J., B. Sattary, A. Khataee and S. Joo. 2015. Removal of a hazardous heavy metal from aqueous solution using functionalized graphene and boron nitride nanosheets: Insights from simulations. J. Mol. Graph. Model 61: 13-20.

Azamat, J., A. Khataee and S. Joo. 2016. Separation of copper and mercury as heavy metals from aqueous solution using functionalized boron nitride nanosheets: A theoretical study. J. Mol. Struct. 1108: 144-149.

Azamat, J. 2017. Removal of nickel (II) from aqueous solution by graphene and boron nitride nanosheets. J. Water Environ. Nanotechnol. 2: 26-33.

Azamat, J. and A. Khataee. 2017. Molecular dynamics simulations of removal of cyanide from aqueous solution using boron nitride nanotubes. Comput. Mater. Sci. 128: 8-14.

Azamat, J. and B. Hazizadeh. 2018. Removal of Cd (II) from water using carbon, boron nitride and silicon carbide nanotubes. Membrane Water Treat. 9: 63-68.

Babuponnusami, A. and K. Muthukumar. 2012. Removal of phenol by heterogenous photo electro Fenton-like process using nano-zero valent iron. Sep. Purif. Technol. 98: 130-135.

Baghbanzadeh, M., D. Rana, C.Q. Lan and T. Matsuura. 2016. Effects of inorganic nano-additives on properties and performance of polymeric membranes in water treatment. Sep. Purif. Rev. 45(2): 141-167.

Beall, G.H. and L.R. Pinckney. 1999. Nanophase glass-ceramics. J. Am. Ceramic Soc. 82(1): 5-16.

Bhattacharya, S., A.B. Gupta, A. Gupta and A. Pandey. 2018. Introduction to water remediation: Importance and methods. pp. 3-8. *In*: S. Bhattacharya, A.B. Gupta, A. Gupta and A. Pandey [eds.]. Water Remediation, Springer.

Bocker, C., A. Herrmann, P. Loch and C. Rüssel. 2015. The nano-crystallization and fluorescence of terbium doped $Na_2O/K_2O/CaO/CaF_2/Al_2O_3/SiO_2$ glasses. J. Mater. Chem. 3(10): 2274-2281.

Cai, W., Y. Hu, J. Chen, G. Zhang and T. Xia. 2012. Synthesis of nanorod-like mesoporous γ-Al_2O_3 with enhanced affinity towards Congo red removal: Effects of anions and structure-directing agents. Cryst. Eng. Comm. 14(3): 972-977.

Chen, B., Q. Ma, C. Tan, T.T. Lim, L. Huang and H. Zhang. 2015. Carbon-based sorbents with three-dimensional architectures for water remediation. Small 11(27): 3319-3336.

Chen, X. and S.S. Mao. 2007. Titanium dioxide nanomaterials: Synthesis, properties, modifications, and applications. Chem. Rev. 107(7): 2891-2959.

Cheng, N., J. Tian, Q. Liu, C. Ge, A.H. Qusti, A.M. Asiri, et al. 2013. Au-nanoparticle-loaded graphitic carbon nitride nanosheets: Green photocatalytic synthesis and application toward the degradation of organic pollutants. ACS Appl. Mater. Inter. 5(15): 6815-6819.

Chronakis, I. 2005. Novel nanocomposites and nanoceramics based on polymer nanofibers using electrospinning process—a review. J. Mater. Process. Technol. 167(2-3): 283-293.

Cui, Y., M. Li, H. Wang, C. Yang, S. Meng and F. Chen. 2018. *In-situ* synthesis of sulfur doped carbon nitride microsphere for outstanding visible light photocatalytic Cr(VI) reduction. Sep. Purifi. Technol. 199: 251-259.

Dadigala, R., R. Bandi, B.R. Gangapuram and V. Guttena. 2017. Carbon dots and Ag nanoparticles decorated g-C_3N_4 nanosheets for enhanced organic pollutants degradation under sunlight irradiation. J. Photochem. Photobio. A: Chem. 342: 42-52.

Das-Gupta, D. and K. Doughty. 1988. Polymer-ceramic composite materials with high dielectric constants. Thin Solid Films 158(1): 93-105.

Fan, Q., J. Liu, Y. Yu, S. Zuo and B. Li. 2017. A simple fabrication for sulfur doped graphitic carbon nitride porous rods with excellent photocatalytic activity degrading RhB dye. Appl. Surf. Sci. 391: 360-368.

Feng, J., X. Hu, P.L. Yue, H.Y. Zhu and G.Q. Lu. 2003. Degradation of azo-dye orange II by a photoassisted Fenton reaction using a novel composite of iron oxide and silicate nanoparticles as a catalyst. Indust. Eng. Chem. Res. 42(10): 2058-2066.

Formo, E., E. Lee, D. Campbell and Y. Xia. 2008. Functionalization of electrospun TiO_2 nanofibers with Pt nanoparticles and nanowires for catalytic applications. Nano Letters 8: 668-672.

Fujishima, A. and K. Honda. 1972. Electrochemical photolysis of water at a semiconductor electrode. Nature 238(5358): 37-38.

Fujishima, A., T.N. Rao and D.A. Tryk. 2000. Titanium dioxide photocatalysis. J. Photochem. Photobio. C: Photochem. Rev. 1(1): 1-21.

Ganiyu, S.O., E.D. Hullerbusch, M. Cretin, G. Esposito and M.A. Oturan. 2015. Coupling of membrane filtration and advanced oxidation process for removal of pharmaceutical residues: A critical review. Sep. Purif. Technol. 156(3): 891-914.

García, A.M., J.M. Villora, D.A. Moreno, C. Ranninger, P. Callejas and M.F. Barba. 2003. Heavy metals bioremediation from polluted water by glass-ceramic materials. J. Am. Ceramic Soc. 86(12): 2200-2202.

Genz, A., A. Kornmüller and M. Jekel. 2004. Advanced phosphorus removal from membrane filtrates by adsorption on activated aluminium oxide and granulated ferric hydroxide. Water Res. 38(16): 3523-3530.

Ghasemian, E. and Z. Palizban. 2016. Comparisons of azo dye adsorptions onto activated carbon and silicon carbide nanoparticles loaded on activated carbon. Int. J. Environ. Sci. Technol. 13(2): 501-512.

Ginebra, M.P., E. Fernández, F.C. Driessens and J.A. Planell. 1999. Modeling of the hydrolysis of α-tricalcium phosphate. J. Am. Ceramic Soc. 82(10): 2808-2812.

Goel, P. 2006. Water pollution: causes, effects and control, New Age International. India.

Graves, J., D. Pletcher, R. Clarke and F. Walsh. 1992. The electrochemistry of Magnéli phase titanium oxide ceramic electrodes Part II: Ozone generation at ebonex and ebonex/lead dioxide anodes. J. Appl. Electrochem. 22(3): 200-203.

Gupta, V.K., A. Fakhri, S. Rashidi, A.A. Ibrahim, M. Asif and S. Agarwal. 2017. Optimization of toxic biological compound adsorption from aqueous solution onto silicon and silicon carbide nanoparticles through response surface methodology. Mater. Sci. Eng.: C. 77: 1128-1134.

Habibi-Yangjeh, A. and A. Akhundi. 2016. Novel ternary g-C_3N_4/Fe_3O_4/Ag_2CrO_4 nanocomposites: Magnetically separable and visible-light-driven photocatalysts for degradation of water pollutants. J. Mol. Catal. A: Chem. 415: 122-130.

Halder, J.N. and M.N. Islam. 2015. Water pollution and its impact on the human health. J. Environ. Human. 2(1): 36-46.

Halmi, M. and S.A. Ahmad. 2014. Chemistry, biochemistry, toxicity and pollution of molybdenum: A mini review. J. Biochem. Microbio. Biotech. 2(1): 1-6.

Hashimoto, K., H. Irie and A. Fujishima. 2005. TiO_2 photocatalysis: A historical overview and future prospects. Jap. J. Appl. Phy. 44(12R): 8269.

Haug, R., M. Weinmann, J. Bill and F. Aldinger. 1999. Plastic forming of preceramic polymers. J. Euro. Ceramic Soc. 19(1): 1-6.

Helmer, R., I. Hespanhol and W.H. Organization. 1997. Water Pollution Control: A Guide to The Use of Water Quality Management Principles. E & FN Spon, London.

Hua, M., S. Zhang, B. Pan, W. Zhang, L. Lv and Q. Zhang. 2011. Heavy metal removal from water/wastewater by nanosized metal oxides: A review. J. Hazar. Mater. 211-212: 317-331.

Janowska, K., F.E.B. Coelho, F. Deganello, G. Magnacca and V. Boffa. 2018. Catalytic activity of doped SrFeO$_3$-δ perovskite-type oxide ceramics for degradation of water pollutants. 2nd Nordic Conf. Ceramic Glass Technol.

Jin, Q., S. Zhang, T. Wen, J. Wang, P. Gu, G. Zhao, et al. 2018. Simultaneous adsorption and oxidative degradation of Bisphenol a by zero-valent iron/iron carbide nanoparticles encapsulated in N-doped carbon matrix. Environ. Pollut. 243: 218-227.

Johnson, B.B., A.V. Ivanov, O.N. Antzutkin and W. Forsling. 2002. 31P nuclear magnetic resonance study of the adsorption of phosphate and phenyl phosphates on γ-Al_2O_3. Langmuir. 18(4): 1104-1111.

Kato, T., K. Ohashi, M. Fuji and M. Takahashi. 2008. Water absorption and retention of porous ceramics fabricated by waste resources. J. Ceramic Soc. Jap. 116(1350): 212-215.

Khataee, A., J. Azamat and G. Bayat. 2016. Separation of nitrate ion from water using silicon carbide nanotubes as a membrane: Insights from molecular dynamics simulation. Comput. Mater. Sci. 119: 74-81.

Khataee, A., G. Bayat and J. Azamat. 2017. Separation of cyanide from an aqueous solution using armchair silicon carbide nanotubes: Insights from molecular dynamics simulations. RSC Adv. 7: 7502-7508.

Khataee, A.R., M.N. Pons and O. Zahraa. 2009. Photocatalytic degradation of three azo dyes using immobilized TiO_2 nanoparticles on glass plates activated by UV light irradiation: Influence of dye molecular structure. J. Hazar. Mater. 168(1): 451-457.

Khataee, A.R., M. Zarei, M. Fathinia and M.K. Jafari. 2011. Photocatalytic degradation of an anthraquinone dye on immobilized TiO_2 nanoparticles in a rectangular reactor: Destruction pathway and response surface approach. Desalination 268(1): 126-133.

Kiani, A., M. Rahmani, S. Manickam and B. Tan. 2014. Nanoceramics: Synthesis, characterization, and applications. J. Nanomater. vol. 2014, Article ID 528348, 2 pages.

Lee, H.D., Y.H. Cho and H.B. Park. 2013. Current research trends in water treatment membranes based on nano materials and nano technologies. Membr. J. 23(2): 101-111.

Lee, K.M., C.W. Lai, K.S. Ngai and J.C. Juan. 2016. Recent developments of zinc oxide based photocatalyst in water treatment technology: A review. Water Res. 88: 428-448.

Lei, C., M. Pi, C. Jiang, B. Cheng and J. Yu. 2017. Synthesis of hierarchical porous zinc oxide (ZnO) microspheres with highly efficient adsorption of congo red. J. Colloid Interface Sci. 490: 242-251.

Lei, W., D. Portehault, D. Liu, S. Qin and Y. Chen. 2013. Porous boron nitride nanosheets for effective water cleaning. Nat. Commun. 4(1): 1777.

Lian, G., X. Zhang, H. Si, J. Wang, D. Cui and Q. Wang. 2013. Boron nitride ultrathin fibrous nanonets: One-step synthesis and applications for ultrafast adsorption for water treatment and selective filtration of nanoparticles. ACS Appl. Mater. Interfaces. 5(24): 12773-12778.

Lim, H., J. Lee, S. Jin, J. Kim, J. Yoon and T. Hyeon. 2006. Highly active heterogeneous Fenton catalyst using iron oxide nanoparticles immobilized in alumina coated mesoporous silica. Chem. Commun. (4): 463-465.

Linsebigler, A.L., G. Lu and J.T. Yates. 1995. Photocatalysis on TiO_2 surfaces: Principles, mechanisms and selected results. Chem. Rev. 95(3): 735-758.

Liqiang, J., S. Xiaojun, X. Baifu, W. Baiqi, C. Weimin and F. Honggang. 2004. The preparation and characterization of La doped TiO_2 nanoparticles and their photocatalytic activity. J. Solid State Chem. 177(10): 3375-3382.

Liu, F., J. Yu, X. Ji and M. Qian. 2015b. Nanosheet-structured boron nitride spheres with a versatile adsorption capacity for water cleaning. ACS Appl. Mater. Interfaces. 7(3): 1824-1832.

Liu, F., S. Li, D. Yu, Y. Su, N. Shao and Z. Zhang. 2018. Template-free synthesis of oxygen-doped bundlelike porous boron nitride for highly efficient removal of heavy metals from wastewater. ACS Sustain. Chem. Eng. 6(12): 16011-16020.

Liu, X., C. Niu, X. Zhen, J. Wang and X. Su. 2015a. Novel approach for synthesis of boehmite nanostructures and their conversion to aluminum oxide nanostructures for remove congo red. J. Colloid Interface Sci. 452: 116-125.

Lu, F. and D. Astruc. 2018. Nanomaterials for removal of toxic elements from water. Coordination Chem. Rev. 356: 147-164.

Maldonado, V.Y., P.J. Espinoza-Montero, C.A. Rusinek and G.M. Swain. 2018. Analysis of Ag (I) biocide in water samples using anodic stripping voltammetry with a boron-doped diamond disk electrode. Analytical Chem. 90(11): 6477-6485.

Maliyekkal, S.M., K.P. Lisha and T. Pradeep. 2010. A novel cellulose-manganese oxide hybrid material by in situ soft chemical synthesis and its application for the removal of Pb (II) from water. J. Hazard. Mater. 181: 986-995.

Margha, F.H., M.S. Abdel-Wahed and T.A. Gad-Allah. 2015. Nanocrystalline Bi_2O_3-B_2O_3-(MoO_3 or V_2O_5) glass-ceramic systems for organic pollutants degradation. Ceramics International 41(4): 5670-5676.

Maria, A. 2003. The costs of water pollution in India. Conference on Market Development of Water and Waste Technologies through Environmental Economics.

Miao, X., Z. Ji, J. Wu, X. Shen, J. Wang, L. Kong, et al. 2017. g-C_3N_4/AgBr nanocomposite decorated with carbon dots as a highly efficient visible-light-driven photocatalyst. J. Colloid Interface Sci. 502: 24-32.

Mills, A. and S. Le Hunte. 1997. An overview of semiconductor photocatalysis. J. Photochem. Photobio. A: Chem. 108(1): 1-35.

Nagpal, M. and R. Kakkar. 2019. Use of metal oxides for the adsorptive removal of toxic organic pollutants. Sep. Purif. Technol. 211: 522-539.

Naslain, R. and F. Langlais. 1986. CVD-processing of ceramic-ceramic composite materials. pp. 145-164. *In*: M.A. Boston [ed.]. Tailoring Multiphase and Composite Ceramics, Springer.

Noll, K.E. 1991. Adsorption Technology for Air and Water Pollution Control, CRC Press.

Ojha, D.P., M.K. Joshi and H.J. Kim. 2017. Photo-Fenton degradation of organic pollutants using a zinc oxide decorated iron oxide/reduced graphene oxide nanocomposite. Ceramics International 43(1, Part B): 1290-1297.

Ollis, D.F. 2000. Photocatalytic purification and remediation of contaminated air and water. Comptes Rendus de l'Académie des Sciences-Series IIC-Chem 3(6): 405-411.

Özkan, A., M.H. Özkan, R. Gürkan, M. Akcay and M. Sökmen. 2004. Photocatalytic degradation of a textile azo dye, sirius gelb GC on TiO_2 or Ag-TiO_2 particles in the absence and presence of UV irradiation: The effects of some inorganic anions on the photocatalysis. Chemistry 163(1-2): 29-35.

Panizza, M., C. Bocca and G. Cerisola. 2000. Electrochemical treatment of wastewater containing polyaromatic organic pollutants. Water Res. 34(9): 2601-2605.

Pawar, M., S. Topcu Sendoğdular and P. Gouma. 2018. A brief overview of TiO_2 photocatalyst for organic dye remediation: Case study of reaction mechanisms involved in Ce-TiO_2 photocatalysts system. J. Nanomater. 2018: 5953609.

Peng, F.F. and P. Di. 1994. Removal of arsenic from aqueous solution by adsorbing colloid flotation. Indust. Eng. Chem. Res. 33(4): 922-928.

Prabhakar, V., T. Bibi, P. Vishnu and T. Bibi. 2013. Nanotechnology future tools for water remediation. Inter. J. Emerging Technol. Adv. Eng. 3(7): 54-59.

Pretto, M., A.L. Costa, E. Landi, A. Tampieri and C. Galassi. 2003. Dispersing behavior of hydroxyapatite powders produced by wet-chemical synthesis. J. Am. Ceramic Soc. 86(9): 1534-1539.

Qu, X., P.J.J. Alvarez and Q. Li. 2013. Applications of nanotechnology in water and wastewater treatment. Water Res. 47(12): 3931-3946.

Rafiq, Z., R. Nazir, S. Durre, M.R. Shah and S. Ali. 2014. Utilization of magnesium and zinc oxide nano-adsorbents as potential materials for treatment of copper electroplating industry wastewater. J. Environ. Chem. Eng. 2(1): 642-651.

Rajaram, T. and A. Das. 2008. Water pollution by industrial effluents in India: Discharge scenarios and case for participatory ecosystem specific local regulation. Futures 40(1): 56-69.

Ramasundaram, S., H.N. Yoo, K.G. Song, J. Lee, K.J. Choi and S.W. Hong. 2013. Titanium dioxide nanofibers integrated stainless steel filter for photocatalytic degradation of pharmaceutical compounds. J. Hazar. Mater. 258-259: 124-132.

Rasalingam, S., R. Peng and R.T. Koodali. 2014. Removal of hazardous pollutants from wastewaters: Applications of TiO_2-SiO_2 mixed oxide materials. J. Nanomater. vol. 2014, Article ID 617405, 42 pages, 2014.

Ren, Z., G. Zhang and J.P. Chen. 2011. Adsorptive removal of arsenic from water by an iron-zirconium binary oxide adsorbent. J. Colloid Interface Sci. 358(1): 230-237.

Safajou, H., H. Khojasteh, M. Salavati-Niasari and S. Mortazavi-Derazkola. 2017. Enhanced photocatalytic degradation of dyes over graphene/Pd/TiO_2 nanocomposites: TiO_2 nanowires versus TiO_2 nanoparticles. J. Colloid Interface Sci. 498: 423-432.

Saravanakumar, K., R. Karthik, S.M. Chen, J. Vinoth Kumar, K. Prakash and V. Muthuraj. 2017. Construction of novel Pd/CeO_2/g-C_3N_4 nanocomposites as efficient visible-light photocatalysts for hexavalent chromium detoxification. J. Colloid Interface Sci. 504: 514-526.

Sartor, J.D. and G.B. Boyd. 1972. Water Pollution Aspects of Street Surface Contaminants. US Government Printing Office.

Sayilkan, F., M. Asilturk, S. Sener, S. Erdemoğlu, M. Erdemoğlu and H. Sayilkan. 2007. Hydrothermal synthesis, characterization and photocatalytic activity of nanosized TiO_2 based catalysts for Rhodamine B degradation. Turkish J. Chem. 31(2): 211-221.

Schulman, L.J., E.V. Sargent, B.D. Naumann, E.C. Faria, D.G. Dolan and J. Wargo. 2002. A human health risk assessment of pharmaceuticals in the aquatic environment. Human Eco. Risk Assessment 8(4): 657-680.

Schwarzenbach, R.P., T. Egli, T.B. Hofstetter, U. von Gunten and B. Wehrli. 2010. Global water pollution and human health. Annual Rev. Environ. Resources. 35(1): 109-136.

Schweitzer, L. and J. Noblet. 2018. Water contamination and pollution. Green Chem. Elsevier. 261-290.

Shahwan, T., S. Abu Sirriah, M. Nairat, E. Boyaci, A.E. Eroğlu, T.B. Scott, et al. 2011. Green synthesis of iron nanoparticles and their application as a Fenton-like catalyst for the degradation of aqueous cationic and anionic dyes. Chem. Eng. J. 172(1): 258-266.

Shahzad, A., K. Rasool, W. Miran, M. Nawaz, J. Jang, K.A. Mahmoud, et al. 2018. Mercuric ion capturing by recoverable titanium carbide magnetic nanocomposite. J. Hazar. Mater. 344: 811-818.

Shin, S., H. Yoon and J. Jang. 2008. Polymer-encapsulated iron oxide nanoparticles as highly efficient Fenton catalysts. Catal. Commun. 10(2): 178-182.

Shinozaki, K., T. Honma, K. Oh-Ishi and T. Komatsu. 2011. Morphology of CaF_2 nanocrystals and elastic properties in transparent oxyfluoride crystallized glasses. Optical Mater. 33(8): 1350-1356.

Sigmund, W.M., N.S. Bell and L. Bergström. 2000. Novel powder-processing methods for advanced ceramics. J. Am. Ceramic Soc. 83(7): 1557-1574.

Singh, B., G. Kaur, P. Singh, K. Singh, J. Sharma, M. Kumar, et al. 2017. Nanostructured BN-TiO_2 composite with ultra-high photocatalytic activity. New J. Chem. 41(20): 11640-11646.

Singh, G., S. Kumar, S.K. Sharma, M. Sharma, V.P. Singh and R. Vaish. 2019a. Antibacterial and photocatalytic active transparent TiO_2 crystallized CaO-BaO-B_2O_3-Al_2O_3-TiO_2-ZnO glass nanocomposites. J. Am. Ceramic Soc. 102(6): 3378-3390.

Singh, G., S. Kumar, M. Sharma and R. Vaish. 2019b. Transparent CaF_2 surface crystallized CaO-$2B_2O_3$ glass possessing efficient photocatalytic and antibacterial properties. J. Am. Ceramic Soc. 102(9): 5127-5137.

Singh, P., G. Kaur, K. Singh, B. Singh, M. Kaur, M. Kaur, et al. 2018a. Specially designed B_4C/SnO_2 nanocomposite for photocatalysis: Traditional ceramic with unique properties. Appl. Nanosci. 8(1): 1-9.

Singh, P., G. Kaur, K. Singh, M. Kaur, M. Kumar, R. Meena, et al. 2018b. Nanostructured boron carbide (B_4C): A bio-compatible and recyclable photo-catalyst for efficient wastewater treatment. Materialia 1: 258-264.

Sreeprasad, T.S., S.M. Maliyekkal, K.P. Lisha and T. Pradeep. 2011. Reduced graphene oxide-metal/metal oxide composites: Facile synthesis and application in water purification. J. Hazard. Mater. 186: 921-931.

Srivastava, A., O. Srivastava, S. Talapatra, R. Vajtai and P. Ajayan. 2004. Carbon nanotube filters. Nat. Mater. 3(9): 610-614.

Srivastava, V., Y.C. Sharma and M. Sillanpaa. 2015. Green synthesis of magnesium oxide nanoflower and its application for the removal of divalent metallic species from synthetic wastewater. Ceramics International. 41(5, Part B): 6702-6709.

Su, C. and R.W. Puls. 2001. Arsenate and arsenite removal by zerovalent iron: Effects of phosphate, silicate, carbonate, borate, sulfate, chromate, molybdate, and nitrate, relative to chloride. Environ. Sci. Technol. 35(22): 4562-4568.

Su, Y., H. Cui, Q. Li, S. Gao and J.K. Shang. 2013. Strong adsorption of phosphate by amorphous zirconium oxide nanoparticles. Water Res. 47(14): 5018-5026.

Subramanian, V., E.E. Wolf and P.V. Kamat. 2004. Catalysis with TiO_2/Gold nanocomposites. Effect of metal particle size on the Fermi level equilibration. J. Am. Chem. Soc. 126(15): 4943-4950.

Suresh, P., J.J. Vijaya and L.J. Kennedy. 2014. Photocatalytic degradation of textile-dyeing wastewater by using a microwave combustion-synthesized zirconium oxide supported activated carbon. Mater. Sci. Semiconductor Processing. 27: 482-493.

Suwarnkar, M., R. Dhabbe, A. Kadam and K. Garadkar. 2014. Enhanced photocatalytic activity of Ag doped TiO_2 nanoparticles synthesized by a microwave assisted method. Ceramics International 40(4): 5489-5496.

Takami, K., A. Nakajima, T. Watanabe and K. Hashimoto. 2004. Preparation and reflectivity of self-organized nanograded SiO_2 / TiO_2 / PMMA thin films. J. Ceramic Soc. Japan. 112(1303): 138-142.

Tan, K.B., M. Vakili, B.A. Horri, P.E. Poh, A.Z. Abdullah and B. Salamatinia. 2015. Adsorption of dyes by nanomaterials: Recent developments and adsorption mechanisms. Sep. Purif. Technol. 150: 229-242.

TenHuisen, K.S. and P.W. Brown. 1999. Phase evolution during the formation of α-tricalcium phosphate. J. Am. Ceramic Soc. 82(10): 2813-2818.

Tesh, S.J. and T.B. Scott. 2014. Nano-composites for water remediation: A review. Adv. Mater. 26(35): 6056-6068.

Theron, J., J. Walker and T. Cloete. 2008. Nanotechnology and water treatment: Applications and emerging opportunities. Critical Rev. Microbio. 34(1): 43-69.

Torres, J., L. Gonzatto, G. Peinado, C. Kremer and E. Kremer. 2014. Interaction of molybdenum (VI) oxyanions with + 2 metal cations. J. Sol. Chem. 43(9-10): 1687-1700.

Trivedi, P., L. Axe and T.A. Tyson. 2001. XAS studies of Ni and Zn sorbed to hydrous manganese oxide. Environ. Sci. Technol. 35(22): 4515-4521.

Vakifahmetoglu, C. 2011. Fabrication and properties of ceramic 1D nanostructures from preceramic polymers: A review. Adv. Appl. Ceramics. 110(4): 188-204.

Vakifahmetoglu, C., D. Zeydanli and P. Colombo. 2016. Porous polymer derived ceramics. Mater. Sci. Eng. R: Reports 106: 1-30.

Velichkova, F., C. Julcour-Lebigue, B. Koumanova and H. Delmas. 2013. Heterogeneous Fenton oxidation of paracetamol using iron oxide (nano) particles. J. Environ. Chem. Eng. 1(4): 1214-1222.

Verbinnen, B., C. Block, D. Hannes, P. Lievens, M. Vaclavikova, K. Stefusova, et al. 2012. Removal of molybdate anions from water by adsorption on zeolite-supported magnetite. Water Env. Res. 84(9): 753-760.

Vinu, R. and G. Madras. 2010. Environmental remediation by photocatalysis. J. Indian Inst. Sci. 90: 189-230.

Vunain E., S.B. Mishra, A.K. Mishra and B.B. Mamba. 2017. Nanoceramics: Fundamentals and advanced perspectives. pp. 1-20. In: A. Mishra [ed.]. Sol-gel Based Nanoceramic Materials: Preparation, Properties and Applications. Springer, Cham.

Wang, C.T. 2007. Photocatalytic activity of nanoparticle gold/iron oxide aerogels for azo dye degradation. J. Non-Cryst. Solids. 353(11): 1126-1133.

Wang, Y., Y. Chen, C. Liu, F. Yu, Y. Chi and C. Hu. 2017. The effect of magnesium oxide morphology on adsorption of U(VI) from aqueous solution. Chem. Eng. J. 316: 936-950.

Wang, Y.H., S.H. Lin and R.S. Juang. 2003. Removal of heavy metal ions from aqueous solutions using various low-cost adsorbents. J. Hazar. Mater. 102(2-3): 291-302.

Wang, Z., C. Fang and M. Megharaj. 2014. Characterization of iron-polyphenol nanoparticles synthesized by three plant extracts and their Fenton oxidation of azo dye. ACS Sustain. Chem. Eng. 2(4): 1022-1025.

Warren, C. 1971. Biology and Water Pollution Control. Peter Doudoroff, W.B. Saunders Company, Philadelphia, Pennsylvania.

Weidner, E. and F. Ciesielczyk. 2019. Removal of hazardous oxyanions from the environment using metal-oxide-based materials. Materials 12(6): 927.

Xie, J., D. Jiang, M. Chen, D. Li, J. Zhu and C. Yan. 2010. Preparation and characterization of monodisperse Ce-doped TiO_2 microspheres with visible light photocatalytic activity. Colloids and Surf. A: Physicochem. Eng. Aspects. 372: 107-114.

Xu, H., Z. Gan, W. Zhou, Z. Ding and X. Zhang. 2017. A metal-free 3C-SiC/g-C_3N_4 composite with enhanced visible light photocatalytic activity. RSC Adv. 7(63): 40028-40033.

Xu, N., Z. Shi, Y. Fan, J. Dong, J. Shi and M. Hu. 1999. Effects of particle size of TiO_2 on photocatalytic degradation of methylene blue in aqueous suspensions. Indust. Eng. Chem. Res. 38(2): 373-379.

Xu, N., C. Christodoulatos and W. Braida. 2006. Adsorption of molybdate and tetrathiomolybdate onto pyrite and goethite: Effect of pH and competitive anions. Chemosphere 62(10): 1726-1735.

Ying, Y., Y. Liu, X. Wang, Y. Mao, W. Cao, P. Hu, et al. 2015. Two-dimensional titanium carbide for efficiently reductive removal of highly toxic chromium (VI) from water. ACS Appl. Mater. Interfaces. 7(3): 1795-1803.

Yun, H.J., H. Lee, J.B. Joo, W. Kim and J. Yi. 2009. Influence of aspect ratio of TiO_2 nanorods on the photocatalytic decomposition of formic acid. J. Phys. Chem. C. 113(8): 3050-3055.

Zeydanli, D., S. Akman and C. Vakifahmetoglu. 2018. Polymer-derived ceramic adsorbent for pollutant removal from water. J. Am. Ceramic Soc. 101(6): 2258-2265.

Zhan, S., D. Chen, X. Jiao and C. Tao. 2006. Long TiO_2 hollow fibers with mesoporous walls: Sol-gel combined electrospun fabrication and photocatalytic properties. J. Phys. Chem. B. 110(23): 11199-11204.

Zhang, R., W. Wan, L. Qiu, Y. Wang and Y. Zhou. 2017. Preparation of hydrophobic polyvinyl alcohol aerogel via the surface modification of boron nitride for environmental remediation. Appl. Surf. Sci. 419: 342-347.

Zhang, X., G. Lian, S. Zhang, D. Cui and Q. Wang. 2012. Boron nitride nanocarpets: Controllable synthesis and their adsorption performance to organic pollutants. Cryst. Eng. Comm. 14(14): 4670-4676.

Zhao, X., L. Lv, B. Pan, W. Zhang, S. Zhang and Q. Zhang. 2011. Polymer-supported nanocomposites for environmental application: A review. Chem. Eng. J. 170(2-3): 381-394.

Zubir, N.A., C. Yacou, X. Zhang and J.C. Diniz da Costa. 2014. Optimisation of graphene oxide-iron oxide nanocomposite in heterogeneous Fenton-like oxidation of acid orange 7. J. Environ. Chem. Eng. 2(3): 1881-1888.

13

Advances in the Use of Metal Oxide Nanocomposites for Remediation of Aqueous Pollutants: Adsorption and Photocatalysis of Organics and Heavy Metals

Michel Franco Galvão Pereira[1], Georgenes Marcelo Gil da Silva[1], Pedro Henrique Novaes Cardoso[1], Carlos Yure Barbosa Oliveira[2], Victor Nascimento de Souza Leão[1] and Evando Santos Araújo[1,*]

13.1 INTRODUCTION

The increase in industrial production associated with the accelerated growth of the world population in recent decades is one of the main factors that contribute to the pollution of the physical and biological components on Earth.

In particular, the scarcity of large scale effective and low-cost procedures of discard and treatment of industrial waste (such as by-products from the production of medicines, organic dyes, pesticides and heavy metals) contributes to the contamination worsening of aquatic environments and potable water for animal and vegetable consumption (Tchounwou et al. 2012).

These substances negatively interfere in the life cycle of species, and in the availability and quality of vital elements for living beings, due to its toxic and carcinogenic components resulted from the decomposition of its molecules (Karpinska and Kotowska 2019). As an example, Table 13.1 provides a description of various heavy metals, their main sources of contamination and their negative effects on human health (WHO 2017a, b, Carolin et al. 2017).

Although the advanced water treatment technologies, such as sedimentation, coagulation, aerobic activated sludge-based treatment, nitrification-denitrification and phosphorus removal have been widely used after conventional primary and secondary treatments, they have some limitations and drawbacks, such as high energy consumption, carbon emission, excess sludge discharge and

[1] Department of Materials Science, Federal University of São Francisco Valley, Juazeiro, Brazil.
[2] Department of Fishing Resources and Aquaculture, Rural Federal University of Pernambuco, Recife, Brazil.
* Corresponding author: evando.araujo@univasf.edu.br

considerable additional cost. In addition, many water treatment techniques used by developing countries (including chlorination, boiling and solar disinfection) are ineffective in removing some contaminants, such as heavy metals (Kwaadsteniet et al. 2013, Presura and Robescu 2017).

TABLE 13.1 Effects of heavy metal on human health and possible sources of contamination

Heavy Metal	Effects on Human Health	Most Common Source of Contamination/Contact Form
Arsenic (As)	Skin damage	Natural form/skin contact
	Circulatory system problems	Electronic industries/oral contact
Cadmium (Cd)	Kidney damage	Natural form/skin contact
	Emergence of cancer cells	Various chemical industries/skin and oral contact
Chromium (Cr)	Allergic dermatites	Natural form/skin contact
	Diarrhea, nausea and	Steel industry/body's airways contact
Copper (Cu)	Gastrointestinal problems	Natural form/oral contact
	Liver and kidney damage	Domestic plumbing systems/oral contact
Lead (Pb)	Kidney damage	Lead-based products/oral contact
	Sleep disorders, abdominal pain, reduction of neural development	Domestic plumbing systems/body's airways contact
Mercury (Hg)	Kidney damage	Petrochemical and pharmaceutical industry; volcanic activities; evaporation of water bodies; electronic devices/oral and body's airways contact
	Damage to the nervous system	

Source: (WHO 2017a, b, Carolin et al. 2017).

The search for new technologies and functional materials for water treatment that are able to meet current needs, with lower cost and environmental impact, is an open field of research today (Chen et al. 2018, Moga et al. 2018, Carolin et al. 2017). In this sense, metal oxide nanostructures (materials that respond to external stimulus – light, electric potential, pressure, among others) have gained increasing attention from academic researchers and industrial developers as potential materials to produce efficient and environmentally friendly devices for the remediation of organic- and heavy metals-contaminated water (Ganachari et al. 2019, Yang et al. 2017). Currently, among the techniques for removing contaminants using metal oxides, adsorption and heterogeneous photocatalysis stand out. These applications are directly related to metal oxide nanostructures properties, such as high chemical and structural stability, high surface area/volume ratio, excellent electrical/electronic bulk properties, surface interactions with contaminant molecules, and the possibility of reusing the material in new decontamination cycles (Yaqoob et al. 2020).

In addition, metal oxides can form nanocomposites with other oxides and other functional materials such as conducting polymers and carbon-based materials, in order to ensure greater surface area/volume, less electron-hole recombination and better performance in adsorption, photocatalysis and sensing, compared to the results obtained with the use of pure oxides (Tan et al. 2013, Kang et al. 2019). These nanostructured materials allow for a range of applications and it is considered an active research object today.

This chapter presents an exploratory bibliographic review in order to present the fundamentals and advances in the use of metal oxide nanocomposites-based photocatalysts and adsorbents for removal of aqueous pollutants, such as organics and heavy metals. A bibliometric analysis was proposed in order to show the evolution of this research topic in the literature over the past two decades. In addition, important scientific contributions to the area are presented and discussed based on this analysis.

13.2 FUNDAMENTALS OF METAL OXIDE NANOCOMPOSITES APPLIED TO PHOTOCATALYSIS AND ADSORPTION

Metal oxides belong to the class of semiconductor materials. These materials have chemical structures formed by the change of charges between oxygen atoms (highly electronegative) and atoms of metallic elements. By definition, semiconductors have a structure of electronic bands with the valence band (filled) separated from the conduction band (unfilled) by a bandgap with sufficient energy (around 1.0-3.0 eV) to maintain them with characteristics of electrical insulators at 0 Kelvin temperature (Böer and Pohl 2014).

These materials present characteristics of conductors when they are excited by external environmental stimulus (such as light, pressure, electrical potential difference, heat, etc.) with sufficiently greater energy than the gap energy, so that electrons (e^-) of the valence band can migrate to the conduction band (consequently generating active holes (h^+) in the valence band) (Böer and Pohl 2014). In other words, their electrical and electronic properties are directly dependent on their structure of electronic bands.

In this context, metal oxide nanostructures (structures with dimensions of the order of 10^{-9} m) stand out in relation to other materials due to their synergistic combination between high superficial area/volume ratio and tunable optical, magnetic, electrical, mechanical, thermal, catalytic and photochemical characteristics (Tong et al. 2012, Dey 2018). These properties, combined with its excellent chemical, structural and environmental stability, ensure them a variety of emerging applications, such as the production of electrochemical, optical and gas sensors, fuel cells, solar cells, piezoelectric devices, supercapacitors, adsorbents, and photocatalysts, among others (Dey 2018, Diao and Wang 2018, Lira-Cantu 2018).

Metal oxide-based nanocomposites are materials resulting from the interaction between two or more metal oxides, or between these and other functional nanomaterials. Conducting polymers and carbon-based materials are some of the examples that have improved the properties of the produced nanocomposites. The main objective of this interaction is to provide enhanced attributes to the resulting material in comparison to those presented by the oxides applied individually, such as greater chemical/structural stability, lower bandgap to minimize electron/hole recombination and greater contact area (Araújo et al. 2015, 2016).

The possibility of producing metal oxide-based nanocomposites with controlled size, morphology and composition, and interaction with other structures improve the mentioned characteristics of these materials and opens new perspectives for environmental applications, such as more efficient photocatalysis and adsorption processes used for decontamination of aqueous media.

Photocatalysis is a photochemical process of the advanced oxidative processes class for the formation of hydroxyl HO·, a free radical that is highly oxidizing and degrades polluting organic substances in the water. These processes are established from a natural or artificial light stimulus on the material, in which the HO· is generated from the water molecules on the metal oxide surface.

The action of a metal oxides-based photocatalyst in an aqueous medium can be summarized as follows: first, the semiconductor is exposed to a light source, with enough energy to activate it. As a consequence, photons are absorbed with energy greater than the potential energy of the bandgap (usually very positive). This process causes an electron transfer from the valence band to the conduction band, leaving a hole (h^+) in the valence band. These holes generate HO· free radicals from the water adsorbed on the metal oxide surface, which induce the organic contaminant oxidation (Araújo et al. 2015). In short, the competition between the capacity for interfacial charge transfer and the recombination of the e^-/h^+ pair is a relevant factor for the photocatalytic efficiency of these materials.

The adsorption is another process that can be activated by metal oxides, with potential applications in the remediation of contaminated aqueous environments. This process involves the accumulation of molecules of contaminating substances (named adsorbates) at the liquid-solid interface between water and the semiconductor surface (which assumes the function of adsorbent material). In this configuration, the physical and chemical properties of both adsorbates and adsorbents are relevant factors for the efficiency of the process efficiency. The adsorption can present physisorption type interaction when the capture of molecules occurs by physical interaction with the adsorbent surface. Similarly, chemisorption occurs when this is a chemical nature interaction. Although, in the first case, the interaction between adsorbate and adsorbent occurs by van der Walls forces and is easily reversible, in chemisorption the attraction between these agents occurs through strong chemical bonds that ensure greater stability to the interaction.

Figure 13.1 illustrates potential metal oxide-based nanocomposites for the removal of organic substances and heavy metals from water by chemisorption and physisorption, and highlights relevant factors in these processes, such as solubility, hydrophobicity, size and surface charge.

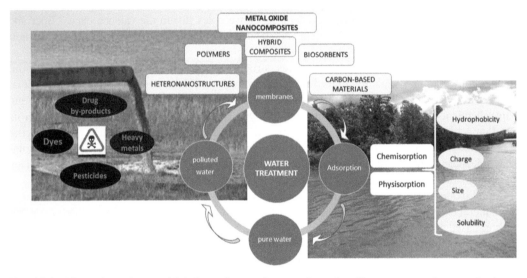

Fig. 13.1   Illustrative scheme with information on the use of metal oxide nanocomposites as adsorbents for contaminating materials in water and the main related variables of the physisorption and chemisorption processes. (*Source*: Authors' files.)

In summary, adsorption allows the solute (contaminant) to be removed from the solution and be accumulated on the semiconductor surface. In this sense, the process efficiency is directly dependent on the pH and aqueous system temperature, the pollutant material concentration, and adsorbent material's choice with surface ability to interact with the contaminant (Arora 2019).

The great attraction by the adsorption techniques for the water contaminant removal refers to its high selectivity at molecular level, combined with the low generation of residues and the possibility of reusing the adsorbent (Kumar et al. 2014).

13.3   BIBLIOMETRIC STUDY

To a brief characterization of the research field on metal oxides nanocomposites for remediation of aqueous pollutants, a data search was performed in the Scopus platform (14 April 2020). The search was carried out using "metal oxide" AND "nanocomposite" AND "pollutant removal" present in the title, abstract or keywords of the documents. This search resulted in 501 documents. The collected data were organized and (or) analyzed in the following ways: (i) trends in publications over the years; (ii) main journals, their Impact Factor and CiteScore; (iii) transitions in the keywords over the years; and (iv) word cloud concept of the main aspects related to the topic. These pollutants were evaluated individually and, in groups of two, three or four pollutants. Analyses were performed using the RStudio® (version 3.5.2) and VOSviewer (version 1.6.14) softwares.

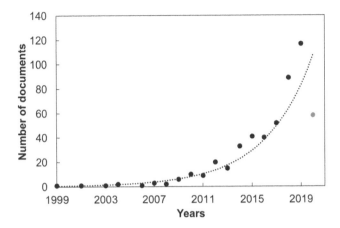

Fig. 13.2   Publishing trend in the field on metal oxide nanocomposites for remediation of aqueous pollutants observed between 1999 and 2020. (*Source:* Authors' files.)

Figure 13.2 shows the trend in quantity of documents published between 1999 and 2020 on metal oxides nanocomposites for remediation of aqueous pollutants. The first ten years of this research field was a period with low variation in the number of published documents (between 1 and 3 per year). In the last decade, it was possible to observe an increase in publications, intensified at the last 4 years, linked to the need to develop water decontamination solutions in the face of the growing evolution in global environmental pollution. This behavior has been observed for research in functional nanomaterials, which involve a series of peculiarities, such as the discovery and synthesis of new materials, the development of more efficient characterization methods and the study of new mechanisms and the synergistic interaction between selective components that ensure greater surface area against contaminants.

This occurs for studies with metallic oxides and their nanocomposites in recent years, in the search for selective materials and synthesis techniques that guarantee a more efficient and environmentally friendly water decontamination by organic and heavy metals. The reduction in the number of documents in 2020 is related to the period of this analysis, which was performed in April, although a large number of documents has already been published.

The top five journals that published these documents are shown in Fig. 13.3. This ranking is led by the 'Journal of Hazardous Materials' (57), 'Chemosphere' (44) and 'Environmental Science' and 'Pollution Research' (20). Much of the research on metal oxide nanocomposites for the remediation of aqueous environments is distributed in the numerous newspapers in the area of materials that specifically deal with synthesis, characterization of new materials and their applications in solving current problems in society. The main journal that published on metal oxides for remediation of aqueous pollutants is also the journal with the highest Impact Factor on CiteScore, as it connects the study of basic science of functional materials with the discussion of new and (or) improved techniques to remedy environmental pollution by hazardous materials, such as organics and heavy metals.

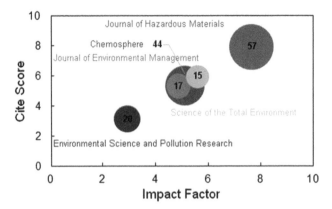

Fig. 13.3 Top five scientific journals that contribute to the topic of metal oxide nanocomposite for remediation of aqueous pollutants. (*Source:* Authors' files.)

A word cloud illustration is presented in Fig. 13.4. It shows the most frequent keywords that emerge in the documents that involve metal oxide nanocomposites for remediation of aqueous pollutants. 'Adsorption' (312), 'Pollutant Removal' (201), 'Water Pollutants' (160), 'Chemical' (149) and '*pH*' (140) were the keywords with the highest number of results in the research documents.

Fig. 13.4 Word cloud of the main keywords that appear in documents of metal oxide nanocomposites for remediation of aqueous pollutants. (*Source:* Authors' files.)

The water chemical pollutants' adsorption shows itself as one of the most used and still promising techniques in comparison to the conventional water decontamination techniques, mainly for its environmentally friendly use and constant development of materials with better surface properties for this application. In addition, the pH of the aqueous environment is one of the main factors to be considered because the activation of surface functional groups of potential nanocomposites is highly dependent on this factor. In this Fig. 13.4, it is possible to examine some words, such as 'ferric compounds', 'iron', 'iron oxide' and 'magnetism', which refer to the recent use of metal oxide nanocomposites based on iron oxide, in order to improve the efficiency of photocatalysis and adsorption processes and the separation of contaminant from water after the response of composites.

High impact water pollutants, such as pharmaceuticals, pesticides, organic dyes and heavy metals have been widely cited in the search because these contaminants represent a large portion of the substances used in industrial processes, which shows that the search for suitable clean-up methods is an open field of research.

An analysis of the keywords' transitions (Fig. 13.5), in the last five years, which appear at least 10 times in the text of these documents, was peformed. Figure 13.5 illustrates this transition in the research field, which represents the advances in the use of metal oxide nanocomposites for remediation of aqueous pollutants. The analysis also shows that graphene has emerged as a new functional material that can interact with metal oxides in order to produce more efficient depollution devices. The pollutant specific search revealed a greater interest in remediation of dyes (373) and heavy metals (333), with some research dealing with both contaminants.

In the investigated researches, the study of mixed metal oxide-based nanocomposites (MMOs) for remediation of aqueous pollutants has been frequent. Titanium dioxide in its allotropic anatase form (anatase TiO_2, ~ 3.2 eV bandgap) has been the most used metal oxide for the production of these nanocomposites, synthesized with other oxides or with other functional materials, in order to improve adsorption and photocatalysis processes (Julkapli et al. 2015). This oxide is non-toxic, and also has a high chemical stability, excellent photoactivity and photostability, is easily UV-activated and has low cost compared to other oxides (Araújo et al. 2015). The anatase form is preferable to the rutile phase (irreversible transition at 600-700°C), mainly because it presents a greater surface area of action when interacting with other materials.

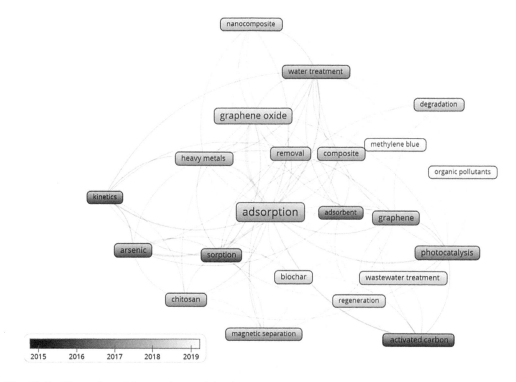

Fig. 13.5 Illustration of keywords-transition in the representative documents of the use of metal oxide nanocomposites for remediation of aqueous pollutants (2015-2019). (*Source:* Authors' files.)

Advances in this field of research involve the incorporation of other metal oxides (such as zinc, tungsten, vanadium, niobium and copper oxides) into TiO_2, in order to increase the efficiency of these processes (Handojo et al. 2020).

Zinc oxide (ZnO) is also widely used in adsorption and photocatalysis processes due to its physico-chemical aspects (biocompatible, non-toxic, very chemically and mechanically stable, insoluble in aqueous environments, ~ 3.1-3.3 band gap, and so on), similar to those of anatase TiO_2. Nanocomposites between TiO_2 and zinc oxide (ZnO) produced by various methods have shown better photocatalytic and adsorption properties, compared to the use of pristine TiO_2 (Upadhyay et al. 2019, Araújo et al. 2015, 2016, Kanjwal et al. 2010). Hybrid TiO_2 nanoparticles have been synthesized from hydrothermal processes, with the incorporation of zinc oxide in the atomic structure (substitution of Ti^{4+} by Zn^{2+}) and on the surface of TiO_2 (Silva et al. 2010, Hanaor and Sorrell 2011, Araújo 2015).

Figure 13.6 presents a micrograph obtained by scanning electron microscopy (with X-ray dispersive energy mapping) of a TiO_2/ZnO nanocomposite sample produced by hydrothermal process.

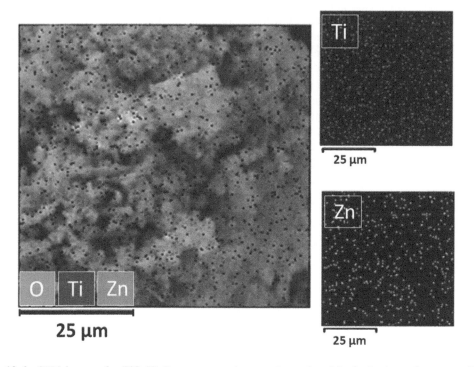

Fig. 13.6 SEM image of a TiO_2/ZnO nanocomposite sample produced by hydrothermal process. EDS mapping (on the right) shows the populations of Ti and Zn atoms distributed throughout the hybrid material. (*Source:* Authors' files.)

The homogeneous distribution of Ti and Zn atoms along the surface of the produced hybrid material is the first indication of a successful synthesis, which can be validated by other characterization techniques such as XRD, Raman, FTIR and BET surface area.

The works report that the TiO_2/ZnO junctions can accelerate the degradation process of dyes, with the mutual chemical process of N-deethylation and cleavage of contaminant molecules. At the level of band theory, electrons' transfer from the zinc oxide conduction band to the titanium dioxide conduction band is reported. As a consequence, there is the photogeneration of holes (h^+) from the TiO_2 valence band to the ZnO valence band. With this configuration, based on the standard photocatalysis mechanisms, the separation of electrical charges is said to occur in an improved way (Upadhyay et al. 2019, Araújo et al. 2015, 2016, Kanjwal et al. 2010).

Araújo's Group developed TiO_2/ZnO hierarchical heteronanostructures as potential materials for photocatalysis processes against organic dyes (Araújo et al. 2016). These nanostructures were synthesized with a sea urchin-like aspect (Fig. 13.7a), from hydrothermal growth of zinc oxide hexagonal nanorods on surface of P25 TiO_2 (3:1 anatase:rutile ratio) nanoparticles, and showed high surface area to dye photodegradation, even with white light (which does not occur with the use of pure TiO_2).

The best decontamination results obtained with Ti/Zn MMOs have been attributed to atomic and surface interactions between the oxides, which reduced the band-gap of the combined material and, consequently, hindered the e^-/h^+ recombination. In addition, the authors proposed the radial growth of zinc crystals on the nanofibers' surface (prepared by electrospinning), initially decorated with TiO_2 particles, and developed photocatalysts with high surface area/volume and better performance (90 percent of dye degradation in aqueous solution after 70 min of exposure to light), when compared to that obtained using the urchin-like nanoparticles. In this last configuration (Fig. 13.7b), the fibers functioned as a polymeric template to stimulate the regular distribution of oxides and prevent aggregation of the particles, which significantly increased the process efficiency.

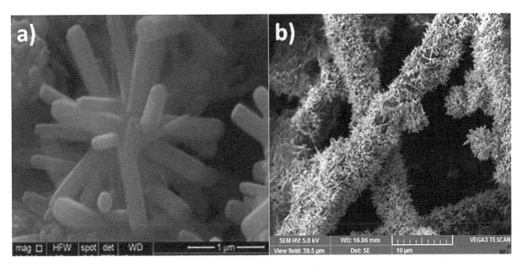

Fig. 13.7 (a) Urchin-like aspect to TiO_2/ZnO hierarchical heteronanostructures, (b) ZnO nanorods radially surrounding the fiber surface decorated with TiO_2 nanoparticles. (*Source:* Authors' files.)

Other works report the interaction of tungstein trioxide (WO_3) with TiO_2, obtained through sintering process, increases the conductivity of the TiO_2/WO_3 system in comparison to pure TiO_2 (Zhang et al. 2018), while the TiO_2 doping with niobium pentoxide (Nb_2O_5) (Yan et al. 2014) or vanadium pentoxide (V_2O_5) (Mandal et al. 2019) delays the phase transition from anatase to rutile. In addition, the combined action between TiO_2, WO_3 and V_2O_5 returns nanocomposites with increased sensibility and phase stability, as reported in the literature (Zhang et al. 2018). These phenomena occur mainly due to the interaction at the atomic level between the atoms of these metals, in which tungsten atoms are set as impurities in the crystalline structure vacancies of Ti and, as a result, return materials with better electronic properties and with greater chemical and structural stability.

In general, the presence of doping metal ions can promote absorption in the visible region, and with a higher surface area, which improves the water cleaning processes described here (Handojo et al. 2020).

Other mixed metal oxides also showed excellent responses against organics and heavy metals in water. As examples of new alternatives to correct these problems, the iron/manganese (Fe/Mn) oxide nanostructures have been used for pharmaceutics, pesticides, organic dyes, and heavy metals adsorption (Zheng et al. 2020, Liang et al. 2020, Zhang et al. 2019a, Singh et al. 2018). These oxides have gained attention due to their high capacity to attract and reduce these substances. In addition, the combined action between Fe and Mn active sites has optimized decontamination processes in aqueous environments due to the greater surface area, resulting from the formation of additional adsorption sites and greater charge mobility, compared to the action of pure oxides.

Other potential MMOs, such as magnesium/aluminum (Mg/Al), silica-alumina (Si-Al), zinc-tin (Zn-Sn) and iron-aluminum (Fe-Al), were also used as adsorbents (Wawrzkiewicz et al. 2017, Kumar et al. 2014, Tsade et al. 2019) and photocatalysts (Yuan et al. 2017) of heavy metals, and return large adsorption, degradation capacity and efficiency, compared to the use of pure oxides and traditional oxides. In addition, scientists also showed that Mg/Al oxide nanocomposite are able to simultaneously degrade multiple metal ions of great environmental interest (Liu et al. 2016).

Mg/Al, Mg/Cerium (Ce) and TiO_2/CeO_2 binary metal oxides (Henych et al. 2019, Manav et al. 2018) are shown as potential materials for the degradation of pesticides because they offer quicker, more efficient and economically viable adsorption and photocatalysis process; Zn/Fe (Di et al. 2017) and $ZnO/ZnFe_2O_4$ MMOs (Di et al. 2019) have been produced to significantly improve the adsorption and photodegradation (and these combined methods) of high-demand pharmaceuticals; Ni/Co/Fe (Pan et al. 2018) and $Zn/Ni/ZnAl_2O_4$ (Zhang et al. 2016) tri-mixed metal oxide nanocomposites showed photocatalytic properties superior to those of P25 TiO_2 for degradation of usual organic dyes.

Another possibility to produce water decontaminating materials includes the interaction of metal oxides with conducting polymers, such as polyaniline (PANI) and polypyrrole (PPy). Conducting polymers (CP) are functional organic materials that stand out from other polymers because they present electrical conductivity. In addition, CP are known for their unique microstructural aspects (highly porous surface and bulk, high surface area/volume ratio and variety of functional surface groups), processability by dispersion, adjustable oxidation-reduction and electrical/electronic properties by organic synthesis from their monomers, and by the dispersion of other electronically active materials, such as metal oxides (Farrel and Kaner 2015). Another differential of these polymers is the possibility of including natural clays, in order to increase the chemical and (or) structural stability of the polymeric matrix.

In addition to the chemical functionalities of their polymeric matrix, CP is used as template for the dispersion of metal oxides and to prevent the aggregation of these particles. Thus, the combined action between oxides and polymer significantly increases the surface area of action against contaminants.

The combination of the CP and metal oxide properties returns a variety of applications in nanotechnology to CP-metal oxide nanocomposites, such as in the development of sensors, supercapacitors, solar cells, organic electronic devices and new efficient adsorbent and photocatalysts materials for the remediation of aqueous environments polluted by organics and heavy metals (Ahmad 2019).

Polyaniline (PANI) is a CP known to have a large amount of superficial functional groups amine and emine. These groups have strong affinity for metal ions and can therefore be used to remove selective organics and heavy metals in aqueous systems. The research field for the development of metal oxide-PANI nanocomposites is open and has returned many innovations in recent years. For example, ZnO-PANI nanocomposite was recently presented as new adsorbent for the removal of heavy metal ions in aqueous solutions (Ahmad 2019). The interaction of zinc nanostructures with PANI returned greater mechanical and thermal resistance, and improved adsorbent properties from the availability of a large quantity and high dispersion of OH functional groups of the oxide and amine groups of polyaniline.

The iron oxide-PANI nanocomposite was prepared with unique morphology, dispersing iron oxide nanowires on rod-like PANI matrix (Dutt et al. 2018). The nanocomposite showed very high and fast adsorption of organic dyes based on electrostatic interaction and with easy magnetic removal of the adsorbent/adsorbate set from water.

Recently, PANI microsphere-Fe_3O_4 magnetic nanocomposite was synthesized as a new adsorbent material (Dutta et al. 2020). The results were very promising with heavy metal adsorption, much higher than that observed in other PANI nanocomposites. The sorption and desorption tests confirmed this PANI-Fe_3O_4 configuration as a sustainable material that can be reused over and over again. In addition, the presence of residual iron in the treated water was noted within the limits allowed by the World Health Organization.

Polypyrrole (PPy) is another CP with great potential to form functional nanocomposites with metal oxides. PPy is synthesized after polymerization of the pyrrole and activated as a conductive polymer after the oxidation process. This polymer is amorphous and has a fractal surface, which also guarantees a high surface area. Its electrical conduction properties are dependent on the conditions and reagents used in the synthesis (Chen et al. 2016, Zhang et al. 2019b).

SiO_2-PPy, Al_2O_3-PPy, and Fe_3O_4-PPy have been used for the adsorption of organic dyes (Chen et al. 2016). The results showed that these nanocomposites have greater affinity with contaminant molecules and new sensory characteristics, directly related to changes in the polymerization process, morphology, Zeta potential and adsorption capacity. For example, Al_2O_3-PPy has shown an adsorption capacity above 130 mg/g, greater than that obtained using the components separately.

Recently, ZnO/PPy-based nanocomposites have returned excellent results of adsorption (capacity of ~ 140 mg/g) and photodegradation of organic dyes (Zhang et al. 2019b, Ahmad et al. 2019). Scientists have shown that this combination can be reused due to excellent chemical and structural stability, and it has great potential for large-scale use.

Other metal oxide nanocomposites with potential use in decontamination of aqueous environments are those prepared from interaction with graphene (Kumar et al. 2019).

In the last decade, graphene has gained very high attention in the scientific community and industry due to its unique physicochemical, electrical and optical properties combined with its honeycomb structure. These materials have high surface area and a wide variety of functional groups, which can be used for several emerging applications, such as in water decontamination. On the other hand, its potential depends strongly on the choice of suitable synthesis techniques. Graphene acts as a photoelectron acceptor in photocatalysis processes, increasing its light absorption range, including the visible range.

The incorporation of metal oxides in its structure provides nanocomposites with a surface area of action even greater than pure graphene, and a chemical interaction by van der Waals forces, which guarantees them to prevent the aggregation of oxides and increase its efficiency.

TiO_2-graphene nanocomposites were studied, and the high capacity of graphene to accept electrons decreased the TiO_2 bandgap by forming an additional valence band (between its original valence and conduction bands), which consequently optimized the photocatalytic activity of the semiconductor (Kiarii et al. 2018). In addition, the interaction of graphene with copper/tin MMO (Cu_2O/SnO_2) and SnO_2 oxide also returned additional valence bands that guaranteed better photodegradation results, when compared to the use of pure TiO_2 (Wang et al. 2013). These results were attributed to the improvement of the electronic coupling and the fast capture of excited electrons from graphene to metal oxides in the composite charge transport. Their unique properties, associated with a variety of applications in environmental remediation, provide an exciting platform of research in the coming years.

13.4 SUMMARY

In this chapter, the main fundamentals about metallic oxides, their composites and the techniques of adsorption and photocatalysis using these semiconductors were presented. The bibliometric study showed the evolution of the use of these materials in the remediation of water over the years. This research also provided the reader a discussion of the latest scientific advances in this topic, presenting carbon-based materials as potential candidates for the development of new nanocomposites with metal oxides for use in water decontamination.

13.5 ACKNOWLEDGMENTS

The authors would like to thank the Brazilian research agencies CNPq, FACEPE, FAPESB and CAPES for the financial support to the projects developed in our research group. In particular, the corresponding author thanks their wife Amara and children Cauã and Duda for their understanding and presence during the writing of this work, amidst the necessary adjustments for the combat of the current COVID-19 global pandemic.

13.6 REFERENCES

Ahmad, N., S. Sultana, S.M. Faisal, A. Ahmed, S. Sabir and M.Z. Khan. 2019. Zinc oxide-decorated polypyr-role/chitosan bionanocomposites with enhanced photocatalytic, antibacterial and anticancer performance. RSC Adv. 9: 41135-41150.

Ahmad, R. 2019. Polyaniline/ZnO Nanocomposite: A Novel Adsorbent for the Removal of Cr(VI) from Aqueous Solution. pp. 1-22. *In*: D. Lucan [ed.]. Advances in Composite Materials Development. IntechOpen, London.

Araújo, E.S., J. Libardi, P.M. Faia and H.P. de Oliveira. 2015. Hybrid ZnO/TiO$_2$ loaded in electrospun polymeric fibers as photocatalyst. J. Chem. 2015: 1-10.

Araújo, E.S., B.P. da Costa, R.A.P. Oliveira, J. Libardi, P.M. Faia and H.P. Oliveira. 2016. TiO$_2$/ZnO hierarchical heteronanostructures: Synthesis, characterization and application as photocatalysts. J. Environ. Chem. Eng. 4: 2820-2829.

Arora, R. 2019. Adsorption of heavy metals–A review. Mater. Today Proc. 18: 4745-4750.

Böer, K.W. and U.W. Pohl. 2014. Bands and Bandgaps in Solids. pp. 1-60. *In*: K.W. Böer and U. Pohl [eds.]. Semiconductor Physics, Springer, Singapore.

Carolin, C.F., P.S. Kumar, A. Saravanan, G.J. Joshiba and M. Naushad. 2017. Efficient techniques for the removal of toxic heavy metals from aquatic environment: A review. J. Environ. Chem. Eng. 5: 2782-2799.

Chen, J., J. Feng and W. Yan. 2016. Influence of metal oxides on the adsorption characteristics of PPy/metal oxides for methylene blue. J. Colloid. Interface Sci. 475: 26-35.

Chen, J., J. Luo, Q. Luo and Z. Pang. 2018. Wastewater Treatment. De Gruyter, Berlin, Boston.

Dey, A. 2018. Semiconductor metal oxide gas sensors: A review. Mater. Sci. Eng. B. 229: 206-217.

Di, G., Z. Zhu, H. Zhang, J. Zhu, H. Lu, W. Zhang, et al. 2017. Simultaneous removal of several pharmaceuticals and arsenic on Zn-Fe mixed metal oxides: Combination of photocatalysis and adsorption. Chem. Eng. J. 328: 141-151.

Di, G., Z. Zhu, Q. Huang, H. Zhang, J. Zhu, Y. Qiu, et al. 2019. Targeted modulation of g-C$_3$N$_4$ photocatalytic performance for pharmaceutical pollutants in water using ZnFe-LDH derived mixed metal oxides: Structure-activity and mechanism. Sci. Total Environ. 650: 1112-1121.

Diao, F.Y. and Y.Q. Wang. 2018. Transition metal oxide nanostructures: Premeditated fabrication and applications in electronic and photonic devices. J. Mater. Sci. 53: 4334-4359.

Dutt, S., T. Vatsa and P.F. Siril. 2018. Synthesis of polyaniline-magnetite nanocomposites using swollen liquid crystal templates for magnetically separable dye adsorbent applications. New J. Chem. 42: 5709-5719.

Dutta, S., S.K. Manna, S.K. Srivastava, A.K. Gupta and M.K. Yadav. 2020. Hollow polyaniline microsphere/ Fe$_3$O$_4$ nanocomposite as an effective adsorbent for removal of arsenic from Water. Sci. Rep. 10: 4982.

Ganachari, S.V., L. Hublikar, J.S. Yaradoddi and S.S. Math. 2019. Metal Oxide Nanomaterials for Environmental Applications. pp. 2357-2368. *In*: L. Martínez, O. Kharissova and B. Kharisov [eds.]. Handbook of Ecomaterials. Springer, Singapure.

Hanaor, D.A.H. and C.C. Sorrell. 2011. Review of the anatase to rutile phase transformation. J. Mater. Sci. 46: 855-874.

Handojo, L., N.A. Ikhsan, R.R. Mukti and A. Indarto. 2020. Indarto. Nanotechnology for remediations of agrochemicals. pp. 535-567. *In*: M. Narasimha and V. Prasad [eds.]. Agrochemicals Detection, Treatment and Remediation. Butterworth-Heinemann, London.

Henych, J., P. Janoš, M. Kormund, J. Tolasz and V. Štengl. 2019. Reactive adsorption of toxic organophosphates parathion methyl and DMMP on nanostructured Ti/Ce oxides and their composites. Arabian. J. Chem. 12: 4258-4269.

Julkapli, N.M., S. Bagheri and A.T. Yousefi. 2015. TiO$_2$ hybrid photocatalytic systems: Impact of adsorption and photocatalytic performance. Rev. Inorg. Chem. 35: 151-178.

Kang, X., S. Liu, Z. Dai, Y. He, X. Song and Z. Tan. 2019. Titanium dioxide: From engineering to applications. Catalysts 9: 1-32.

Kanjwal, M.A., N.A.M. Barakat, F.A. Sheikh, S.J. Park and H.Y. Kim. 2010. Photocatalytic activity of ZnO-TiO_2 hierarchical nanostructure prepared by combined electrospinning and hydrothermal techniques. Macromol. Res. 18: 233-240.

Karpinska, J. and U. Kotowska. 2019. Removal of organic pollution in the water environment. Water 11: 02017.

Kiarii, E.M., K.K. Govender, P.G. Ndungu and P.P. Govender. 2018. Recent advances in titanium dioxide/graphene photocatalyst materials as potentials of energy generation. Bull. Mater. Sci. 41: 1e14.

Kumar, K.Y., H.B. Muralidhara, Y.A. Nayaka and J. Balasubramanyam. 2014. Low-cost synthesis of mesoporous Zn(II)–Sn(II) mixed oxide nanoparticles for the adsorption of dye and heavy metal ion from aqueous solution. Desalin. Water Treat. 52: 22-24.

Kumar, R., M.O. Ansari, M.A. Barakat and J. Rashid. 2019. Graphene/metal oxide-based nanocomposite as photocatalyst for degradation of water pollutants. pp. 221-240. In: M. Jawaid, A. Ahmad and D. Lokhat [eds.]. Graphene-Based Nanotechnologies for Energy and Environmental Applications. Elsevier, Amsterdam.

Kwaadsteniet, M., P. Dobrowsky, A.V. Deventer, W. Khan and T. Cloete. 2013. Domestic rainwater harvesting: Microbial and chemical water quality and point-of-use treatment systems. Water Air Soil Pollut. 224: 1629-1637.

Liang, M., S. Xu, Y. Zhu, X. Chen, Z. Deng, L. Yan, et al. 2020. Preparation and characterization of Fe-Mn binary oxide/mulberry stem biochar composite adsorbent and adsorption of Cr(VI) from aqueous solution. Int. J. Environ. Res. Public Health 17: 676(1-16).

Lira-Cantu, M. 2018. The Future of Semiconductor Oxides in Next-Generation Solar Cells. Elsevier, Amsterdã.

Liu, X.J., H.Y. Zeng, S. Xu, C.R. Chen, Z.Q. Zhang and J.Z. Du. 2016. Metal oxides as dual-functional adsorbents/catalysts for $Cu^{+2}/Cr(VI)$ adsorption and methyl orange oxidation catalysis. J. Taiwan Inst. Chem. E. 60: 414-422.

Manav, N., V. Dwivedi and A.K. Bhagi. 2018. Degradation of DDT, a pesticide by mixed metal oxides nanoparticles. p. 93-99. In: V. Parmar, P. Malhotra and D. Mathur [eds.]. Green Chemistry in Environmental Sustainability and Chemical Education. Springer, Singapore.

Mandal, R.K., S. Kundu, S. Sain and S.K. Pradhan. 2019. Enhanced photocatalytic performance of V_2O_5-TiO_2 nanocomposites synthesized by mechanical alloying with morphological hierarchy. New J. Chem. 43: 2804-2816.

Moga, I.C., I. Ardelean, O.G. Donţu, C. Moisescu, N. Băran, G. Petrescul, et al. 2018. Materials and technologies used in wastewater treatment. IOP Conference Series: Materials Science and Engineering 374: 1-6.

Pan, D., S. Ge, J. Zhao, Q. Shao, L. Guo, X. Zhang, et al. 2018. Synthesis, characterization and photocatalytic activity of mixed-metal oxides derived from NiCoFe ternary layered double hydroxides. Dalton Trans. 47: 9765-9778.

Presura, E. and L. Robescu. 2017. Energy use and carbon footprint for potable water and wastewater treatment. Proceedings of the International Conference on Business Excellence 11: 191-198.

Silva, S.S., F. Magalhães and M.T.C. Sansiviero. 2010. ZnO/TiO_2 semiconductor nanocomposites. Photocatalytic tests. Quim Nova 33: 85-89.

Singh, N.H., K. Kezo, A. Debnath and B. Saha. 2018. Enhanced adsorption performance of a novel Fe-Mn-Zr metal oxide nanocomposite adsorbent for anionic dyes from binary dye mix: Response surface optimization and neural network modeling. Appl. Organomet. Chem. 32: e4165.

Tan, L., W. Ong, S. Chai and A.R. Mohamed. 2013. Reduced graphene oxide-TiO_2 nanocomposite as a promising visible-light-active photocatalyst for the conversion of carbon dioxide. Nanoscale Res. Lett. 8: 465(1-9).

Tchounwou, P.B., C.G. Yedjou, A.K. Patlolla and D.J. Sutton. 2012. Heavy metal toxicity and the environment. Experientia. Suppl. 101: 133-164.

Tong, H., S. Ouyang, Y. Bi, N. Umezawa, M. Oshikiri and J. Ye. 2012. Nano-photocatalytic materials: possibilities and challenges. Adv. Mater. 24: 229-251.

Tsade, H., B. Abebe and H.C.A. Murthy. 2019. Nano sized Fe-Al oxide mixed with natural maize cob sorbent for lead remediation. Mater. Res. Express 6: 085043.

Upadhyay, G.K., J.K. Rajput, T.K. Pathak, V. Kumar and L.P. Purohit. 2019. Synthesis of ZnO:TiO_2 nanocomposites for photocatalyst application in visible light. Vacuum 160: 154-163.

Wang, Z., Y. Du, F. Zhang, Z. Zheng, X. Zhang, Q. Feng, et al. 2013. Photocatalytic degradation of pendimethalin over Cu_2O/SnO_2/graphene and SnO_2/graphene nanocomposite photocatalysts under visible light irradiation. Mater. Chem. Phys. 140: 373e81.

Wawrzkiewicz, M., M. Wiśniewsk, A. Wołowicz, V.M. Gun'ko and V.I. Zarko. 2017. Mixed silica-alumina oxide as sorbent for dyes and metal ions removal from aqueous solutions and wastewaters. Microporous Mesoporous Mater. 250: 128-147.

World Health Organization - WHO. 2017a. Guidelines for Drinking-Water Quality. 4th edition. World Health Organization, Geneva.

World Health Organization - WHO. 2017b. Progress on Drinking Water, Sanitation and Hygiene. World Health Organization, Geneva.

Yan, J., G. Wu, N. Guan and L. Li. 2014. Nb_2O_5/TiO_2 heterojunctions: Synthesis strategy and photocatalytic activity. Appl. Catal. B. 152-153: 280-288.

Yang, Y., S. Niu, D. Han, T. Liu, G. Wang and Y. Li. 2017. Progress in developing metal oxide nanomaterials for photoelectrochemical water splitting. Adv. Energy Mater. 7: 1700555.

Yaqoob, A.A., T. Parveen, K. Umar and M.N.M. Ibrahim. 2020. Role of nanomaterials in the treatment of wastewater: A review. Water 12: 1-30.

Yuan, X., Q. Jing, J. Chen and L. Li. 2017. Photocatalytic Cr(VI) reduction by mixed metal oxide derived from ZnAl layered double hydroxide. Appl. Clay Sci. 43: 168-174.

Zhang, L., C. Dai, X. Zhang, Y. Liu and J. Yan. 2016. Synthesis and highly efficient photocatalytic activity of mixed oxides derived from ZnNiAl layered double hydroxides. Trans. Nonferrous Met. Soc. China 26: 2380-2389.

Zhang, L., X. Liu, X. Huang, W. Wang, P. Sun and Y. Li. 2019a. Adsorption of Pb^{2+} from aqueous solutions using Fe-Mn binary oxides-loaded biochar: Kinetics, isotherm and thermodynamic studies. Environ. Technol. 40: 1853-1861.

Zhang, M., L.Y. Zhao and Z. Yu. 2019b. Fabrication of zinc oxide/polypyrrole nanocomposites for brilliant green removal from aqueous phase. Arab. J. Sci. Eng. 44: 111-121.

Zhang, Q., Y. Wu, L. Li and T. Zuo. 2018. Sustainable approach for spent V_2O_5-WO_3/TiO_2 catalysts management: Selective recovery of heavy metal vanadium and production of value-added WO_3-TiO_2 photocatalysts. ACS Sustainable Chem. Eng. 6: 12502-12510.

Zheng, Q., J. Hou, W. Hartley, L. Ren, M. Wang, S. Tu, et al. 2020. As(III) adsorption on Fe-Mn binary oxides: Are Fe and Mn oxides synergistic or antagonistic for arsenic removal? Chem. Eng. J. 389: 124470.

14

Nanostructured Ceramics for Air Pollution Control: Removal of Gaseous Pollutants and Pathogenic Organisms

Elham Farouk Mohamed[1,*] and Gamal Awad[2]

14.1 ENVIRONMENTAL AIR POLLUTION PROBLEM

14.1.1 What is the Meaning of Air Pollution and What Causes it?

Definition: Simply, air pollution can be defined as the presence of toxic compounds, including both chemical substances and biological organisms, which decrease the air quality and its increase cause irreparable damage to Earth.

Causes: There are various causes of environmental pollution, including human activities, burning of fossil fuels, vehicle exhaust and industries' exhaust. Air pollutants include carbon dioxide (CO_2), nitric oxide (NO), volatile organic compounds (VOCs), suspended particles, and also airborne pathogenic microorganisms (Mohamed 2017).

14.1.2 What is the Solution?

Solution: New strategies, including innovative synthesis and methods, which are able to overcome such dangerous pollution problems are highly required (Mohamed et al. 2015, 2016a, b). Nanotechnologies, including nanostructure ceramics, give a perfect solution in pollutant sensing and controlling different environmental pollutants.

14.2 APPLICATION SCALES OF NANOSTRUCTURED CERAMICS

Nanomaterial and nanoceramic terms refer to materials fabricated from ultrafine particles of diameter less than 100 nm. Nanoceramics are recognized as inorganic, heat resistant, and non-metallic solids. The most important ceramic material systems are metal oxides, carbides, borides and nitrides. With innovation headway, the understanding degree of organization, structure, properties, and utilization of nanoparticles have improved in wide scale of energy, environmental, catalytic, and health materials' applications. As of late, the creation of ceramic composites with nanosized highlights is getting increasing consideration, as they exhibit superior mechanical properties or potential functional features as regard to traditional materials, whereas the cermaic

[1] Air Pollution Department, Environmental Research Division, National Research Centre, 33 EL Bohouth St., Dokki, Giza, P.O.12622, Egypt.

[2] Chemistry of Natural and Microbial Products Department, Pharmaceutical and Drug Industries Research Division, National Research Centre, 33 EL Bohouth St., Dokki, Giza, P.O.12622, Egypt.

* Corresponding author: elham_farouk0000@yahoo.com

nanoparticle (NP) is a very light, strong, flexible, durable, and extremely heat and chemically stable (Palmero 2015). As an outcome, structural ceramic nanocomposite materials are progressively utilized in various applications, for example, in aviation, automotive, drug, computers, catalyst parts, environmental remediation, capacitors, coatings, dense ceramics, batteries, and energy employments. The recent findings reveal the excellent potential of ceramic nanofibers to be used for supporting various catalysts, sensors, filtration media, magnetic parts, electronic devices, and biomedical fields. Esfahani et al. (2017) summarized recent nanoceramic materials according to their applications in various fields (Table 14.1).

TABLE 14.1 Nanoceramic products and their applications

Application	Ceramic Nanomaterials
Catalysts	TiO_2, V_2O_5, ZnO, SnO_2, $CdTiO_3$, Nb_2O_5, Gd_2O_3
Filtration	TiO_2, Al_2O_3, Clay, Fe_3O_4, $SrFe_{12}O_{19}$
Sensors	SnO_2, ZnO, TiO_2, CeO_2, NiO, $LaMnO_3$
Electro-magnetic devices	ZnO, $BaFe_{12}O_{19}$, $CaCu_3Ti_4O_{12}$, ZrO_2, La_2CuO_4
Biomedical	CaO, SiOC, TiO_2, ZnO
Batteries	SiO_2, Al_2O_3, SnO_2, GeO_2, $BaTiO_3$, $LaCoO_3$

An assortment of nanostructured ceramic materials are progressively being utilized for environment contamination control, application in the field of health, and to make sustainable power source less expensive and increasingly proficient (Fig. 14.1).

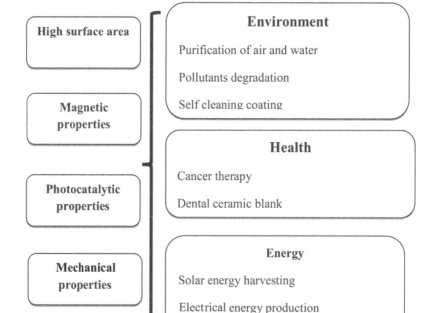

Fig. 14.1 Fundamental properties of nanostructured ceramic materials and its application fields exhibit exclusive properties such as high surface area, photocatalytic activity, high mechanical properties, and high magnetic properties

14.3 NANOSTRUCTURED CERAMICS AND ENVIRONMENTAL POLLUTION CONTROL

This section provides the novel strategies for the application of different nanostructured ceramics for pollutants degradation, as well as outlining future integration of separation and advanced oxidation photocatalysis. These pollutants may be microorganisms, organic and inorganic in origin within water and air. Nanoceramics are composed of inorganics such as hydroxyapatite, zirconia (ZrO_2) and others, that are synthesized by sol-gel, sintering, and laser ablation (Moreno et al. 2012). There are several methods for nanoparticle coating onto ceramic membrane (Kim and Bruggen 2010): (a) electrophoretic deposition: charged nanoparticle deposits onto alumina, (b) 3-aminopropytriethoxysilane (APTES) molecule as a linker between the nanoparticle and the ceramic support, and (c) organic binder (e.g. phytic acid) for doping the nanoparticle on ceramic membrane surface using layer by layer approach (Fig. 14.2). Ceramic nanoparticles have been widely studied for environmental pollutants' removal, separation and sensing, due to nanostructured ceramics' treatment technology having high performance characteristics and long working life even in harsh polluted effluents (Singh 2006).

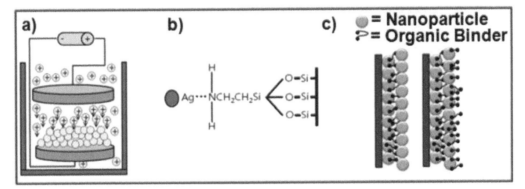

Fig. 14.2 Illustration of three methods for nanoparticle coatings onto ceramic membrane: (a) electrophoretic deposition, charged TiO_2 nanoprticle deposits onto alumina, (b) 3-amino propytriethoxysilane (APTES) molecule as a linker between Ag nanoparticles and ceramic support, (c) organic binder (e.g. phytic acid) for doping Fe_2O_3 nanoparticles on ceramic membrane surface using layer by layer methods (from Kim and Bruggen 2010)

This chapter focuses on the different areas in which nanostructured ceramics are recently used for the air pollutants' treatment. These areas can be divided into three main axes, as follows:

First axe: Nanostructured ceramics and pollutants sensing.

Second axe: Nanostructured ceramics membranes and pollutants separation.

Third axe: Nanostructured ceramics and pollutants removal.

14.4 NANOSTRUCTURED CERAMICS AND POLLUTANTS SENSING

On a very basic level, a sensor is an investigative device that changes over the reaction of transducer to a quantifiable sign, when presented to a pollutant. Successful detection of contaminants and large scale sensor applications helped to improve sensitive, economic, energy consumption and sensor impact throughout the years. Scientific progress and the innovation of advanced nanotechnologies, particularly the nanostructured ceramic material manufacturing technology, helped to improve the sensors for air pollution control. Application of nanostructured ceramics based sensors for gas detection showed an increase in their sensitivity and a decrease in temperature as compared to those prepared with conventional materials (Lamas et al. 2010, Naik et al. 2015). These nanosensors have an excellent sensing performance for detection of different environmental pollutants. Semiconducting substances, such as ZnO, SnO$_2$, TiO$_2$, MoO$_3$, WO$_3$, have an ability to determine lower (ppm) and higher concentrations of gaseous contaminants (Ding et al. 2009). Additionally, Kulshreshth et al. (2017) proposed the advantages that explain the excellent performance of this nanobiosensor, as described in Fig. 14.3.

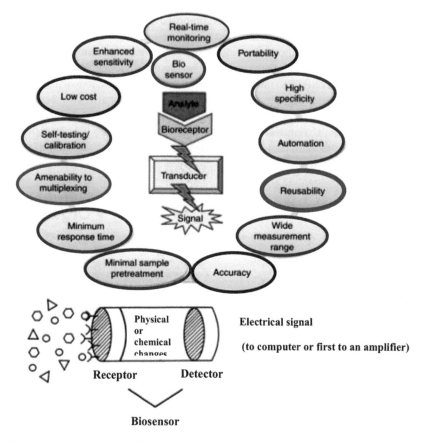

Fig. 14.3   Illustration of the unique advantages of nanosensors (from Kulshreshth et al. 2017)

The structure of gas sensors based on metal oxide nanostructures consists of three sections: (1) detecting film, (2) electrodes and (3) radiator. Metal oxide nanostructures as a thin film change their resistance when they are exposed to pollutants. Couple of electrodes are utilized to quantify the film resistance. The gaseous sensors are typically supplied with a warmer; hence, they are controlled remotely to reach an optimal temperature. Several ceramic nanofibers such as: Fe$_2$O$_3$, TiO$_2$, SnO$_2$, WO$_3$, ZnO, TiO$_2$/ZnO and Cu-doped-SnO$_2$ have been effectively used as sensor to detect various gases with improved detection limit, with notable models including NO$_2$, CO, NH$_3$, H$_2$, and toluene (Zheng et al. 2009, Daia et al. 2011). In addition to their use as sensing substances, they have also been described as excellent materials to support other sensing nanoparticles. As depicted in Fig. 14.4, ceramics in the form of cylinders or wafers are mainly used as supports for fixing sensing materials. In the ceramic tube apparatus, a heating wire is put in the interior tube, whereas in the ceramic wafer apparatus, a heating paste is involved in the posterior side.

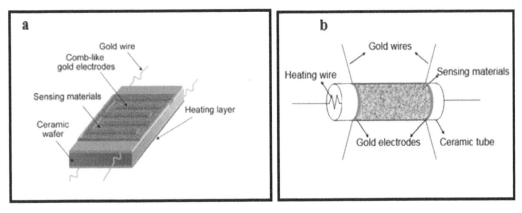

Fig. 14.4 A schematic of (a) device structure based on ceramic wafer substrate, (b) device structure based on ceramic tube substrate (from Sun et al. 2012)

Hongyan et al. (2013) reported SnO_2-ZnO/polyaniline composite thick film for NO_2 detection (Fig. 14.5). It was cited that the composite sensor had high selectivity and response to low concentration of NO_2 gas and also high temperature stability over a long operational period.

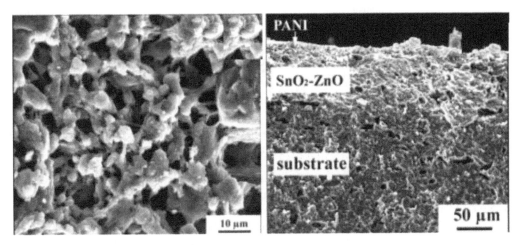

Fig. 14.5 SEM images of SnO_2-ZnO/polyaniline (PANI) composite thick film for NO_2 detection (From Hongyan et al. 2013)

As depicted in Fig. 14.6, graphite doped calcium hydroxyapatite (GHAP) nanoceramic composites are prepared for sensor material. Sensing ability of this nanoceramic sensor is utilized to detect alcoholic vapors such as propanol and ethanol at lower concentrations (100 ppm). Moreover, doping of the nanoceramic materials such as calcium hydroxyapatite enhances the detection of volatile pollutant (VOCs). The novel pattern of these sensor materials is recording separately all the vapors present in the ambient air (Anjum et al. 2018).

On the other side, the environmental pathogens (viral and bacterial) which are rapidly transmittable and intensively affect human health and both agriculture and animals must not be overlooked. Moreover, microbial pathogens are one of the main causes of death in both provincial and urban settings. Inaccurate pathogens detection leads to incorrect problem management. Dependable strategies for fast, specific discovery of airborne pathogens are basic to guarantee the wellbeing of nourishment supplies and to analyze irresistible, bacterial infections accurately.

Conventional and limited laboratory methods were used for the detection and identification of pathogen microorganisms in water and airborne pathogen. However, such conventional techniques have several drawbacks including the difficulty in sample collection, the intensely low pathogen concentration found in aerosol samples and also unsuitable for monitoring (Sanvicens et al. 2009). Therefore, great researches in this field are focused on the need to develop reliable, rapid, sensitive and specific techniques, lower analysis time and low cost, and multi-analyte detection. Many research groups seek this objective, since current diagnostic methods, including biosensors, don't satisfy this requirement. Nanoparticles' incorporation into biosensor systems increases the speed and sensitivity of the detecting methods. We provide in this section a general overview of the progress of NP-based biosensors for detection of pathogenic microorganisms. Nanosensors are devices which have been created by coordinating different parts. Nanobiosensors include biological probes, signal transducers and enhancers. Nanobiosensors for microbe detection in the field mainly include biological substances like enzymes and nucleic acids, attached to the transducer (Sinha et al. 2020).

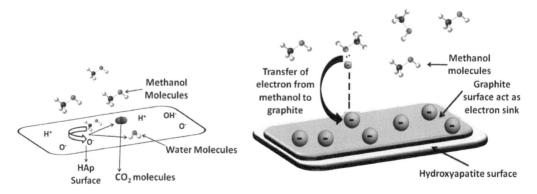

Fig. 14.6 A schematic of graphite doped hydroxyapatite nanoceramic (GHAP): sensor material detects all vapors individually, when collectively present in the ambient air (from Anjum et al. 2018)

A toxin is a poison discharged by cyanobacteria found in the aquatic biological system and causes food contamination (Zhang et al. 2018). Coupling of bi-metallic nanosystem sensors composed of Au-NF (gold nanoflower) and Ag-NP (silver nanoparticle) results in an electromagnetic force more sensitive for microcystins detection, especially in water. The development of these systems might give a convenient decision for biosensors, which should be used for qualitative and quantitative determinations (Zhao et al. 2015). Several researchers cited that the electric biosensors are simple, efficient, and low cost and have an effective performance in detection of several microorganisms (Tombelli et al. 2000). A list of nanofiber ceramic and their ceramic precursors, morphology and application as sensor and nanobiosensor are given in Table 14.2.

TABLE 14.2   Nanoceramic products and their applications as sensors

Ceramic Fiber	Ceramic Precursor(s)	Application	Morphology of Fiber	References
SnO_2	Tin acetate	Gas sensing	Regular fibrillar structure	Kim et al. 2014
SnO_2 doped Al	$SnCl_2 \cdot 2H_2O$, $Al(NO_3)3 \cdot 9H_2O$	Hydrogen sensor	Bead shape fibers sintered after heat treatment	Xu et al. 2011
WO_3	WCl_6	NO_2 gas responses	Short fiber	Giancaterini et al. 2016
ZnO	Zinc acetate dehydrate	Biosensors	Straight, fluffy surface	Stafiniak et al. 2011
NiO	$Ni(NO_3)_2$	Gas sensor, Catalyst	Sintered particles, or lamellar after sintering	Ristic et al. 2014

14.5 NANOSTRUCTURED CERAMIC MEMBRANES AND POLLUTANT SEPARATION

Before moving to detailed explanation of the application of nanostructured ceramic membranes for pollutants separation, let us describe the concept of membrane separation and filtration. A membrane can be defined as a physical hurdle made of a semi-permeable material permitting some components to go through while keeping others back. The permeate is the part of the feed that passes through the membrane and the retentate is the bit of the feed rejected by the membrane (Fig. 14.7). The main thrust over the membrane can be the difference in concentration, temperature or electrochemical potential (Mulder 2013).

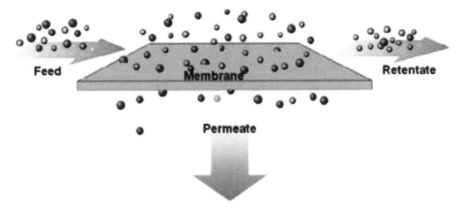

Fig. 14.7 Schematic illustration of gas separation mechanism of a membrane (i) a feed passes through the membrane (ii) a retentate is the portion of feed rejected by the membrane and (iii) the permeate is the portion of the feed that passes through the membrane (from: Schmeling et al. 2010)

In recent decades, membranes consisting of ceramic nanofibers have been created and successfully used in many industrial areas because of their thermal stability and solvent resistance. Furthermore, these membranes have a higher potential to eliminate particulate matters and small particles present in the air and water (Zhang et al. 2011). Nanoceramic membrane is a class of inorganic membrane produced from inorganic materials which are commonly based on oxide ceramics such as γ-Alumina, TiO_2, ZrO_2 or the mixed oxides frameworks of these materials (Meyn 2011). The latter are also hydrophilic due to the presence of hydroxyl groups on the surface (Ren et al. 2015). Nanoceramic membranes having multilayered structure perform separation using the principle of sieving through continuous channels of the porous layers. The separation efficiency is primarily based on the difference in pore size of nanoporous materials and particle or molecular size of the separating species, while the physicochemical interaction between the two types of surfaces governs the performance of the membranes. When gas mixture is introduced to the membrane module, a selective permeation occurs due to adsorption of the gas components into membrane surface and diffuse through the membrane pore (Amanipour et al. 2016). Solubility and diffusivity are different for various gas components (Kutty et al. 2011). Further investigation by Lysenko et al. (2011) demonstrated that the expensive gas helium can be separated from natural gas with the use of filters made from ceramics with open porosity; the nanosize of silicon dioxide powder was developed in the cermic filters. Several researchers fabricate nano-composite ceramic membranes for hydrogen gas and carbon dioxide separation (Nwogu et al. 2016). The permeability of hydrogen through ceramic membrane synthesized by commercial titania coated by alumina oxide via a sol-gel method was higher than N_2, CO and CO_2 (Ahmad et al. 2006). In previous studies, Song et al. (2008) cited that carbon nanofibers could effectively eliminate SO_2 from mixture gases. The penetration of different gases like carbon dioxide, nitrogen, methane, helium, argon and propane through nano-composite ceramic membranes consisting of a ceramic support and a zeolite layer membrane at varying pressures was reported (Shehu et al. 2015).

14.6   NANOSTRUCTURED CERAMICS AND POLLUTANTS REMOVAL

One of the most important environmental methods is the pollutants removal from different ecosystems. Among the innovative nanomaterials, nano-ceramics is one of the most efficient nanomaterials used for various pollutants removal. The integration of nanoceramic into environmental remediation field was investigated by several researchers (Zhou et al. 2019). Photocatalytic ceramic membranes are considerably used for industrial wastewater remediation. It was reported that the efficiency of TiO_2/Al_2O_3 composite membrane under ultraviolet (UV) irradiation for organic pollutants removal was over 96%. On the other hand, photocatalysis is considered as a promising method to decrease a number of air pollutants such as NO_x and VOCs. In this manner, photocatalytically dynamic substances can be added to the asphalt surface and constructing materials (Boonen and Beeldens 2014). The air decontamination using photocatalysis includes various strides: firstly, activation of TiO_2 by UV irradiation. Accordingly, the organic pollutants present in the air are oxidized and fixed at the outside surface of the pavement material. Finally, these pollutants could be removed from the surface by washing with water or by the rain (Fig. 14.8).

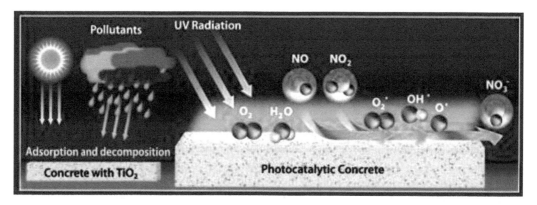

Fig. 14.8   Schematic of photocatalytic air purifying pavement (from: Boonen and Beeldens 2014)

New, self-cleaning innovation utilizing titania (TiO_2) photocatalysts has significant applications in the evacuation of indoor air contaminations. Super hydrophilic TiO_2-coated glass substrates provide a new way to clean indoor air pollutants (Ganesh et al. 2012). High removal rate of formaldehyde gas was achieved by using the immobilized titania (TiO_2) nanoparticles on glass substrates (Leong et al. 2019).

Consequently, nanoceramic innovation field shows noticeable features for advanced and multifunctional treatment that can enhance air pollution performance via different techniques of adsorption, degradation and filtration. Additionally, to expand the application of nanoceramic structural materials, their performance, activity, reuse and stability are of great importance in socioeconomic impacts and the environment, which represents significant technological and design challenges. In brief, comparison between some nanoceramic applications in pollutant remediation is summarized in Table 14.3.

TABLE 14.3 Comparison between different nanomaterials' efficiency for air pollutant removal

Ceramic Fiber	Application	Morphology of Fiber	References
Al_2O_3	Surface adsorption	Straight	Kim et al. 2014
$BiFeO_3$	Photocatalytic activity	Composed of nanoparticles together	Wang et al. 2013
$CdTiO_3$	Removal of industrial pollutants and toxic wastes	Smooth and uniform surface	Hassan et al. 2014
Silicon oxycarbide (SiOC) doped Ag	Antibacterial activity, Gas permeability	Straight, Ag inside the fibers	Guo et al. 2015
Mn_2O_3 and Mn_3O_4	Catalysis ion exchange adsorption biosensors wastewater treatment super capacitors	Straight 3D porous random	Rafei 2015
Nb_2O_5	Photocatalysis applications	Non-woven mat	Leindecker et al. 2014

The current challenges in the pollution remediation sector are the removal of airborne microorganisms from the bioaerosols. In particulate, the airborne pathogen such as viruses, bacteria and fungi are answerable for the chronic communicable diseases. These bioaerosols are kept and accumulated on the air conditioning systems due to the increase in the moisture. Therefore, numerous advanced technologies are trying to find a solution for airborne pathogens' removal. Conventional disinfection processes like chlorination can produce chloro-organic substances that are highly carcinogenic (Gamage and Zhang 2010). Other traditional disinfection methods such as high-efficiency particulate air (HEPA) filters, activated carbon based filters, ozonation and UV irradiation have several drawbacks such as limitation and also generation of mutants colonies and toxic by-products (Robertson et al. 2005). In comparison, nanoceramic photocatalytic innovation materials are capable of overcoming a wide range of pathogenic microorganisms in the air and water, and offers several environmental and practical advantages over the traditional disinfection methods. Broad researches had been reported in the photocatalytic evaluation of diverse microbial and organic contaminations by using several nanoceramic catalysts (Mccullagh et al. 2007, Gamage and Zhang 2010). In indoor air environments, addition of transition metals to TiO_2 improves its photocatalysis performance against the airborne pathogens in indoor air environment (Guillard et al. 2008). A detailed description of the TiO_2 photo disinfection killing of several pathogens in the air has been broadly studied in order to understand the mechanism of the photocatalytic disinfection process. It was reported that the bacterial disinfection of TiO_2 photocatalysis is probably due to the damage of the cell membrane or the oxidative attack of internal cellular species, substantially, leading to cell death (Nadtochenko et al. 2006). In the pioneering study of Guillard et al. (2008), the avian influenza virus (A/H5N2) was observed to be inactive in gaseous phase by photocatalytic systems. Additionally, a great promising application of visible light and doped catalysts was investigated to inactivate various bacteria in indoor air (Pal et al. 2007). Numerous photocatalytic systems for pathogenic air disinfection using polyester supports based catalyst, membrane systems, silver nanoparticles catalysts and CuO catalysts were reported (Mejia et al. 2010, Dunnill et al. 2011). There are several studies proving the high antimicrobial activity of ZnO nanoparticles. In this regard, Jin et al. (2019) reported that the synthesized ZnO nanoceramics had an excellent antibacterial activity against *E. coli* under UV irradiation. To the best of our knowledge, incorporation of Ag nanoparticles to different types of nanoceramic catalysts is playing a crucial and synergetic role in the airborne pathogenic purifications (Tsai et al. 2011). In this assessment, Manna et al. (2015)

reported that the Ag/ZnO on cotton fabrics shows high antibacterial activities and are considered as self-cleaning flexible materials with visible-light photocatalysis. For environmental cleaning from microorganisms and viruses, Rafi et al. (2015) demonstrated that iron oxide is another example of nanoceramic materials that can be used for bacteria inactivation. Additionally, Wang et al. (2018) reported that an iron oxide nanowire-based filter showed significant inactivation efficiency towards *S. epidermidis* (Gram-positive), a common bacterial species of indoor bio-aerosol.

Consequently, we can conclude that the different types of nanoceramic materials can be used as promising photocatalytic systems for microorganism's disinfection. Their application could also be extended to industrial and clinical use as effective and safe systems.

14.7   FUTURE OF HOLLOW CERAMIC NANOFIBERS

Hollow nanofibers (HNFs) are advanced innovation category of nanomaterials that have largely gained interest from the scientific community for their unique open structure. Several applications had been reported for the most attractive features of HNFs, which having long length, high porosity, and large surface area (Homaeigohar et al. 2017, Awad and Mohamed 2019). This chapter focused on the hollow ceramic nanofibers and their application in various fields like air pollution control, catalysis, energy technology and gas sensors, which were reported to become the most promising research areas (Kadir et al. 2014). For example, Zhao et al. (2010) prepared hollow TiO_2 nanofibers (Fig. 14.9). The photocatalytic activity of fibers was utilized for acetaldehyde gas degradation. The multicomposite of TiO_2 HNFs provokes an incident light multi-reflection and an internal pollutant gas capture.

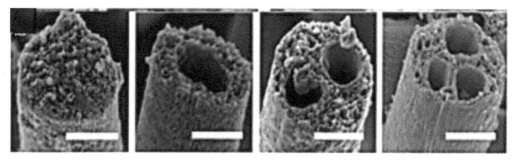

Fig. 14.9   SEM pictures of prepared TiO_2 nanofibers with 4 channels from left to right (0, 1, 2, and 3) (from: Zhao et al. 2010)

One of the greatest challenges in the field of environmental photocatalytic remediation is the enhancement of sunlight utilization. In this regard, Yang et al. (2017) investigated the possibility of narrowing the TiO_2 band gap to increase its photocatalytic application in visible light via loading of Pt nanoparticles onto hollow TiO_2 nanofibers. Figure 14.10 represents the Pt/TiO_2 HNFs' preparation and various photodegradation mechanisms of orange II under visible light.

Additionally, gas sensing is one of the significant application approaches of ceramic HNFs. Among sensing materials, metal oxides such as SnO_2, ZnO, TiO_2, and CuO are broadly used in gas sensors owing to their excellent properties such as response and recovery times and stability (Lu et al. 2019, Wang et al. 2019). Zinc oxide HNFs revealed higher adsorption efficiency for the heavy metal elimination. Additionally, they are used as sensors for several gases such as H_2, NO_2, H_2S, CO, and NH_3.

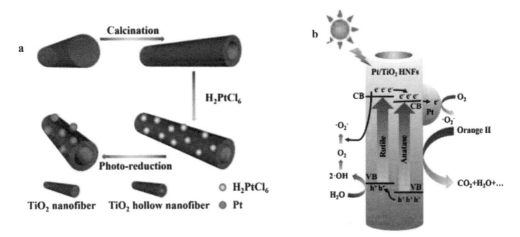

Fig. 14.10 (a) The schematic diagram of the whole preparation of Pt/TiO$_2$-HNFs, (b) Orange II photodegradation under visible light using different photocatalysts (from: Yang et al. 2017)

14.8 CONCLUSIONS

In this chapter, we primarily gave a quick look of environmental air pollution problem including the definition, causes and different proposed solutions. Furthermore, we discussed in brief the current progress in nanoceramic engineering approaches, including several recent application fields such as automotive, drug, computers, coatings, dense ceramics, sensors, fuel cells, batteries, cosmetic and energy employments. Then, we discussed in detail the recent progress towards the use of nanoceramics innovation in the field of environmental remediation, particularly for air pollutants removal and airborne pathogenic disinfection. Finally, this chapter can give valuable information to planners and engineers for designing different filter devices to improve the indoor air quality control system in the future by using nanoceramic structured materials.

14.9 REFERENCES

Ahmad, L., R. Othman and F. Idrus. 2006. Synthesis and characterization of nano-composite alumina-titania ceramic membrane for gas separation. J. Am. Ceram. Soc. 89(10): 3187-3193.

Amanipour, M., G. Babakhani, J. Towfighi and A. Zamaniyan. 2016. Evaluation of a tubular nano-composite ceramic membrane for hydrogen separation in methane steam reforming reaction. RSC Adv. 6: 84276-84283.

Anjum, R., N. Narwade, A. Bogle and S. Khairnar. 2018. Graphite doped hydroxyapatite nanoceramic: Selective alcohol sensor. Nano-Struct. Nano-Objects 14: 98-105.

Awad, G. and E.F. Mohamed. 2019. Immobilization of P450 BM3 monooxygenase on hollow nanosphere composite: Application for degradation of organic gases pollutants under solar radiation lamp. Appl. Catal. B: Environ. 253: 88-95.

Boonen, E. and A. Beeldens. 2014. Recent photocatalytic applications for air purification in belgium. Coating 4: 553-573.

Daia, Y., W. Liub, E. Formoa, Y. Sunc and Y. Xia. 2011. Ceramic nanofibers fabricated by electrospinning and their applications in catalysis, environmental science, and energy technology. Polym. Adv. Technol. 22: 326-338.

Ding, B., M. Wang, J. Yu and G. Sun. 2009. Gas sensors based on electrospun nanofibers. Sensors 9(3): 1609-1624.

Dunnill, W., K. Page, A. Aiken, S. Noimark, G. Hyett, A. Kafizas, et al. 2011. Nanoparticulate silver coated-titania thin films—photooxidative destruction of stearic acid under different light sources and antimicrobial effects under hospital lighting conditions. J. Photoch. Photobio. 220(2): 113-123.

Esfahani, H., R. Jose and S. Ramakrishna. 2017. Electrospun ceramic nanofiber mats today: Synthesis, properties, and applications. Materials (Basel) 10(11): 1238.

Gamage, J. and Z. Zhang. 2010. Review article applications of photocatalytic disinfection. Int. J. Photoenergy 2010: 764870.

Ganesh, A., S. Nair, K. Raut, M. Walsh and S. Ramakrishna. 2012. Photocatalytic superhydrophilic TiO_2 coating on glass by electrospinning. RSC Adv. 2: 2067-2072.

Giancaterini, L., M. Emamjomeh, A. Marcellis, E. Palange and A. Resmini. 2016. The influence of thermal and visible light activation modes on the NO_2 response of WO_3 nanofibers prepared by electrospinning. Sens Actuator B. 229: 387-395.

Guillard, C., H. Bui, C. Felix, V. Moules, B. Lina and P. Lejeune. 2008. Microbiological disinfection of water and air by photocatalysis. C R Chim. 11(1-2): 107-113.

Guo, A., M. Roso, P. Colombo, J. Liu and M. Modesti. 2015. *In situ* carbon thermal reduction method for the production of electrospun metal/SiOC composite fibers. J. Mater. Sci. 50: 2735-2746.

Hassan, S., M. Amna and M. SeobKhil. 2014. Synthesis of high aspect ratio $CdTiO_3$ nanofibers via electrospinning: Characterization and photocatalytic activity. Ceram. Int. 40: 423-427.

Homaeigohar, S., Y. Davoudpour, Y. Habibi and M. Elbahr. 2017. The electrospun ceramic hollow nanofibers. Nanomaterials 7(11): 383-415.

Hongyan, X., X. Chen, J. Zhang, J. Wang, B. Cao and D. Cui. 2013. NO_2 gas sensing with SnO_2-ZnO/PANI composite thick film fabricated from porous nanosolid. Sens. Actuators B: Chem. 176: 166-173.

Jin, S.-E., J.E. Jin, W. Hwang and S.W. Hong. 2019. Photocatalytic antibacterial application of zinc oxide nanoparticles and self-assembled networks under dual UV irradiation for enhanced disinfection. Int. J. Nanomedicine 14: 1737-1751.

Kadir, A., R. Li, Z. Sadek, A.Z. Abdul Rani, R. Zoolfakar, A.S. Field, et al. 2014. Electrospun granular hollow SnO_2 nanofibers hydrogen gas sensors operating at low temperatures. J. Phys. Chem. C. 118: 3129-3139.

Kim, H., J. Yoo, H. Kwak, J. Jung, Y. Kim, H. Park, et al. 2014. Characterization and application of electrospun alumina nanofibers. Nanoscale Res. Lett. 9: 44.

Kim, J. and B. Van der Bruggen. 2010. The use of nanoparticles in polymeric and ceramic membrane structures: Review of manufacturing procedures and performance improvement for water treatment. Environ. Poll. 158(7): 2335-2349.

Kulshreshth, M., D. Shrivastava and P. Singh Bisen. 2017. Contaminant sensors: Nanotechnology-based Contaminant Sensors. pp. 573-628. *In*: Alexandru Grumezescu [ed.]. Chapter 14 in Nanobiosensors Nook. Elsiever.

Kutty, P.V., S. Dasgupta and S. Bandyopadhyay. 2011. Soft chemical synthesis of nanosized zinc aluminate spinel from the thermolysis of different organic precursors. Mater. Sci. Poland 29: 121-126.

Lamas, G., F. Bianchetti, M. Cabezas and N. Walsöe de Reca. 2010. Nanostructured ceramic materials: Applications in gas sensors and solid-oxide fuel cells. J. Alloys Compd. 495(2): 548-551.

Leindecker, C., K. Alves and P. Bergmann. 2014. Synthesis of niobium oxide fibers by electrospinning and characterization of their morphology and optical properties. Ceram. Int. 40: 16195-16200.

Leong, K., Q. Lee, A. Kumar, C. Sim and S. Pichiah. 2019. Facile technique for the immobilization of TiO_2 nanoparticles on glass substrates for applications in the photocatalytic self-cleaning of indoor air pollutants. Malaysian J. Anal. Sci. 23(1): 90-99.

Lu, Z., Q. Zhou, Z. Wei, L. Xu, S. Peng and W. Zeng. 2019. Front. Synthesis of hollow nanofibers and application on detecting SF_6 decomposing products. Mater 6: 1-7.

Lysenko, I., Y. Trufanov and P. Bardakhanov. 2011. Gas filtration and separation with nano-size ceramics. Thermophys. Aeromech. 18(2): 273-280.

Manna, J., S. Goswami, N. Shilpa, N. Sahu and R.K. Rana. 2015. Biomimetic method to assemble nanostructured Ag/ZnO on cotton fabrics: Application as self-cleaning flexible materials with visible-light photocatalysis and antibacterial activities. ACS Appl. Mater. Interfaces 7(15): 8076-8082.

Mccullagh, C., C. Robertson, W. Bahnemann and J. Robertson. 2007. The application of TiO_2 photocatalysis for disinfection of water contaminated with pathogenic microorganisms: A review. Res. Chem. Intermed. 33(3-5): 359-375.

Mejia, I., G. Restrepo, M. Marin, R. Sanjines, C. Pulgarin and E. Mielczarski. 2010. Magnetron-sputtered Ag surfaces. New evidence for the nature of the Ag ions intervening in bacterial inactivation. Appl. Mater. Interfaces 2: 230-235.

Meyn, T. 2011. NOM removal in drinking water treatment using dead-end ceramic microfiltration. (Ph.D.), norwegian university of science and technology.

Mohamed, E.F., M.A. El-Hashemy, N.M. Abdel-Latif and W.H. Shetaya. 2015. Production of sugarcane bagasse-based activated carbon for formaldehyde gas removal from potted plants exposure chamber. J. Air Waste Manag. Assoc. 65: 1413-1420.

Mohamed, E.F., S.A. Sayed Ahmed, N.M. Abdel-Latif and A. Mekawy. 2016a. Air purifier devices based on adsorbents produced from valorization of different environmental hazardous materials for ammonia gas control. RSC Adv. 6: 57284-57292.

Mohamed, E.F., G. Awad, C. Andriantsiferana and A. El-Diwany. 2016b. Biofiltration technology for the removal of toluene from polluted air using *Streptomyces griseus*. Environ. Technol. 37(10): 1197-1207.

Mohamed, E.F. 2017. Nanotechnology: Future of environmental air pollution control. Environ. Manag. Sustain. Dev. 6: 429-454.

Moreno-Vega, I., T. Gomez-Quintero, R.-E. Nunez-Anita, L.-S. Acosta-Torres and V. Castaño. 2012. Polymeric and ceramic nanoparticles in biomedical applications. J. Nanotechnol. 2012: 1-12.

Mulder, J. 2013. Basic Principles of Membrane Technology. Springer Science and Business Media, Netherlands, Kluwer Academic Publishers Dordrecht, Boston, London.

Nadtochenko, V., C. Pulgarin, P. Bowen and J. Kiwi. 2006. Laser spectroscopy of the interaction of bacterial wall membranes and E. coli with TiO_2. J. Photoch. Photobio. A. 181: 401-404.

Naik, T., G. Christopher, T. Ivan, P. Jawwad and D. Binions. 2015. Environmental sensing semiconducting nanoceramics made using a continuous hydrothermal synthesis pilot plant. Sens Actuators B: Chem. 217: 136-145.

Nwogu, C., E. Anyanwu and E. Gobin. 2016. An initial investigation of a nano-composite silica ceramic membrane for hydrogen gas separation and purification. Int. J. Hydrogen. Energ. 41(19): 8228-8235.

Pal, A., O. Pehkonen, E. Yu and B. Ray. 2007. Photocatalytic inactivation of gram-positive and gram-negative bacteria using fluorescent light. J. Photoch. Photobio. A 186(2-3): 335-341.

Palmero, P. 2015. Structural ceramic nanocomposites: A review of properties and powders' synthesis methods. *Nanomaterials* 5: 656-696.

Rafei, L. 2015. Optimization of the electrospinning parameters of Mn_2O_3 and Mn_3O_4 nanofibers. Ceram. Int. 41: 12065-12072.

Rafi, M.M., K. Syed Zameer Ahmed, K. Prem Nazeer, D. Siva Kumar and M. Thamilselvan. 2015. Synthesis, characterization and magnetic properties of hematite (a-Fe_2O_3) nanoparticles on polysaccharide templates and their antibacterial activity. Appl. Nanosci. 5: 515-520.

Ren, C., H. Fang, J. Gu, L. Winnubst and C. Chen. 2015. Preparation and characterization of hydrophobic alumina planar membranes for water desalination. J. Eur. Ceram. Soc. 35(2): 723-730.

Ristić, M., M. Marciuš, Ž. Petrović and S. Musić. 2014. Dependence of NiO microstructure on the electrospinning conditions. Ceram. Int. 40(7): 10119-10123.

Robertson, C., J. Robertson and A. Lawton. 2005. A comparison of the effectiveness of TiO_2 photocatalysis and UVA photolysis for the destruction of three pathogenic micro-organisms. J. Photoch. Photobio. A. 175(1): 51-56.

Sanvicens, N., C. Pastells, N. Pascual and P. Marco. 2009. Nanoparticle-based biosensors for detection of pathogenic bacteria. Anal. Chem. 28(11): 1243-1252.

Schmeling, N., R. Konietzny, D. Sieffert, P. Rölling and C. Staudt. 2010. Functionalized copolyimide membranes for the separation of gaseous and liquid mixtures. Beilstein. J. Org. Chem. 6(1): 789-800.

Senthil, T. and S. Anandhan. 2014. Structure-property relationship of sol-gel electrospun zno nanofibers developed for ammonia gas sensing. J. Colloid Interface Sci. 432: 285-296.

Shehu, H., E. Okon and E. Gobina. 2015. The Use of Nano-composite Ceramic Membranes for Gas Separations. Proceedings of the World Congress on Engineering 2015 Vol II WCE 2015, July 1 - 3, 2015, London, U.K.

Singh, R. 2006. Hybrid Membrane Systems for Water Purification: Technology, Systems Design and Operations, Elsevier Science, USA.

Sinha, R., S. Dwivedi, A. Kumar and P. Srivastava. 2020. Materials in bio-sensing of water pollutants. pp. 187-211. *In*: D. Pooja, P. Kumar, P. Singh and S. Patil [eds.]. Sensors in Water Pollutants Monitoring: Role of Material. Advanced Functional Materials and Sensors. Springer, Singapore.

Song, X., Z. Wang, Z. Li and C. Wang. 2008. Ultrafine porous carbon fibers for SO_2 adsorption via electrospinning of polyacrylonitrile solution. J. Colloid Interface Sci. 327(2): 388-392.

Stafiniak, A., B. Boratynsk, B. Korczyc, A. Szyszka, M. Krasowska, J. Prazmowska, et al. 2011. A novel electrospun ZnO nanofibers biosensor fabrication. Sens. Actuator. B. 160: 1413-1418.

Sun, F., S. Liu, L. Meng, Y. Liu, Z. Jin, T. Kong, et al. 2012. Metal oxide nanostructures and their gas sensing properties: A review. Sensors 12: 2610-2631.

Tombelli, S., M. Mascini, L. Braccini, M. Anichini and A.P. Turner. 2000. Coupling of a DNA piezoelectric biosensor and polymerase chain reaction to detect apolipoprotein E polymorphisms. Biosens. Bioelectron. 15(7-8): 363-370.

Tsai, T.-T., W.-P. Sung and W. Song. 2011. Identification of indoor airborne microorganisms and their disinfection with combined nano-Ag/TiO_2 photocatalyst and ultraviolet light. Environ. Eng. Sci. 28(9): 635-642.

Wang, D., B. Zhu, X. He, Z. Zhu, G. Hutchins, P. Xu, et al. 2018. Iron oxide nanowire-based filter for inactivation of airborne bacteria. Environ. Sci.: Nano 5: 1096-1106.

Wang, J.X., Q. Zhou, Z. Lu, Z. Wei and W. Zeng. 2019. Gas sensing performances and mechanism at atomic level of Au-MoS_2 microspheres. Appl. Surf. Sci. 490: 124-136.

Wang, W., N. Chi, Y. Li, W. Yan, X. Li and C. Shao. 2013. Electrospinning of magnetical bismuth ferrite nanofibers with photocatalytic activity. Ceram. Int. 39: 3511-3518.

Xu, X., J. Sun, H. Zhang, X. Wang, B. Dong, T. Jiang, et al. 2011. Effects of Aldopingon SnO_2 nanofibers in hydrogen sensor. Sens Actuator B. 160: 858-863.

Yang, Z., J. Lu, W. Ye, C. Yu and Y. Chang. 2017. Preparation of Pt/TiO_2 hollow nanofibers with highly visible light photocatalytic activity. Appl. Surf. Sci. 392: 472-480.

Zhang, W., W. Ji, O. Toprakci, Y. Liang and M. Alcoutlabi. 2011. Electrospun nanofiber-based anodes, cathodes, and separators for advanced lithium-ion batteries. Polym. Rev. 51: 239-264.

Zhang, W., Q. Liu, Z. Guo and J. Lin. 2018. Practical application of aptamer-based biosensors in detection of low molecular weight pollutants in water sources. Molecules 23(2): 344.

Zhao, T., Z. Liu, K. Nakata, S. Nishimoto, T. Murakami, Y. Zhao, et al. 2010. Multichannel TiO_2 hollow fibers with enhanced photocatalytic activity. J. Mater. Chem. 20: 5095-5099.

Zhao, Y., X. Yang, H. Li, Y. Luo, R. Yu, L. Zhang, et al. 2015. Au nanoflower-Ag nanoparticle assembled SERS-active substrates for sensitive MC-LR detection. Chem. Comm. 51(95): 16908-16911.

Zheng, W., Z. Li, H. Zhang, W. Wang, Y. Wang and C. Wang. 2009. Electrospinning route for D-Fe_2O_3 ceramic nanofibers and their gas sensing properties. Mater. Res. Bull. 44: 1432-1436.

Zhou, X., M. Wang, D. Yan, Q. Li and H. Chen. 2019. Synthesis and performance of high efficient diesel oxidation catalyst based on active metal species-modified porous zeolite BEA. J. Catal. 379: 138-146.

15

Nanostructured Oxide Ceramic Materials for Applications in the Field of Humidity Sensors

Florin Tudorache

15.1 INTRODUCTION

In last decades, theoretical and experimental research on the functional properties of oxide ceramic materials based on the unique electrical, electro-optical, ferroelectric, piezoelectric, ferromagnetic, magneto-resistive and humidity sensor properties has increased to a great extent. In many research laboratories, nanostructured oxide ceramic materials are given special attention in order to improve their physical and chemical parameters.

In the context of the increasingly accentuated development and evolution of technology, the role those nanostructured oxides ceramic materials have acquired has been essential, due to their possibility of application in fields such as sensors, electronics and telecommunications, medicine, genetic engineering, robotics, etc. The demand of devices based on oxide ceramic materials and their importance has increased considerably due to the relatively simple and less polluting technological process of obtaining, and the lower production costs compared to the technologies of obtaining other ceramic materials like carbides, nitrides, etc. This fact, but also their electrical, mechanical and thermal properties, have made nanostructured oxide ceramic materials of increasing interest in the field of humidity sensors and to impose themselves as materials with multiple functionalities (Søgaard et al. 2011, Cristobal et al. 2009, Abazari et al. 2014, Manikandan et al. 2020a).

It is a known fact that in any substance, the structure of individual molecules and their structural order are key factors that determine the macroscopic properties of the material (Kittel 1966, Belik and Yi 2014, Oldham and Milnes 1964, Tudorache 2018, Andoulsi et al. 2012). Therefore, the importance of the possibility of controlling the structure on the molecular level in case somebody wants to obtain certain oxide ceramic materials with special yields is easy to understand. Investigation made on the oxides ceramic materials as sensitive to moisture has led to the conclusion that the electrical properties of the these are determined not only by their composition or structure, but also by the appearance of secondary phases, thermal treatments, structural defects, contact between interfaces, etc. (Kolesnikova and Kuz'mich 2015, Abdallah et al. 2019, Price et al. 2014, Tudorache et al. 2012, Kuru 2020, Manikandan et al. 2020b).

The oxide ceramic sensors can detect humidity based on the principle of measuring the variation of electrical resistance and electrical capacity by absorbing water vapor. The change of the electrical

Research Centre on Advanced Materials and Technologies, Institute for Interdisciplinary Research, Science Research Department, "Alexandru Ioan Cuza" University of Iasi, Blvd. Carol I No. 11, Iasi 700506, Romania.
Institute for Interdisciplinary Research, Science Research Department, Research Centre on Advanced Materials and Technologies, "Alexandru Ioan Cuza" University of Iasi, Blvd. Carol I No. 11, Iasi 700506, Romania
E-mail: florin.tudorache@uaic.ro

resistance of the electrical capacity of the sensor can be explained by the fact that the water molecules act as donors and acceptors and, consequently, the electrical conductivity alteration will take place. The ability of ceramic oxides based sensors to manifest sensitivity in the presence of water molecules depends on the interaction of water molecules with the surface of the sensor element, that is, the reactivity of its surface. Reactivity is a parameter that depends on surface characteristics such as the chemical composition of the surface, as well as the electronic and morphological structures (Yadav et al. 2016, Gao et al. 2013, Banerjee et al. 2015, Aydin et al. 2019, Andoulsi et al. 2016, Berry et al. 1989, Bhargav et al. 2015a, 2014a, Buscail et al. 1995). These characteristics are greatly affected by the preparation's technology and the material's composition.

From the point of view of the chemical composition, the ceramic materials are classified into oxides, nitrides, carbides and silicates of some metals or semi-metals. Nanostructured oxide ceramic materials can be obtained in the form of amorphous materials (glass), monocrystalline and polycrystalline materials, powders, thin layers, films, composites, depending on the field of application. The first ceramic oxide materials that have been investigated are tin oxide (SnO), indium oxide (In_2O_3), zinc oxide (ZnO) and copper oxide (CuO). In search of alternative materials, these dominant ceramic oxide materials with a simple composition were then joined by other ceramic oxide materials which had two or more oxides in the compositional structure (Sora et al. 2012, Vaingankar et al. 1997). Many oxides of transitional metals present interesting semiconductor properties and are widely used in practical applications (Weingart et al. 2012, Shi et al. 2014, Bhargav et al. 2014b, Kuribayashi et al. 2016, Andoulsi et al. 2018, Bhargav et al. 2015b).

Because of the oxygen vacancies that are present in the structure of the oxide ceramic materials, most of them are sensitive to humidity. Oxygen vacancies, more precisely the interstitial atoms that are contained in the crystalline structure, cause the ceramic oxides to be classified into two categories: n-type or p-type semiconductors.

The reaction of the oxide ceramic material with the humidity vapours in the environment causes changes in its electrical conductivity. Thus, an increase in the electrical conductivity of the oxide ceramic material may occur if the adsorption of oxygen molecules from the moisture vapour plays the role of acceptor or a decrease of the electrical conductivity when the adsorption of the oxygen molecules plays the role of donor. The principle is based on the reversible reaction of atmospheric oxygen with the oxygen vacancies in the compound network and the reduction of the electron concentration. This reaction generates various oxygen species (O_{2-}, O^- or O^{2-}) that react differently with the oxide ceramic material.

The results of these studies had, as main purpose, the elaboration of the principles of operation and construction of devices that have found applications in the most diverse fields of modern science and technology: sensory, electronic, automatic, telecommunications, etc.

The effect of a humidity sensor is to modify the electrical properties (electrical resistivity, electrical conductivity, dielectric constant) of a material when it is exposed to moisture vapour. There are two types of mechanisms underlying the moisture sensitivity of some ceramic oxide materials. The first one is the ionic type, which uses the increase of ionic conductivity due to water adsorption or capillary condensation in micropores (Rezlescu et al. 2004a, Arai et al. 1983, Rezlescu et al. 2004b) and the second one is the electronic type, which uses electronic conductivity change due to water chemisorption (Nitta et al. 1983, Rezlescu et al. 2004c, Rai and Thakur 2016). Chemisorption accompanies the dissociation of water molecules with the formation of hydroxyl OH groups on the surface of oxides. So basically, when the water molecules are adsorbed on the semiconductor oxides, the electrical conductivity increases or decreases, as the oxides are of type n or p. This means that the electrons are apparently transferred from the water molecules to the oxides. The seeming electron exchange from water molecules to oxides can occur through one of the following reactions (Shimizu et al. 1986, Rezlescu et al. 2004d):

1. non-dissociative adsorption with the assignment of a single electron

$$H_2O(g) \leftrightarrow OH_0^-(ad) + e^- \qquad (1)$$

2. dissociative adsorption with the transfer of two electrons

$$H_2O(g) + O_0^{2-} + V_0^{2-} \leftrightarrow 2OH_0^-(ad) + 2e^- \qquad (2)$$

in which V_0^{2-} is an oxygen vacation that captures two electrons.

A good humidity sensor must satisfy several major requirements: sensitivity, high humidity selectivity, short response time, reproducibility, and thermal and chemical stability.

Sensitivity is the main characteristic of the device known as the variation of the conductivity (resistivity) of the oxide ceramic material exposed to moisture vapors. In order to improve the sensitivity, it is of great interest to work with the material which is most sensitive to moisture vapour. Considering that the sensitivity reactions occur mainly on the surface of the sensor material, controlling the size of the semiconductor grains is a major requirement for the qualities of the sensor.

Selectivity is the characteristic by which a sensor element can detect the presence of moisture vapour on certain humidity ranges between 0-97% RH. In this case, it is possible to modify the characteristics of the sensor material to make it more sensitive at a certain humidity range. This could be achieved by modifying some technological process parameters or introducing additions or substitutions into the stoichiometric composition of the basic sensor material.

Stability is the characteristic by which a sensor will give the same results, under the same experimental conditions, after operating with it for a long extent of time. Improving stability can be achieved by performing heat treatments for high temperature calcinations.

The microstructure and electrical conductivity are the main parameters on which the sensor effect depends in a semiconductor oxide ceramic material. The control over the microstructure (phase purity, grain size, shape, porosity) is determined by the choice of precursors (oxides, salts), by the preparation method and by the technological factors involved; it is also the essential factor in obtaining the best humidity sensitivity. A small size of the granulations involves increasing the electrical resistivity by expanding the surface of the granulations in relation to their volume. Also, the doping of ceramic oxides with metal ions can contribute to the increase of the electrical resistivity, which is desirable in the materials used in the field of humidity sensors (Geoffrey et al. 2014, Gholizadeh 2019, Kundu et al. 2019a, Gosavi and Biniwale 2013, Bhat et al. 2013).

Because they have high electrical resistivity, chemical stability and are cheap to obtain, oxide ceramic materials (ferrites) are of high interest both theoretically and practically, having applicability in many fields, such as magnetic storage devices, microwaves, high density recording media, gas sensors, humidity sensors, high quality filters, etc. The use of ceramic oxide materials in environmental monitoring offers potential advantages for their integration in devices with various applications in the field of humidity sensors.

Many reports from the specialized literature refer to a series of ortoferrites which can be used for electronic or humidity sensors (Fan et al. 2013, Kundu et al. 2019b, Li and Wang 2015, Popa and Moreno 2011).

The recently conducted researches (Li et al. 2017, Markova et al. 2013, Naseem et al. 2015, Parrino et al. 2016) have shown that various ceramic oxides, such as pure lanthanum orthoferrite, or in various combinations with metal oxides, may be an alternative to materials used in the field of humidity sensors.

That is why this book chapter refers to the study on the influence of the technological factors on the structural and electrical properties of humidity sensor of the lanthanum orthoferrite ceramic materials in polycrystalline state, made by using the technology of the solid state ceramic method.

15.2   PREPARATION OF OXIDE CERAMIC MATERIALS BY THE SOLID-STATE CERAMIC METHOD

Nanostructured oxide ceramic materials are recognized as essential in obtaining the performance humidity sensors. The achieving technique can reproduce the composition of the nanostructured oxide material and requires a small number of steps in the technological flow. Consequently, nanostructured oxide ceramic materials are fundamental and require an investigation of their role in moisture vapour adsorption and detection. There are numerous procedures to obtain sensing materials from ceramic oxide compounds, some general and other specific to sensors. Oxides ceramic materials obtained using solid state phase reaction method technology have the following advantages: low production cost, friendly to the environment, long life, good physical and chemical stability.

Systematic studies by various research groups on a large number of ceramic oxide materials have shown that the variation of electrical conductivity in the presence of moisture vapour in the air is a common phenomenon for oxides and does not belong to a specific class (Manikandan et al. 2018, Gaikwad et al. 2015, Yang et al. 2018, Yu et al. 2019). The ceramic oxide materials used as humidity sensors are classified into two categories: i) resistive sensors-are those semiconductor oxide materials that have predominantly sensitivity to moisture given by the resistive component (electrical resistance), and ii) capacitive sensors-presenting sensitivity to humidity, mainly to the capacitive component (electrical capacity). If an oxide material has electrical resistivity values in the range $10^3 \div 10^8$ Ωm, then it can be used as a humidity sensor (Park and Park 2013, Manikandan et al. 2019, Jaiswal et al. 2020, Kumar et al. 2019).

Most ceramic materials having metallic oxide precursors crystallize in two crystalline systems with a compact arrangement: *the cubic lattice with centred faces* and *the compact hexagonal lattice*. In these types of networks, the chemical bonding is achieved by pooling the valence electrons that belong simultaneously to all the crystal atoms.

The samples were prepared by the solid phase reaction technology, following the solid state ceramic method, which is one of the most commonly used methods of obtaining the oxide ceramic materials, as it has low production costs and is highly productive. The preparation of the compound occurs through the solid phase reaction, at high temperature, between the oxides of the metals, to form the desired compound. The stages of preparation of a semiconductor oxide material (ferrite) by the classical ceramic method comprise the following technological steps shown in Fig. 15.1. In the following, we will present the experimental results regarding the obtaining and investigation of the structural and electrical properties of the lanthanum orthoferrite.

The oxides taken in the proportions corresponding to the chemical formula, were weighed using an analytical balance, La_2O_3 (manufacturer CarlRoth), Fe_2O_3 (manufacturer Sigma-Aldrich) having a purity higher than 99.9% and were mixed wet for 6 hours in a ball mill Retsch model PM 100. The mixed oxides in powder form were dried at about 150°C, resulting in fine, submicron powders that were pressed into disk form with a 6 mm diameter and a thickness between $1 \div 1.5$ mm. For pressing, a Carver model hydraulic press applying three tons pressing force was used. After pressing, the samples were subjected for 1 hour to a pre-sintering heat treatment performed at 500°C with the aim of eliminating volatile substances.

To investigate the influence of the sintering temperature on the structural and electrical properties of the samples, they were placed in an electric furnace where they were sintered for 5 hours at temperatures of 800°C, 900°C 1000°C and 1100°C and the samples presented in Fig. 15.2 were obtained. The sintering temperature and the cooling were very carefully controlled because they determine the final oxygen content of the sintered sample, and this content has a considerable effect on the electrical properties (Rezlescu et al. 2008, 2005). In order to avoid the thermal shocks, which can lead to the appearance of internal mechanical stresses, the cooling of the samples after each heat treatment was slow, with the oven cooling after disconnecting from the electric network.

During the sintering treatment takes place the diffusion of the ions from the crystalline networks of the oxides and their organization in the crystalline network of the target compound.

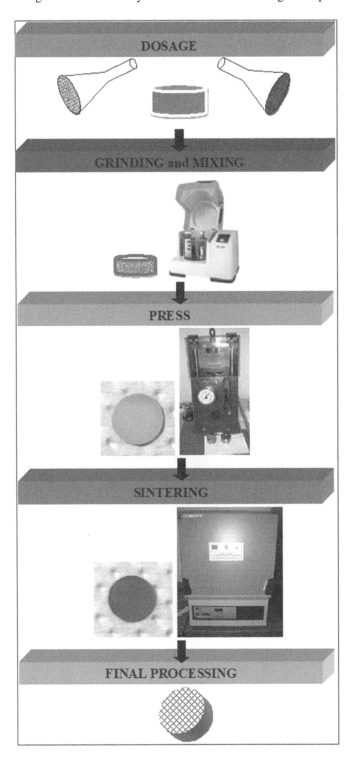

Fig. 15.1 Steps for preparation of the ferrites by the solid-state ceramic method

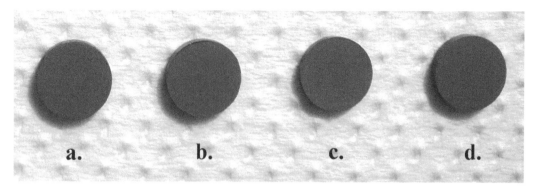

Fig. 15.2 Samples from oxide powders pressed and sintered for 5h at temperatures: (a) 800°C, (b) 900°C (c) 1000°C, (d) 1100°C

In the image shown in Fig. 15.2, we can see that after the sintering treatment carried out at 800°C and 900°C for 5 hours, the samples show a slight change in the colour, and also a small compaction of the diameter. Following the sintering treatment at 1000°C and 1100°C, respectively, the samples changed not only their colour marking, but also the diameter, becoming more compact.

After performing the sintering treatment, the disk-shaped samples were mechanically polished on both sides to thicknesses between 1-1.5 mm, and on both sides of each sample a silver layer with a thickness of 120 nm was deposited as an electrode contact, as in Fig. 15.3.

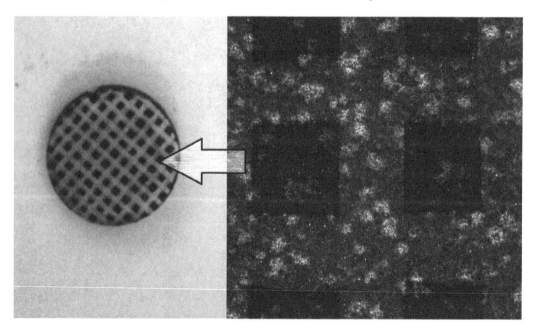

Fig. 15.3 Silver porous contact electrodes, deposited on the surface of the samples

In the image presented in Fig. 15.3, we can easily observe that the electrodes are sieve-type and equidistant, to facilitate the easy penetration of moisture vapour on the surface and in the volume of the samples.

15.3 COMPOSITIONAL AND PHASE CHARACTERIZATION OF OXIDE CERAMIC MATERIALS

The structure and morphology of the ceramic oxide material obtained are important parameters for optimizing the properties of interest in humidity sensors. From the mass and geometric dimensions of the samples obtained after each sintering treatment, the relative volume contraction $\Delta V/V_0$ was determined, the density, $\rho = m/v$ and the porosity p, the values of the obtained parameters being shown in Table 15.1.

TABLE 15.1 Physical parameters of the analysed samples

Sample	Sintering Temperature $T_{sint}[°C]$	Volumic Contraction $\Delta V/V_0$ [%]	Bulk Density ρ [g/cm³]	Porosity p [%]
LaFeO₃	800	1.53	2.86	46
LaFeO₃	900	1.67	2.88	44
LaFeO₃	1000	1.86	2.94	38
LaFeO₃	1100	1.95	2.98	33

The values from Table 15.1 show that after the sintering treatment at 800°C, the sample has the highest porosity value, and the sintered sample at 1100°C has the lowest porosity value.

From the point of view of morphology, the ceramic oxide materials used as sensors can be divided into compact materials and porous materials. In the case of the compact ones, the reaction that determines the behaviour of the sensor takes place only at the surface; in the case of the porous ones, the reaction takes place in the whole volume. Therefore, the porosity of the ceramic oxide material is one of the key factors that determine the humidity sensitivity of the electrical resistivity of a humidity sensor.

The structural properties of the samples were investigated using X-ray diffraction (XRD) and scanning electron microscopy (SEM). The phase analysis presented in Fig. 15.4 was determined using X-ray diffraction at room temperature on sample fragments, using Cu K_α radiation (λ = 1.5405 Å), employing a Shimadzu LabX XRD-6000 diffractometer, which applies a focusing scheme of Bragg-Brentano type.

Figure 15.4 shows the submicronic structure of sintered lanthanum orthoferrite at different treatment temperatures, determined by X-ray diffraction, which shows the peak intensities of the reflection peaks at (311), (222), (022), (400), (220), (310), (312), (440), (531) and (622) planes. From the position of the peaks, we can determine the inter-planar distance, the parameters of the elementary cell and the crystalline phase of the investigated compound. The diffractograms present in Fig. 15.4 show that the samples are multi-phase, and lanthanum orthoferrite (LaFeO₃) and lanthanum oxide (La₂O₃) are present in all samples in certain proportions, but as the sintering temperature increases, the La₂O₃ content decreases. It is recommended that in the case of ceramic oxide materials used as applications in the field of sensors to perform heat treatments in the range of temperatures 800 ÷ 900°C, to avoid densification or the appearance of secondary phases, because these can influence the physico-chemical properties of materials. Decreasing the content of La₂O₃ phases contributes to the increase of the internal electrical resistance in the case of oxide ceramic materials.

The surface morphology of the investigated samples was analysed using Atomic Force Microscopy (AFM). Figure 15.5 shows the images obtained on a surface of 3.3 μm and a height of 1.5 μm. 3D atomic force microscopy images were obtained at room temperature using a Park System NX10 non-contact working microscope with an NCHR tip, AC160TS.

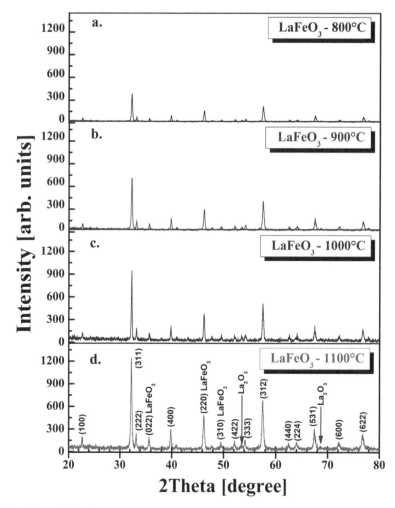

Fig. 15.4 X-ray diffraction patterns of lanthanum orthoferrites sintered at various temperatures

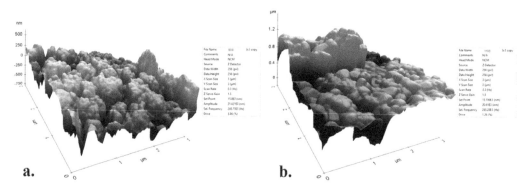

Fig. 15.5 AFM micrographs of Lanthanum orthoferrite samples sintered at: a) 800 °C; b) 1100°C

From the analysis of the atomic force microscopy images performed on the sintered lanthanum orthoferrite samples at temperatures of 800°C and 1100°C, a series of information can be obtained regarding the growth of the crystallites and the surface uniformity. The colour distribution on the analysed surface takes into account the height as well as the local inclination of the surface to highlight the features of the relief. The analysis of the images shown in Fig. 15.5a indicates that the sample sintered at 800°C has a fine granulation, high intergranular porosity and smaller crystallite size compared to the sample presented in Fig. 15.5b, which was sintered at 1100°C. Similar results were reported by Ranieri et al. (2015), Rezlescu et al. (2008), and Tudorache et al. (2008). Through the analysis of the AFM micrographs shown in Fig. 15.5, we observe that the surface of the sample sintered at 800°C is rough (open porosity) compared to the surface of the sample sintered at 1100°C (which is smooth, closed pores), which allows easy penetration of moisture vapours that modify the electrical resistivity of the oxide ceramic material. Also, from Fig. 15.5 it is easy to observe that the sintered sample at 1100°C is much denser than the sintered sample at 800°C; this information is also supported by the dielectric investigations shown in Fig. 15.7.

SEM micrographs were performed at room temperature in order to observe the morphology, shape and tendency of agglomeration of the crystallites, on the fractures of the samples, using a Tescan Vega II SBH electron microscope. The investigations carried out with the help of the electron microscope provide us with detailed information regarding the inter-granular, intragranular pores, structural defects and the homogeneity of the samples. In order to correctly interpret an image obtained with the scanning electron microscope, we must have as much information as possible about the phenomena that may occur when the electron beam interacts with the material from which the sample is made.

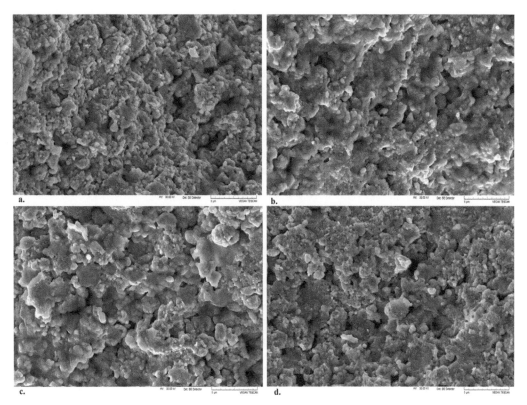

Fig. 15.6 SEM micrographs of Lanthanum orthoferrites samples sintered at: (a) 800°C, (b) 900°C, (c) 1000°C, (d) 1100°C

The micrographs that can be seen in Fig. 15.6 shows us that each sample is characterized by a typical porous structure with intergranular pores, more specifically, spaces between crystallites.

This kind of granulation and agglomeration of samples is used to ensure the electrical properties and sensitivity to moisture vapours. From the SEM micrographs, it can be observed that performing the sintering treatment at high temperatures involves a decrease of the intragranular porosity by increasing the size of the crystallites, thus a densification of the samples. Similar results were reported by Rameshkumar et al. (2019), Mukhopadhyay and Basu (2014), and Hessien et al. (2016). The densification of the samples involves a decrease of the electrical resistivity of the samples, shown in Fig. 15.7b and an increase of the relative electrical permittivity shown in Fig. 15.7a.

In the case of the lanthanum orthoferrite samples sintered at temperatures of 800°C and 900°C, respectively, the small size of the crystallites confers a high electrical resistivity which is in correlation with the graphical representation in Fig. 15.7b. In the case of the lanthanum orthoferrite samples sintered at temperatures of 1000°C and 1100°C, respectively, an increase of the crystallite size is observed, which implies the decrease of the electrical resistivity.

15.4 INVESTIGATION OF ELECTRICAL PROPERTIES WITH POSSIBLE APPLICATIONS IN THE FIELD OF HUMIDITY SENSORS

It is known that the electrical resistivity of an oxide ceramic material is mainly governed by the composition, microstructure, homogeneity and the electrical resistivity of the edges between the crystallites. At the same time, experiments have shown that the microstructure can strongly influence the change of the value of the electrical resistivity at the absorption of the moisture vapour from the environment. In the case of oxides ceramic materials, the mechanism for the detection of the humidity level in the environment is based on the absorption of water vapour into the sample, which involves changing of the electrical resistivity (in the case of resistive sensors) or electrical capacitance (in the case of capacitive sensors). Porosity is the key factor that determines the moisture sensitivity of the electrical resistivity of an oxide ceramic material (Rezlescu et al. 2009, Tudorache et al. 2009).

In the case of semiconductor ceramic oxide materials in the electrical conduction mechanism, two types of charge carriers participate: vacancies and free electrons (Kittel 1966, Kelly and Groves 1970). The property of an oxide material to conduct the electric current is characterized by the electrical conductivity, σ, determined by the relation (Kittel 1966, Oldham and Milnes 1964):

$$J = \sigma \cdot E \tag{1}$$

where j represents the current density and E represents the intensity of the electric field.

In the case of semiconductors, due to the fact that there are two types of charge carriers, the dependence between the electrical conductivity and the concentration of the charge carriers participating in the conduction is given by the relation (Kittel 1966, Oldham and Milnes 1964, Kelly and Groves 1970):

$$\sigma = n \cdot e \cdot (\mu_p + \mu_n) \tag{2}$$

where n represents the concentration of electrons, e is the elementary electric charge, μ_p represent the mobility of the gaps and μ_n represents the mobility of the electrons.

It is desirable that the value of the electrical resistivity of the sensor element be of the order 10^7 Ωm in a dry environment and have an exponential decrease up to 10^3 Ωm when the relative humidity increases. If the value of the electrical resistivity is very high, then it is difficult to measure. However, if the value of the electrical resistivity is too low, it will no longer be able to detect the presence of moisture vapour.

The frequency dependence study of relative dielectric permittivity and electrical resistivity was performed using a Wayne Kerr 6500P High Frequency LCR Meter at room temperature, in the absence of humidity, and is shown in Figs. 15.7a and 15.7b. It is known that the dielectric constant and the electrical resistivity of a semiconductor oxide ceramic material are mainly governed by the composition, microstructure and the resistivity of the crystallites' edges, so it is very important to control, step by step, the method and technological process of preparation.

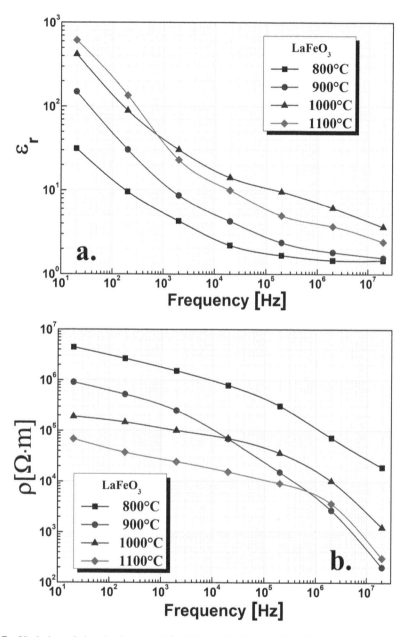

Fig. 15.7 Variation of electrical permittivity (a) electrical resistivity, (b) in the absence of humidity

Variation of relative electrical permittivity versus frequency is characterized by space charge polarization and it is in agreement with Koop's theory. According to Koop's theory, the synthesized ceramic oxide semiconductor is considered to be two conducting layers which are well conducting and poor conducting grains (Virlan et al. 2020, Tudorache et al. 2013, Virlan et al. 2017). High value for relative electrical permittivity is achieved at low frequency because electrons are piled or displaced at poor conducting grains which lead to maximum space charge of polarization.

The high values, at low frequencies, of the relative electrical permittivity presented in Fig. 15.7a, included in the range $\varepsilon' = 20\text{-}600$, confirm that the lanthanum orthoferrite exhibits a semiconductor behaviour (Bhargav et al. 2014b, Bhat et al. 2013, Rai and Thakur 2015, Berchmans et al. 2018, Idrees et al. 2011) due to the jump of the electric charge enhancers positioned in the octahedral networks, between the metallic ions Fe^{2+} and Fe^{3+} from the composition of the oxide ceramic material. The electron displacement between Fe^{2+} and Fe^{3+} ions cannot follow alternate field which results in low space charge polarization (Andoulsi et al. 2016, Bhargav et al. 2015b). Following the dielectric studies performed on the samples, it is found that in the frequency range $2 \times 10^1 - 2 \times 10^7$ Hz all the samples have an exponential decrease in the relative electrical permittivity, this decrease being faster in the low frequency range. From the representation shown in Fig. 15.7b, it can be seen that the increase of the treatment temperature implies a densification of the lanthanum orthoferrites and the decrease of the electrical resistivity. However, small sizes of crystallites offer large surface grain boundaries which increase the value of electrical resistivity.

The electronic device containing the oxide ceramic material (ferrite) has been introduced into a thermostatic enclosure, where using saturated solutions of various salts, various levels of relative humidity can be obtained. Table 15.2 shows the relative humidity values above saturated salts for various temperatures.

TABLE 15.2 The relative humidity at equilibrium of saturated salts for different temperatures

Salt	Temperature [°C]				
	5	**10**	**15**	**20**	**25**
LiCl	11.3%	11.3%	11.3%	11.3%	11.3%
$MgCl_2$	33.6%	33.5%	33.3%	33.1%	32.8%
K_2CO_3	43.1%	43.1%	43.1%	43.2%	43.2%
NaBr	63.5%	62.2%	60.7%	59.1%	57.6%
NaCl	75.7%	75.7%	75.6%	75.7%	75.3%
KCl	87.7%	86.8%	85.9%	85.1%	84.3%
K_2SO_4	98.5%	98.2%	97.9%	97.6%	97.3%

The electrical resistance of the samples was measured using a Wayne Kerr 6500P High Frequency LCR Meter at room temperature in alternating current at the frequency of 1 kHz. The results of the research carried out on the samples under the action of humidity are presented in Fig. 15.8.

From Fig. 15.8a, we can see that in the humidity range 0-60% RH, the samples show a slight increase in the value of the relative electrical permittivity, and in the high humidity fields, above 60% RH, the accentuated increase of the relative permittivity variation is observed. The largest variation of the relative permittivity is found in the sintered lanthanum orthoferrite at 800°C.

Figure 15.8b shows the typical p-type semiconductor behaviour of lanthanum orthoferrite, and the electrical resistivity starts to change significantly from relative humidity values exceeding 60% RH. It is observed that in all samples, regardless of the sintering temperature, the value of the electrical resistivity decreases by one to two orders of magnitude when the humidity increases from 11% RH to 97% RH. In this aspect, the lanthanum orthoferrite sintered at 800°C and 900°C represents the largest variation of the electrical resistivity under the action of humidity. From the graphical variation of electrical resistivity under humidity conditions, we can suggest that lanthanum orthoferrite material can be used in the field of humidity sensors.

As a result of the sensitivity investigations to the humidity vapours carried out on the lanthanum orthoferrites, it can be noticed that the electrical resistivity shows a much greater variation compared

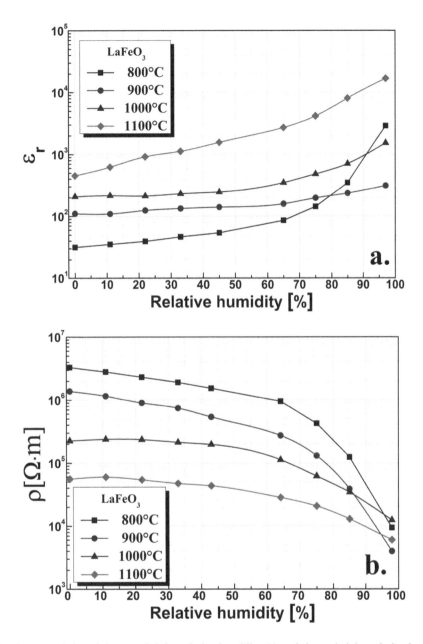

Fig. 15.8 Characteristics of the permittivity-relative humidity (a) and the resistivity-relative humidity (b) of lanthanum orthoferrites

to the relative permittivity so that the lanthanum orthoferrite can be used as a resistive type sensor in the field of humidity sensors. Figure 15.9 shows that about 90% of the total change in electrical resistivity is achieved in the first 180 seconds, when the relative humidity has changed from 0% to 97% RH. The response time of the electrical resistance of the lanthanum orthoferrite to the relative humidity change from 0% to 97% RH was also determined and found to be dependent on the sintering temperature as shown in Fig. 15.9.

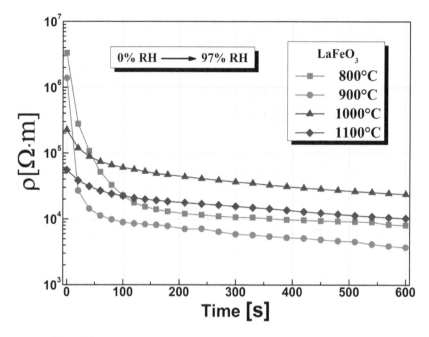

Fig. 15.9   Response time characteristic of lanthanum orthoferrites

In the case of the lanthanum orthoferrite sintered at temperatures of 800°C and 900°C (see Fig. 15.9), the variation of the electrical resistivity when the relative humidity increases is higher than in the case of the same ferrite sintered at temperatures of 1000°C and 1100°C, which can be explained by a higher sensitivity of the first two samples. From the Fig. 15.9, it is possible to appreciate the response time, meaning the time required for the electrical resistivity to reach a stable value when the relative humidity ranged from 0% to 97% RH is about 5 minutes.

15.5   SUMMARY

Within this book chapter was presented the influence of the sintering temperature in the technological process of obtaining nanostructured oxide ceramic materials (ferrites) with applications in the field of humidity sensors. The study of nanostructured oxide ceramic materials was determined by the necessity of obtaining a ferrite with the highest electrical resistivity, with an adequate porosity, low production cost, good physical and chemical stability and the sintering temperatures as low as necessary in the field of humidity sensors.

Samples of $LaFeO_3$ orthoferrite were sintered at temperatures of 800°C, 900°C, 1000°C and 1100°C for 5 hours. It has been found that as the sintering temperature increases, the samples become denser and their colour changes markedly.

X-ray diffraction studies have shown that all samples are multi-phase, and as the sintering temperature increase, the content of the La_2O_3 phase decreases. The studies carried out using the Atomic Force Microscopy and Scanning Electron Microscopy showed significant morphological changes by performing sintering treatment at temperatures of 800°C, 900°C, 1000°C and 1100°C, regarding the shape, size, roughness and porosity of the samples.

The dielectric studies, without humidity influence, performed on the samples show an exponential decrease in the electrical permittivity, which confirms the semiconductor character of the lanthanum orthoferrites. It was also found that the electrical resistivity of the lanthanum orthoferrite decreases with increasing sintering temperature.

It has been found that the high porosity value of over 40% and the sintering temperature of 800°C and 900°C of lanthanum orthoferrite play a large role in the sensitivity to humidity.

Response time study for the samples shows that in the case of lanthanum orthoferrite, it was to be found depending on the sintering temperature. This result is influenced by the increase of the specific surface area and the pore volume which play an important role in water vapours' absorption.

In conclusion, from the point of view of the sensitivity to humidity and the possibility of measuring the different humidity levels through the variation of the electrical resistivity, the sintered lanthanum orthoferrite at temperatures of 800°C and 900°C can be successfully used as a humidity sensor.

15.6 REFERENCES

Abazari, R., S. Sanati and L.A. Saghatforoush. 2014. A unique and facile preparation of lanthanum ferrite nanoparticles in emulsion nanoreactors: Morphology, structure, and efficient photocatalysis. Mater. Sci. Semicond. Process. 25: 301-306.

Abdallah, F.B., A. Benali, M. Triki, E. Dhahri, K. Nomenyo and G. Lerondel. 2019. Investigation of structural, morphological, optical and electrical properties of double-doping Lanthanum ferrite. J. Mater. Sci.: Mater. Electron. 30(4): 3349-3358.

Andoulsi, F.R., K.H.-Naifer and M. Férid. 2012. Structural and electrical properties of calcium substituted lanthanum ferrite powders. Powder Technol. 230: 183-187.

Andoulsi, F.R., K.H.-Naifer and M. Férid. 2016. Influence of zinc incorporation on the structure and conductivity of lanthanum ferrite. Ceram. Int. 42: 1373-1378.

Andoulsi, F.R., N. Sdiri, K.H.-Naifer and M. Férid. 2018. Effect of temperature on the electrical properties of lanthanum ferrite. Spectrochim. Acta Part A. 205: 214-220.

Arai, H., S. Ezaki, Y. Shimizu, O. Shippo and T. Seiyama. 1983. Semiconductive humidity senzor of perovskite-type oxides. Proc. Int. Meeting Chem. Sens., Sept. 19-22 Fukuoka, 393-398.

Aydin, C., H. Aydin, M. Taskin and F. Yakuphanoglu. 2019. A novel study: The effect of graphene oxide on the morphology, crystal structure, optical and electrical properties of lanthanum ferrite based nano electroceramics synthesized by hydrothermal method. J. Nanosci. Nanotechnol. 19: 2547-2555.

Banerjee, K., J. Mukhopadhyay, M. Barman and R.N. Basu. 2015. Effect of 'A'-site non stoichiometry in strontium doped lanthanum ferrite based solid oxide fuel cell cathodes. Mater. Res. Bull. 72: 306-315.

Belik, A.A. and W. Yi. 2014. High-pressure synthesis, crystal chemistry and physics of perovskites with small cations at the A site. J. Phys.: Condens. Matter. 26: 163201.

Berchmans, L.J., R. Sindhu, S. Angappan and C.O. Augustin. 2008. Effect of antimony substitution on structural and electrical properties of $LaFeO_3$. J. Mater. Process. Technol. 207: 301-306.

Berry, F.J., S. Jobsen and S.L. Jones. 1989. An *in-situ* iron-57 mossbauer spectroscopic investigation of the reduction of iron-cerium oxide catalysis in different gaseous reducing agents. Hyperfine Interact. 46: 613-618.

Bhargav, K.K., S. Ram, S.B. Majumder. 2014a. Physics of the multi-functionality of lanthanum ferrite ceramics. J. Appl. Phys. 115: 204109.

Bhargav, K.K., A. Maity, S. Ram and S.B. Majumder. 2014b. Low temperature butane sensing using catalytic nano-crystalline lanthanum ferrite sensing element. Sens. Actuators B. 195: 303-312.

Bhargav, K.K., S. Ram and S.B. Majumder. 2015a. The role of catalytic cobalt-modified lanthanum ferrite nano-crystals in selective sensing of carbon monoxide. J. Mater. Sci. 50: 644-651.

Bhargav, K.K., S. Ram, S.B. Majumder. 2015b. Small polaron conduction in lead modified lanthanum ferrite ceramics. J. Alloys Compd. 638: 334-343.

Bhat, I., S. Husain, W. Khan and S.I. Patil. 2013. Effect of Zn doping on structural, magnetic and dielectric properties of $LaFeO_3$ synthesized through sol-gel auto-combustion process. Mater. Res. Bull. 48: 4506-4512.

Buscail, H., C. Courty and J.-P. Larpin. 1995. Effects of ceria coatings on pure iron oxidation. Comparison with extra low carbon steels. J. Phys. IV Proceedings, EDP Science 05(C7): C7-375-C7-380.

Cristobal, A.A., P.M. Botta, P.G. Bercoff and J.M. Porto Lopez. 2009. Mechanosynthesis and magnetic properties of nanocrystalline $LaFeO_3$ using different iron oxides. Mater. Res. Bull. 44: 1036-1040.

Fan, K., H. Qin, L. Wang, L. Ju and J. Hu. 2013. CO_2 gas sensors based on $La_{1-x}Sr_xFeO_3$ nanocrystalline powders. Sens. Actuators B. 177: 265-269.

Gaikwad, V.M., P. Uikey and S.A. Acharya. 2015. Study of multi-functionality of lanthanum ferrite ($LaFeO_3$). Solid State Physics AIP Conf. Proc. 1665: 140046-1-140046-3.

Gao, B., J. Deng, Y. Liu, Z. Zhao, X. Li, Y. Wang, et al. 2013. Mesoporous $LaFeO_3$ catalysts for the oxidation of toluene and carbon monoxide. Chin. J. Catal. 34: 2223-2229.

Geoffrey, L., I.I. Beausoleil, P. Price, D. Thomsen, A. Punnoose, R. Ubic, et al. 2014. Thermal expansion of alkaline-doped lanthanum ferrite near the neel temperature. J. Am. Ceram. Soc. 97: 228-234.

Gholizadeh, A. 2019. The effects of A/B-site substitution on structural, redox and catalytic properties of lanthanum ferrite nanoparticles. J. Mater. Res. Technol. 8: 457-466.

Gosavi, P.V. and R.B. Biniwale. 2013. Catalytic preferential oxidation of carbon monoxide over platinum supported on lanthanum ferrite ceria catalysts for cleaning of hydrogen. J. Power Sources 222: 1-9.

Hessien, M.M., G.A.M. Mersal, Q. Mohsen and D. Alosaimi. 2017. Structural, magnetic and sensing properties of lanthanum ferrite via facile sol gel oxalate precursor route, J. Mater. Sci.: Mater. Electron. 28(5): 4170-4178.

Idrees, M., M. Nadeem, M. Atif, M. Siddique, M. Mehmood and M.M. Hassan. 2011. Origin of colossal dielectric response in $LaFeO_3$. Acta Mater. 59: 1338-1345.

Jaiswal, A.K., S. Sikarwar, S. Singh, K.K. Dey, B.C. Yadav and R.R. Yadav. 2020. Fabrication of nanostructured magnesium ferrite polyhedrons and their applications in heat transfer management and gas/humidity sensors. J. Mater. Sci.: Mater. Electron. 31: 80-89.

Kelly, A. and G.W. Groves. 1970. Crystallography and Crystal Deffects, Longman Group Ltd, London, UK.

Kittel, C. 1966. Introduction to Solid State Physics. Wiley, New-York.

Kolesnikova, I.G. and Yu. V. Kuz'mich. 2015. Characteristics of nanopowders of Yttrium ferrite $Y_3Fe_5O_{12}$ depending on the conditions of their preparation. Russ. J. Inorg. Chem. 60: 147-150.

Kumar, K., U. Kumar, M. Singh and B.C. Yadav. 2019. Synthesis and characterizations of exohedral functionalized graphene oxide with iron nanoparticles for humidity detection. J. Mater. Sci.: Mater. Electron. 30: 13013-13023.

Kundu, S.K., D.K. Rana, L. Karmakar, D. Das and S. Basu. 2019a. Enhanced multiferroic, magnetodielectric and electrical properties of Sm doped lanthanum ferrite nanoparticles. J. Mater. Sci.: Mater. Electron. 30: 10694-10710.

Kundu, S.K., D.K. Rana, A. Banerjee, D. Das and S. Basu. 2019b. Influence of manganese on multiferroic and electrical properties of lanthanum ferrite nanoparticles. Mater. Res. Express 6: 085032.

Kuribayashi, K., R. Takahashi, Y. Inatomi, S. Ozawa and M.S. Vijaya Kumar. 2016. Containerless processing of metastable multiferroic composite in R-Fe-O system (R: Rare-earth element). Int. J. Microgravity Sci. Application 33: 330215.

Kuru, T.Ş. 2020. Synthesis and investigation of structural, dielectric, impedance, conductivity and humidity sensing properties of Cr^{3+}-substituted Mg-Zn ferrite nanoparticle. App. Phys. A: Mater. Sci. Process. 126(6): 419.

Li, S. and X. Wang. 2015. Synthesis of different morphologies lanthanum ferrite ($LaFeO_3$) fibers via electro spinning. Optik 126: 408-410.

Li, Z., W. Zhang, C. Yuan and Y. Su. 2017. Controlled synthesis of perovskite lanthanum ferrite nanotubes with excellent electrochemical properties. RSC Adv. 7: 12931-12937.

Manikandan, V., Monika Singh, B.C. Yadav and J.C. Denardin. 2018. Fabrication of lithium substituted copper ferrite ($Li-CuFe_2O_4$) thin film as an efficient gas sensor at room temperature. J. Sci.: Adv. Mater. Devices 3: 145-150.

Manikandan, V., S. Sikarwar, B.C. Yadav, S. Vigneselvan, R.S. Mane, J. Chandrasekaran, et al. 2019. Rapid humidity sensing activities of lithium-substituted copper-ferrite ($Li-CuFe_2O_4$) thin films. Mater. Chem. Phys. 229: 448-452.

Manikandan, V., M. Singh, B.C. Yadav, R.S. Mane, S. Vigneselvan, J. Ali Mirzaei, et al. 2020a. Room temperature LPG sensing properties of tin substituted copper ferrite ($Sn-CuFe_2O_4$) thin film. Mater. Chem. Phys. 240: 122265.

Manikandan, V., A. Mirzaei, S. Sikarwar, B.C. Yadav, S. Vigneselvan, A. Vanithadand, et al. 2020b. The rapid response and high sensitivity of a ruthenium-doped copper ferrite thin film ($Ru-CuFe_2O_4$) sensor. RSC Adv. 10: 13611-13615.

Markova-Velichkova, M., T. Lazarova, V. Tumbalev, G. Ivanov, D. Kovacheva, P. Stefanov, et al. 2013. Complete oxidation of hydrocarbons on $YFeO_3$ and $LaFeO_3$ catalysts. Chem. Eng. J. 231: 236-244.

Mukhopadhyay, J. and R.N. Basu. 2014. Morphologically architectured spray pyrolyzed lanthanum ferrite-based cathodese A phenomenal enhancement in solid oxide fuel cell performance. J. Power Sources 252: 252-263.

Naseem, S., W. Khan, B.R. Singh and A.H. Naqvi. 2015. Room temperature optical and dielectric properties of Sr and Ni doped lanthanum ferrite nanoparticles. Solid State Physics AIP Conf. Proc. 1665: 050048-1-050048-3.

Nitta, T., F. Fukushima and T. Matsuo. 1983. Water vapor gas senzor using ZrO_2 – MgO ceramic body. Proc. 1st Int. Meeting Chem. Sens. Sept. 19-22 Fukuoka, 387-392.

Oldham, W.G. and A.G. Milnes. 1964. Interface states in abrupt semiconductor heterojunctions. Solid State Electron. 7: 153-165.

Park, H.J. and J.Y. Park. 2013. A promising high-performance lanthanum ferrite-based composite cathode for intermediate temperature solid oxide fuel cells. Solid State Ionics 244: 30-34.

Parrino, F., E. García-Lopez, G. Marcì, L. Palmisano, V. Felice, I.N. Sora, et al. 2016. Cu-substituted lanthanum ferrite perovskites: Preparation, characterization and photocatalytic activity in gas-solid regime under simulated solar light irradiation. J. Alloys Compd. 682: 686-694.

Popa, M. and J.M. Calderon Moreno. 2011. Lanthanum ferrite ferromagnetic nanocrystallites by a polymeric precursor route. J. Alloys Compd. 509: 4108-4116.

Price, P.M., E. Rabenberg, D. Thomsen, S.T. Misture and D.P. Butt. 2014. Phase transformations in calcium-substituted lanthanum ferrite. J. Am. Ceram. Soc. 97: 2241-2248.

Rai, A. and A.K. Thakur. 2015. The enhancement in dielectric and magnetic property in Na and Mn co substituted lanthanum ferrite. Int. Conf. Condens. Matter Appl. Phys. (ICC 2015) AIP Conf. Proc. 1728: 020304-1-020304-4.

Rai, A. and A.K. Thakur. 2016. Multifunctionality of nanocrystalline Lanthanum ferrite. Int. Conf. Condens. Matter Appl. Phys. (ICC 2015) AIP Conf. Proc. 1728: 020491-1-020491-4.

Rameshkumar, R., T. Ramachandran, K. Natarajan, M. Muralidharan, F. Hamed and V. Kurapati. 2019. Fraction of rare-earth (Sm/Nd)-lanthanum ferrite-based perovskite ferroelectric and magnetic nanopowders. J. Electron. Mater. 48: 1694-1703.

Ranieri, M.G.A., M. Cilense, E.C. Aguiar, C.C. Silva, A.Z. Simões and E. Longo. 2015. Electrical behavior of chemically grown lanthanum ferrite thin films. Ceram. Int. 42: 2234-2240.

Rezlescu, N., E. Rezlescu, C.-L. Sava, F. Tudorache and P.D. Popa. 2004a. Effects of some ionic substitutions on sintering, structure and humidity sensitivity of MgCu ferrite. Phys. Status Solidi. A. 20: 17-25.

Rezlescu, N., E.. Rezlescu, F. Tudorache and P.D. Popa. 2004b. Effects of repalcing Fe by La or Ga in $Mg_{0.5}Cu_{0.5}Fe_2O_4$. Humidity sensitivity. J. Magn. Magn. Mater. 272: E1821-E1822.

Rezlescu, N., E. Rezlescu, C.-L. Sava, F. Tudorache and P.D. Popa. 2004c. On the effects of Ga^{3+} and La^{3+} ions in MgCu ferrite: Humidity-sensitive electrical conduction. Cryst. Res. Technol. 39: 548-557.

Rezlescu, N., E. Rezlescu, F. Tudorache and P.D. Popa. 2004d. MgCu nanocrystalline ceramic with La^{3+} and Y^{3+} ionic substitution used as humidity sensor. J. Optoelectron. Adv. Mater. 6: 695-698.

Rezlescu, N., E. Rezlescu, P.D. Popa and F. Tudorache. 2005. A model of humidity sensor with a Mg-based ferrite. J. Optoelectron. Adv. Mater. 7: 907-910.

Rezlescu, N., F. Tudorache, E. Rezlescu and P.D. Popa. 2008. The effect of the additives and sintering temperature on the structure and humidity sensitivity of a spinel ferrite. J. Optoelectron. Adv. Mater. 10: 2386-2389.

Rezlescu, N., E. Rezlescu, F. Tudorache and P.D. Popa. 2009. Gas sensing properties of porous Cu-, Cd- and Zn-ferrites. Rom. Rep. Phys. 61: 223-234.

Shi, C., H. Qin, M. Zhao, X. Wang, L. Li and J. Hu. 2014. Investigation on electrical transport, CO sensing characteristics and mechanism for nanocrystalline $La_{1-x}Ca_xFeO_3$ sensors. Sens. Actuators B. 190: 25-31.

Shimizu, Y., H. Okada and A. Arai. 1986. Ceramic humidity sensor of lanthanum titanium porous glass. Proc. 2nd Int. Meeting Chem. Sens. July 7-10 Bordeaux, 380-383.

Sora, I.N., T. Caronna, F. Fontana, César de Julián Fernández, A. Caneschi and M. Green. 2012. Crystal structures and magnetic properties of strontium and copper doped lanthanum ferrites. J. Solid State Chem. 191: 33-39.

Søgaard, M., A. Bieberle-Hütter, P.V. Hendriksen, M. Mogensen and H.L. Tuller. 2011. Oxygen incorporation in porous thin films of strontium doped lanthanum ferrite. J. Electroceram. 27: 134-142.

Tudorache, F., E. Rezlescu, P.D. Popa and N. Rezlescu. 2008. Study of some simple ferrites as reducing gas sensors. J. Optoelectron. Adv. Mater. 10: 1889-1893.

Tudorache, F., N. Rezlescu, N. Tudorache, A.M. Catargiu and M. Grigoraş. 2009. Polyaniline and polythiophene-based gas sensors. Optoelectron. Adv. Mater. Rapid Commun. 3: 379-382.

Tudorache, F., P.D. Popa, F. Brinza and S. Tascu. 2012. Structural investigations and magnetic properties of $BaFe_{12}O_{19}$ crystals. Acta Phys. Pol. A 121(1): 95-97.

Tudorache, F., P.D. Popa, M. Dobromir and F. Iacomi. 2013. Studies on the structure and gas sensing properties of nickel-cobalt ferrite thin films prepared by spin coating. J. Mater. Sci. Eng. B. 178: 1334-1338.

Tudorache, F. 2018. Investigations on microstructure, electrical and magnetic properties of copper spinel ferrite with WO_3 addition for applications in the humidity sensors. Superlattices Microstruct. 116: 131-140.

Vaingankar, A.S., S.G. Kulkarni and M.S. Sagare. 1997. Humidity sensing using soft ferrites, Proceedings of the 7th International Conference on Ferrites, September 3-6, 1996, Bordeaux, France, J. Phys. IV, Colloque C1. 7: C1-155-56.

Virlan, C., F. Tudorache, A. Pui. 2017. Increased sensibility of mixed ferrite humidity sensors by subsequent heat treatment, Int. J. Appl. Ceram. Technol. 14(6): 1174-1182.

Virlan, C., F. Tudorache and A. Pui. 2020. Tertiary NiCuZn ferrites for improved humidity sensors: A systematic study, Arabian J. Chem. 13: 2066-2075.

Weingart, C., N. Spaldin and E. Bousquet. 2012. Noncollinear magnetism and single-ion anisotropy in multiferroic perovskites. Phys. Rev. B. 86: 094413.

Yadav, A.K., R.K. Singh and P. Singh. 2016. Fabrication of lanthanum ferrite based liquefied petroleum gas sensor. Sens. Actuators B. 229: 25-30.

Yang, Y., Y. Li, Y. Jiang, M. Zheng, T. Hong, X. Wu, et al. 2018. The electrochemical performance and CO_2 reduction mechanism on strontium doped lanthanum ferrite fuel electrode in solid oxide electrolysis cell. Electrochim. Acta. 284: 159-167.

Yu, T.-F., C.-W. Chang, P.-W. Chung and Y.-C. Lin. 2019. Unsupported and silica-supported perovskite-type lanthanum manganite and lanthanum ferrite in the conversion of ethanol. Fuel Process. Technol. 194: 106117.

16

Applications of Smart Ceramics in Nano/ Micro Sensors and Biosensors

A.A. Jandaghian and O. Rahmani[*]

16.1 INTRODUCTION

Piezoelectricity is an electromechanical interaction that exhibits a coupling between electrical and mechanical states and pyroelectricity is a coupling between electrical and thermal states. The word "piezo" comes from a Greek word that means pressure. The piezoelectric phenomenon was discovered in 1800 by French physicists Jacques and Pierre Curie brothers (Jaffe 2012). It was mentioned that when piezoelectric materials experience mechanical deformation, an electric potential difference could be generated (direct effect) and while an electric field is applied, it leads to mechanical strain or stress (converse effect). Due to their unique properties, these materials can be used in a wide range of applications such as actuators and sensors (Masson and Perriard 2019, Masson et al. 2019, Dosch et al. 1992, Dhuri and Seshu 2007, Kumar and Narayanan 2007, Li et al. 2008a, Zhao et al. 2007, Kumar and Narayanan 2008, Sirohi and Chopra 2000). In general, using piezoelectric materials as actuators to control vibration is not appropriate. Instead, synthetic polycrystalline ceramic materials, for example, lead zirconate titanate (PZT), can be made to show piezoelectric properties significantly. There is an excellent coupling between electrical and mechanical in PZT ceramics and also it is comparatively easy to produce these materials. The fundamental behavior of piezoelectric materials is schematically demonstrated in Fig. 16.1, in two forms, laminar (e.g. polymeric piezoelectrics) and axial (e.g. piezoceramics).

Vasques and Dias Rodrigues (Vasques and Dias Rodrigues 2006) presented a numerical investigation relating to the active vibration control of smart piezoelectric beams. Komijani et al. (Komijani et al. 2013) studied free vibration of functionally graded piezoelectric material (FGPM) beams with rectangular cross-sections in pre/post-buckling states.

An analytical model of the dynamic behavior of an electromechanical piezoelectric bimorph cantilever harvester connected with an AC-DC circuit based on the Euler-Bernoulli beam theory and Hamiltonian theorem was presented by Wang and Meng (Wang and Meng 2013). Poizat and Benjeddou (Poizat and Benjeddou 2006) presented a work to assess and evaluate some well-established analytical and numerical modeling techniques of classical extension bimorphs using general purpose finite element (FE) codes. Maurini et al. (Maurini et al. 2006) investigated various numerical methods for modal analysis of stepped piezoelectric beams modeled by the Euler-Bernoulli beam theory. Sun et al. (Sun et al. 2001) investigated the effect of the debonding on vibration control of beams with piezoelectric sensors and actuators. Wang et al. (Wang et al. 2018b) studied a new sandwich-type piezoelectric transducer that can be stimulated to create the

Smart Structures and New Advanced Materials Laboratory, Department of Mechanical Engineering, University of Zanjan, Zanjan, Iran.
[*] Corresponding author: omid.rahmani@znu.ac.ir

coupled longitudinal and bending vibrations by applying two electrical signals with a shifted phase. Duan et al. studied (Duan et al. 2005) the free vibration analysis of piezoelectric coupled annular plates using the Kirchhoff and Mindlin plate models. Liu et al. (Liu et al. 2002) presented an analytical model for free vibration analysis of piezoelectric coupled moderately thick circular plate based on Mindlin's plate theory for the cases where electrodes on the piezoelectric layers are shortly connected. Transient vibration and harmonic vibration of circular plate and functionally graded plate integrated with two uniformly piezoelectric material were studied by Jandaghian et al. and Jafari et al. (Jandaghian et al. 2013, 2014, Jafari et al. 2014, Jandaghian and Jafari 2012). Ip and Tse (Ip and Tse 2001) investigated the optimal position and orientation of a piezoelectric patch actuator for improving the controllability of isotropic plates against vibration with an emphasis on controlling the first five modes. The characteristic frequencies of infinite piezoelectric plates vibrating between the grounded electrodes of a plane parallel condenser were investigated by Lawson (Lawson 1942). Batra et al. (Batra et al. 1996) found the optimum location and size of the piezoceramics (PZT) region on which a minimum voltage is applied to suppress the motion of the plate. Ebrahimi and Rastgoo (Ebrahimi and Rastgoo 2008) studied nonlinear free vibration of a thin annular functionally graded plate integrated with two uniformly distributed piezoelectric (PZT4) material on the top and bottom surfaces of the FG plate based on Kirchhoff plate theory. Askari Farsangi and Saidi presented (Askari Farsangi and Saidi 2012) an analytical solution for free vibration analysis of moderately thick functionally graded rectangular plates coupled with piezoelectric layers. Active vibration control of functionally graded piezoelectric material (FGPM) plate by piezoelectric material component as actuator and based on the classical laminated plate theory was carried out by Li et al. (Li et al. 2019b). Free vibration of a laminated hollow circular cylinder with traction free surfaces was studied by Paul and Nelson (Paul and Nelson 1996). Ding et al. (Ding et al. 2003) studied transient responses of piezoelectric hollow cylinders for axisymmetric plane strain problems. Chauhan et al. (Chauhan et al. 2019) studied piezoelectric circular cylindrical shells under axisymmetric external pressure and applied electrical voltage in radial direction. Static behavior of functionally graded material (FGM) cylindrical shells with simply supported edges embedded piezoelectric layers was studied by Alibeigloo (Alibeigloo 2009).

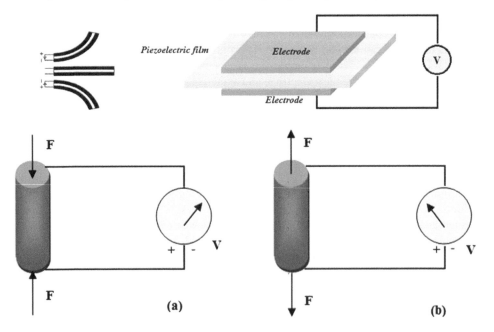

Fig. 16.1   (a) Schematic of piezoelectric effect in laminar format, (b) Schematic of piezoelectric effect in axial format

In the field of materials science, smart materials are one of the new developments that include piezoelectric and piezomagnetic materials. These materials are called magnetoelectroelastic composites that are able to convert energy from one type to the other (among magnetic, electrical and mechanical energies). They also show a magnetoelectric effect, which is not seen in single-phase piezoelectric or piezomagnetic materials. The first report on the coupling effect of magnetoelectric in piezoelectric-piezomagnetic composites was reported by Suchtelen (Van Suchtelen 1972). More exploration of magneto-electro effect of $BaTiO_3$-$CoFe_2O_4$ composites was done by Van den Boomgaard et al. (Van den Boomgaard et al. 1976, Van den Boomgaard et al. 1974). Qing et al. (Qing et al. 2005), Lee (Lee 1991) and He (He 2001) studied basic theories and solution methods of magneto-electro-elastic materials. Annigeri et al. (Annigeri et al. 2007) studied the vibration of multiphase and layered magneto-electro-elastic beam. Milazzo (Milazzo 2013) presented a new one-dimensional model for the dynamic problem of magneto-electro-elastic generally laminated beams. Huang et al. (Huang et al. 2010, 2007) carried out an analytical solution for plane stress problem of generally anisotropic magneto-electro-elastic beams and analytical and semi-analytical solutions for anisotropic functionally graded magneto-electro-elastic beams under an arbitrary load. A general analytical solution for the transient analysis of a magneto-electro-elastic bimorph beam was obtained by Milazzo et al. (Milazzo et al. 2009) based on the Timoshenko beam theory. Vinyas and Kattimani (Vinyas and Kattimani 2017) presented a three-dimensional finite element (FE) formulation for a multilayered magneto-electro-elastic (MEE) beam in a thermal environment. Zhang et al. (Zhang et al. 2020) investigated nonlinear vibration and bending of magneto-electro-elastic laminated beams in thermal environments. Many studies have been conducted about free vibration of magneto-electro-elastic plates, such as exact solutions for three-dimensional, magneto-electro-elastic, and simply-supported rectangular plates under static loadings (Pan 2001), nonlinear free vibration of symmetric magneto-electro-elastic laminated rectangular plates with simply supported boundary condition (Razavi and Shooshtari 2015), free vibration problem of two-dimensional magneto-electro-elastic laminates (Ramirez et al. 2006a), semi-analytical solutions for free vibration of simply supported and multilayered magneto-electro-elastic plates (Xin and Hu 2015), and analysis of free vibration of magneto-electro-elastic laminated plates by using the state-vector approach (Chen et al. 2007). Alaimo et al. (Alaimo et al. 2014) found a finite element formulation to analyze large deflections in magneto-electro-elastic multilayered plates. Chen et al. (Chen et al. 2014) studied free vibration of multilayered magneto-electro-elastic plates under mixed clamped/free lateral boundary conditions. To investigate functionally graded magneto-electro-elastic plates, Bhangale and Ganesan (Bhangale and Ganesan 2006) studied free vibration of functionally graded magneto-electro-elastic plates by semi-analytical finite element method. Active control of nonlinear vibrations of the functionally graded magnet-electro-elastic plate was investigated by Kattimani and Ray (Kattimani and Ray 2015). Pan and Han (Pan and Han 2005) presented an exact solution for the multilayered rectangular plate made of functionally graded, anisotropic, and linear magneto-electro-elastic materials. Ramirez et al. (Ramirez et al. 2006b) obtained natural frequencies of orthotropic magneto-electro-elastic graded composite plates using a discrete layer model with two various approaches. Li et al. (Li et al. 2008b) studied the problem of a functionally graded magneto-electro-elastic circular plate under a uniform load. Static problem of double curved shells made of functionally graded magneto-electro-elastic materials subjected to the electrical, mechanical, and magnetic load was investigated by Wu and Tsai (Wu and Tsai 2007). Bhangale and Ganesan (Bhangale and Ganesan 2005) studied free vibration of functionally graded materials magneto-electro-elastic cylindrical shells. Tsai and Wu (Tsai and Wu 2008) and Wu and Tsai (Wu and Tsai 2010) studied vibration of double curved shells made of functionally graded magneto-electro-elastic shells with simply supported boundary conditions and under open and closed-circuit surface conditions.

Recently, due to fast growth in the flexile electronics and artificial intelligence (Koo et al. 2018, Wang et al. 2018a), wearable and moveable piezoelectric sensors which can directly transmit the generated signal to the sensible electrical information without any additional complicated tools have attracted growing consideration (Nie et al. 2017, Ma et al. 2017, Mannsfeld et al. 2010). Gao et al. (Gao et al. 2020) developed a wearable sensor based on the flexible graphite flakes and MnO_2 nanowires for the useful application in a healthy system and human-computer interaction. Hwang et al. studied new progresses of pliable piezoelectric thin-film harvesters and nanosensors for use in biomedical fields (Hwang et al. 2015). Recent growth in nanotechnology and great mechanical and electrical properties of piezoelectric materials and also excellent mechanical, electrical and magnetic properties of MEE composite materials have captured significant attention of many scientists. In recent years, the MEE and piezoelectric nanomaterials and their nanostructures such as nanowires, nanobeams, nanoplates and nanoshells have become an attractive research subject. Dynamic behavior of these nanostructures plays a key role in the design of intelligent nanodevices. However, the classical continuum elasticity, which is a scale-independent theory, is not able to consider the size-dependent effect. Thus, modified continuum models are proposed to consider the intrinsic parameters correlating micro/nano-structure with macro-structure constitutive equation. Ke and his co-authors extended the nonlocal theory to the piezoelectric and MEE nanostructures (Ke and Wang 2012, 2014). Since then and until now, the nonlocal continuum theory has been applied to model piezoelectric and MEE nanostructures such as nanobeams (Jandaghian and Rahmani 2015, 2016, Ke and Wang 2012, 2014, Ke et al. 2012, Ansari and Gholami 2016, Arani et al. 2014, Ansari et al. 2016, Ebrahimi et al. 2019b, Kunbar et al. 2020, Ebrahimi et al. 2019a), nanoplates (Ansari and Gholami 2017, Jamalpoor et al. 2017, Ke et al. 2015, 2014, Liu et al. 2013, 2018, Zhu et al. 2019, Li et al. 2019a), and nanoshells (Arefi 2020, Arefi and Rabczuk 2019, Wang et al. 2019, Ke et al. 2014a, Sahmani and Aghdam 2018, Ke et al. 2014c, Ebrahimi et al. 2019a, Farajpour et al. 2017) by many researchers.

16.2 BASIC EQUATION

In this section, the equations, which explain electro-mechanical and magneto-electro-mechanical properties of piezoelectric and magneto-electro-elastic materials, are introduced. Based on the IEEE standard, it is supposed that these materials are linear. It is obvious that piezoelectric and magneto-electro-elastic materials show a linear behavior at small mechanical load, small electric fields and at small magnetic fields levels. Nonetheless, significant nonlinearity may be seen if they are subjected to high mechanical load, high electric charge or high magnetic field level. In this chapter, the linear behavior of piezoelectric and magneto-electro-elastic materials is considered.

16.2.1 Piezoelectric Constitutive Equations

For a linear piezoelectric material, the describing electromechanical equations can be written as (Moheimani and Fleming 2006, Meitzler et al. 1988, Fuller et al. 1996):

$$\varepsilon_i = S_{ij}^E \sigma_j + d_{li} E_l,$$
$$D_l = \xi_{ik}^\sigma E_k + d_{li} \sigma_i, \tag{1}$$

where $i, j = 1, 2, ..., 6$ and $l, k = 1, 2, 3$ indicate various directions in the material coordinate system, as displayed in Fig. 16.2 Equation (1) can be re-written in the other form that is frequently used for sensing applications:

$$\varepsilon_i = S_{ij}^D \sigma_j + g_{li} D_l,$$
$$E_l = \beta_{ik}^\sigma D_k + g_{li} \sigma_i \tag{2}$$

where in the eqs. (1) and (2), ε_i is strain (m/m); σ_i is stress (N/m^2); S is coeficients of compliance (m^2/N); E_l and D_l are applied electric field (V/m) and electric displacement (C/m^2), respectively; ξ, d, g and β are permittivity (F/m), piezoelectric strain constants (m/V), piezoelectric constants (m^2/C) and impermitivity component (m/F), respectively. Moreover, the superscripts E, D and σ show that measurements have been taken at zero electric field, zero electric displacement, and zero stress.

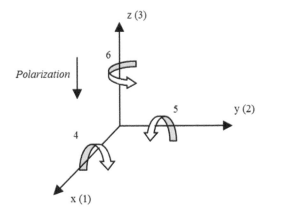

Fig. 16.2 Directions of deformation

no	Axis
1	x
2	y
3	z
4	Shear around x
5	Shear around y
6	Shear around z

Also, equations (1) and (2) can be presented in other forms which are commonly used to describe the coupling between mechanical and electrical variables. Table 16.1 displays the other forms of fundamental relations in piezoelectric materials (Ikeda 1996). In the Table 16.1 d, e, g, and h are four piezoelectric coefficients; β and ε are dielectric impermeability and dielectric constant, respectively.

TABLE 16.1   Types of piezoelectric relation (Ikeda 1996)

Independent Variable	Type	Piezoelectric Relation	From
S, D	Extensive	$T = c^D S - hD$ $E = -hS + \beta^S D$	from-h
T, E	Intensive	$S = s^T T + dE$ $D = dT + \varepsilon^T E$	from-d
T, D	Mixed	$S = s^D T + gD$ $E = -gT + \beta^T D$	from-g
S, E	Mixed	$T = c^E S - eE$ $D = eS + \varepsilon^S E$	from-e

16.2.2   Magneto-Electro-Elastic Constitutive Equations

For a linear magneto-electro-elastic material, the describing electromechanical equations can be written as

$$\sigma_{ij} = c_{ijkl}\varepsilon_{kl} - e_{mij}E_m - q_{nij}H_n - \lambda_{ij}\Delta T,$$
$$D_i = e_{jkl}\varepsilon_{kl} + s_{im}E_m + d_{in}H_n + p_i\Delta T, \tag{3}$$
$$B_i = q_{ikl}\varepsilon_{kl} + d_{im}E_m + \mu_{in}H_n + \beta_i\Delta T,$$

in which σ_{ij}, ε_{ij}, E_i, D_i, H_i and B_i are the stress, strain, electric field, electric displacement, magnetic field, and magnetic induction, respectively; λ_{ij} and ΔT are the thermal moduli and temperature change, respectively; c_{ijkl}, s_{im}, e_{mij}, q_{nij}, μ_{ij}, d_{ij}, p_i and β_i are the elastic, dielectric constants, piezoelectric, piezomagnetic, magnetic, magnetoelectric, pyroelectric and pyromagnetic constants, respectively.

In addition to equation (3), there are other types of constitutive equations, which express the coupling effect among magnetic, elastic, and electric variables (Table 16.2) (Soh et al. 2005).

TABLE 16.2 Constitutive relations of magneto-electro-elastic materials

Independent Variables	Constitutive Relations	Thermodynamic Potentials
ε, E, H	$\sigma = c\varepsilon - e^T E - \tilde{e}^T H$ $D = e\varepsilon + \kappa E + \alpha H$ $B = \tilde{e}\varepsilon + \alpha E + \mu H$	$\Theta_1 = \dfrac{1}{2}(c\varepsilon^2 - \kappa E^2 - \mu H^2)e\varepsilon E - \tilde{e}\varepsilon H - \alpha E H$
ε, D, H	$\sigma = c\varepsilon - h^T D - \tilde{e}^T H$ $E = -h\varepsilon + \beta D - \zeta H$ $B = \tilde{e}\varepsilon + \zeta D + \mu H$	$\Theta_1 = \Theta_2 + DE$
ε, E, B	$\sigma = C\varepsilon - e^T E - \tilde{h}^T B$ $D = e\varepsilon + \kappa E + \eta B$ $H = -\tilde{h}\varepsilon - \eta E + \nu B$	$\Theta_3 = \Theta_1 + BH$
ε, D, B	$\sigma = c\varepsilon - h^T D - \tilde{h}^T B$ $E = -h\varepsilon + \beta D - \lambda B$ $H = \tilde{h}\varepsilon - \lambda D + \nu B$	$\Theta_4 = \Theta_1 + DE + BH$
σ, E, B	$\varepsilon = s\sigma + d^T E + \tilde{g}^T E$ $D = d\sigma + \kappa E + \eta B$ $H = -\tilde{g}\sigma - \eta E + \nu B$	$\Theta_6 = \Theta_1 + BH - \sigma\varepsilon$
σ, D, H	$\varepsilon = s\sigma + g^T D + \tilde{d}^T H$ $E = -g\sigma + \beta D - \zeta H$ $B = \tilde{d}\sigma + \zeta D + \mu H$	$\Theta_7 = \Theta_1 + DE - \sigma\varepsilon$
σ, E, H	$\varepsilon = s\sigma + d^T E + \tilde{d}^T H$ $D = d\sigma + \kappa E + \alpha H$ $B = \tilde{d}\sigma + \alpha E + \mu H$	$\Theta_8 = \Theta_1 - \sigma\varepsilon$

Recent growth in nanotechnology resulted in the progress of micro/nano-electromechanical systems (MEMS and NEMS), which have captured significant attention of scientists in many fields. Duo to high sensitivity, excellent accuracy, and fast response, micro/nano-electromechanical systems are widely used in sensing of physical quantities like molecular mass detector (Jensen et al. 2008), spin (Budakian et al. 2006), and thermal variation (Snyder et al. 2007). To carry out these demands, it is important to decrease the sizes of sensors because it improves the sensitivity efficiency, increases the strength of sensors, and results in more functions that can be integrated into a miniature lab-on-a-chip device.

The nano-sensors are classified into six major groups based on the parameters that are distinguished by them: mechanical, electrical, optical, magnetic, chemical, and thermal (Arash et al. 2015, Wang and Arash 2014). In recent years, with the development of nanotechnology, attention has been concentrated on the replacement of conventional sensors and actuators with smart ceramics such as nano/micro piezoelectric and magneto-electro-elastic. Nanoscale piezoelectric and magneto-electro-elastic transducers can overcome the restrictions of common smart transducers such as rigidity and are consequently not appropriate for adhesion onto electronic skins and wearable devices. Also, to detect very small mechanical input signals like strain, pressure, force, etc. smart transducers with considerably modified sensitivities are required. Dimensions of systems, especially in sensors, have a significant effect on the specifications of the system. At the nanoscale, despite the geometric dependences that can be investigated in classical physics, the existence of quantum effects can lead to considerable size effects. Consequently, significant advantages may occur at the nanoscale. Also, at nano dimensions, surface forces become more important than volume forces and the atoms at the surface become an important part of the whole number of atoms, especially in entirely two dimensional and one layer materials such as single-layer graphene (Falconi 2019). There are many advantages of nanoscale smart materials such as the ability to resist bigger deformations, small voltage, fast speed, greater piezoelectric coefficient, small power, efficient material engineering, low weight, etc. (Falconi 2019).

One of the most important applications of NEMS and MEMS is medicine and biology. For example, accurate measurement of pressure in the body can accurately diagnose the disease and thus effectively treat it. This measurement can be included for blood flow, environment pressure in the cranial cavity, uterus, renal system or other parts. Because of the small size and low power usage, MEMS and NEMS sensors are especially appropriate for this aim. Medical instruments that are made base on MEMS technology are now being expanded widely in *in vivo* and point-of-care usage. Thus, for continuous monitoring of human health conditions *in vivo* sensors are used. The first use of MEMS instruments in medical applications was in the 1970s with the appearance of the silicon-micromachined single-use blood pressure sensors (Blaser et al. 1971). After that, rapid growth has occurred in a number of MEMS instruments for medical applications. An *in vivo* medical device is an instrument, which is planned to work within the body. *In vivo* sensors are used in many applications such as detection of electric signals in the brain, drug delivery, reducing severe pain, measurement of intraocular pressure, blood pressure sensing, monitoring PH in blood, etc. (Schneider and Stieglitz 2004, Hu and Wilson 1997, Ha et al. 2012, Ryu et al. 2007). Electromagnetic energy can transform into mechanical displacements in *in vivo* piezoelectric nanotransducers, hence they are used in biomedical applications such as nano-scissors, nano-pumps and nanoheaters (Falconi et al. 2006, 2007).

16.3 MODELING OF NANOSENSORS

At nanoscale, in addition to experimental methods that are powerful and costly, there are other methods to model nanostructures such as continuum mechanics and molecular dynamics (MD) (Mustapha and Zhong 2010, Wang et al. 2005, Fakhrabadi et al. 2013, Wang et al. 2013). The classical continuum model, which is a scale-independent theory, fails to explain a size-dependent effect. Thus, modified continuum theories are suggested to consider the intrinsic parameters correlating micro/nano-structure with macro-structure constitutive equation. Modified continuum mechanics theories have less computation than other methods such as MD. Also, their formulations are easier than other methods such as multiscale methods. These profits have led to modified continuum theories that are used to study mechanical behavior of nanostructures. For example, recently many researchers applied these theories in the study of buckling, bending and vibration (Kitipornchai et al. 2005, He et al. 2005, Jandaghian and Rahmani 2017, Rahmani and Jandaghian 2015, Jandaghian and Rahmani 2015, 2016c). In classical continuum mechanics, it is assumed that nanomaterials have continuous structures. In this theory, the stress state at a given point is dependent uniquely on the strain state at the same point. The classical continuum model, which is a scale-independent theory, is not able to explain a size-dependent effect. Thus, modified continuum theories are suggested to consider the intrinsic parameters correlating micro/nano-structure with the macro-structure constitutive equation. The most well-known theories in the studies are surface elasticity, gradient elasticity theories, modified coupled theory and nonlocal elastic theories. Among these, nonlocal elasticity theory was presented by Eringen (Eringen 1972, 1983) to consider the scale effect in elasticity by assuming the stress at a reference point to be a function of strain field at every point of the body. These theories have information relating to the forces between atoms (long-range interaction), and the internal length scale. In analysis of nanostructures, nonlocal elasticity theory has gained considerable attention among researches as compared to the other theories due to its simplicity. The nonlocal theory was developed to analyze the bending, buckling, linear and nonlinear vibrations, postbuckling, and wave propagation of nanostructures, such as nanobeams and nanoplates.

According to Eringen's theory, the stress state at a reference point x in the body is regarded to be dependent not only on the strain state at x but also on the strain states at all other points x' of the body. The constitutive equation with zero body force can be written as

$$(1 - (e_0a)^2 \nabla^2)\, \sigma = C : \varepsilon \tag{4}$$

where ∇^2 is Laplacian operator, e_0 is a constant defined experimentally and proper to each material, a is the internal (e.g. lattice parameter, granular size) characteristic length of the nanostructures and e_0a is called the nonlocal parameter, ε and σ are the strain and stress tensors, respectively, and C is the elasticity tensor. After the nonlocal elasticity theory was presented, numerous studies began to use this theory to model nanostructures such as carbon nanotubes, graphene sheets, nanobeams and nanoplates (Mohammadimehr et al. 2010, Jandaghian and Rahmani 2017, Thai 2012, Şimşek 2014, Hosseini-Hashemi et al. 2015).

It can be seen that by setting the nonlocal parameter to zero, i.e. $e_0a = 0$, the nonlocal constitutive relation (4) can be converted to the local constitutive equation. In many studies, a range of $0 \le e_0a \le 2nm$ has been considered for the analysis of nanostructures. It should be noted that to justify the usability of nonlocal models, the nonlocal parameter must be calibrated during a verification method such as molecular simulation or experimental method.

16.3.1 Nonlocal Elasticity Theory for Beam

For Euler-Bernoulli nanobeam, the nonlocal equation of motion is written as (Reddy 2007).

$$EI \frac{\partial^4 w}{\partial x^4} + \rho A \left(\frac{\partial^2 w}{\partial t^2} - (e_0a)^2 \frac{\partial^2 w}{\partial x^2} \right) = 0, \tag{5}$$

where E is the elastic modulus of the beam; w is transverse displacements; ρ is the mass density; A is the area of the beam's cross-section and I denotes the second moment of area of the beam's cross-section.

The Euler-Bernoulli beam model is appropriate for thin nanobeams such as carbon nanotubes (CNTs) and nanowires. Because this model does not consider the transverse shear deformation, it is not adequate for thick beams. Thus, to take into account the effects of shear deformation and rotational inertia, the nonlocal Timoshenko beam theory was developed. The nonlocal equation of motion of Timoshenko beam model is written as (Reddy 2007).

$$K_s AG \frac{\partial}{\partial x}\left(\phi + \frac{\partial w}{\partial x} \right) + (e_0a)^2 \rho A \frac{\partial^4 w}{\partial x^2 \partial t^2} - \rho A \frac{\partial^2 w}{\partial t^2} = 0,$$

$$EI \frac{\partial^2 \phi}{\partial x^2} - K_s AG\left(\phi + \frac{\partial w}{\partial x} \right) - \rho I \frac{\partial^2 \phi}{\partial t^2} + (e_0a)^2 \rho I \frac{\partial^4 \phi}{\partial x^2 \partial t^2} = 0, \tag{6}$$

where G is the shear modulus, K_s is Timoshenko shear coefficient or shear correction factor and ϕ is rotation. Equations (5) and (6) are based on small deformations where there is a linear relation between strain and deformations. In large deformation, nonlinear relation between strain and deformations should be considered. Therefore, to study large deformation problems, Von Karman's theory that is simple for formulation and calculation is used. The nonlinear relation between strain and displacement in Timoshenko beam theory is given as (Yang et al. 2010).

$$\varepsilon_{xx} = \frac{\partial u}{\partial x} + \frac{1}{2}\left(\frac{\partial w}{\partial x} \right)^2 + z \frac{\partial \phi}{\partial x}, \qquad \gamma_{xz} = \frac{\partial w}{\partial x} + \phi, \tag{7}$$

where u is the axial displacement. The nonlinear equations of Timoshenko nanobeams based on the nonlocal elasticity theory are written as (Yang et al. 2010).

$$EA\left(\frac{\partial^2 u}{\partial x^2}+\frac{\partial w}{\partial x}\frac{\partial^2 w}{\partial x^2}\right)=\rho A\frac{\partial^2}{\partial t^2}\left[u-(e_0\alpha)^2\frac{\partial^2 u}{\partial x^2}\right],$$

$$K_s GA\left(\frac{\partial^2 w}{\partial x^2}+\frac{\partial \phi}{\partial x}\right)+L_1-(e_0\alpha)^2 L_2$$

$$=\rho A\frac{\partial^2}{\partial t^2}\left[w-(e_0\alpha)^2\frac{\partial^2 w}{\partial x^2}\right]-\rho A(e_0\alpha)^2\frac{\partial^2}{\partial t^2}\left[\frac{\partial^2 u}{\partial x^2}-(e_0\alpha)^2\frac{\partial^4 u}{\partial x^4}\right] \quad (8)$$

$$EI\frac{\partial^2 \phi}{\partial x^2}-K_s GA\left(\phi+\frac{\partial w}{\partial x}\right)=\rho I\frac{\partial^2}{\partial t^2}\left[\phi-(e_0\alpha)^2\frac{\partial^2 \phi}{\partial x^2}\right],$$

where

$$L_1=EA\left[\frac{\partial^2 u}{\partial x^2}\frac{\partial w}{\partial x}+\frac{3}{2}\left(\frac{\partial w}{\partial x}\right)^2\frac{\partial^2 w}{\partial x^2}+\frac{\partial u}{\partial x}\frac{\partial^2 w}{\partial x^2}\right]$$

$$L_2=EA\left(\frac{\partial^4 u}{\partial x^4}\frac{\partial w}{\partial x}+3\frac{\partial^3 u}{\partial x^3}\frac{\partial^2 w}{\partial x^2}+3\frac{\partial^2 u}{\partial x^2}\frac{\partial^3 w}{\partial x^3}+\frac{\partial u}{\partial x}\frac{\partial^4 w}{\partial x^4}\right) \quad (9)$$

$$+EA\left[3\left(\frac{\partial^2 w}{\partial x^2}\right)^3+9\frac{\partial w}{\partial x}\frac{\partial^2 w}{\partial x^2}\frac{\partial^3 w}{\partial x^3}+\frac{3}{2}\left(\frac{\partial w}{\partial x}\right)^2\frac{\partial^4 w}{\partial x^4}\right]$$

The studies on the vibration of nanobeams show that all natural frequencies are decreased as the nonlocal parameter increases and this reduction is more obvious at higher frequencies (Thai 2012, Rahmani and Pedram 2014, Eltaher et al. 2013). The results obtained by molecular dynamic simulations show that natural frequencies of nanobeams obtained from the nonlocal theory are in good agreement with the results of molecular dynamic simulations (Arash and Ansari 2010, Zhang et al. 2009).

16.3.1.1 Nonlocal Elasticity Theory for Piezoelectric Nanobeam

Nanobeams are one of the most important nanostructures, which are widely used in the nanodevices with piezoelectric, and therefore understanding the vibrational behavior of these nanostructures is of key importance in the design of nanodevices with piezoelectric such as nanosensors and nanoresonators. A schematic sketch of a piezoelectric nanobeam is shown in Fig. 16.3. It is assumed the piezoelectric material is homogeneous. Based on the theory of nonlocal piezoelectricity, the stress tensor and the electric displacement at a reference point depend not only on the strain components and electric-field components at the same position but also on all other points of the body. The nonlocal constitutive relation for the piezoelectric material is then obtained as (Eringen 1983, Jandaghian and Rahmani 2016a)

$$\sigma_{ij}=\int_{-h/2}^{h/2}\alpha(|x-\dot{x}''''''|,\ e_0 a/l)\,[c_{ijkl}\varepsilon_{kl}(x')-\lambda_{ij}\Delta T]\,dV(x') \quad (10)$$

$$D_i=\int_{-h/2}^{h/2}\alpha(|x-\dot{x}''''''|,\ e_0 a/l)\,[c_{ijkl}\varepsilon_{kl}(x')+\Xi_{ij}E(x')+p_i\Delta T]\,dV(x') \quad (11)$$

$$E_i=-\phi_{,i} \quad (12)$$

where Ξ_{ij}, λ_{ij}, p_i denote the dielectric constants, thermal moduli and pyroelectric constants, respectively; ΔT is the variations of temperature and ϕ is the electric potential. According to (Eringen 1983) Eqs. (10) and (11) can be re-written as

$$\sigma_{ij} - (e_0 a)^2 \nabla^2 \sigma_{ij} = C_{ijkl}\varepsilon_{kl} - e_{lij}E_l - \lambda_i \Delta T, \tag{13}$$

$$D_i - (e_0 a)^2 \nabla^2 D_i = e_{ijkl}\varepsilon_{kl} + \Xi_{il}E_l + p_i \Delta T \tag{14}$$

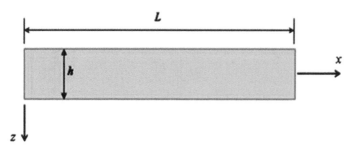

Fig. 16.3   Schematic of piezoelectric nanobeam

For the piezoelectric nanobeam, model distribution of the electric potential is necessary. The electric potential at any point of the piezoelectric should satisfy the Maxwell equation. Therefore, the following electric potential is assumed (Ke and Wang 2012, Ke et al. 2012).

$$\Phi(x, z, t) = -\cos(\beta z)\phi(x, t) + \frac{2z}{h}V_0, \tag{15}$$

where $\beta = \pi/h$; ϕ is the spatial electric potential change along the x-direction and V_0 is the initial applied electric potential. The governing equations of piezoelectric nanobeams based on Euler-Bernoulli beam theory can be obtained as (Jandaghian and Rahmani 2016a).

$$D\frac{\partial^4 w_0}{\partial x^4} + F_{31}\frac{\partial^2 \phi}{\partial x^2} - \rho h \frac{\partial^2}{\partial t^2}\left[w_0 - (e_0 a)^2 \frac{\partial^2 w_0}{\partial x^2}\right] = 0 \tag{16}$$

$$X_{11}\frac{\partial^2 \phi}{\partial x^2} - F_{31}\frac{\partial^2 w_0}{\partial x^2} - X_{33}\phi = 0 \tag{17}$$

where

$$D = \int_{-h/2}^{h/2} \bar{C}_{11} z \, dz, \quad F_{31} = \int_{-h/2}^{h/2} \bar{e}_{31}\beta\sin(\beta z)\, z \, dz, \tag{18}$$

$$X_{11} = \int_{-h/2}^{h/2} \bar{\Xi}_{11}\cos^2(\beta z) dz, \quad X_{33} = \int_{-h/2}^{h/2} \bar{\Xi}_{33}\beta^2\sin^2(\beta z) dz,$$

$$F_{31} = \int_{-h/2}^{h/2} \bar{e}_{31}\beta\sin(\beta z)\, z \, dz \tag{19}$$

$$\bar{C}_{11}(z) = C_{11}(z) - \frac{C_{13}^2(z)}{C_{33}(z)}, \quad \bar{e}_{31}(z) = e_{31}(z) - \frac{e_{33}(z)C_{13}(z)}{C_{33}(z)},$$

$$\bar{\Xi}_{11}(z) = \Xi_{11}(z), \quad \bar{\Xi}_{33}(z) = \Xi_{33}(z) - \frac{e_{33}^2(z)}{C_{33}(z)} \tag{20}$$

The nonlocal equation of motion of Timoshenko piezoelectric nanobeams is written as

$$k_s A_{44}\left(\frac{\partial^2 w}{\partial x^2}+\frac{\partial \Psi}{\partial x}\right)-k_s E_{15}\frac{\partial^2 \phi}{\partial x^2}=I_1\frac{\partial^2}{\partial t^2}\left[w-(e_0\alpha)^2\frac{\partial^2 w}{\partial x^2}\right], \tag{21}$$

$$D_{11}\frac{\partial^2 \Psi}{\partial x^2}-k_s A_{44}\left(\frac{\partial w}{\partial x}+\Psi\right)+F_{31}\frac{\partial \phi}{\partial x}+k_s E_{15}\frac{\partial \phi}{\partial x}=I_3\frac{\partial^2}{\partial t^2}\left[\Psi-(e_0\alpha)^2\frac{\partial^2 \Psi}{\partial x^2}\right], \tag{22}$$

$$F_{31}\frac{\partial \Psi}{\partial x}+E_{15}\left(\frac{\partial^2 w}{\partial x^2}+\frac{\partial \Psi}{\partial x}\right)+X_{11}\frac{\partial^2 \phi}{\partial x^2}-X_{33}\phi=0 \tag{23}$$

where

$$D_{11}=\int_{-h/2}^{h/2}c_{11}z^2 dz,\ \ A_{44}=\int_{-h/2}^{h/2}c_{44}dz,\ \ F_{31}=\int_{-h/2}^{h/2}e_{31}\beta\sin(\beta z)zdz, \tag{24}$$

$$E_{15}=\int_{-h/2}^{h/2}e_{15}\cos(\beta z)dz,$$

$$X_{11}=\int_{-h/2}^{h/2}\varepsilon_{11}\cos^2(\beta z)dz,\ \ X_{33}=\int_{-h/2}^{h/2}\varepsilon_{33}[\beta\sin(\beta z)]^2 dz, \tag{25}$$

The nonlinear equations of piezoelectric Timoshenko nanobeams based upon the nonlocal elasticity theory are written as

$$A_{11}\left(\frac{\partial^2 u}{\partial x^2}+\frac{\partial w}{\partial x}\frac{\partial^2 w}{\partial x^2}\right)=I_1\frac{\partial^2}{\partial t^2}\left[u-(e_0\alpha)^2\frac{\partial^2 u}{\partial x^2}\right], \tag{26}$$

$$k_s A_{44}\left(\frac{\partial^2 w}{\partial x^2}+\frac{\partial \Psi}{\partial x}\right)-k_s E_{15}\frac{\partial^2 \phi}{\partial x^2}+\Sigma_1-(e_0 a)^2\Sigma_2=I_1\frac{\partial^2}{\partial t^2}\left[w-(e_0\alpha)^2\frac{\partial^2 w}{\partial x^2}\right], \tag{27}$$

$$D_{11}\frac{\partial^2 \Psi}{\partial x^2}-k_s A_{44}\left(\frac{\partial w}{\partial x}+\Psi\right)+F_{31}\frac{\partial \phi}{\partial x}+k_s E_{15}\frac{\partial \phi}{\partial x}=I_3\frac{\partial^2}{\partial t^2}\left[\Psi-(e_0\alpha)^2\frac{\partial^2 \Psi}{\partial x^2}\right], \tag{28}$$

$$F_{31}\frac{\partial \Psi}{\partial x}+E_{15}\left(\frac{\partial^2 w}{\partial x^2}+\frac{\partial \Psi}{\partial x}\right)+X_{11}\frac{\partial^2 \phi}{\partial x^2}-X_{33}\phi=0 \tag{29}$$

where

$$\Sigma_1=A_{11}\left[\frac{\partial^2 u}{\partial x^2}\frac{\partial w}{\partial x}+\frac{3}{2}\left(\frac{\partial w}{\partial x}\right)^2\frac{\partial^2 w}{\partial x^2}+\frac{\partial u}{\partial x}\frac{\partial^2 w}{\partial x^2}\right], \tag{30}$$

$$\Sigma_2=A_{11}\left(\frac{\partial^4 u}{\partial x^4}\frac{\partial w}{\partial x}+3\frac{\partial^3 u}{\partial x^3}\frac{\partial^2 w}{\partial x^2}+3\frac{\partial^2 u}{\partial x^2}\frac{\partial^3 w}{\partial x^3}+\frac{\partial u}{\partial x}\frac{\partial^4 w}{\partial x^4}\right)$$

$$+A_{11}\left[3\left(\frac{\partial^2 w}{\partial x^2}\right)^3+9\frac{\partial w}{\partial x}\frac{\partial^2 w}{\partial x}\frac{\partial^3 w}{\partial x^3}+\frac{3}{2}\left(\frac{\partial w}{\partial x}\right)^2\frac{\partial^4 w}{\partial x}\right] \tag{31}$$

To the best of the authors' knowledge, there are no theoretical, molecular dynamic and experimental results on the nonlocal piezoelectric nanobeams available in the literature. Similar to the nanobeams, an increase in the nonlocal parameter results in the reduction of the piezoelectric nanobeams' natural frequencies. Because of the existence of the nonlocal effect, the stiffness of the nanostructures decreases and consequently the values of the natural frequencies decline (Jandaghian and Rahmani 2016b, Ke and Wang 2012, Ke et al. 2012).

16.3.1.2 Nonlocal Elasticity Theory for Magneto-Electro-Elastic Nanobeam

Magneto-electro-elastic (MEE) composite materials are a new kind of smart materials that show an instinct electro-magneto-mechanical coupling effects. Due to the tendency to miniaturize device and with the quick development of nanotechnology, the MEE nanomaterials such as $BiTiO_3$-$CoFe_2O_4$, $BiFeO_3$, and their nanostructures received considerable attention (Martin et al. 2008, Prashanthi et al. 2012). Figure 16.4 shows the schematic of MEE nanobeam.

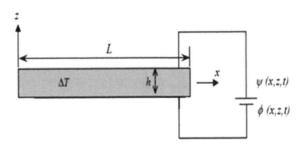

Fig. 16.4 Schematic of magneto-electro-elastic nanobeam

The fundamental equations of a nonlocal MEE material with zero body force is written as (Jandaghian and Rahmani 2016b).

$$\sigma_{ij} = \int_V \alpha \left(|x - \acute{x}''''''|, e_0 a/l\right) [c_{ijkl}\varepsilon_{kl}(x') - e_{mij}E_m(x') - q_{nij}H_n(x') - \lambda_{ij}\Delta T] dV(x') \tag{32}$$

$$D_i = \int_V \alpha \left(|x - \acute{x}''''''|, e_0 a/l\right) [e_{ijkl}\varepsilon_{kl}(x') + s_{mij}E_m(x') + d_{ni}H_n(x') + p_i\Delta T] dV(x') \tag{33}$$

$$B_i = \int_V \alpha \left(|x - \acute{x}''''''|, e_0 a/l\right) [q_{ijkl}\varepsilon_{kl}(x') + d_{mi}E_m(x') + \mu_{ni}H_n(x') + \beta_i\Delta T] dV(x') \tag{34}$$

where D_i, E_i, B_i and H_i are electric displacement, electric field, magnetic induction, and magnetic field and displacement components, respectively; c_{ijkl}, e_{mij}, s_{im}, q_{nij}, d_{in}, μ_{in}, p_i and β_i are the elastic, piezoelectric, dielectric constants, piezomagnetic, magnetoelectric, magnetic, pyroelectric, and pyromagnetic constants, respectively. Based on Eringen (Eringen 1983), the differential type of equations (27)-(29) can be expressed as

$$\sigma_{ij} - (e_0 a)^2 \nabla^2 \sigma_{ij} = c_{ijkl}\varepsilon_{kl} - e_{mij}E_m - q_{nij}H_n + \lambda_{ij}\Delta T, \tag{35}$$

$$D_i - (e_0 a)^2 \nabla^2 D_i = e_{jkl}\varepsilon_{kl} + s_{im}E_m + d_{in}H_n + p_i\Delta T, \tag{36}$$

$$B_i - (e_0 a)^2 \nabla^2 B_i = q_{ikl}\varepsilon_{kl} + d_{im}E_m + \mu_{in}H_n + \beta_i\Delta T \tag{37}$$

To satisfy Maxwell's equations, the electric and magnetic potential are considered as (Ke and Wang 2014, Jandaghian and Rahmani 2016b, Wang 2002)

$$\phi = -\cos(\beta z)\phi_E(t, x) + \frac{2z}{h}V_0, \tag{38}$$

$$\psi = -\cos(\beta z)\psi_H(t, x) + \frac{2z}{h}A_0, \tag{39}$$

in which ϕ_E is spatial variation of the electric potential and ψ_H is spatial variation of the magnetic potential in the x-direction; A_0 and V_0 are the initial applied magnetic and electric potential, respectively. The nonlocal governing equations for the MEE Timoshenko nanobeam are obtained as (Jandaghian and Rahmani 2016b).

$$k_s A_{44} \left(\frac{\partial^2 w}{\partial x^2} + \frac{\partial \theta_x}{\partial x} \right) - k_s \left(E_{15} \frac{\partial^2 \phi}{\partial x^2} + Q_{15} \frac{\partial^2 \psi}{\partial x^2} \right) = I_0 \left(\frac{\partial^2 w}{\partial t^2} - (e_0 a)^2 \frac{\partial^4 w}{\partial x^2 \partial t^2} \right) \tag{40}$$

$$D_{11} \frac{\partial^2 \theta_x}{\partial x^2} - k_s A_{44} \left(\theta_x + \frac{\partial w}{\partial x} \right) + (E_{31} + k_s E_{15}) \frac{\partial \phi_E}{\partial x} + (Q_{31} + k_s Q_{15}) \frac{\partial \psi_H}{\partial x}$$

$$= I_2 \left(\frac{\partial^2 \theta_x}{\partial t^2} - (e_0 a)^2 \frac{\partial^4 \theta_x}{\partial x^2 \partial t^2} \right) \tag{41}$$

$$E_{31} \frac{\partial \theta_x}{\partial x} + E_{15} \left(\frac{\partial^2 w}{\partial x^2} + \frac{\partial \theta_x}{\partial x} \right) + X_{11} \frac{\partial^2 \phi_E}{\partial x^2} + Y_{11} \frac{\partial^2 \psi_H}{\partial x^2} - X_{33} \phi_E - Y_{33} \psi_H = 0 \tag{42}$$

$$Q_{31} \frac{\partial \theta_x}{\partial x} + Q_{15} \left(\frac{\partial^2 w}{\partial x^2} + \frac{\partial \theta_x}{\partial x} \right) + Y_{11} \frac{\partial^2 \phi_E}{\partial x^2} + T_{11} \frac{\partial^2 \psi_H}{\partial x^2} - Y_{33} \phi_E - T_{33} \psi_H = 0 \tag{43}$$

in which

$$D_{11} = \int_{-h/2}^{h/2} \bar{C}_{11} z^2 dz, (A_{11}, A_{44}) = \int_{-h/2}^{h/2} (\bar{C}_{11}, \bar{C}_{44}) dz, (I_0, I_2) = \int_{-h/2}^{h/2} \rho(z^0, z^2) dz,$$

$$(E_{31}, Q_{31}) = \int_{-h/2}^{h/2} (\bar{e}_{31}, \bar{q}_{31}) \beta \sin(\beta z) z dz, (E_{15}, Q_{15}) = \int_{-h/2}^{h/2} (\bar{e}_{15}, \bar{q}_{15}) \cos(\beta z) dz, \tag{44}$$

$$(X_{11}, Y_{11}, T_{11}) = \int_{-h/2}^{h/2} (\bar{s}_{11}, \bar{d}_{11}, \bar{\mu}_{11}) \cos^2(\beta z) dz,$$

$$(X_{33}, Y_{33}, T_{33}) = \int_{-h/2}^{h/2} (\bar{s}_{33}, \bar{d}_{33}, \bar{\mu}_{33}) \beta^2 \sin^2(\beta z) dz.$$

$$\bar{C}_{11} = C_{11} - \frac{C_{13}^2}{C_{33}}, \bar{C}_{44} = C_{44}, \bar{e}_{31} = e_{31} - \frac{C_{13} e_{33}}{C_{33}}, \bar{e}_{15} = e_{15}, \bar{q}_{31} = q_{31} - \frac{C_{13} q_{33}}{C_{33}},$$

$$\bar{s}_{11} = s_{11}, \bar{s}_{33} = s_{33} - \frac{e_{33}^2}{C_{33}}, \bar{d}_{11} = d_{11}, \bar{d}_{33} = d_{33} + \frac{q_{33} e_{33}}{C_{33}}, \bar{\mu}_{11} = \mu_{11}, \bar{\mu}_{33} = \mu_{33} + \frac{q_{33}^2}{C_{33}} \tag{45}$$

$$\bar{q}_{15} = q_{15}, \bar{\lambda}_1 = \lambda_1 - \frac{C_{13} \lambda_3}{C_{33}}, \bar{p}_3 = p_3 + \frac{\lambda_3 e_{33}}{C_{33}}, \bar{\beta}_3 = \beta_3 + \frac{\beta_3 q_{33}}{C_{33}}.$$

k_s and θ_x are shear correction coefficient and the rotation of beam cross-section, respectively. For the nonlinear vibration of the MEE Timoshenko nanobeam, the nonlocal equations based on the von Kármán theory are expressed as

$$A_{11}(U_{,xx} + W_{,x}W_{,xx}) = I_0(\ddot{U} - (e_0\alpha)^2 \ddot{U}_{,xx}), \tag{46}$$

$$k_s A_{44}(\Theta_{x,x} + W_{,xx}) - \kappa_s(E_{15}\phi_{E,xx} + Q_{15}\psi_{H,xx}) + Z_1 - (e_0\alpha)^2 Z_2 + F - (e_0\alpha)^2 F_{,xx}$$
$$= I_0(\ddot{W} - (e_0\alpha)^2 \ddot{W}_{,xx}), \tag{47}$$

$$D_{11}\Theta_{x,xx} - k_s A_{44}(\Theta_x + W_{,x}) + (E_{31} + \kappa_s E_{15})\phi_{E,x} + (Q_{31} + \kappa_s Q_{15})\phi_{E,x} = I_2(\ddot{\Theta}_x - (e_0\alpha)^2 \ddot{\Theta}_{x,xx}), \tag{48}$$

$$E_{31}\Theta_{x,x} + E_{15}(W_{,xx} + \Theta_{x,x}) + X_{11}\phi_{E,xx} + Y_{11}\psi_{H,xx} - X_{33}\phi_E - Y_{33}\psi_H = 0, \tag{49}$$

$$Q_{31}\Theta_{x,x} + Q_{15}(W_{,xx} + \Theta_{x,x}) + Y_{11}\phi_{E,xx} + T_{11}\psi_{H,xx} - Y_{33}\phi_E - T_{33}\psi_H = 0 \tag{50}$$

where

$$Z_1 = A_{11}\left(U_{,xx}W_{,x} + U_{,x}W_{,xx} + \frac{3}{2}W_{,x}^2 W_{,xx}\right), \tag{51}$$

$$Z_2 = A_{11}(U_{,xxxx}W_{,x} + 3U_{,xxx}W_{,xx} + 3U_{,xx}W_{,xxx} + U_{,x}W_{,xxxx}) + A_{11}$$

$$\left(3W_{,xx}^2 + 9W_{,x}W_{,xx}W_{,xxx} + \frac{3}{2}W_{,x}^2 W_{,xxxx}\right) \tag{52}$$

As mentioned for piezoelectric nanobeams, in spite of the significance of designing the MEE nanostructures, there is no literature available for the nanodevices with MEE composite materials based on the nonlocal theory. The MEE nonlocal nanobeams› natural frequencies are lesser than that of the MEE classical nanobeams and reduce by rising of the small-scale parameter. Also, the external magnetic and electric potential have a considerable effect on the natural frequencies of MEE nanobeams (Jandaghian and Rahmani 2016b, Ke and Wang 2014).

16.3.2   Nonlocal Elasticity Theory for Plate

Micro/Nanoplates are the key structures that are extensively used in nanosensors and Micro and Nano Electro-Mechanical Systems (MEMS/NEMS) (Li et al. 2003). To explain the motion of plates, several theories have been developed. The most frequently used are classical plate theory (CLPT or Kirchhoff) and Mindlin-Reissner or first-order shear deformation theory (FSDT).

For CLPT nanoplate, the nonlocal equation is given as (Pradhan and Kumar 2011).

$$D\nabla^4 w = (1 - (e_0\alpha)^2 \nabla^2)\left[-I_0 \frac{\partial^2 w}{\partial t^2} + I_2\left(\frac{\partial^4 w}{\partial x^2 \partial t^2} + \frac{\partial^4 w}{\partial y^2 \partial t^2}\right)\right], \tag{53}$$

where

$$D = \frac{Eh^3}{12(1-v^2)}, \quad I_i = \int_{-h/2}^{h/2} \rho h^i dz \tag{54}$$

h and v are thickness and Poisson's ratios of the nanoplate, respectively. For FSTD nanoplate, the nonlocal equations of motion are expressed as

$$\frac{Eh}{1-v^2}u_{,xx} + \frac{Eh}{4(1+v)}u_{,yy} + \frac{(1-3v)Eh}{4(1-v^2)}v_{,xy} = I_1\ddot{u} + I_2\ddot{\psi}_x - \mu[I_1(\ddot{u}_{xx} + \ddot{u}_{yy}) + I_2(\ddot{\psi}_{x_{xx}} + \ddot{\psi}_{x_{yy}})] \tag{55}$$

$$\frac{(1-3v)Eh}{4(1-v^2)}u_{,xy} + \frac{Eh}{4(1+v)}v_{,xx} + \frac{Eh}{1-v^2}v_{,yy} = I_1\ddot{v} + I_2\ddot{\psi}_y - \mu[I_1(\ddot{v}_{xx} + \ddot{v}_{yy}) + I_2(\ddot{\psi}_{y_{xx}} + \ddot{\psi}_{y_{yy}})] \tag{56}$$

$$\frac{K_s}{4}\frac{Eh}{1+v}w_{,xx} + \frac{K_s}{4}\frac{Eh}{1+v}w_{,yy} + \frac{K_s}{4}\frac{Eh}{1+v}\psi_{x,x} + \frac{K_s}{4}\frac{Eh}{1+v}\psi_{y,y} = I_1 w - \mu[I_1(\ddot{w}_{xx} + \ddot{w}_{yy})] \tag{57}$$

$$-\frac{K_s}{4}\frac{Eh}{1+v}w_{,x}+D\psi_{x,xx}+\frac{(1+v)D}{4}\psi_{x,yy}-\frac{K_s}{4}\frac{Eh}{1+v}\psi_x+\frac{(1-3v)D}{4}\psi_{x,xy}$$

$$=I_2\ddot{u}+I_3\ddot{\psi}_x-\mu[I_2(\ddot{u}_{xx}+\ddot{u}_{yy})+I_3(\ddot{\psi}_{xx}+\ddot{\psi}_{X_{yy}})] \tag{58}$$

$$-\frac{K_s}{4}\frac{Eh}{1+v}w_{,y}+\frac{(1-3v)D}{4}\psi_{x,xy}+\frac{(1-v)D}{4}\psi_{y,xx}+D\psi_{y,yy}-\frac{K_s}{4}\frac{Eh}{1+v}\psi_y$$

$$=I_2\ddot{v}+I_3\ddot{\psi}_y-\mu[I_2(\ddot{v}_{xx}+\ddot{v}_{yy})+I_3(\ddot{\psi}_{y_{xx}}+\ddot{\psi}_{y_{yy}})] \tag{59}$$

where I_1, I_2 and I_3 are the inertia terms; u, v and w are the displacements along x, y and z directions, respectively; ψ_x and ψ_y are the rotations of normal to the mid-surface about the x and y-axis, respectively.

As it is seen for nanobeams, with rise in the nonlocal parameter, the natural frequencies of the nanoplates fall off. In addition, with the growth of nanoplate's length, the natural frequencies go up and close to the local frequencies, which implies a reduction in the effect of the small-scale parameter (Aksencer and Aydogdu 2011, Hosseini-Hashemi 2013, Pradhan and Phadikar 2009).

16.3.2.1 Nonlocal Elasticity Theory for Piezoelectric Nanoplate

Recently, due to the excellent mechanical, electrical and thermal properties of piezoelectric nanostructures, tendency to miniaturize device and with the quick development of nanotechnology, the piezoelectric nanomaterials such as PZT, ZnO, GaN, BaTiO$_3$ and their nanostructures received considerable attention. The governing equations of the piezoelectric nanoplates are presented in this section based on the nonlocal theory for the mentioned piezoelectric materials.

The poling direction of the piezoelectric nanoplate is along to z-axis (Fig. 16.5). As mentioned before, the distribution of the electric potential should satisfy the Maxwell equation. Hence, the following electric potential is assumed (Wang 2002, Quek and Wang 2000)

$$\Phi(x,z,t)=-\cos(\beta z)\phi(x,y,t)+\frac{2z}{h}V_0 \tag{60}$$

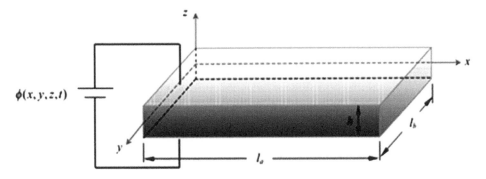

Fig. 16.5 Schematic of piezoelectric nanoplate

According to the Kirchhoff plate theory, the governing equations of piezoelectric nanoplates can be obtained as (Liu et al. 2013).

$$-D_{11}\left(\frac{\partial^4 w}{\partial x^4}+\frac{\partial^4 w}{\partial y^4}\right)-2D_{12}\frac{\partial^4 w}{\partial x^2\partial y^2}-4D_{13}\frac{\partial^4 w}{\partial x^2\partial y^2}+F_{31}\left(\frac{\partial^2\phi}{\partial x^2}+\frac{\partial^2\phi}{\partial y^2}\right)$$

$$=(1-(e_0 a)^2\nabla^2)I_0\ddot{w} \tag{61}$$

$$X_{11}\left(\frac{\partial^2 \phi}{\partial x^2}+\frac{\partial^2 \phi}{\partial y^2}\right)-F_{31}\left(\frac{\partial^2 w}{\partial x^2}+\frac{\partial^2 w}{\partial y^2}\right)-X_{33}\phi = 0 \tag{62}$$

where,
$$(D_{11}, D_{12}, D_{66})=\int_{-h/2}^{h/2}(\bar{C}_{11},\bar{C}_{12},\bar{C}_{66})z^2 dz, \quad \bar{C}_{66}=\frac{1}{2}(\bar{C}_{11}-\bar{C}_{12})$$

The governing equations of the rectangular piezoelectric nanoplate using the Mindlin plate and the nonlocal theory are written as (Ke et al. 2015).

$$k_s A_{44}\left(\frac{\partial^2 W}{\partial x^2}+\frac{\partial^2 W}{\partial y^2}+\frac{\partial \Psi_x}{\partial x}+\frac{\partial \Psi_y}{\partial y}\right)-k_s E_{15}\left(\frac{\partial^2 \Phi}{\partial x^2}+\frac{\partial^2 \Phi}{\partial y^2}\right)$$
$$=[1-(e_0\alpha)^2\nabla^2]I_1\frac{\partial^2 W}{\partial t^2}, \tag{63}$$

$$D_{11}\frac{\partial^2 \Psi_x}{\partial x^2}+D_{12}\frac{\partial^2 \Psi_y}{\partial x\partial y}+D_{66}\left(\frac{\partial^2 \Psi_y}{\partial x\partial y}+\frac{\partial^2 \Psi_x}{\partial y^2}\right)+E_{31}\frac{\partial \phi}{\partial x}-k_s A_{44}\left(\frac{\partial W}{\partial x}+\Psi_x\right)$$
$$+k_s E_{15}\frac{\partial \Phi}{\partial x}=[1-(e_0\alpha)^2\nabla^2]I_3\frac{\partial^2 \Psi_x}{\partial t^2}, \tag{64}$$

$$D_{12}\frac{\partial^2 \Psi_x}{\partial x\partial y}+D_{11}\frac{\partial^2 \Psi_y}{\partial y^2}+D_{66}\left(\frac{\partial^2 \Psi_y}{\partial x^2}+\frac{\partial^2 \Psi_x}{\partial x\partial y}\right)+E_{31}\frac{\partial \Phi}{\partial y}-k_s A_{44}\left(\frac{\partial W}{\partial y}+\Psi_y\right)$$
$$+k_s E_{15}\frac{\partial \Phi}{\partial y}=[1-(e_0\alpha)^2\nabla^2]I_3\frac{\partial^2 \Psi_y}{\partial t^2}, \tag{65}$$

$$E_{15}\left(\frac{\partial^2 W}{\partial x^2}+\frac{\partial^2 W}{\partial y^2}+\frac{\partial \Psi_x}{\partial x}+\frac{\partial \Psi_y}{\partial y}\right)+X_{11}\left(\frac{\partial^2 \Phi}{\partial x^2}+\frac{\partial^2 \Phi}{\partial y^2}\right)+E_{31}\left(\frac{\partial \Psi_x}{\partial x}+\frac{\partial \Psi_y}{\partial y}\right)-X_{33}\Phi = 0 \tag{66}$$

The nonlinear equations of piezoelectric Mindlin nanoplates based on the nonlocal theory and von Kármán theory are written as (Liu et al. 2018)

$$A_{11}\left(\frac{\partial^2 V}{\partial y^2}+\frac{\partial W}{\partial y}\frac{\partial^2 W}{\partial y^2}\right)+A_{12}\left(\frac{\partial^2 U}{\partial x\partial y}+\frac{\partial W}{\partial x}\frac{\partial^2 W}{\partial x\partial y}\right)$$
$$+A_{66}\left(\frac{\partial^2 U}{\partial x\partial y}+\frac{\partial^2 V}{\partial x^2}+\frac{\partial^2 W}{\partial x^2}\frac{\partial W}{\partial y}+\frac{\partial W}{\partial x}\frac{\partial^2 W}{\partial x\partial y}\right)=L_{nol}(I_1\ddot{V}) \tag{67}$$

$$k_s A_{44}\left(\frac{\partial^2 W}{\partial x^2}+\frac{\partial^2 W}{\partial y^2}+\frac{\partial \Psi_x}{\partial x}+\frac{\partial \Psi_y}{\partial y}\right)-k_s E_{15}\left(\frac{\partial^2 \Phi}{\partial x^2}+\frac{\partial^2 \Phi}{\partial y^2}\right)+Z_1+Z_2=L_{nol}(I_1\ddot{W}) \tag{68}$$

$$D_{11}\frac{\partial^2 \Psi_x}{\partial x^2}+D_{12}\frac{\partial^2 \Psi_y}{\partial x\partial y}+D_{66}\left(\frac{\partial^2 \Psi_y}{\partial x\partial y}+\frac{\partial^2 \Psi_x}{\partial y^2}\right)+E_{31}\frac{\partial \phi}{\partial x}-k_s A_{44}\left(\frac{\partial W}{\partial x}+\Psi_x\right)$$
$$+k_s E_{15}\frac{\partial \Phi}{\partial x}=L_{nol}(I_1\ddot{\Psi}_x) \tag{69}$$

$$D_{12} \frac{\partial^2 \Psi_x}{\partial x \partial y} + D_{11} \frac{\partial^2 \Psi_y}{\partial y^2} + D_{66} \left(\frac{\partial^2 \Psi_y}{\partial x^2} + \frac{\partial^2 \Psi_x}{\partial x \partial y} \right) + E_{31} \frac{\partial \phi}{\partial y} - k_s A_{44} \left(\frac{\partial W}{\partial y} + \Psi_y \right)$$

$$+ k_s E_{15} \frac{\partial \Phi}{\partial y} = L_{nol} (I_1 \ddot{\Psi}_y) \tag{70}$$

$$E_{15} \left(\frac{\partial^2 W}{\partial x^2} + \frac{\partial^2 W}{\partial y^2} + \frac{\partial \Psi_x}{\partial x} + \frac{\partial \Psi_y}{\partial y} \right) + X_{11} \left(\frac{\partial^2 \Phi}{\partial x^2} + \frac{\partial^2 \Phi}{\partial y^2} \right) + E_{31} \left(\frac{\partial \Psi_x}{\partial x} + \frac{\partial \Psi_y}{\partial y} \right) - X_{33} \Phi = 0, \tag{71}$$

where

$$L_{nol} = [1 - (e_0 a)^2 \nabla^2], \tag{72}$$

$$Z_1 = \left[A_{11} \left(\frac{\partial U}{\partial x} + \frac{1}{2} \left(\frac{\partial W}{\partial x} \right)^2 \right) + A_{12} \left(\frac{\partial V}{\partial y} + \frac{1}{2} \left(\frac{\partial W}{\partial y} \right)^2 \right) \right] L_{nol} \left(\frac{\partial^2 W}{\partial x^2} \right)$$

$$+ \left[A_{11} \left(\frac{\partial^2 U}{\partial x^2} + \frac{\partial W}{\partial x} \frac{\partial^2 W}{\partial x^2} \right) + A_{12} \left(\frac{\partial^2 V}{\partial x \partial y} + \frac{\partial W}{\partial y} \frac{\partial^2 W}{\partial x \partial y} \right) \right] L_{nol} \left(\frac{\partial W}{\partial x} \right)$$

$$+ \left[A_{12} \left(\frac{\partial U}{\partial x} + \frac{1}{2} \left(\frac{\partial W}{\partial x} \right)^2 \right) + A_{11} \left(\frac{\partial V}{\partial y} + \frac{1}{2} \left(\frac{\partial W}{\partial x} \right)^2 \right) \right] L_{nol} \left(\frac{\partial^2 W}{\partial y^2} \right)$$

$$+ \left[A_{12} \left(\frac{\partial^2 U}{\partial x \partial y} + \frac{\partial W}{\partial x} \frac{\partial^2 W}{\partial x \partial y} \right) + A_{11} \left(\frac{\partial^2 V}{\partial y^2} + \frac{\partial W}{\partial x} \frac{\partial^2 W}{\partial y^2} \right) \right] L_{nol} \left(\frac{\partial W}{\partial y} \right) \tag{73}$$

$$Z_2 = 2 A_{66} \left(\frac{\partial U}{\partial y} + \frac{\partial V}{\partial x} + \frac{\partial W}{\partial x} \frac{\partial W}{\partial y} \right) L_{nol} \left(\frac{\partial^2 W}{\partial x \partial y} \right)$$

$$+ A_{66} \left(\frac{\partial^2 U}{\partial x \partial y} + \frac{\partial^2 V}{\partial x^2} + \frac{\partial W}{\partial x} \frac{\partial^2 W}{\partial x \partial y} \right) L_{nol} \left(\frac{\partial W}{\partial y} \right)$$

$$+ A_{66} \left(\frac{\partial^2 U}{\partial y^2} + \frac{\partial^2 V}{\partial x \partial y} + \frac{\partial^2 W}{\partial x \partial y} \frac{\partial W}{\partial y} + \frac{\partial W}{\partial x} \frac{\partial^2 W}{\partial y^2} \right) L_{nol} \left(\frac{\partial W}{\partial x} \right) \tag{74}$$

Similar to the nanoplate, rising in small-scale parameter reduces the natural frequencies of piezoelectric nanoplates. This is because the stiffness of the nanostructures lessens with increasing the nonlocal parameter and therefore the natural frequencies decline. Moreover, the natural frequencies of piezoelectric nanoplates are sensitive to the change of the applied electric voltage. As the external voltage varies, the natural frequencies also change, and depending on the applied voltage, the natural frequencies of piezoelectric nanoplates may increase or decrease (Ke et al. 2015, Liu et al. 2013, 2018).

16.3.2.2 Nonlocal Elasticity Theory for Magneto-Electro-Elastic Nanoplate

Magneto-Electro-Elastic nanomaterials have better magnetic, electric, mechanical, and physical properties than magneto-electro-elastic bulk composite materials and exhibit more superior magnetoelectric coupling, and therefore are widely used in many applications such as nanoelectronics, mass nanosensors, switchable photovoltaics and NEMS (Kim et al. 2009, Khanmirza et al. 2017, Hosseini et al. 2018). Based on the nonlocal theory, the governing equations of the magneto-electro-elastic nanoplate are presented in this section for the mentioned magneto-electro-elastic materials. It is important to note that the Maxwell equation should be satisfied in the analysis of both macroscale and nanoscale MEE structures. To satisfy Maxwell's equations, the electric and magnetic potential are considered as (Ke and Wang 2014, Jandaghian and Rahmani 2016b, Wang 2002).

$$\phi = -\cos(\beta z)\phi_E(x, y, t) + \frac{2z}{h}V_0,$$

(75)

$$\psi = -\cos(\beta z)\psi_H(x, y, t) + \frac{2z}{h}A_0,$$

(76)

In the classical plate theory, the governing equations of MEE nanoplates are given as (Ke et al. 2014b).

$$-D_{11}\frac{\partial^4 W}{\partial x^4} - 2D_{12}\frac{\partial^4 W}{\partial x^2 \partial y^2} - D_{11}\frac{\partial^4 W}{\partial y^4} - 4D_{66}\frac{\partial^4 W}{\partial x^2 \partial y^2} + E_{31}\left(\frac{\partial^2 \phi_E}{\partial x^2} + \frac{\partial^2 \phi_E}{\partial y^2}\right)$$

$$+ Q_{31}\left(\frac{\partial^2 \psi_M}{\partial x^2} + \frac{\partial^2 \psi_M}{\partial y^2}\right) = (1 - (e_0 a)^2 \nabla^2)I_0\frac{\partial^2 W}{\partial t^2},$$

(77)

$$X_{11}\left(\frac{\partial^2 \phi_E}{\partial x^2} + \frac{\partial^2 \phi_E}{\partial y^2}\right) + Y_{11}\left(\frac{\partial^2 \psi_M}{\partial x^2} + \frac{\partial^2 \psi_M}{\partial y^2}\right) - E_{13}\left(\frac{\partial^2 W}{\partial x^2} + \frac{\partial^2 W}{\partial y^2}\right) - X_{33}\phi_E - Y_{33}\psi_M = 0,$$

(78)

$$Y_{11}\left(\frac{\partial^2 \phi_E}{\partial x^2} + \frac{\partial^2 \phi_E}{\partial y^2}\right) + T_{11}\left(\frac{\partial^2 \psi_M}{\partial x^2} + \frac{\partial^2 \psi_M}{\partial y^2}\right) - Q_{13}\left(\frac{\partial^2 W}{\partial x^2} + \frac{\partial^2 W}{\partial y^2}\right) - Y_{33}\phi_E - T_{33}\psi_M = 0.$$

(79)

where

$$\{D_{11}, D_{12}, D_{66}\} = \left\{\frac{\bar{C}_{11}h^3}{12}, \frac{\bar{C}_{12}h^3}{12}, \frac{\bar{C}_{66}h^3}{12}\right\},$$

$$\{E_{31}, Q_{31}\} = \int_{-h/2}^{h/2} \{\bar{e}_{31}, \bar{q}_{31}\}\beta z \sin(\beta z)\,dz,$$

$$\{X_{11}, Y_{11}, T_{11}\} = \int_{-h/2}^{h/2} \{\bar{s}_{11}, \bar{d}_{11}, \bar{\mu}_{11}\}\cos^2(\beta z)\,dz,$$

$$\{X_{33}, Y_{33}, T_{33}\} = \int_{-h/2}^{h/2} \{\bar{s}_{33}, \bar{d}_{33}, \bar{\mu}_{33}\}[\beta \sin(\beta z)]^2\,dz.$$

(80)

The geometrically nonlinear governing equations of motion of MEE nanoplates based on the FSDT and nonlocal elasticity theory are given as (Ansari and Gholami 2016).

$$A_{11}\left(\frac{\partial^2 u}{\partial x^2} + \frac{\partial w}{\partial x}\frac{\partial^2 w}{\partial x^2}\right) + A_{12}\left(\frac{\partial^2 v}{\partial x \partial y} + \frac{\partial w}{\partial y}\frac{\partial^2 w}{\partial x \partial y}\right)$$

$$+ A_{66}\left(\frac{\partial^2 u}{\partial y^2} + \frac{\partial^2 v}{\partial x \partial y} + \frac{\partial^2 w}{\partial x \partial y}\frac{\partial w}{\partial y} + \frac{\partial w}{\partial x}\frac{\partial^2 w}{\partial y^2}\right)$$

(81)

$$= I_0(1 - (e_0\alpha)^2 \nabla^2)\frac{\partial^2 u}{\partial t^2},$$

$$A_{66}\left(\frac{\partial^2 u}{\partial x \partial y}+\frac{\partial^2 v}{\partial x^2}+\frac{\partial^2 w}{\partial x^2}\frac{\partial w}{\partial y}+\frac{\partial w}{\partial x}\frac{\partial^2 w}{\partial x \partial y}\right)+A_{12}\left(\frac{\partial^2 u}{\partial x \partial y}+\frac{\partial w}{\partial x}\frac{\partial^2 w}{\partial x \partial y}\right)$$

$$+A_{22}\left(\frac{\partial^2 v}{\partial y^2}+\frac{\partial w}{\partial y}\frac{\partial^2 w}{\partial y^2}\right)=I_0(1-(e_0\alpha)^2\nabla^2)\frac{\partial^2 v}{\partial t^2}, \tag{82}$$

$$k_s A_{55}\left(\frac{\partial \psi_x}{\partial x}+\frac{\partial^2 w}{\partial x^2}\right)+k_s A_{44}\left(\frac{\partial \psi_y}{\partial y}+\frac{\partial^2 w}{\partial y^2}\right)-k_s E_{15}\frac{\partial^2 \phi_E}{\partial x^2}-k_s E_{24}\frac{\partial^2 \phi_E}{\partial y^2}$$

$$-k_s Q_{15}\frac{\partial^2 \psi_H}{\partial x^2}-k_s Q_{24}\frac{\partial^2 \psi_H}{\partial y^2}+(1-(e_0\alpha)^2\nabla^2)(Z_1+Z_2) \tag{83}$$

$$=I_0(1-(e_0\alpha)^2\nabla^2)\frac{\partial^2 w}{\partial t^2},$$

$$D_{11}\frac{\partial^2 \psi_x}{\partial x^2}+D_{12}\frac{\partial^2 \psi_y}{\partial x \partial y}+D_{66}\left(\frac{\partial^2 \psi_x}{\partial y^2}+\frac{\partial^2 \psi_y}{\partial x \partial y}\right)-k_s A_{55}\left(\psi_x+\frac{\partial w}{\partial x}\right)$$

$$+(k_s E_{15}+E_{31})\frac{\partial \phi_E}{\partial x}+(k_s Q_{15}+Q_{31})\frac{\partial^2 \psi_H}{\partial x} \tag{84}$$

$$=I_2(1-(e_0\alpha)^2\nabla^2)\frac{\partial^2 \psi_x}{\partial t^2}$$

$$D_{66}\left(\frac{\partial^2 \psi_x}{\partial x \partial y}+\frac{\partial^2 \psi_y}{\partial x^2}\right)+D_{12}\frac{\partial^2 \psi_x}{\partial x \partial y}+D_{22}\frac{\partial^2 \psi_y}{\partial y^2}-k_s A_{44}\left(\psi_y+\frac{\partial w}{\partial y}\right)$$

$$+(k_s E_{24}+E_{32})\frac{\partial \phi_E}{\partial y}+(k_s Q_{24}+Q_{32})\frac{\partial^2 \psi_H}{\partial y} \tag{85}$$

$$=I_2(1-(e_0\alpha)^2\nabla^2)\frac{\partial^2 \psi_y}{\partial t^2},$$

$$E_{15}\left(\frac{\partial \psi_x}{\partial x}+\frac{\partial^2 w}{\partial x^2}\right)+E_{24}\left(\frac{\partial \psi_y}{\partial y}+\frac{\partial^2 w}{\partial y^2}\right)+E_{31}\frac{\partial \psi_x}{\partial x}+E_{32}\frac{\partial \psi_y}{\partial y}+X_{11}\frac{\partial^2 \phi_E}{\partial x^2}$$

$$+X_{22}\frac{\partial^2 \phi_E}{\partial y^2}+Y_{11}\frac{\partial^2 \psi_H}{\partial x^2}+Y_{22}\frac{\partial^2 \psi_H}{\partial y^2}-X_{33}\phi_E-Y_{33}\psi_H=0, \tag{86}$$

$$Q_{15}\left(\frac{\partial \psi_x}{\partial x}+\frac{\partial^2 w}{\partial x^2}\right)+Q_{24}\left(\frac{\partial \psi_y}{\partial y}+\frac{\partial^2 w}{\partial y^2}\right)+Q_{31}\frac{\partial \psi_x}{\partial x}+Q_{32}\frac{\partial \psi_y}{\partial y}+Y_{11}\frac{\partial^2 \phi_E}{\partial x^2}$$

$$+Y_{22}\frac{\partial^2 \phi_E}{\partial y^2}+T_{11}\frac{\partial^2 \psi_H}{\partial x^2}+T_{22}\frac{\partial^2 \psi_H}{\partial y^2}-Y_{33}\phi_E-T_{33}\psi_H=0, \tag{87}$$

where

$$Z_1=\left\{A_{11}\left[\frac{\partial u}{\partial x}+\frac{1}{2}\left(\frac{\partial w}{\partial x}\right)^2\right]+A_{12}\left[\frac{\partial v}{\partial y}+\frac{1}{2}\left(\frac{\partial w}{\partial y}\right)^2\right]\right\}\frac{\partial^2 w}{\partial x^2}+2A_{66}\left(\frac{\partial u}{\partial y}+\frac{\partial v}{\partial x}+\frac{\partial w}{\partial x}\frac{\partial w}{\partial y}\right)\frac{\partial^2 w}{\partial x \partial y}$$

$$+\left\{A_{12}\left[\frac{\partial u}{\partial y}+\frac{1}{2}\left(\frac{\partial w}{\partial y}\right)^2\right]+A_{22}\left[\frac{\partial v}{\partial y}+\frac{1}{2}\left(\frac{\partial w}{\partial y}\right)^2\right]\right\}\frac{\partial^2 w}{\partial y^2}, \tag{88}$$

$$Z_2 = \left\{ A_{66} \left(\frac{\partial^2 u}{\partial y^2} + \frac{\partial^2 v}{\partial x \partial y} + \frac{\partial^2 w}{\partial x \partial y} \frac{\partial w}{\partial y} + \frac{\partial w}{\partial x} \frac{\partial^2 w}{\partial y^2} \right) + A_{11} \left(\frac{\partial^2 u}{\partial x^2} + \frac{\partial w}{\partial x} \frac{\partial^2 w}{\partial x^2} \right) \right.$$

$$\left. + A_{12} \left(\frac{\partial^2 v}{\partial x \partial y} + \frac{\partial w}{\partial y} \frac{\partial^2 w}{\partial x \partial y} \right) \right\} \frac{\partial w}{\partial x} + \left\{ A_{66} \left(\frac{\partial^2 u}{\partial x \partial y} + \frac{\partial^2 v}{\partial x^2} + \frac{\partial^2 w}{\partial x^2} \frac{\partial w}{\partial y} + \frac{\partial w}{\partial x} \frac{\partial^2 w}{\partial x \partial y} \right) \right.$$

$$\left. + A_{12} \left(\frac{\partial^2 u}{\partial x \partial y} + \frac{\partial w}{\partial x} \frac{\partial^2 w}{\partial x \partial y} \right) + A_{22} \left(\frac{\partial^2 v}{\partial y^2} \frac{\partial w}{\partial y} \frac{\partial^2 w}{\partial y^2} \right) \right\} \frac{\partial w}{\partial y} \qquad (89)$$

As mentioned before, the natural frequency of MEE nanoplates decrease with the rise in the nonlocal parameter. By setting the magnetoelectric, piezomagnetic, and magnetic constants to zero (i.e. $d_{ij} = q_{ij} = \mu_{ij} = 0$), the MEE nanoplate converted to the nonlocal piezoelectric and also MEE nanoplate converted to the nonlocal elastic nanoplate if we put $d_{ij} = q_{ij} = \mu_{ij} = e_{ij} = s_{ij} = 0$. The natural frequencies of MEE nanoplates are completely sensitive to the external magnetic, electric and mechanical loading, whereas it is not sensitive to the thermal loading.

16.3.3 Nonlocal Elasticity Theory for Shell

There are several theories to investigate shells problems. Among these theories, two important theories have been generally used in the studies to analyze shells: classical Kirchhoff-Love shell theory and first-order shear deformation theory (FSDT) (Wang and Varadan 2007, Reddy 2003). Based on the classical shell theory, there are three displacement components along the x, θ, and z axes, represented by u_x, u_y, and u_z, respectively (Ke et al. 2014c).

$$u_x(x, \theta, z, t) = u(x, \theta, t) - z \frac{\partial w}{\partial x}(x, \theta, t),$$

$$u_y(x, \theta, z, t) = v(x, \theta, t) - z \frac{\partial w}{\partial \theta}(x, \theta, t), \qquad (90)$$

$$u_z(x, \theta, z, t) = w(x, \theta, t),$$

The governing equations of shell based on the Kirchhoff-Love shell, with radius R, length L, and thickness h is expressed as (Arash et al. 2015).

$$\frac{\partial N_{xx}}{\partial x} + \frac{1}{R} \frac{\partial N_{x\theta}}{\partial \theta} = \rho h \frac{\partial^2 u}{\partial t^2},$$

$$\frac{1}{R} \frac{\partial N_{\theta\theta}}{\partial \theta} + \frac{\partial N_{x\theta}}{\partial x} + \frac{1}{R^2} \frac{\partial M_{\theta\theta}}{\partial \theta} + \frac{1}{R} \frac{\partial M_{\theta\theta}}{\partial x} = \rho h \frac{\partial^2 v}{\partial t^2}, \qquad (91)$$

$$\frac{\partial^2 M_{xx}}{\partial x^2} + \frac{1}{R^2} \frac{\partial^2 M_{\theta\theta}}{\partial \theta^2} + \frac{2}{R^2} \frac{\partial^2 M_{x\theta}}{\partial x \partial \theta} - \frac{N_{\theta\theta}}{R} = \rho h \frac{\partial^2 w}{\partial t^2},$$

in which N_{xx}, $N_{\theta x}$, $N_{\theta\theta}$ are resultant forces and M_{xx}, $M_{\theta x}$, $M_{\theta\theta}$ are bending moments; x and θ are the longitudinal and angular circumferential coordinates, respectively.

In the first-order shear deformation theory, the displacements u_x, u_y, and u_z at any point in the x, θ, and z axis, respectively, are (Arash et al. 2015).

$$u_x(x, \theta, z, t) = u(x, \theta, t) - z \psi_x(x, \theta, t),$$
$$u_y(x, \theta, z, t) = v(x, \theta, t) - z \psi_\theta(x, \theta, t),$$
$$u_z(x, \theta, z, t) = w(x, \theta, z, t), \qquad (92)$$

ψ_x and ψ_θ are rotations of normal to the mid-surface about the x and y-axis, respectively. The governing equations of a shell based on the Donnell shell theory are expressed as

$$\frac{\partial N_{xx}}{\partial x} + \frac{1}{R}\frac{\partial N_{x\theta}}{\partial \theta} = I_0\ddot{u} + I_1\ddot{\psi}_x,$$

$$\frac{\partial N_{x\theta}}{\partial x} + \frac{1}{R}\frac{\partial N_{\theta\theta}}{\partial \theta} + \frac{Q_{\theta\theta}}{R} = I_0\ddot{v} + I_1\ddot{\psi}_\theta$$

$$\frac{\partial Q_{xx}}{\partial x} + \frac{1}{R}\frac{\partial Q_{\theta\theta}}{\partial \theta} - \frac{N_{\theta\theta}}{R} = I_0\ddot{w},$$

$$\frac{\partial M_{xx}}{\partial x} + \frac{1}{R}\frac{\partial M_{x\theta}}{\partial \theta} + Q_{xx} = I_1\ddot{u} + I_2\ddot{\psi}_x,$$

$$\frac{\partial M_{x\theta}}{\partial x} + \frac{1}{R}\frac{\partial M_{\theta\theta}}{\partial \theta} + Q_{\theta\theta} = I_1\ddot{v} + I_2\ddot{\psi}_\theta, \tag{93}$$

where in equation (93), Q_{xx} and $Q_{\theta\theta}$ are stress resultant; $I_i = \int\limits_{-h/2}^{h2} \rho z^i dz$ is inertia terms.

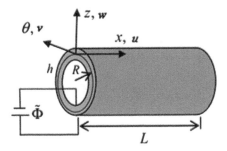

Fig. 16.6 Schematic of a piezoelectric cylindrical shell

The nonlocal cylindrical shell theory, based on the Flugge shell theory (Flügge 2013), is given as (Hu et al. 2008).

$$\frac{\partial^2 u}{\partial x^2} + \frac{1}{2}(1-v)\frac{\partial^2 u}{\partial \theta^2} + \frac{1}{2R}(1+v)\frac{\partial^2 v}{\partial x\partial \theta} - \frac{v}{R}\frac{\partial w}{\partial x}$$

$$+ (1-v^2)\frac{D}{EhR^4}\left(\frac{1}{2}(1-v)\frac{\partial^2 u}{\partial \theta^2} + R^3\frac{\partial^3 w}{\partial x^3} - \frac{R}{2}(1-v)\frac{\partial^3 w}{\partial x\partial \theta^2}\right) \tag{94}$$

$$= \rho h(1-v^2)(1-(e_0\alpha)^2\nabla^2)\frac{\partial^2 u}{\partial t^2}$$

$$\frac{1}{2R}(1-v)\frac{\partial^2 u}{\partial x\partial \theta} + \frac{1}{2}(1-v)\frac{\partial^2 v}{\partial x^2} + \frac{\partial^2 v}{\partial \theta^2} - \frac{\partial w}{\partial \theta}$$

$$+ (1-v^2)\frac{D}{EhR^2}\left(\frac{3}{2}(1-v)\frac{\partial^2 v}{\partial x^2} + \frac{1}{2}(3-v)\frac{\partial^3 w}{\partial x^2\partial \theta}\right) \tag{95}$$

$$= \frac{\rho h}{Eh}(1-v^2)(1-(e_0\alpha)^2\nabla^2)\frac{\partial^2 v}{\partial t^2}$$

$$\frac{v}{R}\frac{\partial u}{\partial x}+\frac{1}{R^2}\frac{\partial v}{\partial \theta}-\frac{w}{R^2}$$

$$-(1-v^2)\frac{D}{EhR^4}\left(R^4\nabla^2\nabla^2 w+R^3\frac{\partial^3 u}{\partial x^3}-\frac{R}{2}(1-v)\frac{\partial^3 u}{\partial x\partial\theta^2}+\frac{R^2}{2}(3-v)\frac{\partial^3 v}{\partial x^2\partial\theta}+w+2\frac{\partial^2 w}{\partial\theta^2}\right) \quad (96)$$

$$=\frac{1}{Eh}(1-v^2)(1-(e_0\alpha)^2\nabla^2)\left(\frac{\partial^2 w}{\partial t^2}\rho h-p\right).$$

16.3.3.1 Nonlocal Elasticity Theory for Piezoelectric Nanoshell

The governing equations of the piezoelectric nanoshells are presented in this section based on the nonlocal theory for the mentioned piezoelectric materials (Fig. 16.6). As mentioned before, the distribution of the electric potential should satisfy the Maxwell equation. Hence, the following electric potential is assumed (Wang 2002, Quek and Wang 2000).

$$\Phi(x,\theta,z,t)=-\cos(\beta z)\phi(x,\theta,t)+\frac{2z}{h}V_0, \quad (97)$$

Based on the Kirchhoff-Love hypothesis, the governing equations of piezoelectric nanoshells can be obtained as (Ke et al. 2014).

$$A_{11}\frac{\partial^2 U}{\partial x^2}+\frac{A_{12}}{R}\left(\frac{\partial^2 V}{\partial x\partial\theta}+\frac{\partial W}{\partial x}\right)+\frac{A_{66}}{R}\left(\frac{1}{R}\frac{\partial^2 U}{\partial\theta^2}+\frac{\partial^2 V}{\partial x\partial\theta}\right)=[1-(e_0\alpha)^2\nabla^2]I_1\frac{\partial^2 U}{\partial t^2} \quad (98)$$

$$\frac{A_{12}}{R}\frac{\partial^2 U}{\partial x\partial\theta}+\frac{A_{11}}{R^2}\left(\frac{\partial^2 V}{\partial\theta^2}-\frac{\partial W}{\partial\theta}\right)+A_{66}\left(\frac{1}{R}\frac{\partial^2 U}{\partial x\partial\theta}+\frac{\partial^2 V}{\partial x^2}\right)-\frac{D_{66}}{R^2}\left(\frac{2\partial^3 W}{\partial x^2\partial\theta}-\frac{\partial^2 V}{\partial x^2}\right)$$

$$-\frac{D_{12}}{R^2}\frac{\partial^3 W}{\partial x^2\partial\theta}-\frac{D_{11}}{R^4}\left(\frac{\partial^3 W}{\partial\theta^3}-\frac{\partial^2 V}{\partial\theta^2}\right)+\frac{E_{31}}{R^2}\frac{\partial\Phi}{\partial\theta}+\frac{Q_{31}}{R^2}\frac{\partial\Psi}{\partial\theta} \quad (99)$$

$$=[1-(e_0\alpha)^2\nabla^2]I_1\frac{\partial^2 V}{\partial t^2},$$

$$-D_{11}\frac{\partial^4 W}{\partial x^4}-\left(\frac{2D_{12}}{R^2}+\frac{4D_{66}}{R^2}\right)\frac{\partial^4 W}{\partial x^2\partial\theta^2}-\frac{D_{11}}{R^4}\frac{\partial^4 W}{\partial\theta^4}+E_{31}\frac{\partial^2\Phi}{\partial x^2}+\frac{E_{31}}{R^2}\frac{\partial^2\Phi}{\partial\theta^2} \quad (100)$$

$$-D_{11}\frac{\partial^4 W}{\partial x^4}-\left(\frac{2D_{12}}{R^2}+\frac{4D_{66}}{R^2}\right)\frac{\partial^4 W}{\partial x^2\partial\theta^2}-\frac{D_{22}}{R^4}\frac{\partial^4 W}{\partial\theta^4}+E_{31}\frac{\partial^2\Phi}{\partial x^2}+\frac{E_{32}}{R^2}\frac{\partial^2\Phi}{\partial\theta^2}$$

$$-\frac{A_{12}}{R}\frac{\partial U}{\partial x}-\frac{A_{22}}{R^2}\left(\frac{\partial V}{\partial\theta}+W\right)+\left(\frac{D_{12}}{R^2}+\frac{2D_{66}}{R^2}\right)\frac{\partial^3 V}{\partial x^2\partial\theta}+\frac{D_{22}}{R^4}\frac{\partial^3 V}{\partial\theta^3} \quad (101)$$

$$=[1-(e_0\alpha)^2\nabla^2]I_1\frac{\partial^2 W}{\partial t^2},$$

$$X_{11}\frac{\partial^2\Phi}{\partial x^2}+X_{22}\frac{\partial^2\Phi}{\partial\theta^2}-X_{33}\Phi-E_{31}\frac{\partial^2 W}{\partial x^2}-\frac{E_{32}}{R^2}\left(\frac{\partial^2 W}{\partial\theta^2}-\frac{\partial V}{\partial\theta}\right)=0 \quad (102)$$

in which

$$\{A_{11}, A_{12}, A_{66}\} = \{\tilde{c}_{11}h, \tilde{c}_{12}h, \tilde{c}_{22}h, \tilde{c}_{66}h\},$$

$$\{D_{11}, D_{12}, D_{66}\} = \left\{\frac{\tilde{c}_{11}h^3}{12}, \frac{\tilde{c}_{12}h^3}{12}, \frac{\tilde{c}_{22}h^3}{12}, \frac{\tilde{c}_{66}h^3}{12}\right\},$$

$$\{E_{31}, E_{32}\} = \int_{-h/2}^{h/2} \{\tilde{e}_{31}, \tilde{e}_{32}\}\beta z \sin(\beta z)\, dz, \quad X_{11} = \int_{-h/2}^{h/2} \tilde{s}_{11} \cos^2(\beta z), \tag{103}$$

$$X_{22} = \int_{-h/2}^{h/2} \tilde{s}_{22}\left[\frac{\cos(\beta z)}{R+z}\right]^2 dz, \quad X_{33} = \int_{-h/2}^{h/2} \tilde{s}_{33}[\beta \sin(\beta z)]^2\, dz,$$

The nonlinear equations of piezoelectric Kirchhoff-Love nanoshells based on the nonlocal theory and von Kármán theory are written (Wang et al. 2019) as

$$A_{11}\left(\frac{\partial^2 v}{R^2 \partial\theta^2} + \frac{\partial w}{R\partial\theta}\frac{\partial^2 w}{R^2 \partial\theta^2} + \frac{1}{R^2}\frac{\partial w}{\partial\theta}\right) + A_{12}\left(\frac{\partial^2 u}{R\partial x\partial y} + \frac{\partial w}{R\partial\theta}\frac{\partial^2 w}{R\partial x\partial\theta}\right)$$

$$+ A_{66}\left(\frac{\partial^2 u}{R\partial x\partial\theta} + \frac{\partial^2 v}{\partial x^2} + \frac{\partial w}{R\partial\theta}\frac{\partial^2 w}{\partial x^2} + \frac{\partial w}{\partial x}\frac{\partial^2 w}{R\partial x\partial\theta}\right) \tag{104}$$

$$= (1 - (e_0\alpha)^2 \nabla^2)\frac{\partial^2 v}{\partial t^2},$$

$$- D_{11}\frac{\partial^4 w}{\partial x^4} - 2D_{12}\frac{\partial^4 w}{R^2 \partial x^2 \partial\theta^2} - D_{11}\frac{\partial^4 w}{R^4 \partial\theta^4} + E_{31}\frac{\partial^2 \Phi}{\partial x^2} + E_{32}\frac{\partial^2 \Phi}{R^2 \partial\theta^2}$$

$$- 4D_{66}\frac{\partial^4 w}{R^2 \partial x^2 \partial\theta^2} + Z_1 + Z_2$$

$$- \frac{1}{R}\left(A_{12}\left(\frac{\partial u}{\partial x} + \frac{1}{2}\left(\frac{\partial w}{\partial x}\right)^2\right) + A_{11}\left(\frac{\partial v}{R\partial\theta} + \frac{1}{2}\left(\frac{\partial w}{R\partial\theta}\right)^2 + \frac{w}{R}\right)\right) \tag{105}$$

$$= (1 - (e_0\alpha)^2 \nabla^2)I_1\frac{\partial^2 w}{\partial t^2},$$

$$X_{11}\frac{\partial^2 \Phi}{\partial x^2} + X_{22}\frac{\partial^2 \Phi}{R^2 \partial\theta^2} - E_{31}\left(\frac{\partial^2 w}{\partial x^2} + \frac{\partial^2 w}{R^2 \partial\theta^2}\right) - X_{33}\Phi = 0 \tag{106}$$

where

$$Z_1 = \left(A_{11}\left(\frac{\partial u}{\partial x} + \frac{1}{2}\left(\frac{\partial w}{\partial x}\right)^2\right) + A_{12}\left(\frac{\partial v}{R\partial\theta} + \frac{1}{2}\left(\frac{\partial w}{R\partial\theta}\right)^2 + \frac{w}{R}\right)\right)(1 - (e_0\alpha)^2 \nabla^2)\frac{\partial^2 w}{\partial x^2}$$

$$+ \left(A_{11}\left(\frac{\partial^2 u}{\partial x^2} + \frac{\partial w}{\partial x}\frac{\partial^2 w}{\partial x^2}\right) + A_{12}\left(\frac{\partial^2 v}{R\partial x\partial\theta} + \frac{\partial w}{R\partial\theta}\frac{\partial^2 w}{R\partial x\partial\theta} + \frac{1}{R}\frac{\partial w}{\partial x}\right)\right)(1 - (e_0\alpha)^2 \nabla^2)\frac{\partial w}{\partial x}$$

$$+ \left(A_{12}\left(\frac{\partial u}{\partial x} + \frac{1}{2}\left(\frac{\partial w}{\partial x}\right)^2\right) + A_{11}\left(\frac{\partial v}{R\partial\theta} + \frac{1}{2}\left(\frac{\partial w}{\partial x}\right)^2 + \frac{w}{R}\right)\right)(1 - (e_0\alpha)^2 \nabla^2)\frac{\partial^2 w}{R^2 \partial y^2} \tag{107}$$

$$+ \left[A_{12} \left(\frac{\partial^2 u}{\partial x^2} + \frac{\partial w}{\partial x} \frac{\partial^2 w}{R \partial x \partial \theta} \right) + A_{11} \left(\frac{\partial^2 v}{R^2 \partial \theta^2} + \frac{\partial w}{\partial x} \frac{\partial^2 w}{R^2 \partial \theta^2} + \frac{1}{R^2} \frac{\partial w}{\partial \theta} \right) \right] (1 - (e_0 \alpha)^2 \nabla^2) \frac{\partial w}{R \partial \theta},$$

$$Z_2 = 2A_{66} \left(\frac{\partial u}{R \partial \theta} + \frac{\partial v}{\partial x} + \frac{\partial w}{\partial x} \frac{\partial w}{R \partial \theta} \right) (1 - (e_0 \alpha)^2 \nabla^2) \frac{\partial^2 W}{R \partial x \partial \theta}$$

$$+ A_{66} \left(\frac{\partial^2 u}{R \partial x \partial \theta} + \frac{\partial^2 v}{\partial x^2} + \frac{\partial^2 w}{\partial x^2} \frac{\partial w}{R \partial \theta} + \frac{\partial w}{\partial x} \frac{\partial^2 w}{R \partial x \partial \theta} \right) (1 - (e_0 \alpha)^2 \nabla^2) \frac{\partial W}{R \partial \theta} \tag{108}$$

$$+ A_{66} \left(\frac{\partial^2 u}{R^2 \partial \theta^2} + \frac{\partial^2 v}{R \partial x \partial \theta} + \frac{\partial^2 w}{R \partial x \partial \theta} \frac{\partial w}{R \partial \theta} + \frac{\partial w}{\partial x} \frac{\partial^2 w}{R^2 \partial \theta^2} \right) (1 - (e_0 \alpha)^2 \nabla^2) \frac{\partial w}{\partial x}.$$

Studies in this field show that natural frequencies of piezoelectric nanoshells decrease when the nonlocal parameter increases. At higher circumferential mode numbers, the size effect is more evident than that of lower circumferential mode numbers. The natural frequencies of piezoelectric nanoshells are sensitive to the external electric potential (Ke et al. 2014, Wang et al. 2019).

16.3.3.2   *Nonlocal Elasticity Theory for Magneto-Electro-Elastic Nanoshells*

Based on the nonlocal theory, the governing equations of the magneto-electro-elastic nanoshells are presented in this section for the mentioned magneto-electro-elastic materials. In the analysis of both macroscale and nanoscale MEE structures, the basic assumption of the electric potential and the magnetic potential is that it should satisfy the Maxwell equation. To satisfy Maxwell's equations, the electric and magnetic potential are considered as (Ke and Wang 2014, Jandaghian and Rahmani 2016a, Wang 2002).

$$\phi(x, \theta, z, t) = -\cos(\beta z) \phi_E(x, \theta, t) + \frac{2z}{h} V_0, \tag{109}$$

$$\psi(x, \theta, z, t) = -\cos(\beta z) \psi_H(x, \theta, t) + \frac{2z}{h} A_0 \tag{110}$$

According to Love's shell theory, the governing equations of MEE nanoshells are given as (Ke et al. 2014).

$$A_{11} \frac{\partial^2 U}{\partial x^2} + \frac{A_{12}}{R} \left(\frac{\partial^2 V}{\partial x \partial \theta} + \frac{\partial W}{\partial x} \right) + \frac{A_{66}}{R} \left(\frac{1}{R} \frac{\partial^2 U}{\partial \theta^2} + \frac{\partial^2 V}{\partial x \partial \theta} \right) = [1 - (e_0 \alpha)^2 \nabla^2] I_1 \frac{\partial^2 U}{\partial t^2} \tag{111}$$

$$\frac{A_{12}}{R} \frac{\partial^2 U}{\partial x \partial \theta} + \frac{A_{11}}{R^2} \left(\frac{\partial^2 V}{\partial \theta^2} - \frac{\partial W}{\partial \theta} \right) + A_{66} \left(\frac{1}{R} \frac{\partial^2 U}{\partial x \partial \theta} + \frac{\partial^2 V}{\partial x^2} \right) - \frac{D_{66}}{R^2} \left(\frac{2 \partial^3 W}{\partial x^2 \partial \theta} - \frac{\partial^2 V}{\partial x^2} \right)$$

$$- \frac{D_{12}}{R^2} \frac{\partial^3 W}{\partial x^2 \partial \theta} - \frac{D_{11}}{R^4} \left(\frac{\partial^3 W}{\partial \theta^3} - \frac{\partial^2 V}{\partial \theta^2} \right) + \frac{E_{31}}{R^2} \frac{\partial \Phi}{\partial \theta} + \frac{Q_{31}}{R^2} \frac{\partial \psi}{\partial \theta} \tag{112}$$

$$= [1 - (e_0 \alpha)^2 \nabla^2] I_1 \frac{\partial^2 V}{\partial t^2},$$

$$+ Q_{31} \frac{\partial^2 \psi}{\partial x^2} + \frac{Q_{31}}{R^2} \frac{\partial^2 \psi}{\partial \theta^2} - \frac{A_{12}}{R} \frac{\partial^2 U}{\partial x} - \frac{A_{22}}{R^2} \left(\frac{\partial V}{\partial \theta} + W \right) + \left(\frac{D_{12}}{R^2} + \frac{4 D_{66}}{R^2} \right) \frac{\partial^3 V}{\partial x^2 \partial \theta^2}$$

$$+ \frac{D_{11}}{R^4} \frac{\partial^3 V}{\partial \theta^3} = [1 - (e_0 \alpha)^2 \nabla^2] I_1 \frac{\partial^2 W}{\partial t^2}, \tag{113}$$

$$X_{11}\left(\frac{\partial^2 \Phi}{\partial x^2} + \frac{\partial^2 \Phi}{\partial \theta^2}\right) + Y_{11}\left(\frac{\partial^2 \Psi}{\partial x^2} - \frac{\partial \Psi}{\partial \theta^2}\right) - X_{33}\Phi - Y_{33}\Psi - E_{31}\frac{\partial^2 W}{\partial x^2} - \frac{E_{31}}{R^2}\left(\frac{\partial^2 W}{\partial \theta^2} - \frac{\partial V}{\partial \theta}\right) = 0,$$

(114)

$$Y_{11}\left(\frac{\partial^2 \Phi}{\partial x^2} + \frac{\partial^2 \Phi}{\partial \theta^2}\right) + T_{11}\left(\frac{\partial^2 \Psi}{\partial x^2} + \frac{\partial^2 \Psi}{\partial \theta^2}\right) - Y_{33}\Phi - T_{33}\Psi - Q_{31}\frac{\partial^2 W}{\partial x^2}$$

$$- \frac{Q_{31}}{R^2}\left(\frac{\partial^2 W}{\partial \theta^2} - \frac{\partial V}{\partial \theta}\right) = 0.$$

(115)

where

$$\{A_{11}, A_{12}, A_{66}\} = \{\tilde{c}_{11}h, \tilde{c}_{12}h, \tilde{c}_{22}h, \tilde{c}_{66}h\},$$

$$\{D_{11}, D_{12}, D_{66}\} = \left\{\frac{\tilde{c}_{11}h^3}{12}, \frac{\tilde{c}_{12}h^3}{12}, \frac{\tilde{c}_{22}h^3}{12}, \frac{\tilde{c}_{66}h^3}{12}\right\},$$

$$\{E_{31}, Q_{31}\} = \int_{-h/2}^{h/2} \{\tilde{e}_{31}, \tilde{q}_{31}\}\beta z \sin(\beta z)\, dz, \quad X_{11} = \int_{-h/2}^{h/2} \tilde{s}_{11} \cos^2(\beta z)\, dz$$

$$X_{22} = \int_{-h/2}^{h/2} \tilde{s}_{22}\left[\frac{\cos(\beta z)}{R+z}\right]^2 dz, \quad X_{33} = \int_{-h/2}^{h/2} \tilde{s}_{33}[\beta \sin(\beta z)]^2\, dz, \quad Y_{11} = \int_{-h/2}^{h/2} \tilde{d}_{11} \cos^2(\beta z)\, dz,$$

$$Y_{22} = \int_{-h/2}^{h/2} \tilde{d}_{22}\left[\frac{\cos(\beta z)}{R+z}\right]^2 dz,$$

(116)

$$Y_{33} = \int_{-h/2}^{h/2} \tilde{d}_{33}[\beta \sin(\beta z)]^2\, dz, \quad T_{11} = \int_{-h/2}^{h/2} \tilde{\mu}_{11} \cos^2(\beta z)\, dz, \quad T_{22} = \int_{-h/2}^{h/2} \tilde{\mu}_{22}\left[\frac{\cos(\beta z)}{R+z}\right]^2 dz$$

$$T_{33} = \int_{-h/2}^{h/2} \tilde{\mu}_{33}[\beta \sin(\beta z)]^2\, dz.$$

Numerical results on MEE nanoshells reveal that the natural frequencies decrease with an increase in the nonlocal parameter. The natural frequencies of MEE nanoshells are sensitive to the external electrical and magnetic potentials and temperature changes. An increase of the length-to-radius ratio leads to the decrease of fundamental frequency of MEE nanoshells.

16.4 SUMMARY

In this chapter, a general overview of the applications of intelligent materials in nano/micro sensors and biosensors was conducted. Recent growth in smart materials and nanotechnology has led to the widespread use of these materials such as piezoelectric and piezomagnetic in micro/nano-devices. However, understanding the mechanical behavior of these micro/nano-devices plays an important role in their design. At nanoscale, in addition to experimental methods that are powerful and costly, there are other methods to model nanostructures such as continuum mechanics and molecular dynamics (MD). The nonlocal elasticity theory gives a useful method for modeling nanostructures, which is able to consider the scale effect in the study of the nanostructures. These benefits have led to the nonlocal continuum modeling being used in many researches to simulate nanostructures.

However, the success of nonlocal continuum model completely depends on the correct value of nonlocal parameter and so calibrating this value is very important. The nano-sensors are classified into six major groups based on the parameters that are distinguished by them: mechanical, electrical, optical, magnetic, chemical, and thermal. In recent years, with the development of nanotechnology, attention has been concentrated on the replacement of conventional sensors and actuators with smart ceramics such as nano/micro piezoelectric and magneto-electro-elastic. Miniature size, compressed structure, low energy usage, lightweight, high sensitivity, and high-performance are advantages of smart ceramics. These advantages have led to the development of biosensors› self-powered systems that have functions such as persistent health monitoring and real-time detection. Apart from the mentioned advantages, there are many challenges in the production of piezoelectric nano-sensors such as the characterization, the modeling, and the synthesis. In summary, although there are many problems related to nanodevices, the existence of advantages of smart ceramics and the rising interest in the combination of these nano-devices with biology can cause the widespread use of smart ceramics in nano/microsensors and biosensors.

16.5 REFERENCES

Aksencer, T. and M. Aydogdu. 2011. Levy type solution method for vibration and buckling of nanoplates using nonlocal elasticity theory. Physica E: Low Dimens. Syst. Nanostruct. 43(4): 954-959.

Alaimo, A., I. Benedetti and A. Milazzo. 2014. A finite element formulation for large deflection of multilayered magneto-electro-elastic plates. Compos. Struct. 107: 643-653.

Alibeigloo, A. 2009. Static analysis of a functionally graded cylindrical shell with piezoelectric layers as sensor and actuator. Smart Mater. Struct. 18(6): 065004.

Annigeri, A.R., N. Ganesan and S. Swarnamani. 2007. Free vibration behaviour of multiphase and layered magneto-electro-elastic beam. J. Sound Vib. 299(1-2): 44-63.

Ansari, R., M. Faraji Oskouie, R. Gholami and F. Sadeghi. 2016. Thermo-electro-mechanical vibration of postbuckled piezoelectric Timoshenko nanobeams based on the nonlocal elasticity theory. Compos. B. Eng. 89: 316-327.

Ansari, R. and R. Gholami. 2016. Size-dependent nonlinear vibrations of first-order shear deformable magneto-electro-thermo elastic nanoplates based on the nonlocal elasticity Theory. Int. J. Appl. Mech. 08(04): 1650053.

Ansari, R. and R. Gholami. 2017. Size-dependent buckling and postbuckling analyses of first-order shear deformable magneto-electro-thermo elastic nanoplates based on the nonlocal elasticity theory. Int. J. Struct. Stab. Dyn. 17(01): 1750014.

Arani, A. Gh., R. Kolahchi and S.A. Mortazavi. 2014. Nonlocal piezoelasticity based wave propagation of bonded double-piezoelectric nanobeam-systems. Int. J. Mech. Mater. Des. 10(2): 179-191.

Arash, B. and R. Ansari. 2010. Evaluation of nonlocal parameter in the vibrations of single-walled carbon nanotubes with initial strain. Physica E: Low Dimens. Syst. Nanostruct. 42(8): 2058-2064.

Arash, B., J.W. Jiang and T. Rabczuk. 2015. A review on nanomechanical resonators and their applications in sensors and molecular transportation. Appl. Phys. Rev. 2(2): 021301.

Arefi, M. and T. Rabczuk. 2019. A nonlocal higher order shear deformation theory for electro-elastic analysis of a piezoelectric doubly curved nano shell. Compos. B. Eng. 168: 496-510.

Arefi, M. 2020. Electro-mechanical vibration characteristics of piezoelectric nano shells. Thin-Walled Struct. 155: 106912.

Askari Farsangi, M.A. and A.R. Saidi. 2012. Levy type solution for free vibration analysis of functionally graded rectangular plates with piezoelectric layers. Smart Mater. Struct. 21(9): 094017.

Batra, R.C., X.Q. Liang and J.S. Yang. 1996. Shape control of vibrating simply supported rectangular plates. AIAA J. 34(1): 116-122.

Bhangale, R.K. and N. Ganesan. 2005. Free vibration studies of simply supported non-homogeneous functionally graded magneto-electro-elastic finite cylindrical shells. J. Sound Vib. 288(1): 412-422.

Bhangale, R.K. and N. Ganesan. 2006. Free vibration of simply supported functionally graded and layered magneto-electro-elastic plates by finite element method. J. Sound Vib. 294(4): 1016-1038.

Blaser, E.M., W.H. Ko and E.T. Yon. 1971. A miniature digital pressure transducer. Paper read at Proc. 24th Annu. Conf. Engineering Medicine and Biology.

Budakian, R., H.J. Mamin and D. Rugar. 2006. Spin manipulation using fast cantilever phase reversals. Appl. Phys. Lett. 89(11): 113113.

Chauhan, S., A. Gupta and Ch. kumar. 2019. A study on buckling of piezoelectric circular cylindrical shell. IOP Conf. Ser. Mater. Sci. Eng. 691: 012018.

Chen, J., H. Chen, E. Pan and P.R. Heyliger. 2007. Modal analysis of magneto-electro-elastic plates using the state-vector approach. J. Sound Vib. 304(3): 722-734.

Chen, J.Y., P.R. Heyliger and E. Pan. 2014. Free vibration of three-dimensional multilayered magneto-electro-elastic plates under combined clamped/free boundary conditions. J. Sound Vib. 333(17): 4017-4029.

Dhuri, K.D. and P. Seshu. 2007. Favorable locations for piezo actuators in plates with good control effectiveness and minimal change in system dynamics. Smart Mater. Struct. 16(6): 2526.

Ding, H.J., H.M. Wang and P.F. Hou. 2003. The transient responses of piezoelectric hollow cylinders for axisymmetric plane strain problems. Int. J. Solids. Struct. 40(1): 105-123.

Dosch, J.J., D.J. Inman and E. Garcia. 1992. A self-sensing piezoelectric actuator for collocated control. J. Intell. Mater. Syst. Struct. 3(1): 166-185.

Duan, W.H., S.T. Quek and Q. Wang. 2005. Free vibration analysis of piezoelectric coupled thin and thick annular plate. J. Sound Vib. 281(1): 119-139.

Ebrahimi, F. and A. Rastgoo. 2008. Free vibration analysis of smart annular FGM plates integrated with piezoelectric layers. Smart Mater. Struct. 17(1): 015044.

Ebrahimi, F., M. Dehghan and A. Seyfi. 2019a. Eringen's nonlocal elasticity theory for wave propagation analysis of magneto-electro-elastic nanotubes. Adv. Nano. Res. 7(1): 1.

Ebrahimi, F., M. Karimiasl and A. Singhal. 2019b. Magneto-electro-elastic analysis of piezoelectric-flexoelectric nanobeams rested on silica aerogel foundation. Eng. Comput. 1-8.

Eltaher, M.A., A.E. Alshorbagy and F.F. Mahmoud. 2013. Vibration analysis of Euler-Bernoulli nanobeams by using finite element method. Appl. Math. Model. 37(7): 4787-4797.

Eringen, A. Cemal. 1972. Nonlocal polar elastic continua. Int. J. Eng. Sci. 10(1): 1-16.

Eringen AC. 1983. On differential equations of nonlocal elasticity and solutions of screw dislocation and surface waves. Int. J. Appl. Phys. 54: 4703-4710.

Fakhrabadi, M.M.S., P. Khoddam Khorasani, A. Rastgoo and M.T. Ahmadian. 2013. Molecular dynamics simulation of pull-in phenomena in carbon nanotubes with Stone-Wales defects. Solid State Commun. 157: 38-44.

Falconi, C., A. D'Amico and Z.L. Wang. 2006. Wireless nanosensors and nanoactuators for *in vivo* biomedical applications. Proceedings of the Eurosensors XX, Goteborg, Sweden.

Falconi, Ch., Arnaldo D'Amico and Zh. Lin Wang. 2007. Wireless joule nanoheaters. Sens. Actuators B Chem. 127(1): 54-62.

Falconi, Ch. 2019. Piezoelectric nanotransducers. Nano Energy 59: 730-744.

Farajpour, A., A. Rastgoo and M.R. Farajpour. 2017. Nonlinear buckling analysis of magneto-electro-elastic CNT-MT hybrid nanoshells based on the nonlocal continuum mechanics. Compos. Struct. 180: 179-191.

Flügge, W. 2013. Stresses in shells: Springer Science and Business Media.

Fuller, Ch. C., Sh. Elliott and Ph. A. Nelson. 1996. Active control of vibration: Academic Press.

Gao, L., K. Cao and X. Hu. 2020. Nano electromechanical approach for flexible piezoresistive sensor. Appl. Mater. Today 18: 100475.

Ha, D., W.N. de Vries, S.W.M. John, P.P. Irazoqui and W.J. Chappell. 2012. Polymer-based miniature flexible capacitive pressure sensor for intraocular pressure (IOP) monitoring inside a mouse eye. Biomed. Microdevices 14(1): 207-215.

He, J.-H. 2001. Variational theory for linear magneto-electro-elasticity. Int. J. Nonlinear Sci. Numer. Simul. 2(4): 309-316.

He, X.Q., S. Kitipornchai and K.M. Liew. 2005. Resonance analysis of multi-layered graphene sheets used as nanoscale resonators. Nanotechnology 16(10): 2086-2091.

Hosseini, M., M.R. Mofidi, A. Jamalpoor and M. Safi Jahanshahi. 2018. Nanoscale mass nanosensor based on the vibration analysis of embedded magneto-electro-elastic nanoplate made of FGMs via nonlocal Mindlin plate theory. Microsyst. Technol. 24(5): 2295-2316.

Hosseini-Hashemi, Sh., M. Zare and R. Nazemnezhad. 2013. An exact analytical approach for free vibration of Mindlin rectangular nano-plates via nonlocal elasticity. Compos. Struct. 100: 290-299.

Hosseini-Hashemi, Sh., R. Nazemnezhad and H. Rokni. 2015. Nonlocal nonlinear free vibration of nanobeams with surface effects. Eur. J. Mech. A Solids. 52: 44-53.

Hu, Y. and G.S. Wilson. 1997a. Rapid changes in local extracellular rat brain glucose observed with an *in vivo* glucose sensor. J. Neurochem. 68(4): 1745-1752.

Hu, Y. and G.S. Wilson. 1997b. A temporary local energy pool coupled to neuronal activity: Fluctuations of extracellular lactate levels in rat brain monitored with rapid-response enzyme-based sensor. J. Neurochem. 69(4): 1484-1490.

Hu, Yan-Gao, Kim Meow Liew, Q. Wang, X.Q. He and B.I. Yakobson. 2008. Nonlocal shell model for elastic wave propagation in single-and double-walled carbon nanotubes. J. Mech. Phys. Solids. 56(12): 3475-3485.

Huang, D.J., H.J. Ding and W.Q. Chen. 2007. Analytical solution for functionally graded magneto-electro-elastic plane beams. Int. J. Eng. 45(2-8): 467-485.

Huang, D.J., H.J. Ding and W.Q. Chen. 2010. Static analysis of anisotropic functionally graded magneto-electro-elastic beams subjected to arbitrary loading. Eur. J. Mech. A Solids. 29(3): 356-369.

Hwang, Geon-Tae, Myunghwan Byun, Chang Kyu Jeong and Keon Jae Lee. 2015. Flexible piezoelectric thin-film energy harvesters and nanosensors for biomedical applications. Adv. Healthc. Mater. 4(5): 646-658.

Ikeda, Takurō. 1996. Fundamentals of Piezoelectricity: Oxford University Press.

Ip, K.H. and P.Ch. Tse. 2001. Optimal configuration of a piezoelectric patch for vibration control of isotropic rectangular plates. Smart Mater. Struct. 10(2): 395.

Jafari, A.A., A.A. Jandaghian and O. Rahmani. 2014. Transient bending analysis of a functionally graded circular plate with integrated surface piezoelectric layers. IJMME. 9(1): 8.

Jaffe, B. 2012. Piezoelectric Ceramics. Vol. 3: Elsevier.

Jamalpoor, A.A., A.A. Ahmadi-Savadkoohi, M. Hosseini and Sh Hosseini-Hashemi. 2017. Free vibration and biaxial buckling analysis of double magneto-electro-elastic nanoplate-systems coupled by a visco-Pasternak medium via nonlocal elasticity theory. Eur. J. Mech. A Solids. 63: 84-98.

Jandaghian, A.A., A.A. Jafari and O. Rahmani. 2014. Vibrational response of functionally graded circular plate integrated with piezoelectric layers: An exact solution. Eng. Solid Mech. 2(2): 119-130.

Jandaghian, A.A. and A.A. Jafari. 2012. Investigating the Effect of Piezoelectric layers on Circular Plates under Forced Vibration.

Jandaghian, A.A., A.A. Jafari and O. Rahmani. 2013. Exact solution for transient bending of a circular plate integrated with piezoelectric layers. Appl. Math. Model. 37(12-13): 7154-7163.

Jandaghian, A.A. and O. Rahmani. 2015. On the buckling behavior of piezoelectric nanobeams: An exact solution. J. Mech. Sci. Technol. 29(8): 3175-3182.

Jandaghian, A.A. and O. Rahmani. 2016a. An analytical solution for free vibration of piezoelectric nanobeams based on a nonlocal elasticity theory. J. Mech. 32(2): 143-151.

Jandaghian, A.A. and O. Rahmani. 2016b. Free vibration analysis of magneto-electro-thermo-elastic nanobeams resting on a Pasternak foundation. Smart Mater. Struct. 25(3): 035023.

Jandaghian, A.A. and O. Rahmani. 2016c. Vibration analysis of functionally graded piezoelectric nanoscale plates by nonlocal elasticity theory. Superlattices Microstruct. 100: 57-75.

Jandaghian, A.A. and O. Rahmani. 2017. Buckling analysis of multi-layered graphene sheets based on a continuum mechanics model. Appl. Phys. A. 123(5): 324.

Jensen, K., K. Kim and A. Zettl. 2008. An atomic-resolution nanomechanical mass sensor. Nat. Nanotechnol. 3(9): 533.

Kattimani, S.C. and M.C. Ray. 2015. Control of geometrically nonlinear vibrations of functionally graded magneto-electro-elastic plates. Int. J. Mech. Sci. 99: 154-167.

Ke, L.L., Y. Sh. Wang and Zh. D. Wang. 2012. Nonlinear vibration of the piezoelectric nanobeams based on the nonlocal theory. Compos. Struct. 94(6): 2038-2047.

Ke, L.L. and Y. Sh. Wang. 2012. Thermoelectric-mechanical vibration of piezoelectric nanobeams based on the nonlocal theory. Smart Mater. Struct. 21(2): 025018.

Ke, L.L., Y.S. Wang and J.N. Reddy. 2014a. Thermo-electro-mechanical vibration of size-dependent piezoelectric cylindrical nanoshells under various boundary conditions. Compos. Struct. 116: 626-636.

Ke, L.L., Y. Sh. Wang, J. Yang and S. Kitipornchai. 2014b. Free vibration of size-dependent magneto-electro-elastic nanoplates based on the nonlocal theory. Acta Mech. Sin. 30(4): 516-525.

Ke, L.L., Y. Sh. Wang, J. Yang and Sritawat Kitipornchai. 2014c. The size-dependent vibration of embedded magneto-electro-elastic cylindrical nanoshells. Smart Mater. Struct. 23(12): 125036.

Ke, L.L. and Y. Sh. Wang. 2014. Free vibration of size-dependent magneto-electro-elastic nanobeams based on the nonlocal theory. Physica E: Low Dimens. Syst. Nanostruct. 63: 52-61.

Ke, L.L., Ch. Liu and Y. Sh. Wang. 2015. Free vibration of nonlocal piezoelectric nanoplates under various boundary conditions. Physica E: Low Dimens. Syst. Nanostruct. 66: 93-106.

Khanmirza, E., A. Jamalpoor and A. Kiani. 2017. Nano-scale mass sensor based on the vibration analysis of a magneto-electro-elastic nanoplate resting on a visco-Pasternak substrate. Eur. Phys. J. Plus. 132(10): 422.

Kim, Chang Hyun, Myung, Yoon, Cho, Yong Jae, Kim, Han Sung, Park, Seong-Hun, Park, Jeunghee, et al. 2009. Electronic structure of vertically aligned Mn-doped $CoFe_2O_4$ nanowires and their application as humidity sensors and photodetectors. J. Phys. Chem. C. 113(17): 7085-7090.

Kitipornchai, S., X.Q. He and K.M. Liew. 2005. Continuum model for the vibration of multilayered graphene sheets. Phys. Rev. B. 72(7): 075443.

Komijani, M., Y. Kiani, S.E. Esfahani and M.R. Eslami. 2013. Vibration of thermo-electrically post-buckled rectangular functionally graded piezoelectric beams. Compos. Struct. 98: 143-152.

Koo, J.H., D. Ch. Kim, H.J. Shim, T.H. Kim and D.H. Kim. 2018. Flexible and stretchable smart display: materials, fabrication, device design, and system integration. Adv. Funct. Mater. 28(35): 1801834.

Kumar, K. Ramesh and S. Narayanan. 2007. The optimal location of piezoelectric actuators and sensors for vibration control of plates. Smart Mater. Struct. 16(6): 2680.

Kumar, K. Ramesh and S. Narayanan. 2008. Active vibration control of beams with optimal placement of piezoelectric sensor/actuator pairs. Smart Mater. Struct. 17(5): 055008.

Kunbar, Laith A. Hassan, Luay Badr Hamad, Ridha A. Ahmed and Nadhim M. Faleh. 2020. Nonlinear vibration of smart nonlocal magneto-electro-elastic beams resting on nonlinear elastic substrate with geometrical imperfection and various piezoelectric effects. Smart. Struct. Syst. 25(5): 619-630.

Lawson, A.W. 1942. The vibration of piezoelectric plates. Phys. Rev. 62(1-2): 71-76.

Lee, PCY. 1991. A variational principle for the equations of piezoelectromagnetism in elastic dielectric crystals. Int. J. Appl. Phys. 69(11): 7470-7473.

Li, Chenlin, Huili Guo, Xiaogeng Tian and Tianhu He. 2019a. Size-dependent thermo-electromechanical responses analysis of multi-layered piezoelectric nanoplates for vibration control. Compos. Struct. 225: 111112.

Li, F.X., R.K.N.D. Rajapakse, D. Mumford and M. Gadala. 2008a. Quasi-static thermo-electro-mechanical behaviour of piezoelectric stack actuators. Smart Mater. Struct. 17(1): 015049.

Li, Jinqiang, Yu Xue, Fengming Li and Yoshihiro Narita. 2019b. Active vibration control of functionally graded piezoelectric material plate. Compos. Struct. 207: 509-518.

Li, X.Y., H.J. Ding and W.Q. Chen. 2008b. Three-dimensional analytical solution for functionally graded magneto-electro-elastic circular plates subjected to uniform load. Compos. Struct. 83(4): 381-390.

Li, Xiaodong, Bharat Bhushan, Kazuki Takashima, Chang-Wook Baek and Yong-Kweon Kim. 2003. Mechanical characterization of micro/nanoscale structures for MEMS/NEMS applications using nanoindentation techniques. Ultramicroscopy 97(1-4): 481-494.

Liu, Chen, Liao-Liang Ke, Yue-Sheng Wang, Jie Yang and Sritawat Kitipornchai. 2013. Thermo-electro-mechanical vibration of piezoelectric nanoplates based on the nonlocal theory. Compos. Struct. 106: 167-174.

Liu, Chen, Liao-Liang Ke, Jie Yang, Sritawat Kitipornchai and Yue-Sheng Wang. 2018. Nonlinear vibration of piezoelectric nanoplates using nonlocal Mindlin plate theory. Mech. Adv. Mater. Struct. 25(15-16): 1252-1264.

Liu, X., Q. Wang and S.T. Quek. 2002. Analytical solution for free vibration of piezoelectric coupled moderately thick circular plates. Int. J. Solids Struct. 39(8): 2129-2151.

Ma, Yanan, Liu, Nishuang, Li, Luying, Hu, Xiaokang, Zou, Zhengguang, Wang, Jianbo, et al. 2017. A highly flexible and sensitive piezoresistive sensor based on MXene with greatly changed interlayer distances. Nat. Commun. 8(1): 1207.

Mannsfeld, Stefan C.B., Benjamin C.K. Tee, Randall M. Stoltenberg, et al. 2010. Highly sensitive flexible pressure sensors with microstructured rubber dielectric layers. Nat. Mater. 9(10): 859-864.

Martin, L.W., S.P. Crane, Y.H. Chu, M.B. Holcomb, M. Gajek, M. Huijben, et al. 2008. Multiferroics and magnetoelectrics: thin films and nanostructures. J. Condens. Matter Phys. 20(43): 434220.

Masson, L., X. Ren and Y. Perriard. 2019. Novel test environment for the development of self-sensing piezoelectric actuators. Paper read at 2019 22nd International Conference on Electrical Machines and Systems (ICEMS).

Masson, Louis and Yves Perriard. 2019. Study of self-sensing actuation strategies for quasi-static piezoelectric actuators. Paper read at 2019 22nd International Conference on Electrical Machines and Systems (ICEMS).

Maurini, C., M. Porfiri and J. Pouget. 2006. Numerical methods for modal analysis of stepped piezoelectric beams. Sound Vib. 298(4): 918-933.

Meitzler, A., H.F. Tiersten, A.W. Warner, D. Berlincourt, G.A. Couqin and F.S. Welsh III. 1988. IEEE Standard on Piezoelectricity. Society.

Milazzo, ALBERTO, CALOGERO Orlando and ANDREA Alaimo. 2009. An analytical solution for the magneto-electro-elastic bimorph beam forced vibrations problem. Smart Mater. Struct. 18(8): 085012.

Milazzo, Alberto. 2013. A one-dimensional model for dynamic analysis of generally layered magneto-electro-elastic beams. J. Sound Vib. 332(2): 465-483.

Mohammadimehr, M., A.R. Saidi, A. Ghorbanpour Arani, A. Arefmanesh and Q. Han. 2010. Torsional buckling of a DWCNT embedded on winkler and pasternak foundations using nonlocal theory. J. Mech. Sci. Technol. 24(6): 1289-1299.

Moheimani, S.O. Reza and Andrew J. Fleming. 2006. Piezoelectric Transducers for Vibration Control and Damping: Springer Science and Business Media.

Mustapha, K.B. and Z.W. Zhong. 2010. Free transverse vibration of an axially loaded non-prismatic single-walled carbon nanotube embedded in a two-parameter elastic medium. Comput. Mater. Sci. 50(2): 742-751.

Nie, B., X. Li and J. Shao. 2017. Flexible and transparent strain sensors with embedded multiwalled carbon nanotubes meshes. ACS Appl. Mater. Interfaces. 9(46): 40681-40689.

Pan, E. 2001. Exact solution for simply supported and multilayered magneto-electro-elastic plates. J. Appl. Mech. 68(4): 608-618.

Pan, E. and F. Han. 2005. Exact solution for functionally graded and layered magneto-electro-elastic plates. Int. J. Eng. Sci. 43(3): 321-339.

Paul, H.S. and Vazhapadi K. Nelson. 1996. Flexural vibration of piezoelectric composite hollow cylinder. The J. Acoust. Soc. Am. 99(1): 309-313.

Poizat, Christophe and Ayech Benjeddou. 2006. On analytical and finite element modelling of piezoelectric extension and shear bimorphs. Comput Struct. 84(22): 1426-1437.

Pradhan, S.C. and J.K. Phadikar. 2009. Nonlocal elasticity theory for vibration of nanoplates. J. Sound Vib. 325(1-2): 206-223.

Pradhan, S.C. and A. Kumar. 2011. Vibration analysis of orthotropic graphene sheets using nonlocal elasticity theory and differential quadrature method. Compos. Struct. 93(2): 774-779.

Prashanthi, K., P.M. Shaibani, A. Sohrabi, T.S. Natarajan and T. Thundat. 2012. Nanoscale magnetoelectric coupling in multiferroic $BiFeO_3$ nanowires. Physica. Status Solidi. (RRL) 6(6): 244-246.

Qing, Guang-Hui, Jia-Jun Qiu and Yan-Hong Liu. 2005. Modified HR mixed variational principle for magnetoelectroelastic bodies and state-vector equation. Appl. Math. Mech. 26(6): 722-728.

Quek, S.T. and Q. Wang. 2000. On dispersion relations in piezoelectric coupled-plate structures. Smart Mater. Struct. 9(6): 859.

Rahmani, O. and O. Pedram. 2014. Analysis and modeling the size effect on vibration of functionally graded nanobeams based on nonlocal Timoshenko beam theory. Int. J. Eng. Sci. 77: 55-70.

Rahmani, O. and A.A. Jandaghian. 2015. Buckling analysis of functionally graded nanobeams based on a nonlocal third-order shear deformation theory. Appl. Phys. A. 119(3): 1019-1032.

Ramirez, F., P.R. Heyliger and E. Pan. 2006a. Free vibration response of two-dimensional magneto-electro-elastic laminated plates. J. Sound Vib. 292(3): 626-644.

Ramirez, Fernando, Paul R. Heyliger and Ernian Pan. 2006b. Discrete layer solution to free vibrations of functionally graded magneto-electro-elastic plates. Mech. Adv. Mater. Struct. 13(3): 249-266.

Razavi, S. and A. Shooshtari. 2015. Nonlinear free vibration of magneto-electro-elastic rectangular plates. Compos. Struct. 119: 377-384.

Reddy, Junuthula Narasimha. 2003. Mechanics of Laminated Composite Plates and Shells: Theory and Analysis: CRC Press.

Reddy, J.N. 2007. Nonlocal theories for bending, buckling and vibration of beams. Int. J. Eng. Sci. 45(2): 288-307.

Ryu, WonHyoung, Zhinong Huang, Fritz B. Prinz, Stuart B. Goodman and Rainer Fasching. 2007. Biodegradable micro-osmotic pump for long-term and controlled release of basic fibroblast growth factor. J. Control. Release. 124(1-2): 98-105.

Sahmani, S. and M.M. Aghdam. 2018. Thermo-electro-radial coupling nonlinear instability of piezoelectric shear deformable nanoshells via nonlocal elasticity theory. Microsyst. Technol. 24(2): 1333-1346.

Schneider, Andreas and Thomas Stieglitz. 2004. Implantable flexible electrodes for functional electrical stimulation. Med. Device Technol. 15(1): 16-18.

Şimşek, Mesut. 2014. Large amplitude free vibration of nanobeams with various boundary conditions based on the nonlocal elasticity theory. Compos. B. Eng. 56: 621-628.

Sirohi, Jayant and Inderjit Chopra. 2000. Fundamental understanding of piezoelectric strain sensors. J. Intell. Mater. Syst. Struct. 11(4): 246-257.

Snyder, Phillip W., Gwangrog Lee, Piotr E. Marszalek, Robert L. Clark and Eric J. Toone. 2007. A stochastic, cantilever approach to the evaluation of solution phase thermodynamic quantities. Proc. Natl. Acad. Sci. 104(8): 2579-2584.

Soh, A.K., Liu and J.X. 2005. On the constitutive equations of magnetoelectroelastic solids. J. Intell. Mater. Syst. Struct. 16(7-8): 597-602.

Sun, Dongchang, Liyong Tong and Satya N. Atluri. 2001. Effects of piezoelectric sensor/actuator debonding on vibration control of smart beams. Int. J. Solids Struct. 38(50): 9033-9051.

Thai, Huu-Tai. 2012. A nonlocal beam theory for bending, buckling, and vibration of nanobeams. Int. J. Eng. Sci. 52: 56-64.

Tsai, Yi-Hwa and Chih-Ping Wu. 2008. Dynamic responses of functionally graded magneto-electro-elastic shells with open-circuit surface conditions. Int. J. Eng. Sci. 46(9): 843-857.

Van den Boomgaard, J., D.R. Terrell, R.A.J. Born and H.F.J.I. Giller. 1974. An *in situ* grown eutectic magnetoelectric composite material. J. Mater. Sci. 9(10): 1705-1709.

Van den Boomgaard, J., A.M.J.G. Van Run and J. Van Suchtelen. 1976. Piezoelectric-piezomagnetic composites with magnetoelectric effect. Ferroelectrics 14(1): 727-728.

Van Suchtelen, J. 1972. Product properties: A new application of composite materials. Philips Res. Rep. 27(1): 28-37.

Vasques, C.M.A. and J. Dias Rodrigues. 2006. Active vibration control of smart piezoelectric beams: Comparison of classical and optimal feedback control strategies. Comput. Struct. 84(22): 1402-1414.

Vinyas, M. and S.C. Kattimani. 2017. Static studies of stepped functionally graded magneto-electro-elastic beam subjected to different thermal loads. Compos. Struct. 163: 216-237.

Wang, C.G., L. Lan, Y.P. Liu, H.F. Tan and X.D. He. 2013. Vibration characteristics of wrinkled single-layered graphene sheets. Int. J. Solids Struct. 50(10): 1812-1823.

Wang, Chunfeng, Chonghe Wang, Zhenlong Huang and Sheng Xu. 2018a. Materials and structures toward soft electronics. Adv. Mat. 30(50): 1801368.

Wang, Hongjin and Qingfeng Meng. 2013. Analytical modeling and experimental verification of vibration-based piezoelectric bimorph beam with a tip-mass for power harvesting. Mech. Syst. Signal. Pr. 36(1): 193-209.

Wang, Liang, Viktor Hofmann, Fushi Bai, Jiamei Jin and Jens Twiefel. 2018b. Modeling of coupled longitudinal and bending vibrations in a sandwich type piezoelectric transducer utilizing the transfer matrix method. Mech. Syst. Signal. Pr. 108: 216-237.

Wang, Q. 2002. On buckling of column structures with a pair of piezoelectric layers. Eng. Struct. 24(2): 199-205.

Wang, Q. and V.K. Varadan. 2007. Application of nonlocal elastic shell theory in wave propagation analysis of carbon nanotubes. Smart Mater. Struct. 16(1): 178-190.

Wang, Quan and Behrouz Arash. 2014. A review on applications of carbon nanotubes and graphenes as nano-resonator sensors. Comput. Mater. Sci. 82: 350-360.

Wang, X., H.K. Yang and K. Dong. 2005. Torsional buckling of multi-walled carbon nanotubes. Mater. Sci. Eng. A. 404(1): 314-322.

Wang, Yanqing, Yunfei Liu and J.W. Zu. 2019. Nonlinear free vibration of piezoelectric cylindrical nanoshells. Appl. Math. Mech. 40(5): 601-620.

Wu, Ch. P. and Y.H. Tsai. 2010. Dynamic responses of functionally graded magneto-electro-elastic shells with closed-circuit surface conditions using the method of multiple scales. European J. Mech. - A/Solids. 29(2): 166-181.

Wu, Chih-Ping and Yi-Hwa Tsai. 2007. Static behavior of functionally graded magneto-electro-elastic shells under electric displacement and magnetic flux. Int. J. Eng. Sci. 45(9): 744-769.

Xin, L. and Zh. Hu. 2015. Free vibration of simply supported and multilayered magneto-electro-elastic plates. Compos. Struct. 121: 344-350.

Yang, J., L.L. Ke and S. Kitipornchai. 2010. Nonlinear free vibration of single-walled carbon nanotubes using nonlocal Timoshenko beam theory. Physica E: Low Dimens. Syst. Nanostruct. 42(5): 1727-1735.

Zhang, X.L., Q. Xu, X. Zhao, Y.H. Li and J. Yang. 2020. Nonlinear analyses of magneto-electro-elastic laminated beams in thermal environments. Compos. Struct. 234: 111524.

Zhang, Y.Y., C.M. Wang and V.B.C. Tan. 2009. Assessment of Timoshenko beam models for vibrational behavior of single-walled carbon nanotubes using molecular dynamics. AAMM. 1(1): 89-106.

Zhao, X., H. Gao and G. Zhang. 2007. Active health monitoring of an aircraft wing with embedded piezoelectric sensor/actuator network: I. Defect detection, localization and growth monitoring. Smart Mater. Struct. 16(4): 1208.

Zhu, J., Z. Lv and H. Liu. 2019. Thermo-electro-mechanical vibration analysis of nonlocal piezoelectric nanoplates involving material uncertainties. Compos. Struct. 208: 771-783.

17

Nanostructured Oxides for Photocatalytic Applications

Suman[1], Surjeet Chahal[1], Ashok Kumar[1],
Anand Kumar[2] and Parmod Kumar[3,*]

17.1 INTRODUCTION

Human culture is truly undermined by the consistently expanding utilisation of regular assets, particularly fresh water, and man is compelled to fix the dirtied water to spare his reality on Earth (Shabani et al. 2019, Liu et al. 2015, Subramonian et al. 2017, Wang and Huang 2016). Human beings are threatened just because of the growth of industries and ever increasing population. Normally, the water pollutants are derived from the sewage effluent of industries and domestic contaminants. Paper industry, textile industry, pharmaceutical industry, dye industry etc. are examples of sewage effluents and detergent, pesticide, pharmaceuticals etc. come under domestic contaminants (Fig. 17.1) (Buthiyappan et al. 2016, Solís et al. 2012, Cundy et al. 2008, Ribeiro et al. 2015). Till now, a significant number of contaminations have been identified and classified as inorganic particles, pathogens, synthetic dyes which are toxic to organisms. Therefore, numerous conventional systems including chemical, physical and biological have been created so as to treat polluted water. For example, sedimentation (Khoufi et al. 2007), coagulation (Jiang 2015, Shi et al. 2007, Zhou et al. 2008), adsorption (Ali 2014, Lata and Samadder 2016), ozonation (Gupta and Suhas 2009), membrane filtration (Avlonitis et al. 2008, Šostar-Turk et al. 2005), reverse osmosis (Al-Bastaki 2004) and biological degradation (Barragán et al. 2007, Rai et al. 2005). Because of complex composition and diverse physico-chemical properties, the contaminants are not degraded by the above said strategies. Thus, traditional techniques have a lot of limitations including high energy consumption, low efficiency and toxic by-products (Mamba and Mishra 2016, Lee and Park 2013, Nasirian and Mehrvar 2016, Oturan and Aaron 2014). So, to treat wastewater, an environmentally friendly advanced oxidation process is followed, which includes sonolysis, Fenton process, and photocatalysis (Zou 2015, Uma et al. 2020, Bustillo-Lecompte et al. 2016, Ghafoori et al. 2014, Hamad et al. 2016, Chen et al. 2019, Kim et al. 2016). Among these, photocatalysis is a good technique that can utilize wide solar spectrum of sunlight, as sunlight is the principal source of energy on Earth.

The increasing demand for clean water has driven the researchers to find new materials that have good physical and chemical properties. To utilize full visible light, semiconductors have gained more attention of researchers due to their unique crystal structure, tunable surface morphology and intrinsic catalytic property (Chong et al. 2010, Di Paola et al. 2012, Xu et al. 2012). In spite of the

[1] Dept. of Physics, Deenbandhu Chhotu Ram University of Science and Technology, Murthal-131039, Haryana, India.
[2] Dept. of Physics, Institute of Integrated and Honors Studies, Kurukshetra University Kurukshetra-136119, Haryana, India.
[3] Dept. of Physics, J.C. Bose University of Science and Technology, YMCA, Faridabad-121006, Haryana, India.
* Corresponding author: parmodphysics@gmail.com

good catalytic property of semiconductors, their short hole diffusion length and recombination of charge carriers reduce degradation efficiency (Qu and Duan 2013, Moniz et al. 2015). So, a catalyst to be used in degradation must fulfil the defined criterion such as appropriate band gap, chemical stability, band gap configuration and effective electron hole pair separation.

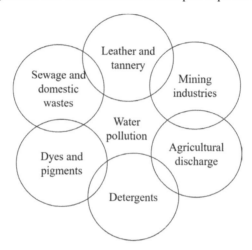

Fig. 17.1 Conclusive outlook of water pollution by various toxic pollutants

Various morphologies of ceramic materials such as nanoflowers, nanowires, nanoring, nanotubes, nanorods etc., have been developed as compared to bulk material in order to enhance the rate capability and cyclic stability of photodegradation. So, nanoscience has totally swapped the past innovation for the following reasons:

1. Nanomaterials have high linking of adsorbing substances, due to their smaller size, porosity and dynamic surface.
2. Nanomaterials have high surface to volume ratio as compared to bulk counterparts, so they provide more active sites to chemical reaction.
3. Nanomaterials as catalyst completely remove most of the contaminants from polluted water.
4. Nanomaterials are non-corrosive in presence of chemicals and water, non-toxic, easily synthesised, inexpensive, and thermally and chemically stable.

Thus, to get better results for wastewater treatment, nanoparticles have become an alternative strategy for removal of organic waste from polluted water.

Nanostructures of various metal oxides have attractive practical applications owing to their novel chemical and physical properties including huge surface area as compared to bulk counterparts, gas affectability, chemical inertness, high photocatalytic, electrochemical activities and biocompatibility (Sharma et al. 2018, 2012, Yu et al. 2019, Mittal et al. 2014). Among various nanosized metal oxides, titanium dioxide (TiO_2, 3.2 eV) (Reda et al. 2020), magnesium oxide (MgO, 7.8 eV) (Zeng et al. 2020), zinc oxide (ZnO, 3.2 eV) (Zhao et al. 2017), cerium oxide (CeO_2, 3.3 eV) (Chahal et al. 2020a) and ferric oxide (α-Fe_2O_3, 2.2 eV) (Warwick et al. 2015) are considered as likely competitors in wastewater evacuation due to their high redox potential, nontoxicity, low cost and higher catalytic activities brought about by size quantum effect. Despite being good photocatalyst in dye degradation, quick electron hole pairs' recombination diminishes the photodegradation performance. Also, the main drawback of TiO_2, MgO, ZnO, CeO_2 being used as catalyst is that they have wide band gap, which can't absorb visible light under sunlight, and are active only under irradiation of UV light (Kumar and Rao 2017). It is notable that sun oriented range comprised around 50% of visible light and just 5% UV light (Chen et al. 2010). Thus, wide band gap materials under sunlight irradiation yield low quantum efficiency. Therefore, these metal oxides need to be modified in order to enhance the degradation efficiency. This aim can be achieved using various strategies including doping of metals/non-metals, core shell heterojunctions, depositing noble metals etc. in these metal oxides (Güy and Özacar 2016, Lassoued et al. 2018b, Nasirian et al. 2017, Chahal et al. 2020b, Shooshtari and Ghazi 2017, Zhang et al. 2019).

A great deal of hard investigations on the synthesis of catalyst (TiO_2, MgO, ZnO, CeO_2, α-Fe_2O_3) nanostructures have been done. For example, nanotubes, nanowires, hollow spheres, nanobelts, nanorods and so forth, and synthesized by various approaches including template-assisted, electrospinning, hydrothermal, electrochemical anodization methods and chemical co-precipitation (Gandha et al. 2016). Among different nanostructures, nanotubes possess high surface area for a given diameter as tubes provide both outside and inside surfaces. The nanostructures are useful in improving the optical properties of metal oxides and have advantages over bare metal oxides. The superior photocatalytic activity of nanostructures are mainly due to the specific morphology and correspondingly large surface area of the prepared nanomaterials as compared to bulk counterparts.

This chapter summarises the various morphologies of metal oxides as catalyst in wastewater treatment using photocatalytic technique and also focuses on better understanding of the degradation mechanism of photocatalytic activity. Finally, summary and future perspectives of metal oxide nanostructures used for photocatalyst are presented and discussed.

17.2 PHOTOCATALYTIC ACTIVITY AND ITS MECHANISM

A catalyst is one whose presence in the chemical reaction can be fastened, without any change in itself. The catalytic property of a material within the sight of light is known as photocatalyst; photocatalysis alludes to the speeding up of the pace of compound response under the nearness of light source by initiating a catalyst. The catalysts used mainly in photodegradation are semiconductors which can generate charge carriers i.e. electrons and holes by absorbing light and the reactions that take place are mainly reduction and oxidation (Ani et al. 2018, Qi et al. 2017, Jack et al. 2015, Augugliaro et al. 2019, Sajjadi and Goharshadi 2017).

17.2.1 Mechanism of Photocatalytic Activity

The essential component (Fig. 17.2) of the photocatalytic mechanism could be represented by following steps:

1. Pollutants get out of the water body and go to surface of catalyst where they are influenced by the morphological properties of the catalyst, for example, porosity, surface area, surface charge etc.
2. With photon excited reaction, the particles of the pollutants get adsorbed on the surface of catalyst.
3. Redox reaction occurs between the adsorbed molecules and the charge molecules produced by photon excitation, which gives CO_2 and H_2O as final by-products on degradation.

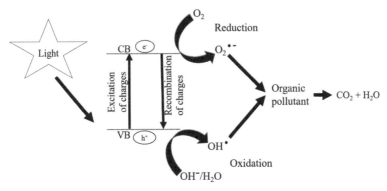

Fig. 17.2 Basic photocatalytic mechanism involved in organic pollutants degradation

17.2.2 Selection of Photocatalyst

Based on the mechanism of photocatalyst mentioned above, five main criteria could be proposed for design of photocatalyst as given below.

1. The semiconductor ought to have a narrow band gap so as to effortlessly excite the charge carriers.
2. The photon ingestion limit of the catalyst ought to be high so more charge transporters can be created.
3. The recombination of charge carriers should be prevented as much as possible so that the efficiency of the catalyst in dye degradation can be increased.
4. The surface area of catalyst should be more so that it can provide maximum reaction sites.
5. The chemical and physical properties of catalyst should be steady with the goal that catalyst can be utilized again.

In order to get the above requirements for the design of the photocatalyst, researchers have developed a lot of techniques, among which surface modification, which is the easiest and is low cost, is discussed in this chapter.

17.3 PARAMETERS AFFECTING PHOTOCATALYTIC DEGRADATION

Various factors influence the photocatalytic efficiency of metal oxide semiconductors such as structure of photocatalyst, concentration of dyes, time of illumination of light, reaction temperature, dissolved oxygen in the reaction medium, pH of the medium, surface area of photocatalyst, initial dye concentration, photocatalyst particle size, temperature of the system, and the intensity of light as appearing in Fig. 17.3 (Mamun et al. 2015, Akpan and Hameed 2009). Many scientists have done studies on degradation of organic pollutants and an optimal condition for photodegradation was concluded.

17.3.1 Impact of Catalyst Particle Size on Dye Degradation

In general, photocatalysis varies with respect to size of nanoparticles, which is known as nanoscale effect. Typically, smaller particles have higher absorption capacity due to increase in surface area to volume proportion as compared to bulk counterparts. Mekasuwandumrong et al. showed that with decrease in particle size, the surface area also reduces and correspondingly photocatalytic activity diminishes and found that ZnO nanoparticles of smaller size have higher photocatalytic efficiency for methylene blue dye degradation (Mekasuwandumrong et al. 2010). Pardeshi and Patil discovered that photocatalytic degradation rate was reliant on particle size as well as on morphology of ZnO nanoparticles and found that photodegradation diminished with increase in particle size (Pardeshi and Patil 2009). Hence, we can conclude that particles in nano-range have more impact on photodegradation as compared to bulk.

17.3.2 Impact of pH Variation on Dye Degradation

The pH impact is associated with surface charged property of catalyst material and provides significant contribution towards dye degradation (Kumar 2017, Shahrodin et al. 2019, Mittal et al. 2016). The surface charge of the semiconductor can be modified in many ways and the modified surface charge affects the reaction rate and mechanism of the catalyst. For example, TiO_2 exhibits an amphoteric character to develop negative or positive charge on its surface (Poulios et al. 2006), and variation in pH can affect the dye molecules in being adsorbed on TiO_2 surface (Wang et al. 2008). Bubacz et al. found that the same pattern between pH and dye degradation rate was followed in their work with methylene blue dye (Bubacz et al. 2010). Similarly, Kansal et al. investigated about the degradation rate of Reactive Black 5 and Reactive Orange 4 and found that degradation process was supported by acidic medium with TiO_2 as catalyst (Kansal et al. 2009). Flores et al. reported different nanostructures of ZnO for methylene blue dye degradation and found that higher photocatalytic activity was achieved at pH 9 compared to other pH values (Flores et al. 2014). In summary, dyes have different activity under photocatalytic reaction. Some dyes are easily degradable at lower pH value while others do so at higher pH scale and all this also depend on nature of pollutants to be degraded. Therefore, it is necessary to study the nature of pollutants, and accordingly determine the correct pH to degrade them photocatalytically.

17.3.3   Impact of Catalyst Loading on Dye Degradation

Catalyst amount is a primary factor that governs the degradation process; if concentration of catalyst is more, then it does not degrade the dye efficiently and if concentration of catalyst is less, then also its degradation efficiency decreases, so an optimum concentration of catalyst is required for significant degradation of dye (Ahmed et al. 2011). Sun et al. pointed out that excess catalyst dose prevents illumination, hydroxyl radicals decrease in catalytic system and accordingly photodegradation reduces (Sun et al. 2008a). Moreover, further increase in catalyst concentration above optimum point causes agglomeration of catalyst nanoparticles and the catalyst surface for photon absorption becomes unavailable (Huang et al. 2008). Krishnakumar et al. inspected the degradation rate of Acid Violet 7 by catalyst dose and observed that maximum dye was degraded by ZnO nanoparticles at an optimum catalyst amount of 2 gL^{-1} and further increase of catalyst concertation reduces the degradation efficiency (Krishnakumar and Swaminathan 2011). Hence, we can conclude that optimum amount of catalyst dose is required in dye degradation. If catalyst concentration is increased, it will cause agglomeration of nanoparticles and less surface of catalyst will be involved in photon absorption and correspondingly degradation efficiency will decrease.

17.3.4   Impact of Catalyst Surface Area on Dye Degradation

Surface area is the crucial factor for dye degradation as large surface provides more active sites to photocatalyst and this can be expanded by developing nanostructures of semiconductor materials like nanoparticles, hollow spheres, spindles, urchins, platelets nanotube, and nanorods. Azeez et al. synthesized TiO_2 nanoparticles via hydrothermal route for methylene blue dye degradation and observed that reduction in particle size yields higher surface area. A maximum surface area of 183.6 m^2g^{-1} was observed at pH 10 and correspondingly maximum 97% dye degradation was achieved at a degradation rate constant of 0.018 min^{-1} under irradiation of UV light (Azeez et al. 2018). Maji et al. examined the rose bengal dye degradation by Fe_2O_3 nanoparticles prepared at different annealing temperatures. Results revealed that sample prepared at 500°C showed the highest dye degradation efficiency of 98% and this maximum dye degradation was achieved corresponding to higher surface area of 79.2 m^2/g. This showed that photodegradation rate decreases with reduction in surface area (Maji et al. 2012). In conclusion, large surface area provides more active sites which participates in photochemical reaction and in turn is helpful in dye degradation.

17.3.5   Impact of Initial Dye Concentration on Degradation

Introductory/initial dye amount focused additionally on a fundamental job in dye degradation, while keeping the fixed measure of catalyst. Sobana et al. observed that increase in dye concentration reduced the molecules adsorbed at surface and correspondingly dye degradation effectiveness of Direct Yellow 4 declines as this likewise diminishes the path length of photon entering in dye arrangement (Sobana et al. 2013). Different scientists' work likewise covered comparable impact of starting dye concentration on photodegradation rate (Giwa et al. 2012, Saggioro et al. 2011, Daneshvar et al. 2003). Neppolian et al. studied the effect of initial dye concentration on the percentage degradation. They differed introductory dyes' focus from 1×10^{-4} to 5×10^{-4} M in

Reactive Blue 4 (RB4) dye, from 4.16×10^{-4} to 1.25×10^{-3} M in Reactive Red 2 (RR2) dye and from 8×10^{-4} to 1.2×10^{-3} M in Reactive Yellow 17 (RY17) dye and with optimum TiO_2 catalyst loading. Author found that degradation rate reduces with increase in concentration of initial dye (Neppolian et al. 2002). In conclusion, the initial dye concentration has more impact on degradation efficiency and activity of catalyst reduces with an increase in initial dye concentration.

17.3.6   Impact of Light Intensity on Dye Degradation

The intensity of light irradiation is an important factor whose amount should be so that the electrons can reach at the conduction band and can participate in redox response of photocatalysis. Under the high light intensity, electron hole pairs predominate over recombination and when light intensity decreases, separation between electron and holes reduces and correspondingly decreases photodegradation rate (Bahnemann 2001). Liu et al. inspected the impact of light intensity in degradation of Acid Yellow 17 under three different light intensities, for example, 1.24 mW/cm^2, 2.04 mW/m^2 and 3.15 mW/m^2 and found that dye degradation rate expanded with increase in intensity of light (Liu et al. 2006). The effect of light intensity on degradation rate was in agreement with other groups also (Sivalingam et al. 2003, So et al. 2002). In conclusion, at high light intensity, the photodegradation is higher and correspondingly, probability of electrons' excitation increases, while at low light intensity the photodegradation efficiency of dyes decreases as charge generation and rate of recombination compete with each other.

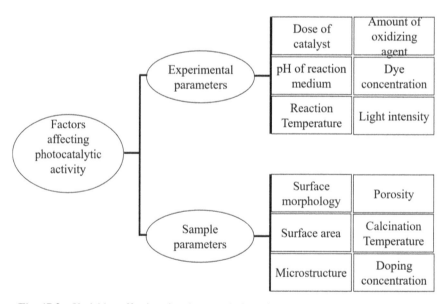

Fig. 17.3   Variables affecting the photocatalytic action of metal oxide semiconductors

17.4   NANOSTRUCTURES OF CERAMIC MATERIALS

Ceramic materials such as metal oxide semiconductors, despite several advantages, were labelled as poor photocatalyst mainly due to short hole diffusion length, charge carriers getting recombined and because they decrease the degradation efficiency of a catalyst. So, surface modification has been potentially used to improve photocatalytic efficiency. The primary parameter which plays an essential role in surface modification is growing of nanostructures of metal oxides. Use of nanostructured ceramics is expected to deliver a better and efficient system for photocatalytic activity due to smaller dimensions as nanoscale measurements give progressive surface territory to volume proportion as compared to bulk counterparts to a catalyst and high surface region supports the exchange of holes and electrons, and diminishes the possibility of recombination rate. So, to have more surface area of different metal oxide semiconductors, morphologies like nanoflowers, nanotubes, nanowires, empty circles, nanobelts, nanorods, nanourchins of different photocatalyst materials (ZnO, Fe_2O_3, TiO_2, MgO) are grown. Different morphological materials as catalyst utilized in dye degradation are outlined in Fig. 17.4. Various techniques have been used by many scientists to prepare metal oxides in nanodimensions. The literature presented here is limited to nanostructured metal oxides of direct relevance to this chapter, i.e. method of preparation. The vast majority of the ordinarily utilized methods for combination of metal oxide nanostructures are given in Table 17.2 with their related dye degradation efficiency.

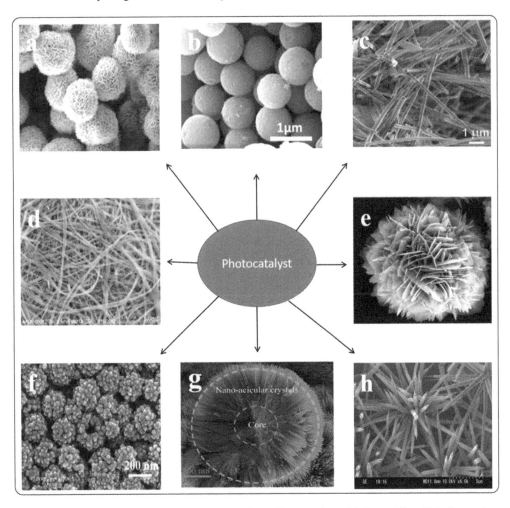

Fig. 17.4 (a) Nano flakes, (b) spheres, (c) nanotubes, (d) nanowires, (e) flower-like, (f) hollow spheres, (g) microspheres, (h) nanorods of photocatalytic materials (Natarajan et al. 2018)
(Reprinted with permission from Elsevier, Journal of Environmental Sciences, vol. 65, pp. 201-222, 2018.)

Semiconductors, for example, zinc oxide (ZnO), cerium oxide (CeO_2), titanium oxide (TiO_2) and magnesium oxide (MgO) exhibit low photodegradation efficiency in the presence of sunlight irradiation. So, relatively wide band gap of these metal oxides limits its absorption range in visible spectrum and hinders its practical applications. To improve degradation efficiency of these metal oxides, nanostructures of various morphology including nanotubes, nanoflowers, nanorods, nanowires, nanotubes, hollow spheres etc. have been created.

17.4.1   ZnO Nanostructures as Photocatalyst

Author (Xu et al. 2009) reported the different nanostructures of ZnO, synthesised via solvothermal method by using different solvents as shown in Fig. 17.5. According to authors, first time they grew various morphologies of ZnO using zinc acetylacetonate as the source of zinc precursor material. The impact of incorporated nanostructures of ZnO, for example, was explored for the phenol dye degradation. FESEM pictures indicated the impact of various solvents on morphology of ZnO and found that cauliflower structure was obtained by using THF as solvent as shown in Fig. 17.5 (a, b). The accumulation of nanoparticles in cauliflower had size 15-25 nm. Size of cauliflower was on micrometre. Figure 17.5 (c, d) indicated the truncated hexagonal conical shape on utilizing decane as solvent and observed that the side length of the top and bottom hexagonal face ranged from 300-400 nm and 700 nm to 1 μm, respectively, and cone height was about 1 μm. When water was used as solvent, rod like structures of ZnO were obtained. Upon closer examination of Fig. 17.5 (e, f), it was observed that some of rods were hollow with diameter ranging from 200-500 nm. Figure 17.5 (g, h) showed that hourglass-like shape could be obtained only when toluene is used as solvent. The bottom side length of hourglass-like was 500 nm and had diameter of around 1.2 μm. When ethanol was used as solvent, rod like nanostructures of ZnO were obtained. Figure 17.5 (i, j) showed that the rods were connected with each other to a centre to form radial shape having diameter around 30 nm and length to be around 250 nm. Spherical shape nanostructures of ZnO were observed on using acetone as solvent. From Fig. 17.5 (k, l), it is clear that spherical shape seems to consist of nanoparticles with size 30-50 nm in diameter and size of the spheres ranged from 750 nm to 1 μm. The photocatalytic activity of ZnO nanostructures was screened for phenol dye degradation and it was compared in terms of first order kinematic rate constant. Nanostructures grown using THF as solvent were found to show highest degradation rate and almost complete phenol dye degradation occurred in 20 min under the irradiation of UV light having 0.1496 min^{-1} as rate constant (k). The rate constant (k) for synthesized nanostructures with different solvents are given in Table 17.1. Nanostructures grown with acetone as solvent had shown lowest degradation constant rate of 0.0385 min^{-1}, but still its degradation rate was more than twice of the commercial ZnO (k = 0.0182 min^{-1}). The cauliflower shaped nanostructures of ZnO synthesized via THF as solvent exhibited excellent photocatalytic activity and it was 9 times more enhanced than commercial ZnO.

TABLE 17.1   ZnO morphologies grown using different solvents with kinematic rate constant values

Photocatalyst Structure	Solvent	Rate Constant (k) min^{-1}
Cauliflower	THF	0.1496
Truncated hexagonal conical	Decane	0.0867
Rod like	Water	0.0651
Hourglass like	Toluene	0.0518
Rod like	Ethanol	0.0468
Spherical	Acetone	0.0385

Similarly, Sun et al. synthesized nanostructures of ZnO such as nanobelt arrays, rod-/comb and thin film like for degradation of methyl orange dye under illumination of UV light source (Sun et al. 2008b). Results demonstrated that 59% and 71% of dye was degraded in 5h by thin film and rod-/comb, respectively, whereas ZnO nanobelt arrays showed excellent degradation property and degraded 94% of dye in 5h. The observed degradation is associated with band gap and surface area of catalyst. They announced that less recombination of charge carriers prompts high synergist efficiency.

Zhang et al. (2014) reported results on ZnO nanorods synthesized at different annealing temperatures (350, 400, 450 and 500°C) to see their effect on decomposition of methylene blue dye. Author showed the impact of different oxygen defects on degradation efficiency of ZnO nanorods. Results indicated that as the length of rod increased, photocatalytic efficiency enhanced correspondingly. It was found that sample synthesized at 350°C showed 99.3% dye decomposition, while 88.1%, 86.0%, and 56.9% degraded by 400, 450 and 500°C samples, respectively, in 80 min under sodium lamp. The maximum dye was degraded by 350°C sample and also, this sample had stability over five recycles showing high reusability performance due to low recombination rate of electron and hole pairs.

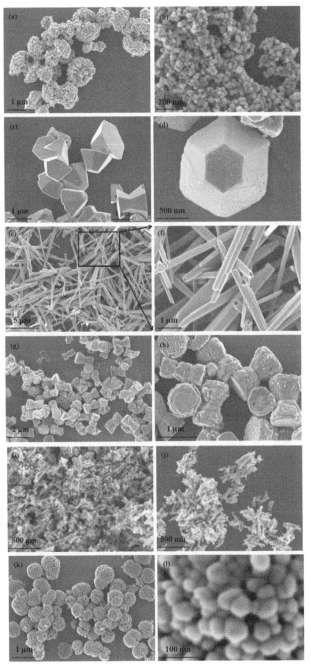

Fig. 17.5 FE-SEM images of as-synthesized ZnO nanomaterials in different solvents: (a, b) THF, (c, d) decane, (e, f) H₂O, (g, h) toluene, (i, j) ethanol, (k, l) acetone (Xu et al. 2009)

(Reprinted with permission from ACS, Journal of Chemistry of Materials, vol. 21, pp. 2875–2885, 2009.)

17.4.2 CeO$_2$ Nanostructures as Photocatalyst

CeO$_2$ has likewise been considered as one of the finest materials in the field to degrade harmful contaminants in water. Wu et al. prepared CeO$_2$ nanorods on K$_{1.33}$Mn$_8$O$_{16}$ nanowires by template method using Sodium dodecyl benzene sulphonate (SDBS) as a surfactant to reduce harmful Cr(VI) to Cr(III) via photoreduction at RT (Wu et al. 2015). UV illumination initiates CeO$_2$–oxalic acid complex to create organic radicals like C$_2$O$^•_4$ from adsorbed oxalic acid on the surface through a ligand-to-metal charge transfer that causes reduction of Cr(VI). Yu et al. prepared single crystalline highly porous hexagonal CeO$_2$ nanosheets to degrade RhB dye. Nanosheets were merely 2.4 nm thick and few micrometerers in length. Highly porous CeO$_2$ nanosheets exhibited superior photocatalytic activity for RhB dye degradation compared to polycrystalline CeO$_2$ nanoparticles. The improvement in efficiency could be accomplished because of more grounded reducibility of energized electrons and inherent properties of single-crystal-like CeO$_2$ nanosheets (Yu et al. 2013).

Sabari Arul et al. synthesized flower-like nanostructures of cerium oxide using solvothermal method to seek enhancement in photocatalytic activity towards degradation of AO7 dye. Up to 65% of AO7 dye was degraded in 10 hr using prepared catalysts under the illumination of UV light having rate constant of 0.259 h^{-1} (Sabari Arul et al. 2013). Qian et al. prepared CeO$_2$ microspheres using bio-template of lotus pollen grains. Highly porous structures of CeO$_2$ assumed an essential job in degradation of MB dye and achieved almost 95% degradation efficiency (Qian et al. 2014). Chahal et al. utilized sol-gel technique to prepare pure CeO$_2$ nanoparticles that convert into grain-like structure on La doping for RB dye degradation under the irradiation of UV light. Authors found that grain-like structures in doped frameworks have higher degradation efficiency than conventional cerium oxide nanoparticles. 8% La doped CeO$_2$ nanostructures showed maximum efficiency of 86.7% in 90 minutes (Chahal et al. 2020c).

17.4.3 TiO$_2$ Nanostructures as Photocatalyst

TiO$_2$ has been considered as the best photocatalyst in pure form among oxide semiconductors. Marien et al. fabricated nanotubes of TiO$_2$ having different lengths that varies from 1.5 μm to 25 μm. Photocatalytic activity for paraquat was carried out to probe the degradation efficiency of synthesized photocatalysts. It was found that the degradation efficiency of nanotubes with a length of 7 μm exhibits maximum degradation efficiency when compared with other lengths (Marien et al. 2016).

Alias et al. prepared nanoflowers-like rutile phase of TiO$_2$ using hydrothermal method to degrade the MB dye solution. MB dye was degraded by 98.95% in 180 minutes using these nanoflowers catalysts under UV light source (Alias et al. 2020). Du et al. prepared different morphologies (rods and flower-like) of TiO$_2$ using solvothermal process by controlling the concentrations of acetic acid and N-Dimethylformamide. 10 ppm solution of RhB dye was taken to observe the photocatalytic test under visible light. Photocatalytic efficiency was enhanced by 5.55 times for flower-like structures calcined at 400°C with molten salt (MS) (Du et al. 2020).

Tian et al. synthesized controlled 3D hierarchical flower-like nanostructured photocatalysts of TiO$_2$ via hydrothermal method. Photocatalytic activity for degradation of phenol dye has been examined under exposure of UV irradiation. 10 ppm phenol dye solution was degraded using different morphologies and it was found that flower-like photocatalysts showed highest efficiency and achieved 86% in 120 min (Tian et al. 2011). Similarly, other results on enhanced photodegradation using different nanostructures of MgO, Fe$_2$O$_3$, TiO$_2$, CeO$_2$ and ZnO were studied by many researchers as given in Table 17.2.

TABLE 17.2 Preparation of ceramic materials having different dimensions and shapes along with their enhanced photodegradation efficiency

Dimension	Shape	Metal Oxide	Synthesis Route	Characterisations	Characteristic of the Obtained Product	Pollutant	Degradation Rate	Ref.
0D	Nanoparticles	α-Fe_2O_3	Chemical co-precipitation	E_g: 2.14 eV, M_s: 27.32 emu/g	The average size of α-Fe_2O_3 particles with irregular spherical shape lies between 19-37 nm	MB	89% decolourisation rate of MB dye was achieved in 140 min by α-Fe_2O_3 nanoparticles calcined at 700°C under visible light irradiation	(Lassoued et al. 2018a)
	Nanoparticles	MgO	Green synthesis	E_g: 4.1758 eV	The spherical shape morphology was observed with the average size of 50 nm	MB	81% dye was degraded in 250 min under UV-lamp (250 W)	(Khan et al. 2020)
	Nanoparticles	ZnO	Green synthesis	–	Leaf extracts (15 ml) used to synthesize nanoparticles having average crystallite size of 0.2 nm; nanoparticles are constituted by spherical particles	MB	98.89% dye was degraded within 50 min under solar irradiation	(Muthuvel et al. 2020)
1D	Nanorods	α-Fe_2O_3	Hydrothermal	Surf. Area: 27.28-96.73 m^2/g	Smooth nanorod shaped with length of 340-850 nm, diameter of 48-83 nm with a pore diameter of 2-8 nm	MO	Electrostatic attraction between MO and Fe_2O_3 nanorod enhanced the photodegradation with 34.5 mg/g removal capacity	(Cha et al. 2011)
	Nanorods	MgO	Solvent-free thermal decomposition	–	Rod-like nanostructures of 100-300 nm length and 20-40 nm diameter	MB	90% degradation was obtained within 180 min after irradiation under visible light	(Salehifar et al. 2016)
	Nanotubes	α-Fe_2O_3	Electrochemical anodization	E_g: 2.04 eV	Well-aligned uniform nanotubes consist of circular and oval shaped open surface, having diameter 50-70 nm, and length of ~ 4 μm	MB	Sonicated clean Fe_2O_3 nanotube illustrates maximum degradation rate in comparison to untreated nanotube hematite under illumination of visible light	(Momeni 2016)
	Nanotubes	CeO_2	Hydrothermal	Surf. Area: 30.2 m^2/g, E_g: 2.9 eV	Tube-like nanostructures with 90 ± 10 nm diameter and 0.6-1.2 m length	Cr(VI)	The reduction of 60 mL, 100 mg/L $K_2Cr_2O_7$ aqueous solution with 50 mg catalyst having 0.05 g oxalic acid. The conversion of Cr(VI) achieved 99.6% in 50 min, under UV light irradiation	(Wu et al. 2015)
	Nanotubes	TiO_2	Ultrasonication	–	Tube Length = 7 μm, Wall thickness = 19.5 nm, Internal diameter = 64 nm	Paraquat	Photocatalytic activity decreases with increase in length of tubes after 7 μm	(Marien et al. 2016)

Contd.

TABLE 17.2 Contd.

	Morphology	Material	Method	Surf. Area / E_g	Structure	Dye	Result	Reference
2D	Nanodisks	ZnO	Chemical hydrolysis method	–	Nanodisks have a uniform average diameter of 200-300 nm and each nanodisk is made of large number of nanocrystallines with 30 ± 10 nm diameter	MO	Complete dye degradation within 120 min in presence of UV light	(Zhai et al. 2012)
	Nanodisks	α-Fe$_2$O$_3$	Hydrothermal	E_g: 2.05 eV	Each nanodisk consists of relatively rough surface with thickness of ~ 60 nm, diameter of ~ 100 nm and consists of 6 to 10 layers	MB	More than 90% dye was degraded in 30 min in visible light	(Huang et al. 2015)
	Nanoplatelets	α-Fe$_2$O$_3$	Hydrothermal	Surf. Area: 69 m^2/g	Randomly oriented and uniform nanoplates having length – 300 nm, width – 100 nm, thickness – 30 nm	MB	47.62% dye was degraded successfully in 4.5 hr after illumination of solar light	(Ayachi et al. 2015)
	Nanosheets	α-Fe$_2$O$_3$	Thermal oxidation	Surf. Area: 70.866 m^2/g	Highly porous nanostructures consisting of dense Fe$_2$O$_3$ nanosheets were found	As, Cr	Chromium can be removed by thermal oxidation of 24 hr upto 100% after 10 min, while arsenic upto 100% after 10 min	(Lin et al. 2014)
	Nanosheets	CeO$_2$	Sol-gel	–	Single-crystalline 2.4 nm thick porous nanosheets of CeO$_2$ with pore size of 3.7 nm	RhB	Porous nanosheets of CeO$_2$ showed much higher degradation efficiency than polycrystalline nanoparticles of CeO$_2$	(Yu et al. 2013)
	Nanosheets	ZnO	By annealing	Surf. Area: 9.04 m^2/g, E_g: 3.05 eV	The width of nanosheets is 0.26-1.34 mm and thickness 58 nm	MO	93.4% dye was degraded by being exposed to {001} surfaces of ZnO nanosheets	(Chen et al. 2015)
3D	Nanodendrites	α-Fe$_2$O$_3$	Reactable ionic liquid	–	Dendritic α-Fe$_2$O$_3$, micropines structure with diameter: 4-6 µm, and branch trunk range: 50 nm to 1.5 µm	MB	90% of MB dye degraded in 150 min under visible light	(Xia et al. 2012)
	Nanoflowers	MgO	Precipitation method	Surf. Area: 142.9 m^2/g	Flower-like particles having diameters of 3.8-6.0 µm	MB	99.8% of MB dye degraded in 90 min under UV light	(Zheng et al. 2019)

Contd.

TABLE 17.2 Contd.

	Morphology	Material	Synthesis	Surface Area / Properties	Structure	Dye	Result	Reference
3D	Nanoflowers	α-Fe$_2$O$_3$	Hydrothermal	–	Flower-like microstructures consist of well-crystalline rods of length: 900 ± 100 nm and diameter: 100 ± 15 nm	RB	Completely degraded RB dye in 35 min under UV light	(Atabaev 2015)
	Nanoflowers	CeO$_2$	Solvothermal	Surf. Area: 71 m^2/g	Average grain size of CeO$_2$ using plane (111) was 20 nm and pore size lies between 2-10 nm	AO7	Degradation by 65% of AO7 in 10 hr on UV illumination	(Sabari Arul et al. 2013)
	Nanoflowers	ZnO	Aqueous solution route	Surf. Area: 25.1617 m^2/g, E_g: 3.23 eV	Flower-like microstructures on accumulation of many interleaving nanosheets with a thickness of around 10 nm and diameter ranged in 1-2 μm	RhB	Complete decolourisation of dye was obtained in 100 min	(Pan et al. 2011)
	Nanoflowers	ZnO	Green synthesis	Surf. Area: 9.18 m^2/g	–	MB	88% dye was degraded in 270 min under solar light illumination	(Vinayagam et al. 2020)
	Nanoflowers	TiO$_2$	Hydrothermal method	Surf. Area: 27.25 m^2/g	–	MB	Highest removal of MB by 98.95%	(Alias et al. 2020)
	Nanoflowers	TiO$_2$	Solvothermal approach	Surf. Area: 58 m^2/g, E_g: 2.86 eV	Flower-like TiO$_2$ structures composed of nanothorns of average diameter 30-50 nm	RhB	Photocatalytic activity for MS calcined sample at 400°C was around 5.5 times higher than that calcined in air under visible light irradiation	(Du et al. 2020)
	Nanoflowers	TiO$_2$	Solvothermal method	Surf. Area: 72 m^2/g	Flowers were composed of nanopetals made from aggregation of nanothorns. Nanopetals were connected to the centre of flowers to form 3D structures	Phenol	Exhibited a 97% degradation efficiency in 120 min under UV light irradiation	(Tian et al. 2011)
	Nano-urchins	α-Fe$_2$O$_3$	Hydrothermal	Surf. Area: ~76.703 m^2/g	Nano-urchins made by composition of many nanoneedles at a centre point. Size of each nanoneedle: 110 ± 30 nm and urchin size: 2 ± 0.5 μm	CR, ER, MB	α-Fe$_2$O$_3$ nano-urchin degraded CR dye by 98%, while ER and MB was degraded 84% and 80%, respectively, under visible light radiation	(Jiao et al. 2015)
	Nanourchins	ZnO	Thermal oxidation	–	–	MB	Largest dye degradation, with urchin like structures of ZnO in comparison to other synthesized needle like structure under UV light irradiation	(Ballesteros-Balbuena et al. 2020)

Contd.

TABLE 17.2 Contd.

3D	Microspheres	CeO_2	Ultrasonic process	Surf. Area: 85.5 cm²/g	Microspheres of diameter: 10 μm to 15 μm, and mesopores on the surface: 2-5 nm	MB	MB dye was degraded 80% in 120 min after irradiation of visible light	(Qian et al. 2014)
	Hollow spheres	ZnO	Hydrothermal method	Surf. Area: 63 m²/g	Diameter of hollow spheres: 10 μm	RhB	Molar ratio of glucose to zinc ions having R = 15 shows maximum photocatalytic activity	(Yu and Yu 2008)
	Hollow spheres	CeO_2	Hydrothermal	Surf. Area: 72 m²/g	–	Cr(VI) and As(V)	Maximum removal capacity of 15.4 mg/g and 22.4 mg/g for Cr(VI) and As(V) respectively. Also, 100% CO conversion was attained at RT	(Cao et al. 2010)
	Hollow spheres	TiO_2	Solvothermal route	Surf. Area: 58 m²/g	Hollow structures of diameter: 0.8-1.0 μm and wall thickness: 40-60 nm	Phenol	99.9% dye degradation was observed using THS-E-600 as catalyst in presence of UV light source for 3 hr	(Shang et al. 2012)
	Hollow spheres	TiO_2	Sol-gel	Surf. Area: 58.3 m²/g	Average pore size: ~ 13.3 nm	RhB	Completely decomposed on 90 min of irradiation with UV light	(Bao et al. 2018)
	Nanospheres	ZnO	Hydrothermal	Surf. Area: 26.58 m²/g, E_g: 3.08 eV	Nanospheres size: 30-70 nm	MO	Complete degradation in 2 hr under UV light, the stability of catalyst was retained by 95% up to three cycles by reusability test.	(Shahi et al. 2018)
	Rose bridal bouquet	TiO_2	Hydrothermal	Surf. Area: 60.98 m²/g	Flower-shaped geometry of size: 1.8-2.0 μm, Individual rose consists of partially rolled sheet-like nanopetals of width around 20-30 nm	MB	99% of MB dye was decolourised in 60 min on UV irradiation	(Nguyen-Phan et al. 2011)

Abbreviations used: MO: Methyl Orange; ER: Eosin Red; MB: Methylene Blue; RhB: Rhodamine B; RB 5: Reactive Black 5; CR: Congo Red; AO7: acid orange 7; As: arsenic; Cr: chromium; Surf. Area: Surface area; M_s: Saturation magnetisation; E_g: Band gap

17.4.4 Ferrites Nanostructures as Photocatalyst

Spinel ferrites are the most studied materials in photocatalysis owing to their narrow band gap. Specially, nanostructures of ferrites have gained much attention in photocatalysis and controlled growth of nanostructures has high impact on degradation efficiency. Dhiman et al. examined the effect of various morphologies of $ZnFe_2O_4$ as catalyst including porous nanorods, nanoparticles, nanoflowers and hollow microspheres as shown in Fig. 17.6. Both cationic (Safranine-O) and anionic (Remazol Brilliant Yellow) dyes were used for degradation. Results indicated that porous nanorods exhibited highest degradation compared to other morphology which was due to high specific surface area of nanorods and slow recombination of electron hole pairs. Photocatalytic degradation followed the order as porous nanorods > nanoparticles > nanoflowers > hollow microspheres (Dhiman et al. 2016).

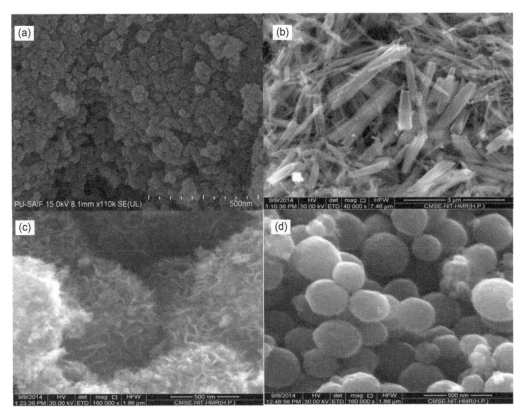

Fig. 17.6 FE-SEM images of $ZnFe_2O_4$ nanostructures (a) Nanoparticles, (b) Porous nanorods, (c) Nanoflowers, (d) Hollow microspheres (Dhiman et al. 2016)

(Reprinted with permission from Elsevier, *Ceramics International, vol. 42, pp. 12594–12605, 2016.*)

Similarly, Li et al. (2011) examined $ZnFe_2O_4$ nanospheres synthesized by solvothermal route without using any surfactant or template for rhodamine B dye degradation. Author observed that almost 100% dye was degraded in 5 hr under Xe lamp irradiation by nanospheres of $ZnFe_2O_4$. This shows that $ZnFe_2O_4$ nanospheres had shown excellent photocatalytic efficiency compared to nanoparticles under same light conditions due to high specific surface area of nanospheres (51.81 m^2g^{-1}). Shen et al. (2013) synthesized $CuFe_2O_4$ nanospheres having diameter around 116 nm via solvothermal route for photocatalytic conversion of benzene. The benzene conversion by $CuFe_2O_4$ nanoparticles was 29.6%, while 44.8% was achieved using $CuFe_2O_4$ nanospheres under same Xe lamp irradiation. Guo et al. (2012) reported the methylene blue dye degradation using colloidal nanocrystal clusters and hollow spheres of $MnFe_2O_4$ and found that band gap values for both colloidal nanocrystal clusters and hollow spheres were 1.68 eV and 1.74 eV, respectively. The results indicated that hollow spheres have superior catalytic efficiency under the illumination of visible light compared to colloidal nanocrystal clusters of $MnFe_2O_4$, as more outer surface area is explored in case of hollow spheres that provides more active sites in redox reaction and correspondingly enhance photocatalytic performance. Raina and Manimekalai (2018) fabricated nanorods of $Co_{0.7}Zn_{0.3}Fe_2O_4$ as catalyst through hydrazine precursor via co-precipitation route. Raina and Manimekalai (2018) examined the various dyes degradation such as Rose Bengal, Methylene blue, Malachite green, Methyl red, Rhodamine B and Congo red dye and found that Congo red and Rose Bengal dye were degraded by 58% within 40 min under solar spectrum. Thus, nanostructures of various ferrites have played an important role in dye degradation.

17.4.5 Effect of Various Metal Ions' Doping on Photocatalyst

Doping of ceramic materials with various metal ions such as Zn^{2+}, Cr^{3+}, Ni^{2+}, Cu^{2+}, Co^{2+}, Fe^{3+} and so on induce a local energy level within photocatalyst band gap and help in increasing life time of charge carriers and correspondingly improve the photocatalytic activity. For example, Singla et al. synthesized parent and zinc doped TiO_2 nanoparticles through sol gel route to investigate the catalyst effect on Methyl Red and Eriochrome Black T dye degradation under the exposure of UV-visible light. Singla et al. (2013) observed that 0.7 mol% Zn-doped TiO_2 have superior degradation rate as compared to pure TiO_2 nanoparticles. Kaur et al. (Kaur et al. 2014) prepared zinc sulphide nanoparticles capped with thioglycerol and doped with copper as catalyst to see their effects on crystal violet dye degradation. Author found that dye degradation reached up to 99% for $Zn_{1-x}Cu_xS$ at $x = 0.03$ in 180 min under illumination of UV light. Suman et al. pointed out the effect of zinc ions on hematite nanoparticles for Rose Bengal dye degradation. Results revealed that Zn 4% showed highest degradation efficiency of 87% in comparison to pure Fe_2O_3 nanoparticles under UV light irradiation in 90 min. In conclusion, doping of various metal ions increases the life time of charge carriers by prohibiting the recombination rate and helps in improving the photocatalytic efficiency (Suman et al. 2020).

17.5 FUTURE PERSPECTIVES

Seeing the vast literature available regarding the photocatalysis development, it appears that nothing is left but still multiple issues regarding the practical application performance of it needs to be sorted out. On a whole, we can work over the following aspects for required push in the development of photocatalysts for their use in treatment of wastewater.

1. More simplified, cost-effective, higher efficiency and environmentally friendly methods are required, as existing methods are complex, expensive, and dangerous to the environment.
2. The catalytic mechanism of these photocatalysts is not clearly understood and more efforts are required to understand it clearly for efficient photocatalysis.
3. When large-scale applications are the ultimate goal, the research focus should be directed towards the use of abundant and noble-metal free co-catalysts and protective layers.
4. Last but not the least, the use of these catalysts in the natural system is limited due to minimum or no recycling ability, small size, and lower mechanical strength of these catalysts.

17.6 SUMMARY

Water pollution is a worldwide crisis and to address this, efforts have been put in developing various photocatalytic processes. Alongside, efforts are also put in the improvement of photocatalytic efficiency of catalysts made with semiconductors. Here, we have tried to compile the multiple approaches used regarding current day development of surface modified semiconductor photocatalysts including the extremely used strategies to nail down the difference in semiconductor band gaps for retarding the recombination of photogenerated electron-hole pairs. These are done to increase the adsorption capacity for visible light, and also to minimize the recombination. We have also tried to summarize the distinct morphologies and the performance of different ceramic materials as catalyst. It is expected that in the coming years, there will be major progress in wastewater treatment through photocatalysis technique by using various naturally occurring and synthesized nanostructured materials.

17.7 ACKNOWLEDGEMENTS

Parmod Kumar acknowledges the financial support from DST under DST-INSPIRE faculty scheme (No. DST/INSPIRE/04/2015/003149) and one of the author Suman would like to thank UGC (UGC-Ref. No.: 1320/(OBC)(CSIR-UGC NET DEC. 2016)) for providing research fellowship. Also, we would like to acknowledge various journals (Ceramics International, Journal of Environmental Sciences and Chemistry of Materials) for giving us permission to reuse images in this chapter.

17.8 REFERENCES

Ahmed, S., M.G. Rasul, R. Brown and M.A. Hashib. 2011. Influence of parameters on the heterogeneous photocatalytic degradation of pesticides and phenolic contaminants in wastewater: A short review. J. Environ. Manage. 92: 311-330.

Akpan, U.G. and B.H. Hameed. 2009. Parameters affecting the photocatalytic degradation of dyes using TiO_2-based photocatalysts: A review. J. Hazard. Mater. 170: 520-529.

Al-Bastaki, N. 2004. Removal of methyl orange dye and Na_2SO_4 salt from synthetic waste water using reverse osmosis. Chem. Eng. Process. Process Intensif. 43: 1561-1567.

Ali, I. 2014. Water treatment by adsorption columns: Evaluation at ground level." Sep. Purif. Rev. 43: 175-205.

Alias, S.S., Z. Harun, F.H. Azhar and S.A. Ibrahim. 2020. Comparison between commercial and synthesised nano flower-like rutile TiO_2 immobilised on green super adsorbent towards dye wastewater treatment. J. Clean. Prod. 251: 119448.

Ani, I.J., U.G. Akpan, M.A. Olutoye and B.H. Hameed. 2018. Photocatalytic degradation of pollutants in petroleum refinery wastewater by TiO_2- and ZnO-based photocatalysts: Recent development. J. Clean. Prod. 205: 930-954.

Atabaev, T.S. 2015. Facile hydrothermal synthesis of flower-like hematite microstructure with high photocatalytic properties. J. Adv. Ceram. 4: 61-64.

Augugliaro, V., G. Palmisano, L. Palmisano and J. Soria. 2019. Heterogeneous Photocatalysis and Catalysis: An Overview of Their Distinctive Features. Elsevier B.V.

Avlonitis, S.A., I. Poulios, D. Sotiriou, M. Pappas and K. Moutesidis. 2008. Simulated cotton dye effluents treatment and reuse by nanofiltration. Desalination 221: 259-267.

Ayachi, A.A., H. Mechakra, M.M. Silvan, S. Boudjaadar and S. Achour. 2015. Monodisperse α-Fe_2O_3 nanoplatelets: Synthesis and characterization. Ceram. Int. 41: 2228-2233.

Azeez, F., E. Al-Hetlani, M. Arafa, Y. Abdelmonem, A.A. Nazeer, M.O. Amin, et al. 2018. The effect of surface charge on photocatalytic degradation of methylene blue dye using chargeable titania nanoparticles. Sci. Rep. 8: 1-9.

Bahnemann, D. 2001. Photocatalytic detoxification of polluted waters. ACS Div. Environ. Chem. Prepr. 41: 929-933.

Ballesteros-Balbuena, M., G. Roa-Morales, A.R. Vilchis-Nestor, V.H. Castrejon-Sanchez, E. Vigueras-Santiago, P. Balderas-Hernandez, et al. 2020. Photocatalytic urchin-like and needle-like ZnO nanostructures synthetized by thermal oxidation. Mater. Chem. Phys. 244: 122703.

Bao, Y., Q.L. Kang, C. Liu and J.Z. Ma. 2018. Sol-gel-controlled synthesis of hollow TiO_2 spheres and their photocatalytic activities and lithium storage properties. Mater. Lett. 214: 272-275.

Barragán, B.E., C. Costa and M.C. Márquez. 2007. Biodegradation of Azo dyes by bacteria inoculated on solid media. Dye. Pigment. 75: 73-81.

Bubacz, K., J. Choina, D. Dolat and A.W. Morawski. 2010. Methylene blue and phenol photocatalytic degradation on nanoparticles of anatase TiO_2. Polish J. Environ. Stud. 19: 685-691.

Bustillo-Lecompte, C.F., S. Ghafoori and M. Mehrvar. 2016. Photochemical degradation of an actual slaughterhouse wastewater by continuous UV/H_2O_2 photoreactor with recycle. J. Environ. Chem. Eng. 4: 719-732.

Buthiyappan, A., A.R. Abdul Aziz and W.M.A. Wan Daud. 2016. Recent advances and prospects of catalytic advanced oxidation process in treating textile effluents. Rev. Chem. Eng. 32: 1-47.

Cao, C.Y., Z.M. Cui, C.Q. Chen, W.G. Song and W. Cai. 2010. Ceria hollow nanospheres produced by a template-free microwave-assisted hydrothermal method for heavy metal ion removal and catalysis. J. Phys. Chem. C. 114: 9865-9870.

Cha, H., S.J. Kim, K.J. Lee, M.H. Jung and Y.S. Kang. 2011. Single-crystalline porous hematite nanorods: Photocatalytic and magnetic properties. J. Phys. Chem. C. 115: 19129-19135.

Chahal, S., A. Kumar and P. Kumar. 2020a. Erbium-doped oxygen deficient cerium oxide: Bi-functional material in the field of spintronics and photocatalysis. Appl. Nanosci. 10: 1721-1733.

Chahal, S., N. Rani, A. Kumar and P. Kumar. 2020b. Electronic structure and photocatalytic activity of samarium doped cerium oxide nanoparticles for hazardous rose bengal dye degradation. Vacuum 172: 109075.

Chahal, S., S. Singh, A. Kumar and P. Kumar. 2020c. Oxygen-deficient lanthanum doped cerium oxide nanoparticles for potential applications in spintronics and photocatalysis. Vacuum 177: 109395.

Chen, C., W. Ma and J. Zhao. 2010. Semiconductor-mediated photodegradation of pollutants under visible-light irradiation. Chem. Soc. Rev. 39: 4206-4219.

Chen, L., X. Zuo, S. Yang, T. Cai and D. Ding. 2019. Rational design and synthesis of hollow $Co_3O_4@Fe_2O_3$ core-shell nanostructure for the catalytic degradation of norfloxacin by coupling with peroxymonosulfate. Chem. Eng. J. 359: 373-384.

Chen, Y., L. Zhang, L. Ning, C. Zhang, H. Zhao, B. Liu, et al. 2015. Superior photocatalytic activity of porous wurtzite ZnO nanosheets with exposed {001} facets and a charge separation model between polar (001) and (001-) surfaces. Chem. Eng. J. 264: 557-564.

Chong, M.N., B. Jin, C.W.K. Chow and C. Saint. 2010. Recent developments in photocatalytic water treatment technology: A review. Water Res. 44: 2997-3027.

Cundy, Andrew B., L. Hopkinson and R.L.D. Whitby. 2008. Use of iron-based technologies in contaminated land and groundwater remediation: A review. Sci. Total Environ. 400: 42-51.

Daneshvar, N., D. Salari and A.R. Khataee. 2003. Photocatalytic degradation of azo dye acid red 14 in water: investigation of the effect of operational parameters. J. Photochem. Photobiol. A. Chem. 157: 111-116.

Dhiman, M., R. Sharma, V. Kumar and S. Singhal. 2016. Morphology controlled hydrothermal synthesis and photocatalytic properties of $ZnFe_2O_4$ nanostructures. Ceram. Int. 42: 12594-605.

Di Paola, A., E. García-López, G. Marcì and L. Palmisano. 2012. A survey of photocatalytic materials for environmental remediation. J. Hazard. Mater. 211-212: 3-29.

Du, M., Q. Chen, Y. Wang, J. Hu and X. Meng. 2020. Synchronous construction of oxygen vacancies and phase junction in TiO_2 hierarchical structure for enhancement of visible light photocatalytic activity. J. Alloys Compd. 830: 154649.

Flores, N.M., U. Pal, R. Galeazzi and A. Sandoval. 2014. Effects of morphology, surface area, and defect content on the photocatalytic dye degradation performance of ZnO nanostructures. RSC Adv. 4: 41099-41110.

Gandha, K., J. Mohapatra, M.K. Hossain, K. Elkins, N. Poudyal, K. Rajeshwarb, et al. 2016. Mesoporous iron oxide nanowires: Synthesis, magnetic and photocatalytic properties. RSC Adv. 6: 90537-90546.

Ghafoori, S., M. Mehrvar and P.K. Chan. 2014. A statistical experimental design approach for photochemical degradation of aqueous polyacrylic acid using photo-fenton-like process. Polym. Degrad. Stab. 110: 492-497.

Giwa, A., P.O. Nkeonye, K.A. Bello, E.G. Kolawole and A.M.F.O. Campos. 2012. Solar photocatalytic degradation of reactive yellow 81 and reactive violet 1 in aqueous solution containing semiconductor oxides. Int. J. Appl. Sci. Technol. 2: 90-105.

Guo, P., G. Zhang, J. Yu, H. Li and X.S. Zhao. 2012. Controlled synthesis, magnetic and photocatalytic properties of hollow spheres and colloidal nanocrystal clusters of manganese ferrite. Colloids Surfaces A Physicochem. Eng. Asp. 395: 168-174.

Gupta, V.K. and Suhas. 2009. Application of low-cost adsorbents for dye removal-a review. J. Environ. Manage. 90: 2313-2342.

Güy, N. and M. Özacar. 2016. The influence of noble metals on photocatalytic activity of ZnO for congo red degradation. Int. J. Hydrogen Energy 41: 20100-20112.

Hamad, D., R. Dhib and M. Mehrvar. 2016. Photochemical degradation of aqueous polyvinyl alcohol in a continuous UV/H_2O_2 process: Experimental and Statistical Analysis. J. Polym. Environ. 24: 72-83.

Huang, M., C. Xu, Z. Wu, Y. Huang, J. Lin and J. Wu. 2008. Photocatalytic discolorization of methyl orange solution by Pt modified TiO_2 loaded on natural zeolite. Dye. Pigment. 77: 327-334.

Huang, Y., D. Ding, M. Zhu, W. Meng, Y. Huang, F. Geng, et al. 2015. Facile synthesis of α-Fe_2O_3 nanodisk with superior photocatalytic performance and mechanism insight. Sci. Technol. Adv. Mater. 16: 14801.

Jack, R.S., G.A. Ayoko, M.O. Adebajo and R.L. Frost. 2015. A review of iron species for visible-light photocatalytic water purification. Environ. Sci. Pollut. Res. 22: 7439-7449.

Jiang, J.Q. 2015. The role of coagulation in water treatment. Curr. Opin. Chem. Eng. 8: 36-44.

Jiao, Y., Y. Liu, F. Qu, A. Umar and X. Wu. 2015. Visible-light-driven photocatalytic properties of simply synthesized α-iron (III) oxide nanourchins. J. Colloid Interface Sci. 451: 93-100.

Kansal, S.K., N. Kaur and S. Singh. 2009. Photocatalytic degradation of two commercial reactive dyes in aqueous phase using nanophotocatalysts. Nanoscale Res. Lett. 4: 709-716.

Kaur, J., M. Sharma and O.P. Pandey. 2014. Structural and optical studies of undoped and copper doped zinc sulphide nanoparticles for photocatalytic. Superlattices Microstruct. 77: 35-53.

Khan, Muhammad Isa, M.N. Akhtar, N. Ashraf, J. Najeeb, H. Munir, T.I. Awan, et al. 2020. Green synthesis of magnesium oxide nanoparticles using dalbergia sissoo extract for photocatalytic activity and antibacterial efficacy. Appl. Nanosci. 10: 2351-2364.

Khoufi, S., F. Feki and S. Sayadi. 2007. Detoxification of olive mill wastewater by electrocoagulation and sedimentation processes. J. Hazard. Mater. 142: 58-67.

Kim, S.E., J.Y. Woo, S.Y. Kang, B.K. Min, J.K. Lee and S.W. Lee. 2016. A facile general route for ternary Fe_2O_3@TiO_2@nanometal (Au, Ag) composite as a high-performance and recyclable photocatalyst. J. Ind. Eng. Chem. 43: 142-149.

Krishnakumar, B. and M. Swaminathan. 2011. Influence of operational parameters on photocatalytic degradation of a genotoxic azo dye acid violet 7 in aqueous ZnO suspensions. Spectrochim. Acta - Part A Mol. Biomol. Spectrosc. 81: 739-744.

Kumar, A. 2017. A review on the factors affecting the photocatalytic degradation of hazardous materials. Mater. Sci. Eng. Int. J. 1: 106-114.

Kumar, S.G. and K.S.R.K. Rao. 2017. Comparison of modification strategies towards enhanced charge carrier separation and photocatalytic degradation activity of metal oxide semiconductors (TiO_2, WO_3 and ZnO). Appl. Surf. Sci. 391: 124-148.

Lassoued, A., M.S. Lassoued, B. Dkhil, S. Ammar and A. Gadri. 2018a. Photocatalytic degradation of methylene blue dye by iron oxide (α-Fe_2O_3) nanoparticles under visible irradiation. J. Mater. Sci. Mater. Electron. 29: 8142-8152.

Lassoued, A., M.S. Lassoued, S. García-Granda, B. Dkhil, S. Ammar and A. Gadri. 2018b. Synthesis and characterization of Ni-doped α-Fe_2O_3 nanoparticles through Co-precipitation method with enhanced photocatalytic activities. J. Mater. Sci. Mater. Electron. 29: 5726-5737.

Lata, S. and S.R. Samadder. 2016. Removal of arsenic from water using nano adsorbents and challenges: A review. J. Environ. Manage. 166: 387-406.

Lee, S.Y. and S.J. Park. 2013. TiO_2 photocatalyst for water treatment applications. J. Ind. Eng. Chem. 19: 1761-1769.

Li, X., Y. Hou, Q. Zhao and L. Wang. 2011. A general, one-step and template-free synthesis of sphere-like zinc ferrite nanostructures with enhanced photocatalytic activity for dye degradation. J. Colloid Interface Sci. 358: 102-108.

Lin, D., B. Deng, S.A. Sassman, Y. Hu, S. Suslov and G.J. Cheng. 2014. Magnetic field assisted growth of highly dense α-Fe_2O_3 single crystal nanosheets and their application in water treatment. RSC Adv. 4: 18621-18626.

Liu, C.C., Y.H. Hsieh, P.F. Lai, C.H. Li and C.L. Kao. 2006. Photodegradation treatment of azo dye wastewater by UV/TiO_2 process. Dye. Pigment. 68: 191-195.

Liu, Y., L. Sun, J. Wu, T. Fang, R. Cai and A. Wei. 2015. Preparation and photocatalytic activity of ZnO/Fe_2O_3 nanotube composites. Mater. Sci. Eng. B. Solid-State Mater. Adv. Technol. 194: 9-13.

Maji, S.K., N. Mukherjee, A. Mondal and B. Adhikary. 2012. Synthesis, characterization and photocatalytic activity of α-Fe_2O_3 nanoparticles. Polyhedron 33: 145-149.

Mamba, G. and A. Mishra. 2016. Advances in magnetically separable photocatalysts: Smart, recyclable materials for water pollution mitigation. Catalysts 6: 1-34.

Mamun, K., R. Asw and K. Fahmida. 2015. Parameters affecting the photocatalytic degradation of dyes using TiO_2: A review. Appl. Water Sci. 7: 1569-1578.

Marien, Cédric B.D., T. Cottineau, D. Robert and P. Drogui. 2016. TiO_2 nanotube arrays: Influence of tube length on the photocatalytic degradation of paraquat. Appl. Catal. B. Environ. 194: 1-6.

Mekasuwandumrong, O., P. Pawinrat, P. Praserthdam and J. Panpranot. 2010. Effects of synthesis conditions and annealing post-treatment on the photocatalytic activities of ZnO nanoparticles in the degradation of methylene blue dye. Chem. Eng. J. 164: 77-84.

Mittal, M., M. Sharma and O.P. Pandey. 2014. UV – visible light induced photocatalytic studies of Cu doped ZnO nanoparticles prepared by Co-precipitation method. Sol. Energy 110: 386-397.

Mittal, M., M. Sharma and O.P. Pandey. 2016. Fast and quick degradation properties of doped and capped ZnO nanoparticles under UV – visible light radiations. Sol. Energy 125: 51-64.

Momeni, M.M. 2016. Influence of top morphology of hematite nanotubes on photo degradation of methylene blue and solar water splitting performance. Mater. Res. Innov. 20: 390-394.

Moniz, S.J.A., S.A. Shevlin, D.J. Martin, Z.X. Guo and J. Tang. 2015. Visible-light driven heterojunction photocatalysts for water splitting: A critical review. Energy Environ. Sci. 8: 731-59.

Muthuvel, A., M. Jothibas and C. Manoharan. 2020. Effect of chemically synthesis compared to biosynthesized ZnO-NPs using solanum nigrum leaf extract and their photocatalytic, antibacterial and *in vitro* antioxidant activity. J. Environ. Chem. Eng. 8: 103705.

Nasirian, M. and M. Mehrvar. 2016. Modification of TiO_2 to enhance photocatalytic degradation of organics in aqueous solutions. J. Environ. Chem. Eng. 4: 4072-4082.

Nasirian, M., C.F. Bustillo-Lecompte and M. Mehrvar. 2017. Photocatalytic efficiency of Fe_2O_3/TiO_2 for the degradation of typical dyes in textile industries: Effects of calcination temperature and UV-assisted thermal synthesis. J. Environ. Manage. 196: 487-498.

Natarajan, S., H.C. Bajaj and R.J. Tayade. 2018. Recent advances based on the synergetic effect of adsorption for removal of dyes from waste water using photocatalytic process. J. Environ. Sci. 65: 201-222.

Neppolian, B., H.C. Choi, S. Sakthivel, B. Arabindoo and V. Murugesan. 2002. Solar/UV-induced photocatalytic degradation of three commercial textile dyes. J. Hazard. Mater. 89: 303-317.

Nguyen-Phan, T.D., E.J. Kim, S.H. Hahn, W.J. Kim and E.W. Shin. 2011. Synthesis of hierarchical rose bridal bouquet- and humming-top-like TiO_2 nanostructures and their shape-dependent degradation efficiency of dye. J. Colloid Interface Sci. 356: 138-144.

Oturan, M.A. and J.J. Aaron. 2014. Advanced oxidation processes in water/wastewater treatment: Principles and applications. A review. Crit. Rev. Environ. Sci. Technol. 44: 2577-2641.

Pan, C., L. Dong, B. Qu and J. Wang. 2011. Facile synthesis and enhanced photocatalytic performance of 3D ZnO hierarchical structures. J. Nanosci. Nanotechnol. 11: 5042-5048.

Pardeshi, S.K. and A.B. Patil. 2009. Effect of morphology and crystallite size on solar photocatalytic activity of zinc oxide synthesized by solution free mechanochemical method. J. Mol. Catal. A. Chem. 308: 32-40.

Poulios, I., M. Kositzi, K. Pitarakis, S. Beltsios and I. Oikonomou. 2006. Photocatalytic oxidation of methomyl in the presence of semiconducting oxides. Int. J. Environ. Pollut. 28: 33-44.

Qi, K., B. Cheng, J. Yu and W. Ho. 2017. Review on the improvement of the photocatalytic and antibacterial activities of ZnO. J. Alloys Compd. 727: 792-820.

Qian, J., Z. Chen, C. Liu, X. Lu, F. Wang and M. Wang. 2014. Improved visible-light-driven photocatalytic activity of CeO_2 microspheres obtained by using lotus flower pollen as biotemplate. Mater. Sci. Semicond. Process. 25: 27-33.

Qu, Y. and X. Duan. 2013. Progress, challenge and perspective of heterogeneous photocatalysts. Chem. Soc. Rev. 42: 2568-2580.

Rai, H.S., M.S. Bhattacharyya, J. Singh, T.K. Bansal, P. Vats and U.C. Banerjee. 2005. Removal of dyes from the effluent of textile and dyestuff manufacturing industry: A review of emerging techniques with reference to biological treatment. Crit. Rev. Environ. Sci. Technol. 35: 219-238.

Raina, O. and R. Manimekalai. 2018. Photocatalysis of cobalt zinc ferrite nanorods under solar light. Res. Chem. Intermed. 44: 5941-5951.

Reda, S.M., M. Khairy and M.A. Mousa. 2020. Photocatalytic activity of nitrogen and copper doped TiO_2 nanoparticles prepared by microwave-assisted sol-gel process. Arab. J. Chem. 13: 86-95.

Ribeiro, A.R., O.C. Nunes, M.F.R. Pereira and A.M.T. Silva. 2015. An overview on the advanced oxidation processes applied for the treatment of water pollutants defined in the recently launched directive 2013/39/ EU. Environ. Int. 75: 33-51.

Sabari Arul, N., D. Mangalaraj and T.W. Kim. 2013. Photocatalytic degradation mechanisms of self-assembled rose-flower-like CeO_2 hierarchical nanostructures. Appl. Phys. Lett. 102: 1-5.

Saggioro, E.M., A.S. Oliveira, T. Pavesi, C.G. Maia, L.F.V. Ferreira and J.C. Moreira. 2011. Use of titanium dioxide photocatalysis on the remediation of model textile wastewaters containing azo dyes. Molecules 16: 10370-10386.

Sajjadi, S.H. and E.K. Goharshadi. 2017. Highly monodispersed hematite cubes for removal of ionic dyes. J. Environ. Chem. Eng. 5: 1096-1106.

Salehifar, N., Z. Zarghami and M. Ramezani. 2016. A facile, novel and low-temperature synthesis of MgO nanorods *via* thermal decomposition using new starting reagent and its photocatalytic activity evaluation. Mater. Lett. 167: 226-229.

Shabani, M., M. Haghighi, D. Kahforoushan and A. Haghighi. 2019. Mesoporous-mixed-phase of hierarchical bismuth oxychlorides nanophotocatalyst with enhanced photocatalytic application in treatment of antibiotic effluents. J. Clean. Prod. 207: 444-457.

Shahi, S.K., N. Kaur, J.S. Shahi and V. Singh. 2018. Investigation of morphologies, photoluminescence and photocatalytic properties of ZnO nanostructures fabricated using different basic ionic liquids. J. Environ. Chem. Eng. 6: 3718-3725.

Shahrodin, N.S.M., J. Jaafar, A.R. Rahmat, N. Yusof, M.H.D. Othman and M.A. Rahman. 2019. Superparamagnetic iron oxide as photocatalyst and adsorbent in wastewater treatment – a review. Micro Nanosyst. 12: 4-22.

Shang, S., X. Jiao and D. Chen. 2012. Template-free fabrication of TiO_2 hollow spheres and their photocatalytic properties. ACS Appl. Mater. Interfaces. 4: 860-865.

Sharma, M., T. Jain, S. Singh and O.P. Pandey. 2012. Photocatalytic degradation of organic dyes under UV – visible light using capped ZnS nanoparticles. Sol. Energy 86: 626-633.

Sharma, M., M. Olutas, A. Yeltik, Y. Kelestemur, A. Sharma and S. Delikanli. 2018. Understanding the journey of dopant copper ions in atomically flat colloidal nanocrystals of CdSe nanoplatelets using partial cation exchange reactions. Chem. Mater. 30: 3265-75.

Shen, Y., Y. Wu, H. Xu, J. Fu, X. Li, Q. Zhao, et al. 2013. Facile preparation of sphere-like copper ferrite nanostructures and their enhanced visible-light-induced photocatalytic conversion of benzene. Mater. Res. Bull. 48: 4216-4222.

Shi, B., G. Li, D. Wang, C. Feng and H. Tang. 2007. Removal of direct dyes by coagulation: The performance of preformed polymeric aluminum species. J. Hazard. Mater. 143: 567-574.

Shooshtari, N.M. and M.M. Ghazi. 2017. An investigation of the photocatalytic activity of nano A-Fe_2O_3/ZnO on the photodegradation of cefixime trihydrate. Chem. Eng. J. 315: 527-536.

Singla, P., M. Sharma, O.P. Pandey and K. Singh. 2013. Photocatalytic degradation of azo dyes using Zn-doped and undoped TiO_2 nanoparticles. Appl. Phys. A. 116: 371-78.

Sivalingam, G., K. Nagaveni, M.S. Hegde and G. Madras. 2003. Photocatalytic degradation of various dyes by combustion synthesized nano anatase TiO_2. Appl. Catal. B. Environ. 45: 23-38.

So, C.M., M.Y. Cheng, J.C. Yu and P.K. Wong. 2002. Degradation of azo dye procion red MX-5B by photocatalytic oxidation. Chemosphere 46: 905-912.

Sobana, N., B. Krishnakumar and M. Swaminathan. 2013. Synergism and effect of operational parameters on solar photocatalytic degradation of an azo dye (direct yellow 4) using activated carbon-loaded zinc oxide. Mater. Sci. Semicond. Process. 16: 1046-1051.

Solís, M., A. Solís, H.I. Pérez, N. Manjarrez and M. Flores. 2012. Microbial decolouration of azo dyes: A review. Process Biochem. 47: 1723-1748.

Šostar-Turk, S., M. Simonič and I. Petrinić. 2005. Wastewater treatment after reactive printing. Dye. Pigment. 64: 147-152.

Subramonian, W., T.Y. Wu and S.P. Chai. 2017. Photocatalytic degradation of industrial pulp and paper mill effluent using synthesized magnetic Fe_2O_3-TiO_2: Treatment efficiency and characterizations of reused photocatalyst. J. Environ. Manage. 187: 298-310.

Suman, S. Chahal, A. Kumar and P. Kumar. 2020. Zn doped α-Fe_2O_3: An efficient material for UV driven photocatalysis and electrical conductivity. Crystals 10: 273.

Sun, J., L. Qiao, S. Sun and G. Wang. 2008a. Photocatalytic degradation of orange G on nitrogen-doped TiO$_2$ catalysts under visible light and sunlight irradiation. J. Hazard. Mater. 155: 312-319.

Sun, T., J. Qui and C. Liang. 2008b. Controllable fabrication and photocatalytic activity of ZnO nanobelt arrays. J. Phys. Chem. C. 112: 715-721.

Tian, G., Y. Chen, W. Zhou, K. Pan, C. Tian, X. Huang, et al. 2011. 3D hierarchical flower-like TiO$_2$ nanostructure: Morphology control and its photocatalytic property. Cryst. Eng. Comm. 13: 2994-3000.

Uma, K., B. KrishnaKumar, G.T. Pan, T.C.K. Yang and J.H. Lin. 2020. Enriched silver plasmon resonance activity on the sonochemical synthesis of ZnO flowers with α-Fe$_2$O$_3$ as an efficient catalyst for photo-fenton reaction and photo-oxidation of ethanol. J. Water Process Eng. 34: 101089.

Vinayagam, R., R. Selvaraj, P. Arivalagan and T. Varadavenkatesan. 2020. Synthesis, characterization and photocatalytic dye degradation capability of calliandra haematocephala-mediated zinc oxide nanoflowers. J. Photochem. Photobiol. B. Biol. 203: 111760.

Wang, C. and Z. Huang. 2016. Controlled synthesis of α-Fe$_2$O$_3$ nanostructures for efficient photocatalysis. Mater. Lett. 164: 194-197.

Wang, N., J. Li, L. Zhu, Y. Dong and H. Tang. 2008. Highly photocatalytic activity of metallic hydroxide/titanium dioxide nanoparticles prepared *via* a modified wet precipitation process. J. Photochem. Photobiol. A. Chem. 198: 282-287.

Warwick, M.E.A., K. Kaunisto, D. Barreca, G. Carraro, A. Gasparotto, C. Maccato, et al. 2015. Vapor phase processing of α-Fe$_2$O$_3$ photoelectrodes for water splitting: An insight into the structure/property interplay. ACS Appl. Mater. Interfaces 7: 8667-8676.

Wu, J., J. Wang, Y. Du, H. Li, Y. Yang and X. Jia. 2015. Chemically controlled growth of porous CeO$_2$ nanotubes for Cr(VI) photoreduction. Appl. Catal. B. Environ. 174-175: 435-444.

Xia, J., H. Liu, X. Cheng, S. Yin, H. Li, H. Xu, et al. 2012. Reactable ionic liquid synthesis and visible-light photocatalytic activity of dendritic ferric oxide hierarchical structures. Micro Nano Lett. 7: 806-809.

Xu, L., Y.L. Hu, C. Pelligra, C.H. Chen, L. Jin, H. Huang, et al. 2009. ZnO with different morphologies synthesized by solvothermal methods for enhanced photocatalytic activity. Chem. Mater. 21: 2875-2885.

Xu, P., G.M. Zeng, D.L. Huang, C.L. Feng, S. Hu, M.H. Zhao, et al. 2012. Use of iron oxide nanomaterials in wastewater treatment: A review. Sci. Total Environ. 424: 1-10.

Yu, J. and X. Yu. 2008. Hydrothermal synthesis and photocatalytic activity of zinc oxide hollow spheres. Environ. Sci. Technol. 42: 4902-4907.

Yu, J., Z. Li, Y. Liao, C. Kolodziej, S. Kuyuldar, W.S. Warren, et al. 2019. Probing the spatial heterogeneity of carrier relaxation dynamics in CH$_3$NH$_3$PbI$_3$ perovskite thin films with femtosecond time-resolved nonlinear optical microscopy. Adv. Opt. Mater. 1901185: 3-8.

Yu, Y., Y. Zhu and M. Meng. 2013. Preparation, formation mechanism and photocatalysis of ultrathin mesoporous single-crystal-like CeO$_2$ nanosheets. Dalt. Trans. 42: 12087-12092.

Zeng, W., Z. Yin, M. Gao, X. Wang, J. Feng, Y. Ren, et al. 2020. *In-situ* growth of magnesium peroxide on the edge of magnesium oxide nanosheets: Ultrahigh photocatalytic efficiency based on synergistic catalysis. J. Colloid Interface Sci. 561: 257-264.

Zhai, T., S. Xie, Y. Zhao, X. Sun, X. Lu, M. Yu, et al. 2012. Controllable synthesis of hierarchical ZnO nanodisks for highly photocatalytic activity. Cryst. Eng. Comm. 14: 1850-1855.

Zhang, R., M. Sun, G. Zhao, G. Yin and B. Liu. 2019. Hierarchical Fe$_2$O$_3$ nanorods/TiO$_2$ nanosheets heterostructure: Growth mechanism, enhanced visible-light photocatalytic and photoelectrochemical performances. Appl. Surf. Sci. 475: 380-388.

Zhang, X., J. Qin, Y. Xue, P. Yu, B. Zhang, L. Wang, et al. 2014. Effect of aspect ratio and surface defects on the photocatalytic activity of ZnO nanorods. Sci. Rep. 4: 4-11.

Zhao, S.W., H.F. Zuo, Y.R. Guo and Q.J. Pan. 2017. Carbon-doped ZnO aided by carboxymethyl cellulose: Fabrication, photoluminescence and photocatalytic applications. J. Alloys Compd. 695: 1029-1037.

Zheng, Y., L. Cao, G. Xing, Z. Bai, J. Huang and Z. Zhang. 2019. Microscale flower-like magnesium oxide for highly efficient photocatalytic degradation of organic dyes in aqueous solution. RSC Adv. 9: 7338-48.

Zhou, Y., Z. Liang and Y. Wang. 2008. Decolorization and COD removal of secondary yeast wastewater effluents by coagulation using aluminum sulfate. Desalination 225: 301-311.

Zou, X.L. 2015. Combination of ozonation, activated carbon, and biological aerated filter for advanced treatment of dyeing wastewater for reuse. Environ. Sci. Pollut. Res. 22: 8174-8181.

Index